Lecture Notes in Computer Science 2128

Edited by G. Goos, J. Hartmanis, and J. van Leeuwen

D0454706

Springer
Berlin
Heidelberg
New York
Barcelona
Hong Kong
London
Milan
Paris
Tokyo

Hartmut Ehrig   Gabriel Juhás
Julia Padberg Grzegorz Rozenberg (Eds.)

Unifying
Petri Nets

Advances in Petri Nets

 Springer

Series Editors

Gerhard Goos, Karlsruhe University, Germany
Juris Hartmanis, Cornell University, NY, USA
Jan van Leeuwen, Utrecht University, The Netherlands

Volume Editors

Hartmut Ehrig
Julia Padberg
Technische Universität Berlin
Institut für Softwaretechnik und Theoretische Informatik
Franklinstr. 28/29, 10587 Berlin, Germany
E-mail:{ehrig/padberg}@cs.tu-berlin.de

Gabriel Juhás
Katholische Universität Eichstätt
Ostenstr. 28, 85072 Eichstätt, Germany
E-mail: gabriel.juhas@ku-eichstaett.de

Grzegorz Rozenberg
Leiden University
Leiden Institute of Advanced Computer Science
Niels Bohrweg 1, 2333 CA Leiden, The Netherlands
E-mail: rozenberg@liacs.nl

Cataloging-in-Publication Data applied for

Die Deutsche Bibliothek - CIP-Einheitsaufnahme

Unifying Petri nets : advances in Petri nets / Hartmut Ehrig ... (ed.). -
Berlin ; Heidelberg ; New York ; Barcelona ; Hong Kong ; London ; Milan ;
Paris ; Tokyo : Springer, 2001
  (Lecture notes in computer science ; Vol. 2128)
  ISBN 3-540-43067-9

CR Subject Classification (1998): F.1, D.1-3, F.3, C.2, J.1, G.2

ISSN 0302-9743
ISBN 3-540-43067-9 Springer-Verlag Berlin Heidelberg New York

Springer-Verlag Berlin Heidelberg New York
a member of BertelsmannSpringer Science+Business Media GmbH

http://www.springer.de

© Springer-Verlag Berlin Heidelberg 2001
Printed in Germany

Typesetting: Camera-ready by author, data conversion by Olgun Computergrafik
Printed on acid-free paper   SPIN 10839998   06/3142   5 4 3 2 1 0

Preface

Since their introduction nearly 40 years ago, research on Petri nets has taken many different directions. Various kinds of Petri net classes, motivated either by theory or applications, with their specific features and analysis methods, have been proposed since then. The fact that Petri nets are widely used and are still considered to be an important research area, demonstrates both the usefulness and the power of this approach. This successful development has led to a very heterogeneous landscape of diverse models, and this in turn has stimulated research on concepts and approaches that provide some (often partial) unification/structuring of this landscape. Since most of these unifying approaches are scattered through the literature, we are convinced that the time is ripe for the publication of a volume comprising the most relevant approaches developed up to now. The title of this volume "Unifying Petri Nets" in the series "Advances in Petri Nets" provides a compact representation of its contents. The goals we hope to achieve by publishing this volume are:

- a stimulation of research in this important area,
- a meaningful comparison of various approaches,
- a cross-fertilization between different approaches,
- a compact presentation of the state of the art.

Although different approaches to unifying Petri nets aim at different goals, there are some common benefits. A uniform approach to Petri nets captures the common concepts of different kinds of Petri nets, such as places, transitions, net structure, and (in the case of high-level nets) data types. General notions, such as firing behavior, invariants, etc., that are essential for all kinds of Petri nets, can be formulated within a unifying approach. In this way these notions become independent of their definition within a specific net class. Results achieved within a unifying approach can often be "naturally" transferred to the net classes captured by this approach.

The volume begins with an introductory paper that presents some of the paradigms underlying the theory of Petri nets.

Part I: Application Oriented Approaches is mainly concerned with an overview of (and recent developments concerning) the German DFG Researcher Group *Applied Petri Net Technology*, where the concept of a "Petri Net Baukasten" has been developed in order to allow a unified access to theory, applications, and tool development in the area of Petri nets.

Part II: Unifying Frameworks presents various mathematical approaches, based on partial algebras, category theory, and rewriting logic, that allow a classification as well as a uniform presentation of various Petri net classes.

Part III: Theoretical Approaches is a collection of contributions investigating more specialized aspects of a uniform theoretical treatment of Petri nets.

We hope that this volume offers new insights and suggests new and important research topics for all readers interested in Petri nets.

April 2001   Hartmut Ehrig
Gabriel Juhás
Julia Padberg
Grzegorz Rozenberg

Table of Contents

"What Is a Petri Net?"
Informal Answers for the Informed Reader

Jörg Desel and Gabriel Juhás

Katholische Universität Eichstätt
Lehrstuhl für Angewandte Informatik
85071 Eichstätt, Germany
{joerg.desel,gabriel.juhas}@ku-eichstaett.de

Abstract. The increasing number of Petri net variants naturally leads to the question whether the term "Petri net" is more than a common name for very different concepts. This contribution tries to identify aspects common to all or at least to most Petri nets. It concentrates on those features where Petri nets significantly differ from other modeling languages, i.e. we ask where the use of Petri nets leads to advantages compared to other languages. Different techniques that are usually comprised under the header "analysis" are distinguished with respect to the analysis aim. Finally, the role of Petri nets in the development of distributed systems is discussed.

1 Introduction

What is a Petri net? Very often, the thesis of Carl Adam Petri [23] written in the early sixties is cited as the origin of Petri nets. However, Petri did of course not use his own name for defining a class of nets. Moreover, this fundamental work does not contain a definition of those nets that have been called Petri nets later on. In fact, there are hundreds of different definitions and extensions in the literature on Petri nets since then. Most authors did not mean to define something completely new when coming up with a new definition. They use the term "Petri net" to express that the basic concept of a notion is the one of Petri nets, no matter how this notion is formulated mathematically or which extensions of standard definitions are used. In this contribution we try to identify central aspects of this basic concept of Petri nets. In other words, we aim at providing characteristics of Petri nets that are common to all existing and future variants. It should be clear that this can only be done in a very subjective manner. So we like to place the following disclaimer at the very beginning of the paper: We do not consider our list of important aspects of Petri nets complete, and for each aspect claimed to be common to all Petri net variants there might exist very reasonable exceptions.

This paper is not an introduction to Petri net theory. Instead, we assume that the readers have some knowledge about Petri nets and preferably even know different Petri net classes. For an overview of Petri net theory we refer to the proceedings of the previous advanced course on Petri nets [25, 26]. The other

H. Ehrig et al. (Eds.): Unifying Petri Nets, LNCS 2128, pp. 1–25, 2001.
© Springer-Verlag Berlin Heidelberg 2001

contributions in this book should also be helpful, although the present paper is meant to be an introductory note to this book. In particular, the work of the "Forschergruppe Petrinetz-Technologie", represented by the papers [35, 17, 13, 36], show how different variants of Petri nets can be subsumed and structured in a unified framework.

There are also examples of modeling notions which do not carry "Petri net" in the name but apparently stem from Petri nets. Among these notions are event-driven process chains (EPCs) [31] (originally called "Ereignisgesteuerte Prozess-ketten" in German), a standard notion for modeling business processes in the framework of the "ARIS-Toolset". The first publications on this model explictly refer to Petri nets. Still, the central idea is the one of Petri nets although there are some significant differences. Another example is given by activity diagrams, a language within the Unified Modeling Language (UML). These diagrams more or less look like Petri nets and have an interpretation which is very similar to Petri nets but have some additional features such as "swim lanes", associating each diagram element to an object. Although people from the UML community insist that activity diagrams have nothing to do with Petri nets, there already exist a number of publications establishing close connections between these two languages [14, 18]. Actually, Petri nets are suggested for a formal semantics of activity diagrams – this notion has evolved to a standard without having any fixed semantics by now. So this paper is about Petri nets and those related formalisms which are based on the same concepts as Petri nets.

Many papers defining or using Petri nets emphasize the following characteristics of the model; Petri nets are a *graphical notion* and at the same time a *precise mathematical notion*. So we take it that these two properties are the most important ones and we devote the following two sections to them. The next important characteristics of Petri nets is described by their *executability*, their *semantics*, their *behavior* or the like. Whereas it seems that the first two characteristic features do not rise any dissension, there is no common agreement what the semantics of a Petri net should look like, i.e., what the behavior of a Petri net formally is. We split the consideration on behavior in two parts; behavior is constituted by the *occurrence rule* – which defines under which conditions a transition is enabled and what happens when it occurs – and by derived formal descriptions of the entire *behavior*, given by the set of occurrence sequences, partially ordered runs or any kind of trees or graphs representing all runs of a net. These parts constitute the topics of sections four and five. *Analysis* of Petri nets is the next important subject, addressed in section six. This term comprises many different concepts; analysis by simulation, by employing structural properties of the net, or by analysis of the exhibited behavior of a net. We distinguish between analysis techniques that automatically provide useful information for a given net (like deadlock-freedom), techniques that automatically verify a given property (like mutual exclusion) and techniques that help in manually proving the correctness of a net with respect to a given specification. The last section is concerned with topics that are not explicitly addressed in most other papers on Petri nets. Each Petri net is a model of a system, if it is not just a counterexample or an illustration of a proof. There are many different languages for modeling systems, most of them not comparable with Petri nets (consider, e.g., models of the architecture, models of the data structure etc.). Therefore we have to be more precise; a Petri net models the behavioral aspects of a system. The same can be said about differential equations. So we should add that the behavior is constituted by discrete events. Again, there are more prominent languages for this task, namely the variety of automata models. The core issue of Petri nets is that they model behavioral aspects of *distributed systems*, i.e., systems with components that are locally separated and communicate which each other. Surprisingly, neither components nor any notion of locality appears with the usual definition of a Petri net. The section on *distributed systems* discusses aspects of this gap.

Each section header is an answer to the question raised at the beginning of the paper.

2   A Graphical Notation

Most modeling languages have graphical notations, and this has good reasons. Models are used as a means to specify concepts and ideas, and to communicate them between humans. Nearly everybody would use some kind of graphics to express his or her understanding of a system, even without using any explicit modeling language. We asked our first semester students to give a model of the enrollment procedure of our university. The result was a very interesting variety of models emphasizing surprisingly many different aspects of the procedure. All these models were supported by graphics. It does not need psychological research to state that graphics employing two dimensions allow for a better understanding of complex structures than one-dimensional text. Since specification of systems and communication of models are the main applications of Petri nets in practice, understandability for humans is among the most crucial quality criteria for modeling languages. Petri nets have a nice graphical representation using only very few different types of elements, which is a good basis for an easy understandability of a model and for the learnability of the language. These two criteria for modeling languages belong to the most important ones recognized in the "Guidelines of Modeling" [3].

Many modeling languages are supported by graphics that possibly abstracts from some details of a model. Petri nets are not only supported by graphics but each Petri net *is* a special annotated graph. One could argue that the annotations of a Petri net are as essential as the graphics. In fact, for some high-level Petri net classes it is possible to represent any model equivalently by a trivial net structure, putting all the information about the model into the annotations of a single place, a single transition and the connecting arcs [32]. In general, often one has to trade off between specification by graphical means and specification by textual means in the annotations. It is a typical feature of Petri nets that the semantics of textual annotations can be given in terms of nets, i.e. of graphs. As an example, consider the low-level unfolding of a high-level Petri net [32]. In

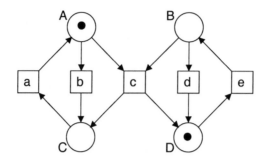

Fig. 1. A picture of a Petri net

this sense, annotations can be viewed as shortcuts for more complex graphical representations, employing, e.g., symmetries of a net. Hence it is justified to claim that a Petri net is a graph.

In the previous paragraphs we confused mathematical graphs with graphical notations. So what is a Petri net, a mathematical object representing the components of a graph or a picture? It is important to notice that by definition the way a net is drawn does not carry any semantic information. This is different for languages such as SADT [28] where it makes an important difference whether an arc touches a node at its right, left, upper or lower side. Also the relative position of Petri net nodes carries no formal information. However, the topology of a drawn Petri net is important from a pragmatic perspective. The modeler might place the elements representing a single system component on a cycle if this helps to understand the net. In this case, additional knowledge about the model and its relation to the system is put in the picture. Alternatively, a tool can calculate a nice way to draw a net; then the figure carries information about the net itself and about some analysis results. So a Petri net picture can be more than a mathematically defined graph. The difference is irrelevant for analysis tools. But it is significant when the net is used as a means for human communication. Even simple models can be drawn in a spaghetti style such that this picture does not help much (compare for example two pictures of the same Petri net in Figures 1 and 2). The topology of a net drawing is an important topic in the context of interchange standards for Petri nets [20]. The exchange information of a picture might contain information about the relative position of the nodes, about their shape etc.

It is often emphasized that Petri nets are bipartite graphs, because each directed arc either leads from a place to a transition or from a transition to a place. This is not exactly true; Petri nets are more than that. In bipartite graphs the two sets of nodes play a symmetric role whereas places and transitions are dual concepts. Exchanging places and transitions leads to a completely different net. The existence of places and transitions and their distinction, is one of the fundamental ingredients of Petri nets. Therefore this formalism is neither primarily based on actions (like data flow diagrams), represented by transitions, nor is it primarily based on states (like automata), represented by places. Instead, the

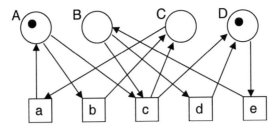

Fig. 2. Another picture of the Petri net from Figure 1

mutual interplay of local activities and local states constitutes the basic components of each net, as will be discussed later in more detail. The equal footing of actions and states is reflected in nets by the definition of places and transitions on the same level. So the following definition is the most common answer to the question "What is a Petri net?"

The usual definition of a "core" Petri net
A Petri net is a directed graph with two kinds of nodes,
interpreted as places and transitions,
such that no arc connects two nodes of the same kind.

Places of a Petri net are usually represented by round graphical objects (circles or ellipses), and transitions by rectangular objects (boxes or squares), as shown in Figures 1 and 2. There is a standard arc type between vertices of different type representing the flow relation, as shown in the figures. This convention makes it easy to guarantee a rough understanding of any Petri net without additional legend. One of the main advantages of the Unified Modeling Language (UML, see [30]) is that it unifies the shape of vertices and arcs in its diagrams that have been used in a contradictory way in different languages before. Likewise, the consistent use of graphical symbols for Petri net objects is one of the main reasons for the worldwide and long standing success of Petri nets. Whenever someone acquainted with Petri nets is confronted with a new variant, the general interpretation of places and transitions does not have to be explained and gives no rise to misunderstandings. So Petri nets play the role of a "Unified Process Language" since a long time.

Sometimes the use of only circles and squares is considered a disadvantage. Instead of circles or squares, special symbols representing the actual type of the represented system component can be used. Branches at transitions or at places can be substituted by special branching nodes. These variants are – among others – implemented in many commercial Petri net tools. The vendors claim that the readability of their models is improved by the graphical extensions. This might sometimes be true, but there is the danger of inconsistency between different products. Moreover, an increasing number of features leads to an increasing number of modeling errors. Someone only familiar with such a specific application-dependent notion can not understand an example net given in another proprietary notation. However, as long as additional graphical notions have a unique and easy translation to traditional Petri net components, a good knowledge about Petri nets will help to understand any such model. In this way, Petri nets – including their graphical representation – can be seen as an "interlingua" for many different related modeling languages.

3   A Precise Mathematical Language

One might say that this answer to the question raised in this paper is a matter of course. However, often the mathematics, and particularly its presentation, is the reason to consider Petri nets difficult for users. As mentioned in the previous section, it is the graphical representation of a net which is actually used in practice and which can easily be understood. So why do we need any further mathematical foundation? And what does it mean for a modeling language to be precise? The answer to these questions concerns two parts: *syntax* and *semantics*.

There are different ways to specify a class of nets syntactically. Well-known examples are restricted classes such as free-choice Petri nets [5] where the local vicinities of net objects are restricted in a characteristic way. Another frequently used possibility is given by the class of all nets that are generated from an initial one using a given set of production rules. Such Petri net grammars can be used for a syntactic formulation of a Petri net class, defining exactly which Petri nets belong to that class.

We consider here another way to deal with the syntax of Petri nets; each Petri net should have a precise syntax. In other words, it should be clear what kind of objects belong to a given Petri net and which objects do not belong to it. This syntax differs for different classes of Petri nets. It turns out that this kind of formal syntax can be more conveniently be given in terms of simple mathematics than in terms of the graphical representations. So for definition purposes, Petri nets are syntactically defined as annotated graphs in a mathematical setting. The usual notions equip tuples of sets, relations and mappings. The following definition shows an example.

The mathematical definition of a place/transition Petri net
A place/transition net is a tuple (S, T, F, M_0, W, K), where
S is the set of places,
T is the set of transitions,
F is the flow relation,
M_0 is the initial marking,
        formally given as a mapping from S to the nonnegative integers,
W maps arcs to positive numbers (arc weights) and
K maps places to positive numbers (capacity restrictions).

These objects have to satisfy some restrictions, such as $M_0(s) \leq K(s)$ for each place s (no capacity restriction is violated initially). All these objects and restrictions are very easy to explain using the graphical representation. For example, "no arc connects two places or two transitions" might be more plausible than the usual expression $F \subseteq (S \times T) \cup (T \times S)$. But all the used objects and restrictions have to be precise enough that an equivalent mathematical formulation can be given in a simple and obvious way and such that there is no doubt how this formulation would look like. The tuple notion induces an order on the objects (in the above example, places before transitions before arcs before ...). This order does not imply valences of the used objects. It is just arbitrarily fixed for convenience sake; the formulation "given a place/transition net (A, B, C, D, E, F)" is the shortest way to define all components of a place/transition net. But the tuple-notion never represents the core idea of a Petri net. When teaching Petri net theory one should be careful not to emphasize this notion too much – it unnecessarily complicates the matter.

When are two Petri nets identical? Using the mathematical definition, the answer is obvious: two nets are identical if and only if all their objects are pairwise identical. This implies in particular that a different graphical representation of a Petri net does not change the Petri net. Conversely, two Petri nets which look the same, i.e. which have identical graphical representations, are not necessarily identical, assuming that the graphical representation does not include the identity of each single element. This is often not exactly what one would like to have. Instead, Petri nets that look the same should sometimes not be distinguished. Imagine for example the net with a single (unmarked) place, a single transition and one arc from the place to the transition (arc-weights, capacities etc. are ignored for this example). Putting this description into mathematics one needs to define a place s, a transition t and an arc (s, t). There is no unique net that matches the above description, because the identity of the place and the identity of the transition is chosen arbitrarily. The net with place s' (where $s \neq s'$), t and arc (s', t) is different to the one defined before. This difference is only meaningful if the net models something; then s and s' model different objects of the system domain. But syntax does not distinguish what is modeled. So, intuitively one is interested in the class of all nets which can be obtained from the original one by consistent renaming. In other words, the syntactical definition of a Petri net comes with the notion of an isomorphism relation.

Isomorphism of Petri nets
Two Petri nets are isomorphic if there are bijections
between their respective sets of objects (places and transitions)
which are respected by all annotations, relations and mappings
that belong to the syntactical definition.

The simple but important distinction between equality and isomorphism of Petri nets is only easily possible on a mathematical level. Intuitively, a single (graphical) Petri net is mathematically given by an isomorphism class of tuples, where each single tuple of the class has the same Petri net as its graphical representation. Isomorphism classes are particularly important for labeled Petri nets, i.e. nets where each element carries a label which establishes the connection to the modeled world. In a labeled Petri net, two distinct places can represent the same object and two distinct transitions the same action. For example, process nets representing partially ordered runs of other Petri nets are labeled Petri nets.

The same run can be represented by many isomorphic process nets (see the next section).

What is the semantics of a Petri net? Taking the original meaning of the word "semantics", the answer should associate objects of a net to objects of the modeled system. Considering also the dynamics of the net, the behavior of the net should correspond to the behavior of the modeled system. In the context of modeling languages, the term "semantics" is used in a different way, usually together with the prefix "formal". A class of Petri nets has one (or several) formal semantics although the world of modeled systems is not considered at all when defining such a class. The formal semantics generically defines the behavior of each Petri net that belongs to the class, i.e. the role of each possible ingredient of a net with respect to behaviour is precisely defined by the semantics. Since Petri nets are defined by mathematics, so are their formal semantics. At this stage, we do not discuss different variants of semantics, because this will be the topic of the next section.

Many modeling notions used in practice do not have a precise semantics. Defining a formal semantics is only possible for a notion possessing a formal syntax. Hence, without explicitly defining the syntax it is impossible to formalize semantics. Some languages do have a formal syntax, with or without a mathematically given description, but no fixed semantics. These notions are frequently called "semi-formal". It is often claimed that semi-formal modeling languages allow more flexibility and are hence better suited for practical applications than formal modeling languages like Petri nets. Moreover, semi-formal models are said to be easier to understand and easier to learn. We claim that the opposite is true. The theory of Petri nets offers classes of nets where specific details of the model are left open. For example, channel/agency nets define only the structure given by places, transitions and arcs together with the interpretations of these elements, but no behavior [24]. Place/transition nets identify all tokens and thus abstract from different token objects. Conflicts, i.e., different mutually exclusively enabled transitions, can be interpreted as incomplete specifications – the vicinity that decides which alternative will be chosen is missing. Most nets abstract from all notions of time. So there are various ways to express different kinds of vagueness. The important point is that it is always very clear which aspects are expressed by the net and which aspects are not. Many modeling notions outside the Petri net world exhibit moreover a kind of meta-vagueness. For these models, it is a matter of interpretation to decide which aspects are represented in the model and which are not. So flexibility concerns not only the model itself but also its interpretation – a feature that we do not consider desirable. Instead, it is much easier to understand a model and also the modeling language if there is a precise understanding about *what* has been modeled and *how* it is modeled. Only a precise mathematical language, such as given by most variants of Petri nets, provides sufficient clarity.

As an example for a semi-formal notion, consider event-driven process chains (EPCs) [31]. This language is a derivative of Petri nets. In the application field of business processes it has emerged to a quasi-standard. The major benefit of EPCs is that they are integrated in a larger context containing additionally a data model and a structure model (the "house of ARIS" [31]). An EPC has three types of nodes, two comparable to places and transitions, and an additional node type for the logical connections AND, OR, and XOR (exclusive or). Not surprisingly, the OR connector raises severe problems. A binary OR-split is interpreted as follows. Either one of the output arcs or both arcs are chosen for forwarding the control. A binary OR-merge cannot be interpreted in such a simple way. After receiving the control from one input arc, either one has to wait for the control from the other arc or one can continue immediately which corresponds to the different possible decisions at the OR-split. This technical problem has led to quite a number of research activities (see e.g. [29]), but there exists no really satisfying solution yet. The problem is that EPCs have no formal semantics. When asking experts in EPC modeling about the correct interpretation of an OR-merge in a difficult example, they come to very vague (and different) answers. Surprisingly, it is often claimed that EPCs are more compact, more appropriate and easier understandable than Petri nets in the application area of business processes. The paper [2] proves that they are not smaller than equivalent Petri nets in general. Nonetheless, the Petri net community should learn from EPCs which kind of concepts and which kind of links between concepts are necessary for successfully selling a modeling language together with an associated tool.

4 A Structured Set of Activities that Remove and Add Tokens

Most Petri net variants are equipped with a notion for behavior. Some variants, however, are not. For example, channel/agency nets do not have an explicit behavioral definition [24]. They are used as a first step when developing a Petri net model. Refinement and completion of a channel/agency net leads to a more detailed model, which can then be equipped with behavior.

In this section, we restrict our considerations to nets that do have a behavior.

In contrast to all automata models and transition systems, a (global) state of a net is not a fundamental concept but it is constituted by local states of all places of the Petri net. States are formally represented by markings. A marking associates a set, multi-set, list etc. of tokens to each place, where tokens are elements of some domain. So a global state is only a derived concept (with the exception that the definition of a Petri net often contains initial or final global states).

Principle of Distributiveness
States are associated to places and thus distributed.
A global state is constituted by all local states.

In most cases the behavior of a net is formulated by means of a rule stating under which conditions a single transition can occur and stating the consequences of its occurrence, the so-called occurrence rule. It is one of the central principles of

Petri nets that both the enabling conditions and the consequences only concern the immediate vicinity of a transition. In other words, if the occurrence of a transition is related to the state of a place then there must be some arc connecting the transition and the place.

1. Principle of Locality
The conditions for enabling a transition, in a certain mode if applicable, only depend on local states of (some) places in its immediate vicinity.

2. Principle of Locality
The occurrence of an enabled transition only changes the local state of (some) places in its immediate vicinity.

We formulated the locality principle in two parts because the relevant sets of places for enabledness and for change in the vicinity of a transition are not necessarily identical. For place/transition nets, all places in the pre-set (i.e., sources of arcs leading to the transition) are relevant for enabling the transition, and places in the post-set (i.e., targets of arcs from the transition) only play a role when capacity restrictions are involved. The state is only changed for places which are either in the pre-set or in the post-set but not both (as long as arc weights are not considered). Moreover, the new state of a place depends on its previous value in place/transition nets, because a token is added. However, the relative change of the state of a place does not depend on its previous state. Given a transition, we can distinguish:

a) places where the local state is relevant for enabling but is not changed (such as read places or inhibitor places),
b) places where the local state is relevant for enabling and is changed by the transition occurrence (places in the pre-set in case of place/transition nets without capacity restrictions), and
c) places where the local state is not relevant for enabling but is changed by the transition occurrence (places in the post-set in case of place/transition nets without capacity restrictions).

Orthogonally, places where the local state is changed by the transition occurrence (cases (b) and (c)) can be divided into:

1) places where the new local state depends on its previous value (places in the pre-set and places in the post-set in case of place/transition nets), and
2) places where the new state does not depend on the previous one (such as places reset by the transition occurrence in case of nets with reset arcs).

Often, the different role of the places is depicted by different arc types such as inhibitor arcs or reset arcs. When talking about the vicinity of a transition, we mean all places connected with the transition by an arbitrary arc.

It might be worth mentioning that the majority of Petri net formalisms considers test-and-set-operations elementary, i.e. reading a local state and changing it depending on the previous value is considered one atomic action. These Petri net formalisms have no difficulty with simultaneous access to different places, even if these places model conditions at different locations. The general paradigm is the one of removing and adding tokens. It can even be phrased as:

The Token Flow Paradigm
Tokens flow with infinite speed from place to place,
sometimes they mutate, join or split in transitions.

Perhaps surprisingly, read actions (case (a)) and write actions (case (c),(2)) are not that usual in the Petri net literature. As explained above, reading the state of a place means that this place is relevant for a respective transition but the state of this place is not changed by the occurrence of the transition. Although concurrent read is an essential operation in most areas of computer science, many semantics of Petri nets do not allow any concurrent access to the tokens of a place (see [11]). Likewise, writing is a central issue in other areas of computer science but there is hardly any corresponding concept in the Petri net literature. Petri nets with reset arcs are an exception, but they model only the special case that a place looses all tokens when the corresponding transition occurs. More generally, writing the local state of a place means changing the state arbitrarily without taking the previous state into account. The only way to model writing with Petri nets is by synchronous removing the old tokens and adding new ones. It needs a special variant of high-level nets to perform arbitrary removing with a single transition, such that the previous local state has no influence on any new state (see [6]).

There are generalizations of the occurrence rule concerning the simultaneous occurrence of many transitions. These variants still obey the principle of locality, because the vicinities of all simultaneous transitions have to be considered.

5 A Compact Way to Specify Behavior

The behavior of a Petri net does not only concern occurrences of single transition but sets of occurring transitions which can be in different relations such as causal relationship, concurrency, choice, or being totally ordered. The behavior can also include intermediate local or global states or the final global state and possible continuations from these states. Different ways to describe the behavior of Petri nets are given by different semantics of the respective Petri net classes.

Given a model of a dynamic system, the behavior of the model should be in a close relationship to the system's behaviour. If the model is executable, i.e. if it has a defined semantics, then runs of the model can be generated. These runs correspond to the runs of the system. Analysis of the model's behaviour yields information about the system's behaviour. In this section we concentrate on the question how to formalize the behavior of a net. Since the behavior is the most interesting aspect of a model, one can phrase this question also as: What kind of behavior is represented by a Petri net?

We will not discuss different semantics in detail. Other contributions to this book are devoted to this topic [4, 11, 21, 22]. Instead, we provide a rough land-scape of different behavioral notions that can be formulated for arbitrary Petri net variants that are equipped with dynamic behaviour, formulated by means of an occurrence rule.

We distinguish different ways to formalize single runs, namely sequences, causal runs and arbitrary partially ordered runs. Orthogonally, we distinguish single runs, tree-like structures representing more than one run, and graphs, representing all runs and taking cyclic behavior into account.

5.1 Runs

Given a Petri net with initial marking, not only a single transition can occur but also sets of transitions, constituting a run. We call the occurrence of a transition in a run an event. In the sequel, runs for different semantics will be sketched. For each semantics, we provide a Petri net notation for its runs.

The behavior of a net is a net
Runs of Petri nets consist of events and pre- and post-conditions
that generate a (partial) order.
Runs can always be represented by nets.

For sequential semantics, representing runs by nets is not usual. Instead, often words or sequences are used to formalize totally ordered runs. Automata-like trees and graphs represent the entire behavior. In this paper, we represent all types of runs by Petri nets. An obvious advantage is that, using this unifying approach, different semantics can be easier related and compared which each other. However, we do not claim that this representation is better readable than alternative graphical or textual representations.

In the sequel, the Petri net modeling a system will be called system net, to avoid confusion.

An *occurrence sequence* describes a sequential view on a single run. In the initial state, some transition can occur yielding a follower state. In this state, again some transition occurs, and so on. Hence the events of an occurrence sequence are totally ordered and can be represented by a sequence of transition names (as the name occurrence sequence suggests): $t_1 t_2 \ldots t_n$ for finite occurrence sequences with n events or $t_1 t_2 t_3 \ldots$ for infinite occurrence sequences. Notice that, for $i \neq j$, t_i and t_j might denote the same transition. Sometimes all intermediate global states are represented as well. However, they do not provide any additional information because each global state can be calculated from the subsequence leading to it and the initial state, using the occurrence rule.

A sequential run can also be conveniently represented by a very simple Petri net, where places represent tokens and transitions represent events. An example is shown in Figure 3. In general, each place in the pre-set of a transition represents a token of the marking enabling that transition, and similarly for post-sets. In this example, the number of tokens is two for all markings, but this is not the case in general. The net representation of an occurrence sequence is unique up to isomorphism.

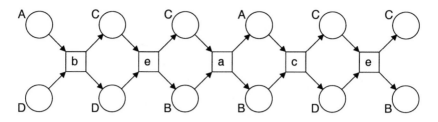

Fig. 3. A Petri net representing the occurence sequence *b e a c e* of the system net from Figure 1

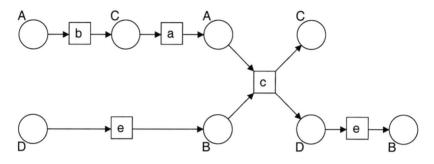

Fig. 4. A process net of the system net from Figure 1

A *process net* is a Petri net representing all events of a run and their mutual causal dependencies. Any such dependency states that a transition can only occur after another transition has occurred. General dependencies are generated by immediate dependencies, stating that a transition occurrence creates a token that is used to enable the other transition. These tokens are represented by places of the process net. Reasons for immediate dependencies are always explicitly modeled in the system net. So there is a close connection between the vicinities of a transition representing an event of a process net and the vicinity of the corresponding transition of the system net. Process nets have specific syntactic restrictions:

- Each place has at most one input transition and at most one output transition, representing the creation and the deletion of a token instance in one single run.
- The places with empty pre-set correspond to the initial token distribution which is given by the initial state.
- The relation "connected by a directed path" is a partial order, i.e., a process net contains no cycles. This is due to the fact that this relation represents the dependency relation, which obviously is acyclic.

Figure 4 shows an example of a process net.

The next semantics under consideration is given by arbitrary *partially ordered runs*. Process nets induce partially ordered sets of events. Occurrence sequences induce totally ordered sets of events. Sometimes arbitrary partial orders which

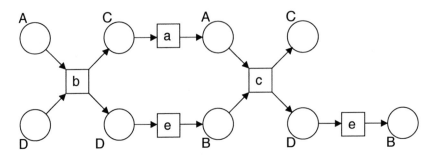

Fig. 5. A Petri net representing the process term $(b \,;(a + e)\,;c\,;e)$ of the system net from Figure 1

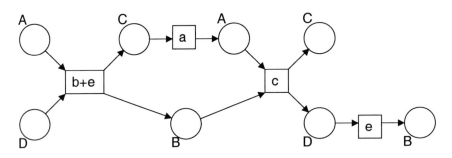

Fig. 6. A Petri net representing a run with the step $\{b, e\}$

define more dependencies than a process net and less dependencies than an occurrence sequence are useful. For example, when a Petri net variant contains timing information then it might be useful to define a relation "later than". This relation can express that an event occurs after another event, even when there is no token that constitutes an explicit dependency between the events.

Another example is given by a so-called process term semantics (see e.g. [11]). A process term such as $(b\,;(a+e)\,;c\,;e)$ is a generalization of the sequence representation of a sequential run. It describes that transitions a and e occur concurrently, both after b, and both before c, which occurs before e. A Petri net representation of this term is given in Figure 5. As discussed in [11], process terms do not have the expressive power to describe arbitrary process nets. However, sets of process terms can be used to specify an arbitrary process net.

Steps represent sets of simultaneous events. Simultaneous occurrences and concurrent occurrences of transition are different in general. Being simultaneous is a transitive relation whereas concurrency is not (in the above example of process nets, the events labeled by b and e are concurrent, the events labeled by e and a are concurrent but the events labeled by b and a are not concurrent). In general, concurrent events can occur in a step but not each step refers to a set of concurrent events. A Petri net representation of the run given by the process net of Figure 4 using the step $\{b, e\}$ is shown in Figure 6.

Since a run can be infinite, all the mathematical objects corresponding to the above representations of a run can be infinite as well.

5.2 Trees

Two different runs can start identically and then proceed differently. A compact representation of these runs contains the common prefix only once and then splits for the different continuations. This representation also explicitly shows after which events there exist alternative continuations (in Petri net theory, alternatives are also called choices or conflicts). This construction can be performed for arbitrary sets of runs and for all representations of runs listed above. Taking the Petri net representation of occurrence sequences and our above example, the occurrence tree of Figure 7 is obtained this way. Notice that this net, seen as a graph, is not really a tree but only a tree-like structure which we call tree by abuse of notation. When markings are represented by single vertices, which is the usual way to draw occurrence sequences, then the resulting graph actually is a tree.

If the reason for constructing an occurrence tree is only to identify the set of reachable markings, then it is not necessary to consider any event leading to a marking that was already identified as reachable before. In our example shown in Figure 8, it suffices to consider the occurrence sequences $e\,b$ and b because the marking reached after $e\,d$ or $b\,a$ is the initial marking, the marking reached after $e\,c$ is also reached after b and the marking reached after $b\,e$ is also reached after $e\,b$ (these are all possible continuations). In other words, we can cut the complete tree after the occurrence of $e\,b$ and after the occurrence of b.

In the example, any sequential construction of the occurrence tree will stop after three events if the above cut criterion is used. In general, a Petri net can have infinitely many different reachable markings. Then there still exists a finite tree-like structure that provides at least some information on the reachable markings: If the above stop criterion is changed to: "stop if a transition that occurred previously produced a marking that is smaller than the one produced by the current transition" then the so-called coverability tree is obtained (see [8]).

Tree-like structures can also be constructed from process nets. The resulting nets are called unfoldings of the system net. Again, cut criteria can be used to obtain finite representations of the behavior. For unfoldings representing all process nets, these criteria are given by cut-off transitions, as defined in [15]. A similar concept can be used for unfoldings obtained from an arbitrary subset of process nets [10].

When process terms do not only have operators for sequential and concurrent composition but moreover allow to express alternatives, then the corresponding Petri net representation is a tree-like structure obtained by glueing common prefixes of their Petri net representations. Likewise, it is not difficult to define a corresponding concept for steps.

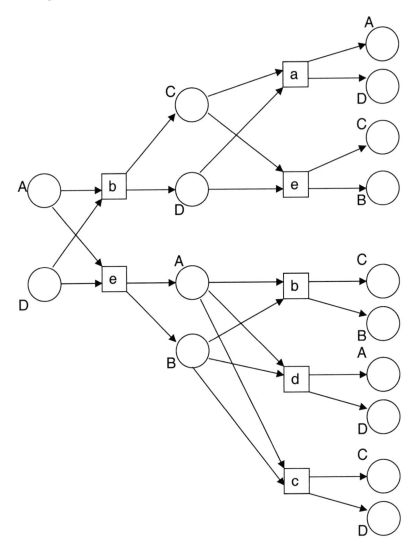

Fig. 7. A Petri net representing an occurrence tree of the system net from Figure 1

5.3 Graphs

In addition to the glueing of common prefixes of runs, one can identify sets of places that represent the same marking, to be explained next. In the previous subsection we suggested to stop the tree construction when the post-set of an event represents a marking that is already represented by the places of the post-set of another event. The next step to obtain graphs is simply performed by adding the new event and drawing arcs from it to all places of the set of places that represent the reached marking. The graph obtained this way is the reach-ability graph of the system net. Actually, the usual definition of a reachability

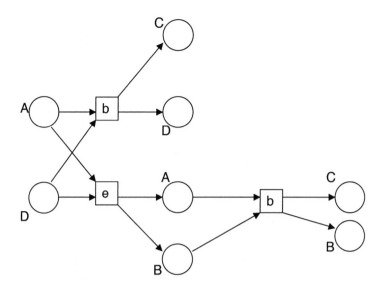

Fig. 8. A Petri net identifying all reachable markings of the system net from Figure 1

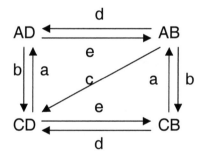

Fig. 9. The reachability graph of the system net from Figure 1

graph employs markings as nodes and transitions as arc labels, see Figure 9. It is not difficult to see that our Petri net notion of reachability graphs is equivalent, see Figure 10.

Similarly, one can construct coverability graphs from coverability trees to obtain a smaller representation of the entire behavior. Steps can also be taken into account in reachability graphs in the obvious way. However, for process nets and other Petri nets describing partially ordered runs or trees there is no obvious way to construct graphs representing the entire behavior. The reason is that markings of these nets are properly distributed. In fact, glueing all places and transitions with respective equal label usually resembles the original net, and in this case nothing is gained by the construction.

For process terms, loops in the corresponding graphs correspond to additional operators for iteration.

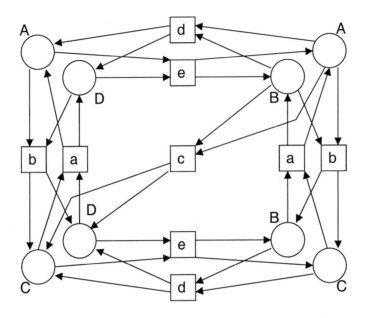

Fig. 10. The Petri net representing the reachability graph of the system net from Figure 1

6   A Formalism Equipped with Analysis Methods

A considerable amount of the huge number of Petri net articles published in the past thirty year is devoted to the analysis of Petri nets. Clearly, here is no space to give a survey of all these results. Instead, we provide a classification of different methods that are often summarized under the word "analysis".

6.1   Simulation

Simulation means creation and investigation of runs. In most cases, not all runs of a model can be generated. Runs might be infinite, and the set of runs might be infinite. But even in the case of finitely many finite runs their number often is too large to allow a complete simulation of all runs of a given Petri net. Therefore, simulation usually considers only a part of the system's behavior. Like testing of programs, simulation can thus only identify undesirable behavior but cannot prove the correctness of a model, as long as not all possible runs of a model are simulated.

Simulation can be performed by playing the token game by hand or even in mind. This procedure is quite error-prone. When finding an undesirable behavior, it is hard to say whether the identified error is due to a design error or to a simulation error. Therefore simulation is usually done by computer tools.

Many simulation tools just offer a visualization of the token game such that the constructed run is represented only implicitly. Other tools create runs which can be represented to the user and also can be input to further analysis. For example, the VIPtool [10] creates process nets which can be analyzed with respect to a given specification.

There are other applications of simulation approaches in the context of performance evaluation, where quantitative measures, e.g. about the average throughput time of a system, are derived in a simulative way.

6.2   Analysis

Analysis in a narrow sense means to gain information about a Petri net model. For example, results of analysis can be the information on deadlock-freedom, liveness, boundedness and the like. Analysis can also yield information which is useful for proof methods. For example, an analysis tool can calculate a set of place invariants (see below).

Analysis of syntactical properties such as the free-choice property [5], strong connectedness etc. are based on the structure of the net, whereas analysis of behavioral properties such as deadlock-freedom usually needs the construction of a tree or graph representing the behavior. For many properties of general Petri nets it can be proven that essentially there does not exist any more efficient way to decide the property [16]. Exceptions only exist for subclasses of Petri nets such as free-choice nets.

Quantitative analysis provides quantitative results. In contrast to simulation, quantitative analysis computes these results from quantitative parameters and attributes associated to a Petri net.

6.3   Verification

The term "Verification" often comes along with the term "specification". Verification finds out whether a given specification holds true. There are numerous ways to formulate a specification. Like analysis, a typical approach to verification is based on the construction of the reachability graph of a Petri net which is then further investigated. More efficient approaches construct reduced reachability graphs that still carry all the information that is relevant for verification. For example, reachability graphs can be reduced by employing symmetries and methods reducing the redundancy which is caused by concurrency. In particular, the so-called stubborn set method [34] has proved to yield significant reductions of reachability graphs without spoiling information about the possible enabledness of transitions. An alternative efficient approach to verification employs unfoldings of nets, i.e. a behavior description based on process nets, see [15].

6.4   Semi-decision Methods

A semi-decision method is a method for verifying a given property which has either the possible answers "yes" and "don't know" or the possible answers "no" and "don't know". Thus, one possible output provides a useful information whereas the other one just states that this method is of no use in the current case.

A well-known example for an efficient semi-decision method for deciding reachability of a marking of a place/transition net is given by the marking equation: For every reachable marking this equation system has a nonnegative integer-valued solution. A marking is proven to be unreachable if it is shown that no such solution exists, whereas the existence of a solution does not prove anything. Weaker but more efficient methods look for arbitrary integral solutions, for nonnegative rational-valued solutions or even for rational-valued solutions. For a discussion of the respective expressive power and the complexity of these approaches, see [9].

6.5   Proof Methods

The most prominent formal concepts for Petri net analysis are place invariants for almost all variants of Petri nets and siphons and traps for place/transition nets. Perhaps surprisingly, these concepts are of little use for analysis in the narrow sense of 6.2. Usually the existence of a specific place invariant is not a property relevant for a user. In general, the number of minimal (non-negative non-zero integral) place invariants, the number of siphons and the number of traps can grow exponentially with the size of the net. So even an enumeration of these objects is not feasible.

Instead, place invariants, siphons and traps can be used very elegantly for proving that a desirable property holds. However, the user has to find the necessary invariants, traps and siphons first. Tools can help to verify that a suggested invariant actually is an invariant of the investigated net. Place invariants have close relations to the semi-decision method based on the marking equation. Namely, there is a place invariant for proving a property if and only if the marking equation has no rational-valued solution. A similar result holds for so-called modulo place-invariants and integral solutions of the marking equation [7]. In this sense, the proof methods based on these concepts can be viewed as nondeterministic semi-decision algorithms; if the right place invariant, siphon, or trap is guessed, then its characteristic property can esily be verified and it can be used to prove the desired property.

Place invariants, siphons and traps are based on simple arguments on assertions; the associated properties are preserved by arbitrary change from a (not necessarily reachable) marking to a follower marking that is allowed by the occurrence rule. Actually, only the changes from reachable markings are relevant. The restricted expressive power is due to the possibility that the property under consideration is not preserved by an unreachable marking change, and hence the argument cannot be used, although it might be preserved for all reachable changes. However, the restriction to reachable changes is not easy because it requires the construction of all reachable markings. In most cases this construction is very time consuming and provides an immediate proof of the desired property without using assertions.

6.6   Validation

Whereas verification checks a model against a given specification, validation checks a model against the modeled system or against desired properties of the system. If the model is not correct then analysis and verification of the model is of little use because in this case the behavior of the system might significantly differ from the behavior of the model.

Validation of a model means to compare the model with either the existing system or a planned system where some properties are known. It can be done on a structural level by comparing all the components (elements and connections) of the model with reality. A further step in validation uses simulation: Simulated runs of the model should correspond to runs of the system and vice versa. Verification and analysis can also be used; when applied to the model, the results should coincide with corresponding properties expected from the system.

The investigation of a system by analysis of its model only makes sense if the model can be assumed to be valid. So it is useful to proceed in two steps: First the above mentioned methods are applied to ensure that the model is correct with respect to the modeled system, i.e., that it is valid. After that, it can be assumed that the model's behavior and the system's behavior are closely related, and further application of the above methods to the model provides information about the system's behavior.

7   A Model of a Distributed System

Complex distributed systems with a large number of connected components exhibit a very complex behavior. Every component might depend in some way on each other component. The set of global states reachable by consecutive transition occurrences often grows exponentially in the size of the system. The central feature of Petri net theory is that

Petri nets can manage the complexity of large systems.

Instead of yielding rapidly growing state spaces, the number of places grows linearly with the size of the modeled system. The reachable states do not have to be represented explicitly but are implicitly given by the many combinations of local states. Instead of explicitly stating all direct or indirect dependencies, only the immediate dependencies are represented – other dependencies follow transitively in runs of the model. It does not matter whether transitions and their vicinities are taken as elementary building blocks, as discussed in Section 4, or whether places and their vicinities representing the relevant actions are considered. The result is the same: a Petri net. This way of modeling has not only the advantage of keeping the complexity of the model manageable, it also resembles the modular structure of the modeled system.

However, in general the single components of a system and their connections cannot be identified in its Petri net model. Petri nets are not equipped with notions for physical distribution, channels, messages or locality (at least, this holds for the most common Petri net variants). The lack of these apparently important concepts is often claimed to be a disadvantage of Petri nets. Other modeling languages are based on local components, have means for communication such as message passing or synchronization and provide elegant ways for composing components, refinement of components etc.

Comparing Petri nets with such notions, it turns out that Petri nets support all these concepts as well. Since Petri nets constitute a very general language, different concepts for locality, refinement, composition and communication can be expressed.

When using Petri nets for modeling distributed systems of a specific kind, this model is easier to understand when its components, the communication between components etc. are easily identified in the model. Since in conventional Petri nets the information about these concepts is lost, it is useful to define languages that are based on Petri nets but restrict to certain macros defining possible building blocks in a given paradigm. Petri nets are general enough such that this kind of macros can be easily defined for different modeling paradigms. On this macro level, it is easy to understand the model from a behavioral view, because the model is still a Petri net. It is also easy to identify components and communication because they are formulated in terms of macros. By definition of the macros, restricting to certain sets of macros ensures that the model obeys the rule for the given paradigm. For example, a model of a message-passing system can not use a macro representing a shared variable. Here are some examples for suitable macros:

A *local component* can be given by a subnet which is connected to other subnets in a very restricted way. Different states of a component can be represented by different places or by a single high-level place (i.e., a place of a high-level Petri net). It is useful to give a graphical representation for the subnets that represent components. In a more compact representation, a single subnet of a high-level net might represent a set of similar components [27].

A *variable* can be represented by a special kind of place that is only connected to transitions that read or write the variable (see Section 4). It is useful to give a special graphical representation for variables.

A *message channel* can be represented by a specific place. Only transitions of components that actually have access to the channel can remove a token. Sometimes a channel is represented as a chain of places and transitions. In this case it is useful to provide a coarser view by a single place that is refined to this chain.

Synchronous communication can only be applied to transitions that model interfaces of components. It is useful to provide a graphical representation for these transitions. When synchronized, two transitions occur together. This can either be defined as part of the semantics or an additional common transition is introduced.

The concept of *Asymmetric Synchronization* means that a transition can only occur together with another transition, which in turn can only occur alone if the first one is not enabled [33, 19, 12]. This concept frequently occurs when modeling modular technical systems. There exist translations to traditional Petri nets. However, the number of transitions in such translation can grow rapidly. Also, a macro notation using special event arcs keeps the nets more readable.

8   Conclusion

This paper presented the author's selection of possible readings of Petri nets, commenting on them from the personal perspective. It was not meant to be technical and attempted no comparisons between models, nor between different variants of nets. Instead, it tried to concentrate on the common grounds of Petri net variants. There would have been hundreds of opportunities to add references to other work but the authors avoided to create an annotated bibliography. So also the selection of pointers to the literature was a (sometimes biased) personal choice.

Some readers my feel that some topics should have been treated in greater detail, or in a more technical fashion. We will end the paper with a couple of links to further readings which we left out because their respective topics concern only a part of the world of Petri nets.

There exists prosperous research on Petri nets equipped with *time*. Time can be associated to transitions, to places, or to arcs. Time can be deterministic, i.e., the occurrence of a transition always lasts the same amount of time, or it can be stochastic. Timed Petri nets and the concept of concurrent runs are not a very good match but they do not totally exclude each other. The major part of research on timed Petri net is considered with performance evaluation, i.e. with the calculation and estimation of throughput time, delays etc. of the modeled systems.

As mentioned at some places above, there are different *levels* of Petri nets – from low-level to high-level. Actually, this dimension allows for many more variants than suggested by these terms. Different high-level Petri nets emphasize a syntactial view or a semantical view or a functional view etc. On the highest level in this classification, a Petri net represents an entire class of models which all satisfy a syntactically given specification. These nets are called algebraic Petri nets. They involve algebraic specifications. Any interpretation of such a specification leads to another concrete Petri net.

There exists very many translations and correspondences between Petri nets and *other formalisms* for concurrent systems, some of them mentioned above. A related topic is the integration of nets and other formalisms. For example, transitions can be inscribed by expressions of a programming language. Then every occurrence of a transition corresponds to a run of the respective program, taking the tokens as input and output values. Other integrating approaches combine Petri nets with formal data models. When Petri nets are used in the process of system design then there is no way of using them totally separated from other methods. So integration concepts, as well as respective tools, are necessary. Although some solutions in this directions have been developed in the last years, we consider this research direction most urgent to further disseminate the very idea of Petri nets in practical applications.

Acknowledgement

We a grateful to anonymous referees for their helpful remarks.

References

1. W.M.P. van der Aalst, J. Desel and A. Oberweis (Eds.) *Business Process Management.* Springer, LNCS 1806, 2000.
2. W.M.P. van der Aalst. Formalization and Verification of Event-Driven Process Chains. *Information and Software Technology,* 41(10):639-650, 1999.
3. J. Becker, M. Rosemann and C. von Uthmann. Guidelines of Business Process Modeling. In [1], pp. 30–49.
4. R. Bruni and V. Sassone. Two Algebraic Process Semantics for Contextual Nets. In this volume.
5. J. Desel and J. Esparza. *Free Choice Petri Nets.* Cambridge Tracs in Theoretical Computer Science 40, Cambridge University Press 1995.
6. J. Desel. How Distributed Algorithms Play the Token Game. In C. Freksa, M. Jantzen and R. Valk (Eds.) *Foundations of Computer Science: Potential–Theory–Cognition,* Springer, LNCS 1337, pp. 297–306, 1997.
7. J. Desel, K.-P. Neuendorf and M.-D. Radola. Proving Nonreachability by Modulo-Invariants. *Theoretical Computer Science,* 153(1-2): 49–64.
8. J. Desel and W. Reisig. Place/Transition Petri Nets. In [25], pp. 122–173.
9. J. Desel. *Petrinetze, lineare Algebra und lineare Programmierung.* Teubner-Texte zur Informatik, Band 26, B. G. Teubner, 1998.
10. J. Desel. Validation of Process Models by Construction of Process Nets. In [1], pp. 110–128.
11. J. Desel, G. Juhás and R. Lorenz. Petri Nets over Partial Algebra. In this volume.
12. J. Desel, G. Juhás and R. Lorenz. Process Semantics and Process Equivalence of NCEM. In S. Philippi (Ed.) *Proc. 7. Workshop Algorithmen und Werkzeuge für Petrinetze AWPN'00,* Fachberichte Informatik, Universität Koblenz-Landau, pp. 7–12, October 2000.
13. C. Ermel and M. Weber. Implementation of Parametrized Net Classes with the Petri Net Kernel. In this volume.
14. R. Eshuis and R. Wieringa. A formal semantics for UML activity diagrams, 2000. Available at http://www.cs.utwente.nl/ eshuis/sem.ps.
15. J. Esparza. Model checking using net unfoldings. *Science of Computer Programming* 23(2): 151–195, 1994.
16. J. Esparza. Decidability and Complexity of Petri Net Problems –An Introduction. In [25], pp. 374–428.
17. M. Gajewsky and H. Ehrig. The PNT-Baukasten and its Expert View. In this volume.
18. T. Gehrke, U. Goltz and H. Wehrheim. Zur semantischen Analyse der dynamischen Modelle von UML mit Petri-Netzen. In E. Schnieder (Ed.): Tagungsband der 6. Fachtagung Entwicklung und Betrieb komplexer Automatisierungssysteme (EKA '99), pp. 547–566, Beyrich, Braunschweig, 1999.

19. H.-M. Hanisch and M. Rausch. Synthesis of Supervisory Controllers Based on a Novel Representation of Condition/Event Systems. In *Proc. IEEE International Conference on Systems Man and Cybernetics*, Vancouver, British Columbia, Canada, October 22-25, 1995.

20. M. Jüngel, E. Kindler and M. Weber. The Petri Net Markup Language. In S. Philippi (Ed.) *Proc. 7. Workshop Algorithmen und Werkzeuge für Petrinetze AWPN'00*, Fachberichte Informatik, Universität Koblenz-Landau, pp. 47–52, October 2000.

21. J. Meseguer, P. Ölveczky, and M.-O. Stehr. Rewriting Logic as a Unifying Framework for Petri Nets. In this volume.

22. J. Padberg and H. Ehrig. Parametrized Net Classes: A uniform approach to net classes. In this volume.

23. C. A. Petri. *Kommunikation mit Automaten*. PhD thesis, Univ. Bonn, 1962.

24. W. Reisig. *A Primer in Petri Net Design*. Springer, 1992.

25. W. Reisig and G. Rozenberg (Eds.). *Lectures on Petri Nets I: Basic Models*. Advances in Petri Nets, Springer, LNCS 1491, 1998.

26. W. Reisig and G. Rozenberg (Eds.). *Lectures on Petri Nets II: Applications*. Advances in Petri Nets, Springer, LNCS 1491, 1998.

27. W. Reisig. Elements of Distributed Algorithms: Modeling and Analysis with Petri Nets. Springer, 1998.

28. D. Ross and K. Schoman. Structured analysis for requirements definition. IEEE Transactions on Software Engineering Vol. SE-3, No. 1, pp. 6-15, 1977.

29. F. J. Rump. *Geschäftsprozeßmanagement auf der Basis ereignisgesteuerter Prozeßketten*. Teubner-Reihe Wirtschaftsinformatik, B. G. Teubner, 1999.

30. J. Rumbaugh, I. Jacobson, G. Booch. *The unified Modeling Language Reference Manual*. Addison-Wesley, 1999.

31. A.-W. Scheer and M. Nüttgens. ARIS Architecture and Reference Models for Business Process Management. In [1] pp. 376–390.

32. E. Smith. Principles of high-level net theory. In [25], pp. 174–210.

33. R. S. Sreenivas and B. H. Krogh Petri Net Based Models for Condition/Event Systems. In *Proceedings of 1991 American Control Conference*, vol. 3, pp. 2899-2904, Boston, MA, 1991.

34. A. Valmari. The State Explosion Problem. In [25], pp. 429-528.

35. H. Weber, H. Ehrig and W. Reisig (Eds.). *Proc. Colloquium on Petri Net Technologies for Modelling Communication Based Systems*. Berlin, October 1999.

36. H. Weber, S. Lembke and A. Borusan. Petri Nets Made Usable: The Petri Net Baukasten for Application Development. In this volume.

The »Petri Net Baukasten«: An Overview[*]

Meike Gajewsky and Hartmut Ehrig

Technical University Berlin
Institute for Communication and Software Technology
{gajewsky,ehrig}@cs.tu-berlin.de

Abstract This paper presents an overview of the »Petri Net Baukasten« developed by the "DFG-Forschergruppe PETRINETZ-TECHNOLOGIE" in Berlin. The »Petri Net Baukasten« provides a unified presentation with different views on theory, application, and tools of Petri nets: The Expert View, the Application Developer View, and the Tool Developer View. All of these views are related to a Common Base, which comprises Petri net notions in a semiformal description. The relations establish so-called Petri net techniques given by a combination of Petri net types, corresponding methodological procedures, formalizations, and tools from the different views. The »Petri Net Baukasten« represents the Petri net techniques in a structured and application-oriented way, which yields an application oriented Petri net technology. It bridges the gap between theory, practice and tools for Petri nets. In this paper the basic concepts are summarized.

1 Introduction

Within the last four decades of research on Petri nets numerous Petri net notions and methods as well as tools and tool environments have evolved. These have been successfully employed in various application areas, such as automatic production, control systems, workflow management etc. In such large scale applications different Petri net variants, called Petri net types, can be employed. A Petri net type represents a Petri net variant including a set of techniques based on that variant like structuring, analysis, and verification techniques. To identify the adequate Petri net types and a method for the employment of Petri nets within the system development process for a specific application domain is still a difficult task. Hence, there is a strong need for a structured access to various Petri net techniques comprising methodological procedures, tool support and formal techniques. A structured presentation of various methods and techniques is called a technology. The strong motivation for such a technology derives from the rich and diverse Petri net theory and its various applications [Rei85,Jen97,Sch99].

---

[*] This work is part of the joint research project "DFG-Forschergruppe PETRINETZ-TECHNOLOGIE" between H. Weber (Coordinator), H. Ehrig (both from the Technical University Berlin) and W. Reisig (Humboldt-Universität zu Berlin), supported by the German Research Council (DFG).

H. Ehrig et al. (Eds.): Unifying Petri Nets, LNCS 2128, pp. 26–53, 2001.
© Springer-Verlag Berlin Heidelberg 2001

In this paper we give an overview over such a technology, called »Petri Net Baukasten«, which has been previously presented in [WER99,DFG99]. The »Petri Net Baukasten« provides a classification of Petri nets and corresponding notions independent of their use in applications, their formalizations, and tool support. The semi-formal classification is given by class diagrams that describe Petri net types and its notions. It is represented using UML and constitutes the base of the »Petri Net Baukasten«, called the Common Base.

These representations of Petri net notions are given in specific views. They concern the use of the Petri net notions within a certain application domain, their formalization in the theory of Petri nets, and their implementation as Petri net tools. Hence, these views are called Application Developer View, Expert View, and Tool Developer View respectively. One of the distinguished application domains of the research group PETRI NET TECHNOLOGY is the area of business processes [DG91,MOS93,ADO00]. Various Petri net techniques provide excellent means for the description of business processes and hence represent a good basis for business process management. These techniques allow the visualization, the formal description, early evaluation as well as verification of business processes. Although developed with special attention to the area of business processes, we claim that the »Petri Net Baukasten« is equally useful for other application domains like the application domain of traffic control systems.

This paper is organized as follows: in Section 2 we define the notion of Petri net technology and show that the »Petri Net Baukasten« is such a technology. In Section 3 to 6 we present an overview of the Common Base and the different views of the »Petri Net Baukasten«. These sections are mainly extracts of the corresponding parts of [WLB01,Gaj99,Web99] and describe the different views on a conceptual level. In Section 7 we discuss installments, maintenance and evolution of the »Petri Net Baukasten«. In the appendix we present the Expert View in more detail. Details of the other views and further results can be found in [WLB01,EW01] of this volume and in [DFG99,Pad99,Deh99,Web99].

Acknowledgments. This work is part of the joint research project "DFG-Forschergruppe PETRINETZ-TECHNOLOGIE" between H. Weber (Coordinator), H. Ehrig (both from the Technical University Berlin) and W. Reisig (Humboldt-Universität zu Berlin), supported by the German Research Council (DFG). The conceptualization of the »Petri Net Baukasten« has been a major task, which has been achieved by an intensive cooperation among all members of the DFG-Forschergruppe. We would like to express our gratitude to all members as documented in [DFG99], which is the basis for main parts of this paper.

2   Petri Net Technology and the »Petri Net Baukasten«

In this section we want to explain the notions PETRI NET TECHNOLOGY and »Petri Net Baukasten« and we will show that the »Petri Net Baukasten« is an application oriented PETRI NET TECHNOLOGY.

2.1 Petri Net Technology

We start with the general definitions of "technology", "method", and "technique" according to WWWebster [MW99].

> A *technology* is in general a manner of accomplishing a task especially using technical processes, methods, or knowledge.
> A *method* is a discipline that deals with the principles and techniques of scientific inquiry. It implies an orderly logical effective arrangement usually in steps.
> A *technique* is a body of technical methods as in a craft or in scientific research.

According to these general definitions we define a Petri net technology to be a Petri net based manner of accomplishing the task of system development using methods for employing Petri net techniques. A method deals with principles of employing Petri net techniques that in general should answer the following questions:

- What kind of Petri net techniques should be used?
  A distinguished set of Petri net techniques constitutes the body of a method.
- How do these Petri net techniques help to accomplish a given task?
  There has to be a method how to use the different Petri net techniques in order to accomplish a given task.
- What support for these Petri net techniques is available?
  Method, formal foundation, and tool support for the chosen Petri net techniques should be named, if available.
- Are these Petri net techniques reliable?
  Those of the chosen Petri net techniques, that have a formal foundation, and the kind of formal consistency, that can be obtained, should be stated.
- How to use the different Petri net techniques?
  Methodological procedures for the use of the different Petri net techniques should be provided, if they are available.

A Petri net technology is called application oriented if it is suitable for system development in different application domains and allows an application domain specific interpretation of Petri net notions.

2.2 The »Petri Net Baukasten«

As mentioned already in the introduction the research group PETRI NET TECHNOLOGY has developed the »Petri Net Baukasten«, documented in [WER99], [DFG99] to enable a more straight forward understanding of Petri net types and the development of Petri net tools, to aid in the application of Petri net types, and to provide a unified represention of the formal definition of Petri net types. The »Petri Net Baukasten« includes a classification concept for Petri nets that serves these purposes. Moreover, the concept of the »Petri Net Baukasten« is governed by the following basic requirements.

- The »Petri Net Baukasten« has to provide different views for different purposes.
- The classification of Petri net types has to be achieved in a semi-formal manner.
- The »Petri Net Baukasten« has to comprise a large variety of Petri net types, which are important for certain application domains.
- The »Petri Net Baukasten« has to provide for each Petri net type a rigorous (i. e. mathematical formal) description to ensure consistent techniques including analysis, structuring and verification.

The »Petri Net Baukasten« supports the application developer, the tool developer and the Petri net expert in their different objectives.

1. The Application Developer View enables an engineer developing an application
   - to use the application-oriented interpretation of Petri net notions,
   - to find the Petri net technique that serves him best in the development of the application,
   - to use the adequate method for developing the application, and
   - to rely on the chosen Petri net technique.
2. The Expert View enables the Petri net expert
   - to define new types and notions in a uniform way,
   - to state properties of all variants of Petri nets in a formal and constructive manner,
   - to transfer results between net types, and
   - to make results and notions available for applications.
3. The Tool Developer View enables tool developers
   - to find appropriate tools,
   - to fit tools together in a prototyping way,
   - to add and change tools and tool references, and
   - to provide tools and modular prototyping of tools.

These three aspects describe the most important views on Petri nets. In order to systematize Petri net notions and to relate the different views a basic classification of Petri nets is provided. This classification includes different Petri net types, like place/transition nets and high-level Petri nets, and corresponding Petri net notions like structuring, verification, and analysis. The Petri net types are classified by attributes and attribute values. The classification is represented in the Common Base and connects the different views in the following way: A Petri net type in the Common Base comprises the notions that belong to a Petri net technique, the Application Developer View provides methodological procedures for the use of certain Petri net types and methods for the system development process, the Expert View represents the mathematical foundations for formal Petri net techniques, and the Tool Developer View provides tools and tool development facilities.

The relationship of the three different views and their Common Base is depicted in Figure 1. The Common Base relates the different views and their representation of the same concepts, namely it describes Petri net notions which are

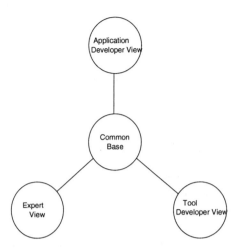

Figure 1. The ≫Petri Net Baukasten≪

- interpreted in the application domain as a part of the Application Developer View,
- formalized using mathematical definitions and propositions in the Expert View, and
- implemented using algorithms and program code in the Tool Developer View.

In the following we present a first overview of the concepts of the Common Base and the three different views which are given in Section 3–6 in more detail. Finally, we show that the ≫Petri Net Baukasten≪ is an application-oriented PETRI NET TECHNOLOGY.

2.3   Common Base

The Common Base, see [Deh99,WLB01], encompasses all Petri net types relevant for a specific application domain, it integrates all these types into a common scheme and defines the relationship between the Petri net types. It is intended that the Common Base is structured as simple as possible and understandable to application developers, tool developers and Petri net experts alike. The structuring concept that is considered to be simple and still powerful enough to capture all the properties of the different Petri net types is a specialization/generalization relationship between the different Petri net types. The Petri net classification in the Common Base is a specialization hierarchy along distinguishing characteristics. This classification is represented in the Common Base using class diagrams of UML. The distinguishing characteristics are given in terms of attributes and attribute values. These describe Petri net notions on a conceptual level. The domain dependent interpretation of these concepts are part of the Application Developer View. The encoding into software belongs to the Tool Developer View and the mathematical formalization of these concepts is part of the Expert View.

2.4   Application Developer View

The Application Developer View, see [Lem99,WLB01], provides the prerequisites for an application-oriented assistance of Petri net based development projects. The planning of development activities in an application domain is supported in the following way: The Application Developer View provides information for the planning of development activities, i. e. methods of employing Petri net techniques. This information supports the application developer to choose the appropriate Petri net techniques, based on tools, methodological procedures and formal Petri net techniques for the development activities. Additionally, modelling activities are supported by sample and standard solutions, and methodological procedures for a chosen Petri net type. The Application Developer View includes information on Petri net notions and Petri net techniques in an application- and problem-oriented way. These so-called assistance concepts yield application domain specific interpretations of Petri net notions of the Common Base.

2.5   Expert View

The Expert View provides the formal foundation of Petri net types in the »Petri Net Baukasten« in terms of a mathematical presentation of the underlying notions and results. The Expert View is given in a structured way. It comprises formal Petri net techniques, abstract Petri net frames, actualizations, and transformations. Each of these notions comprises a coherent and consistent piece of Petri net theory. Formal Petri net techniques, and transformations are directly related to the Common Base. Abstract Petri net frames, actualizations, and abstract transformations describe relation and dependencies of formal Petri net techniques and transformations on a more abstract level. Hence, these yield a uniform description of Petri nets as a foundation of the classification given in the Common Base.

More details on the Expert View can be found in Section 5 and the Appendix.

2.6   Tool Developer View

The main task of the Tool Developer View, for details see [Web99,EW01], is to provide support for tool development. This comprises the management of existing tools, facilities for tool development as well as possibilities to extend tools. Petri net tools support the Petri net based system development. They provide support for editing, simulating, structuring, and analyzing a Petri net variant. The management of existing tools is also important for the rest of the »Petri Net Baukasten«, since it offers tool support for the Petri net types of the Common Base. A Petri net tool corresponds to a Petri net type of the Common Base if it supports the notions comprised by that Petri net type. The Tool Developer View provides support for the development of tools in terms of object-oriented and parameterization concepts, a component-oriented approach for already existing tools and encoded algorithms such that they become easily accessible. The Petri net type used in the encoded algorithm is given in the Common Base. The formal representation of the algorithm is given in the corresponding formal Petri net technique in the Expert View.

2.7   Relation of Common Base with Views

The Common Base is related with the views according to different representations of Petri net notions in the Common Base and the views.

The notions given in the Common Base in an informal way by attributes are represented in the Application Developer View by application-oriented notions within a methodological procedure, explaining the use of these notions in a certain application domain. Within the Tool Developer View these notions are represented either as algorithms or as tools. The Expert View provides consistent formal Petri net techniques for the Petri net type given in the Common Base.

As an example the notion *marking of a Petri net* may be considered

- in the Common Base to be a distribution of tokens over places indicating the state of the net,
- in the Application Developer View to be a representation of documents and business objects in different processes,
- in the Tool Developer View to be a record of places, and
- in the Expert View to be an element of the free commutative monoid over the set of places.

The specialization hierarchy of the Common Base can be used for the navigation with respect to all three views. It allows embedding of new tools or new theoretical results and making them available for practice.

2.8   Petri Net Techniques in the »Petri Net Baukasten«

The »Petri Net Baukasten« provides different methods and Petri net techniques, that serve the task of system development. A method consists of the combination of various Petri net techniques. A Petri net technique is built up by a consistent set of Petri net types, formal Petri net techniques, methodological procedures, and tools. We illustrate a specific method in Figure 2 which is the sequence of three Petri net techniques where transformations between different Petri net techniques are indicated by arrows.

The »Petri Net Baukasten« is an application oriented Petri net technology in the sense of Section 2.1 since

- it establishes the base for a systematic practice for employing Petri nets in different application domains,
- it comprises different ways of the accomplishment of the task of system development,
- it allows an application domain specific interpretation of Petri net notions, and
- it provides well-defined Petri net techniques and methods.

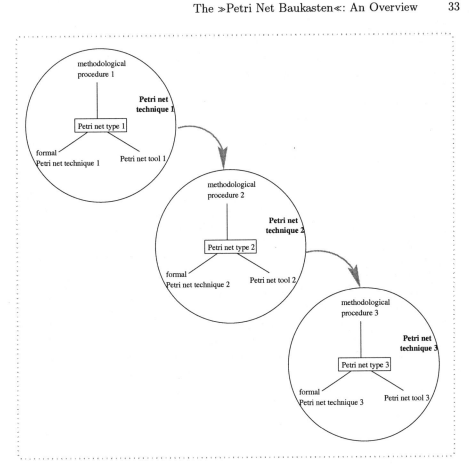

Figure 2. A Petri Net Method

In particular, the general questions concerning Petri net techniques given in Section 2.1 can be answered now in case of the »Petri Net Baukasten«:

- What kind of Petri net techniques should be used?
  Petri net techniques based on Petri net types are given in the Application Developer View for different application domains.
- How do these Petri net techniques help to accomplish a given task?
  The Petri net techniques are combined to a Petri net method (see Figure 2) given in the Application Developer View in order to accomplish a specific task for suitable application domains.
- What support for these Petri net techniques is available?
  A Petri net technique is supported by methodological procedures, formal Petri net techniques, and Petri net tools in the three views of the »Petri Net Baukasten«.

- Are these Petri net techniques reliable?
  There are corresponding consistent formal Petri net techniques in the Expert View, which are the basis for reliability.
- How to use the different Petri net techniques?
  Methodological procedures for each Petri net techniques can be found in the Application Developer View.

3 Common Base

In this section we present an overview over the Common Base of the »Petri Net Baukasten«, which establishes the relation between the different views. A fundamental requirement is that it encompasses all relevant Petri net variants and respresents them in a common scheme in order to define the relationship between these variants. The scheme itself must be simple and easily understandable to anyone working with the »Petri Net Baukasten«, including application developers, tool developers and Petri net experts. Therefore a structuring concept from software engineering is chosen, which is object-oriented. It uses specialization and generalization concepts. Different Petri net variants are related to each other via these relationships. The general scheme of the Common Base is a hierarchy of Petri net variants, called Petri net classification, depicted in Figure 3. Intuitively, the Petri net classification follows similar principles as the construction of class hierarchies in UML.

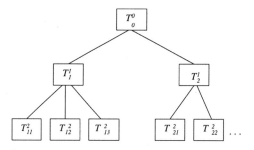

Figure 3. General diagram of the Common Base. T_0^0 represents the most simple Petri net type and $T_{jk...}^i$ represents the specializiation $jk...$ of the degree i

The object-oriented semiformal description used for the »Petri Net Baukasten« introduces Petri net variants as objects, which are called Petri net types in the following. Characteristics of a variant are described as characteristics of an object. The classification of Petri net types is based on the set of characteristics of the Petri net types. Accordingly, Petri net types whose characteristics are a subset of another Petri net type are considered as generalizations of the other type. Vice versa, a Petri net type is a specialization of another Petri net type if it carries a superset of characteristics. Petri net types classified according to this classification scheme are all specializations of the most common Petri net type at the root of this hierarchy. Analogously to the notions in object-oriented techniques, this hierarchy is called inheritance hierarchy. It depicts the inheritance of characteristics from top to bottom. Our inheritance hierarchy in the Common Base is an acylic directed graph, because multiple inheritance is allowed.

The root of our Petri net classification is given by a Petri net type comprising the characteristics: place, transition, (two kinds of) edge, activation, firing, marking, and token. Starting from the root we obtain different specialization paths which each focusses on a particular Petri net property. We distinguish three groups of specialization paths: *elementary, additional,* and *operational* paths. An *elementary* path specifies one of the basic constituents of a Petri net variant, e. g. marking limit (capacities) on places, edge weights, pre-and post-sets of transitions, token colours, etc. Optional characteristics like authorization, time, etc. are described in the *additional* paths. The *operational* path is concerned with available operations like verification, structuring, or analysis.

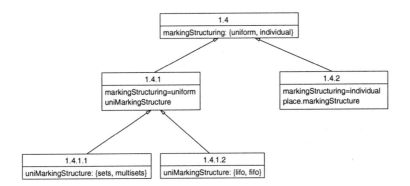

Figure 4. Specialization example for characteristic *Markingstructuring*

An example for specializations is given in Figure 4. It is concerned with the elementary characteristic *MarkingStructuring* introduced in the second specialization level. Petri net type 1.4 is specialized by Petri net type 1.4.x, where the inherited characteristics are more specific and new characteristics are added. More details on the underlying principles for specialization can be found in [WLB01] and an initial installment of about 100 different Petri net types in [DFG99].

Summarizing, the Petri net classification is built from Petri net types with characteristics ordered hierarchically by a specialization/generalization relation. A classification glossary provides a description of the meaning of the different characteristics. Moreover, a Petri net type holds an indicator to Application Developer View, Expert View, or Tool Developer View, if there is a direct relation to the corresponding view. Such a relation exists if the Petri net type is represented in that view. More details about these indicators and the relation of the

Common Base to the different views can be found in the corresponding sections of this paper. For more details see [DFG99,Deh99] and [WLB01] in this volume.

4   Application Developer View

The Application Developer View provides adequate support for application developers within application development projects. For planning and performing the developing activities the Application Developer View offers informations to allow application developers to select suitable

- Petri net techniques,
- development methods, and
- sample solutions.

Therefore, the Application Developer View builds a bridge between practice and Petri net theory. We intend to faciliate understanding of the use of Petri nets for practioners of different application domains. To achieve better understanding we take objectives and aspects of several application domains into account and create a link to corresponding Petri net notions. In [WLB01] it is discussed in more detail how these objectives refer to application-, problem-, method-, solution-, tool-, need-, and multiple-way orientation. The support for application developers is realized by the interpretation of Petri net concepts. Therefore, it is necessary to use technical terms of application domains and map them to Petri net notions in application specific glossaries. A detailed description of the elements of the Application Developer View together with a structural model can be found in [WLB01].

Application domains are of central importance in our »Petri Net Baukasten«, as one of the major aims of the »Petri Net Baukasten« is application-orientation. Application orientation is realized by an application domain specific interpretation of Petri net notions. The description of an application domain may contain some application-oriented aspects, which are relevant entities and questions of a particular application domain. For example, the application domain of *workflow management* contains the aspects according to the workflow process definition metamodel of the Workflow Management Coalition (WfMC), documented in [WfMC00]. This includes workflow application, workflow participants, workflow process activities, workflow relevant data, etc. Application domains may be related to others, e. g. if one is the superdomain. One superdomain of *logistics*, for example, is *business process modeling*.

In the following we will briefly sketch the support for the application developer given by the Application Developer View.

The Application Developer View offers Petri net techniques as defined in Section 2.8: For this purpose, characteristics of the Petri net techniques are identified which are relevant for application developers. These characteristics concern e. g. the application domain, the Petri net type, intended objectives, or tool support. Additionally, a so-called technique-guide offers informal descriptions about the Petri net techniques.

The Application Developer View offers development methods: In our terminology a development method comprises a set of Petri net techniques, given by a Petri net type, its formalization, a corresponding tool, and a methodological procedure. This procedure is given by a sequence of development steps, which lead to a result which is required in that particular method. Moreover, the procedure describes the development activities. In contrast to a process model, a method explicitly comprises techniques, and in particular Petri net types. Each development method supports a specific application domain.

Last, but not least, the Application Developer View offers sample solutions. A sample solution is given by a Petri net model solving specific problems together with a description of the solution. The solution description uses the terminology of the corresponding application domain in order to support understandability and can be considered as a didactical support, how to model a specific problem in an application domain.

Relation to the Common Base. The relation between the Application Developer View and the Common Base is established by application domains, development techniques, sample solutions, and application-oriented aspects. They are either related to a Petri net type in the Common Base (as *Application Domain, Development Technique,* and *Sample Solution*) or to some characteristics of a Petri net type. More details can be found in [WLB01].

5   The Expert View

The Expert View provides a uniform structuring of the theory of Petri nets. It supports Petri net experts in so far as a uniform presentation of new notions is supported, and the transfer of results is faciliated. The major task of the Expert View within the »Petri Net Baukasten« is to provide the formal foundation of the Petri net types in the Common Base in terms of mathematical definitions and results based on the underlying notions.

The Expert View provides the following notions which will be explained below: abstract Petri net frames, formal Petri net techniques, actualiziations, and transformations. Schemes are used for the structured representation of these notions in order to obtain a uniform structure for the presentation of Petri net theory in the Expert View. The relation of the Expert View to the »Petri Net Baukasten« is discussed at the end of this section.

The Role and Structure of the Expert View. The role of the Expert View is to represent mathematical concepts and results on Petri nets in a structured way. The main idea is to represent *formal Petri net techniques*: A formal Petri net technique consists of a core formalism – i.e. the mathematical description of nets – and compatible operations on Petri nets like structuring, analysis, and verification techniques. Furthermore, a formal Petri net technique is the formal foundation of a Petri net type in the Common Base. A formal Petri net technique is one part within a Petri net technique as shown in Figure 2.

In order to describe different formal Petri net techniques in a generic way we use the notion of *abstract Petri net frames*. Due to their high level of abstraction these Petri net frames allow to represent different notions of Petri nets in a uniform way. An abstract Petri net frame defines basic Petri net notions and operations on Petri nets using formal parameters. Actualization of these formal parameters allows to propagate notions and results defined on the abstract level to concrete formal Petri net techniques. Note, that several abstract Petri net frames are considerd.

Transformations between different formal Petri net techniques are used to state similarities, dependencies, refinement or abstractions. Transformations are essential for Petri net based system development when changing from one Petri net technique to another in a development step, depicted as arrows in Figure 2.

Abstract Petri net frames and formal Petri net techniques are related in the following way: The *actualization* of an abstract Petri net frame yields a formal Petri net technique. This is achieved by replacing the abstract mathematical entity (formal parameter) by a concrete mathematical entity (actual parameter).

Transformations are mappings between abstract Petri net frames and formal Petri net techniques respectively. More precisely, transformations allow to transfer notions and operations on Petri nets from one formal technique to another.

Summarizing, the role of the Expert View is to

- present mathematical concepts and results on Petri nets
- provide consistent formal Petri net techniques
- relate different Petri net notions

The structure of the Expert View is depicted in Figure 5 and given in more detail in Section A.1–A.4 in the appendix together with some typical examples. It consists of

- formal Petri net techniques (see Section A.1)
- abstract Petri net frames (see Section A.2)
- actualizations (see Section A.3)
- transformations (see Section A.4)

Schemes. For a structured representation a meta notation, called schemes, is used in the Expert View. A scheme symbolizes a fixed pattern of keywords representing Petri net notions. For abstract Petri net frames, formal Petri net techniques, actualizations, and transformations a corresponding scheme consisting of relevant keywords is introduced.

Instantiation of the corresponding scheme by definitions, facts, theorems, algorihms etc. leads to a specific abstract Petri net frame, formal Petri net technique etc. Instantiations can be either explicit or by reference to a Definition/Fact/Theorem in a corresponding paper. Moreover, it might be annotated by explanations for the sake of readability.

Note, that for an instantiated abstract Petri net frame, formal Petri net technique etc. it is not necessary that every single component of the corresponding scheme is instantiated.

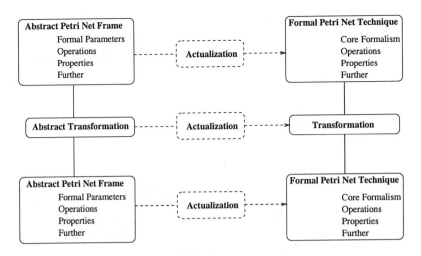

Figure 5. The Structure of the Expert View

Relation of Expert View and »Petri Net Baukasten«. The Expert View presents the formalizations of concepts described in the Common Base in a semiformal way. Each formal Petri net technique in the Expert View is a formal counterpart of a Petri net type in the Common Base. The correspondence is given by using name identity of the Petri net type and the formal Petri net technique.

This correspondence is established as follows. For each attribute of the Petri net type, there is a mathematical description in the corresponding formal Petri net technique. Of course, the relation between attributes and formalization cannot be expected to be strictly one-to-one. Hence, one Petri net type in the Common Base may have various counterparts in the Expert View. This is an advantage for the application or tool developer, because it allows to choose between different formal Petri net techniques formalizing one and the same Petri net type.

A further connection between Expert View and Common Base is given by transformations. In the simple specialization/generalization structure of the Common Base transformations are presented as attributes. Each transformation establishes a relation between two formal Petri net techniques, where a transformation attribute in the Common Base corresponds to a transformation in the Expert View.

Transformations play an important role in the »Petri Net Baukasten« as they allow to combine different formal Petri net techniques to a method as defined in Section 2 (see Figure 2 where the arrows correspond to transformations).

This shows that formal Petri net techniques and transformations in the Expert View are most important for the development activities in the Application Developer View, where the correspondence is established via the Common Base. Last, but not least, the implementation of parameterized net classes, described in [EW01], establishes a direct relation between Expert View and Tool Developer View.

6   Tool Developer View

The task of the tool developer is supported by the Tool Developer View. The support comprises various aspects of tool development:

- management of existing tools,
- facilities for tool development, and
- support for extension of existing tools.

Management of Existing Tools. The Tool Developer View has a tool administration component which stores informations about tools.The administration is based on a form containing all relevant information about a tool. The form contains information concerning e. g. a general description, the underlying Petri net type, the implemented Petri net operations, some technical instructions, information about licences and some evaluation notes. The information captured by this component faciliates the search for an adequate tool. This establishes an easy access to existing tools.

Facilities for Tool Development. For building prototypes the Tool Developer View contains the Petri net kernel (PNK), see [KW99a,KW99b] for details. It yields an infrastructure for building Petri net tools by offering standard functions and a graphical user interface. The PNK is not restricted to a particular Petri net type but covers all relevant Petri net types due to parameterization. The design of the PNK according to [Web99] was driven by the following objectives:

- Implementation of a new algorithm for analysis, simulation, or verification should be faciliated. The tool developer should be free from caring about implementation of parser, graphical interface etc, which is provided by the PNK.
- The net information should be accessible via a simple interface of the PNK, which reflects the typical mathematical notions on Petri nets such as pre- and postsets. The interface should not require the knowledge of a particular software technique, such that an unexperienced programmer should be able to efficiently use the interface within short time.
- It should be easy to integrate several algorithms which have been developed independently and to tailor them to specific application domains.
- The implementation of a tool for newly defined Petri net types should be supported by the PNK — again without implementing additional parse operations or editor functions.

Extension of Existing Tools. Extension of existing tools is a special case of tool development. It may be caused by the incremental development, by further advances of Petri net theory, or in order to fit it to the purpose of a user. This kind of support is covered by the general support for tool development as decsribed above.

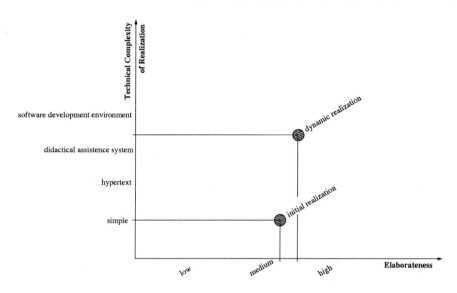

Figure 6. Installments of the »Petri Net Baukasten«

Relation to the Common Base. Petri net tools support one or more tasks within a development method. Usually, they are based on one Petri net type, and support a corresponding Petri net technique. A tool may support modeling, structuring, analyzing, testing, verifying, or transforming Petri nets, provided such operations are available in the Petri net technique. There is a relation between a Petri net type and a tool, if it supports the characteristics of that type.

A direct relation to the Expert View is described in [EW01], where the implementation of one specific abstract Petri net frame, namely parameterized net classes, with the PNK is presented.

7   Conclusion: Installment, Maintenance, and Evolution of the »Petri Net Baukasten«

In this conclusion we discuss the problems of installment, maintenance, and evolution of the »Petri Net Baukasten«.

Installments of the »Petri Net Baukasten«

Installments[1] of the »Petri Net Baukasten« can be accomplished in different ways. There are various ways to apply the »Petri Net Baukasten« in practice which are highly dependent from the intended use.

As shown by the two axes in Figure 6 an installment of the »Petri Net Baukasten« depends on the elaborateness of its contents as well as on the

---

[1] Installment has been called "realization" in previous papers about the »Petri Net Baukasten«. The new term emphasizes the fact, that it is one part of a serial story.

technical complexity. Two installments are considered up to now, the initial installment, which has been called initial realization in [DFG99], and a second installment presented in [DGLW01]. The main ideas of the initial installment can be summarized as follows:

Initial Installment of the »Petri Net Baukasten«:

Common Base

In the initial installment a class diagram representing the classification of Petri net types relevant for business processes has been established and is depicted in [DFG99] using the UML tool Rational Rose. The classification comprises about 100 Petri net types and twelve different specialization paths, that can be distinguished into three categories: elementary, additional, and operational extensions.

Application Developer View

In the initial installment an assistance system is provided on a conceptual level for the support of the application developer. There are two kinds of variants for the assistance that can be distinguished, namely assistance to find a suitable Petri net technique and assistance to find a suitable solution example. Both of these variants are supported by prescriptive, navigating, or descriptive assistance methods. The system architecture sketches the fundamental ideas about the assistance system in [DFG99]. For more detail we refer to the paper [WLB01] in this volume.

Expert View

The structured representation of significant aspects of Petri net theory in the initial installment is based on schemes for abstract Petri net frames, formal Petri net techniques, actualizations, and transformations. Each of these schemes consists of a list of relevant keywords. The initial installment of the Expert View comprises instantiations of all these schemes, e. g. parameterized net classes as an actualization of abstract Petri net frames algebraic high-level nets, coloured Petri nets, elementary nets, place/transition nets, and FunSoft nets as instantiations of formal Petri net techniques.

Petri Net Kernel in the Tool Developer View

The Petri Net Kernel in the initial installment is an object-oriented tool for the fast prototyping of simple Petri net tools. The basic idea of the Petri Net Kernel is the distinction between fixed and variable aspects of Petri nets. Variable aspects can be considered as parameters and allow the automatic generation of Petri net tools by actualization of these parameters. It implements that part of the Tool Developer View that is actually concerned with the tool development. The implementation of parameterized net classes with the Petri Net Kernel is discussed in the paper [EW01] of this volume.

Maintenance of the »Petri Net Baukasten«

Maintenance of the »Petri Net Baukasten« is the general task to update the contents of the initial and other installments within the Common Base and the three views. Specific tasks would be to add new notions or results of the theory of Petri nets into the Expert View, tools or tool descriptions in the Tool Developer View, or methods in the Application Developer View. Maintenance of the Common Base might be necessary as a consequence of the maintenance of some view, and may induce further changes in the other views. It is important to point out that maintenance preserves the structure of each view and the connection to the Common Base. However, different installments require different maintenance scenarios. A collection of various scenarios can also be found in [DFG99].

Evolution of the »Petri Net Baukasten«

In contrast to maintenance of the »Petri Net Baukasten« where the conceptual structure of the Common Base and the different views is preserved, evolution of the »Petri Net Baukasten« allows to change the structure and the technical complexity of the installment as shown in Figure 6. This requires also an update of the contents and leads to a new installment of the »Petri Net Baukasten«. An evolution step towards higher technical complexity is presented in [DGLW01], called second installment. It comprises a database, services operating on this database and explicit access for each group (user interfaces). Each of these components respects the conceptual structure, given by the different views. Moreover, a software architecture as a further refinement has been introduced.

Hopefully, further installments will be provided not only by the research group PETRI NET TECHNOLOGY, but also by other development groups in different application domains. Last, but not least, it is important to point out that the general idea of the »Petri Net Baukasten« is not at all restricted to Petri nets, but can be extended to other kinds of semi-formal and formal specification techniques in the literature as summarized in [EOP99].

References

[Deh99]   J. Dehnert. The Common Base of the Petri Net Baukasten. In Weber et al. [WER99], pages 211–229.

[DFG99]   DFG-Forschergruppe PETRI NET TECHNOLOGY. Initial realization of the »Petri Net Baukasten«. Informatik-Berichte 129, Humboldt-Universität zu Berlin, October 1999.

[DG91]   W. Deiters and V. Gruhn. Software Process Model Analysis Based on FUNSOFT Nets. *Mathematical Modelling and Simulation*, 8, May 1991.

[DGLW01]   J. Dehnert, M. Gajewsky, S. Lembke, and H. Weber. The Petri Net Baukasten: Second Installment. In *Proceedings of ETAPS-UniGra*. Elsevier, 2001. To Appear.

[DHP91]   C. Dimitrovici, U. Hummert, and L. Petrucci. Composition and net properties of algebraic high-level nets. In *Advances of Petri Nets*. Springer Verlag, Lecture Notes in Comp. Science 524, 1991.

[EGW98]   H. Ehrig, M. Gajewsky, and U. Wolter. From Abstract Data Types to Algebraic Development Techniques: A Shift of Paradigms. In *Proc. of Workshop on Algebraic Development Techniques*, pages 1–17. Springer Verlag, Lecture Notes in Comp. Science 1376, 1998.

[EHKP91]   H. Ehrig, A. Habel, H.-J. Kreowski, and F. Parisi-Presicce. Parallelism and concurrency in high-level replacement systems. *Math. Struct. in Comp. Science*, 1:361–404, 1991.

[EM85]   H. Ehrig and B. Mahr. *Fundamentals of Algebraic Specification 1: Equations and Initial Semantics*, volume 6 of *EATCS Monographs on Theoretical Computer Science*. Springer Verlag, Berlin, 1985.

[EOP99]   H. Ehrig, F. Orejas, and J. Padberg. Relevance, integration and classification of specification formalisms and formal specification techniques. In *Proc. FORMS'99, Braunschweig, Germany*, 1999.

[EP97]   H. Ehrig and J. Padberg. A Uniform Approach to Petri Nets. In Ch. Freksa, M. Jantzen, and R. Valk, editors, *Foundations of Computer Science: Potential - Theory - Cognition*. Springer Verlag, Lecture Notes in Comp. Science 1337, 1997.

[EPR94]   H. Ehrig, J. Padberg, and L. Ribeiro. Algebraic High-Level Nets: Petri Nets Revisited. In *Recent Trends in Data Type Specification*, pages 188–206. Springer Verlag, Lecture Notes in Comp. Science 785, 1994.

[ER97]   Hartmut Ehrig and Wolfgang Reisig. An algebraic view on Petri nets. *Bulletin of the EATCS*, 61:52–58, February 1997.

[EW01]   C. Ermel and M. Weber. Implementation of Parameterized Net Classes with the Petri Net Kernel. In H. Ehrig, G. Juhás, J. Padberg, and G. Rozenberg, editors, *Unifying Petri Nets*. Springer Verlag, Advances in Petri Nets, 2001. In this volume.

[Gaj99]   M. Gajewsky. The Expert View of the Petri Net Baukasten. In Weber et al. [WER99], pages 243–265.

[GB92]   J. A. Goguen and R. M. Burstall. Institutions: Abstract Model Theory for Specification and Programming. *Journals of the ACM*, 39(1):95–146, January 1992.

[GE00]   M. Gajewsky and C. Ermel. Transition Invariants in Algebraic High-Level Nets. In A. Ertas, editor, *4th World Conference on Integrated Design and Process Technology*, 1999/2000. CD-ROM, 8 pages.

[Hum89]   U. Hummert. *Algebraische High-Level Netze*. PhD thesis, Technische Universität Berlin, 1989.

[Jen97]   K. Jensen. *Coloured Petri Nets - Basic Concepts, Analysis Methods and Practical Use*, volume 3: Practical Use. Springer Verlag, EATCS Monographs in Theoretical Computer Science edition, 1997.

[KW99a]   Ekkart Kindler and Michael Weber. The Petri Net Kernel: An infrastructure for building Petri net tools. In *Petri Nets '99. 20th International Conference on Application and Theory of Petri Nets. Petri Net Tool Presentations*, Williamsburg, USA, June 1999.

[KW99b]   Ekkart Kindler and Michael Weber. *The Petri Net Kernel. Documentation of the Application Interface. PNK Version 2.0.* Humboldt-Universität zu Berlin, Institut für Informatik, January 1999. http://www.informatik.hu-berlin.de/ kindler/PN-Kern/.

[Lem99]   S. Lembke. The Application Developer View of the Petri Net Baukasten. In Weber et al. [WER99], pages 231–241.

[Lil95]   J. Lilius. *On the Structure of High-Level Nets*. PhD thesis, Helsinki University of Technology, Digital Systems Laoratory, Research Report 33, 1995.

[ML70]   S. Mac Lane. *Categories for the Working Mathematician*. Springer Verlag, Berlin Heidelberg New York, 1970.

[MM90]   J. Meseguer and U. Montanari. Petri Nets are Monoids. *Information and Computation*, 88(2):105–155, 1990.

[MOS93]   T. Mochel, A. Oberweis, and V. Sänger. INCOME/STAR: The Petri net simulation concepts. *Systems Analysis - Modelling - Simulation, Journal of Modelling and Simulation in Systems Analysis*, 13:21–36, 1993.

[MP92]   Zohar Manna and Amir Pnueli. *The Temporal Logic of Reactive and Concurrent Systems, Specification*. Springer Verlag, 1992.

[MW98]   A. Martini and U. Wolter. A systematic study of mappings between institutions. In F. Parisi-Presicce, editor, *Recent Trends in Algebraic Development Techniques*, pages 300–315. 12th International Workshop, WADT'97, Tarquinia, Italy, June 1997, Selected Papers, Springer Verlag, Lecture Notes in Comp. Science 1376, 1998.

[MW99]   Incorporated Merriam-Webster. *WWWebster Dictionary*, 1999. http://www.m-w.com/dictionary.

[Pad96]   J. Padberg. *Abstract Petri Nets: A Uniform Approach and Rule-Based Refinement*. PhD thesis, Technical University Berlin, Shaker Verlag, 1996.

[Pad98a]   J. Padberg. Abstract Petri Nets as a Uniform Approach to High-Level Petri Nets. In *Proc. WADT 98*, pages 240–259. Springer Verlag, Lecture Notes in Comp. Science 1589, 1998.

[Pad98b]   Julia Padberg. Classification of Petri Nets Using Adjoint Functors. *Bulletin of EACTS 66*, 1998.

[Pad99]   J. Padberg. The Petri Net Baukasten: An Application-Oriented Petri Net Technology. In Weber et al. [WER99], pages 191–209.

[PER95]   J. Padberg, H. Ehrig, and L. Ribeiro. Algebraic high-level net transformation systems. *Mathematical Structures in Computer Science*, 5:217–256, 1995.

[PGE98]   J. Padberg, M. Gajewsky, and C. Ermel. Rule-Based Refinement of High-Level Nets Preserving Safety Properties. In E. Astesiano, editor, *Fundamental Approaches to Software Engineering*, pages 221–238. Springer Verlag, Lecture Notes in Computer Science 1382, 1998.

[PHG00]   Julia Padberg, Kathrin Hoffmann, and Maike Gajewsky. Stepwise Introduction and Preservation of Safety Properties in Algebraic High-Level Net Systems. In T. Maibaum, editor, *Fundamental Approaches to Software Engineering*, pages 249–265. Springer Verlag, Lecture Notes in Comp. Science 1783, 2000.

[Rei85]   W. Reisig. *Petri Nets*, volume 4 of *EATCS Monographs on Theoretical Computer Science*. Springer Verlag, 1985.

[Rei91]   W. Reisig. Petri Nets and Algebraic Specifications. *Theoretical Computer Science*, 80:1–34, 1991.

[Sch96]   K. Schmidt. *Symbolische Analysemethoden für algebraische Petri-Netze*. PhD thesis, Humboldt-Universität zu Berlin, 1996.

[Sch99]   E. Schnieder, editor. *Methoden der Automatisierung: Beschreibungsmittel, Modellkonzepte und Werkzeuge für Automatisierungssysteme*. Studium Technik. Vieweg, 1999.

[Vau87]   J. Vautherin. Parallel System Specification with Coloured Petri Nets. In G. Rozenberg, editor, *Advances in Petri Nets 87*, pages 293–308. Springer Verlag, 1987. Lecture Notes in Computer Science 266.

[ADO00]   van der Aalst, W., Desel, J., and Oberweis, A., editors. *Business Process Management - Models, Techniques and Empirical Studies*, Springer Verlag, Lecture Notes in Computer Science 1806, 2000.

[Web99]   M. Weber. The Tool Developer View of the Petri Net Baukasten. In Weber et al. [WER99], pages 267–277.

[WER99]   H. Weber, H. Ehrig, and W. Reisig, editors. *Int. Colloquium on Petri Net Technologies for Modelling Communication Based Systems, Part II: The »Petri Net Baukasten«*. Fraunhofer Gesellschaft ISST, October 1999.

[WLB01]   H. Weber, S. Lembke, and A. Borusan. Improving the Usability of Petri Nets with the »Petri Net Baukasten«. In H. Ehrig, G. Juhás, J. Padberg, and G. Rozenberg, editors, *Unifying Petri Nets*, Springer Verlag Advances in Petri Nets, 2001. In this volume.

[WfMC00]   Workflow Management Coalition. Homepage of the Workflow Management Coalition. http://www.aiim.org/wfmc/mainframe.htm, 2000.

A   Appendix

In this appendix we present more details of the Expert View, described in Section 5 on a conceptual level. For illustration parts of the initial installment (see [DFG99]) of the »Petri Net Baukasten« are presented here.

A.1   Formal Petri Net Techniques

The notion of a formal Petri Net technique comprises a core formalism and compatible operations on Petri nets. The core formalism is a mathematical definition of a class of Petri nets which comprises the definitions of the net structure, data type structure, time, organizational structure, initial marking, and firing. An operation on Petri nets like structuring, analysis, and verification is a function that can be applied to Petri nets of a specific core formalism. Compatibility among these operations and with the core formalism is essential for their combination with the core formalism to a formal Petri net technique. Consequently, these compatibilities are main properties of the formal Petri net technique. Additionally, there might be further theoretical aspects, which are of less importance within the »Petri Net Baukasten«.

The scheme for a formal Petri net technique given below corresponds to this description.

For illustration we present the formal Petri net technique of algebraic high-level nets as an instantiation of the scheme for a formal Petri net technique. We first discuss this technique informally before we give the instantiation of the scheme.

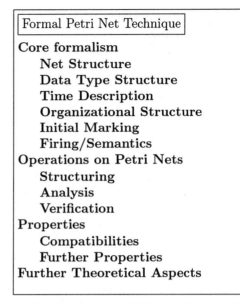

Formal Petri Net Technique
Core formalism
Net Structure
Data Type Structure
Time Description
Organizational Structure
Initial Marking
Firing/Semantics
Operations on Petri Nets
Structuring
Analysis
Verification
Properties
Compatibilities
Further Properties
Further Theoretical Aspects

An *Algebraic High-Level Net (AHL net)* consists — roughly speaking — of a Petri net with inscriptions of an algebraic specification $SPEC$ [EM85] defining the data type part of the net. For AHL nets structuring techniques are formulated within the frame of high-level replacement systems [EHKP91]. Results from the theory of AHL nets [PER95,PGE98,GE00,PHG00] comprise horizontal structuring techniques like union (composition of two nets with respect to a common interface) and fusion (the gluing of places, transitions, or even subnets within a given net), and vertical structuring like rule-based modification and rule-based refinement preserving safety properties in the sense of [MP92]. All of these structuring techniques are pairwise compatible with each other. Furthermore, there are concurrency properties of rule-based modification and rule-based refinement like local confluence and parallelism which are essential for the AHL net technique. On the other hand the notion of T-invariants defined for AHL nets (see [GE00]) is — in general — not compatible with rule-based modification. Therefore, T-invariants are not given in the operations of the formal Petri net technique below, but they are only listed under further theoretical aspects.

We are now going to instantiate the scheme according to the above description.

Formal Petri Net Technique: AHL Nets

Core formalism

   Net Structure

      Algebraic High-Level Net, see [PER95], Def. 4.1

   Data Type Structure

      Specification and SPEC-algebra, see [EM85], Def. 1.14

   Firing/Semantics

      Operational Behaviour of AHL-Nets, see [PER95], Def. 4.2

Operations on Petri Nets
  Structuring
    – Union, see [PER95], Def. 3.5
    – Fusion, see [PER95], Def. 3.1
    – Rule-Based Modification, see [PGE98], Def. 2.3 (Rule and Transformation)
    – Rule-Based Refinement, see [PGE98], Concept 4.1 (Vertical Structuring Technique: Rule-Based Refinement)
Properties
  Compatibilities
    – Fusion and Union, see [PER95,ML70]
    – Rule-Based Modification and Rule-Based Refinement, see [Pad96], Fact 4.3.3 (Induced Q-Transformations and Pushouts in QCAT)
    – Fusion and Rule-Based Modification, see [PER95], Theorem 3.4 (Fusion Theorem)
    – Union and Rule-Based Modification, see [PER95], Theorem 3.7 (Union Theorem)
    – Fusion and Rule-Based Refinement, see [Pad96], Fact 4.5.5 (Fusion is compatible with Q-Transformations)
    – Union and Rule-Based Refinement, see [Pad96], Fact 4.5.10 (Union is compatible with Q-Transformations)
    – Preservation of Safety Properties, [PGE98,PHG00]
  Further Properties
    – Instance of Parametrized Net Classes, see [Pad96], Example 3.5.1 (Algebraic High-Level Nets)
    – Transformations: $Unf : AHL \to PT$, see Section A.4
    – Category AHL-net, see [PGE98], Def. 4.6
    – Category of AHL nets is cocomplete, see [PER95], Theorem 5.10
    – HLR-properties of AHL-net transformation systems, see [PER95], Theorem 5.10
    – Local Church-Rosser Theorem, see [PER95], Theorem 2.11 proof see [EHKP91]
    – Paralellism Theorem, see [PER95], Theorem 2.13, proof see [EHKP91]
Further Theoretical Aspects
  Variants of AHL nets in [Vau87,Hum89,Rei91,DHP91,Lil95,Sch96]
  Sheaf Semantics [Lil95]
  T-Invariants [GE00]

A.2   Abstract Petri Net Frames

Abstract Petri Net frames are uniform descriptions of Petri nets which are based on the concept of parametrization/actualization. Abstract Petri net frames contain formal parameters and a frame body leading to a parameterized description of Petri nets. The frame body describes the invariant aspects that do not change for any actualization of the formal parameters. The abstract Petri net frame comprises structural as well as behavioral aspects in a generic way including operations defined already on the generic level. These operations have to be compatible with each other such that each actualization leads to operations on Petri nets within a consistent formal Petri net technique. These compatibilities are relevant properties for the »Petri Net Baukasten«. Such properties and further theoretical aspects are also presented in an abstract Petri net frame. This leads to the following scheme for abstract Petri net frames:

| Abstract Petri Net Frame |

Formal Parameters
  parameter name: formalization
Frame Body
Operations
Properties
Further Theoretical Aspects

In the following, we present as an instantiation of the scheme the abstract Petri net frame of parameterized net classes [Pad96,EP97,Pad98a]. For the implementation of parameterized net classes within the »Petri Net Baukasten« we refer to [EW01].

Parametrized Net Classes are based on an algebraic representation of Petri nets [MM90,PER95,EGW98,ER97]. Orthogonal parameters for the structure of the net, the structure of the marking (adjoint functors[ML70]), and for the data type specification (institutions [GB92]) can be declared. Horizontal and vertical structuring techniques are defined on the generic level of parametrized net classes where all these structuring techniques are compatible with each other. Thus, actualization of the parameters leads to net classes with compatible structuring techniques. Most of the known and interesting new net classes are obtained by appropriate actualizations, which will be discussed below.

The representation of data type specifications as institutions serves as a uniform framework where a flexible change of specification techniques is supported by institution morphisms [GB92,MW98].

| **Abstract Petri Net Frame: "Parametrized Net Classes"** |

[Pad96,EP97,Pad98a]

Formal parameters
  – Data type structure: specification frames (institutions), see [Pad98a], Definition 2.11 or [Pad96], Definition 3.2.11
    Explanation: Data type part of nets
  – Marking structure (functor): A composite functor from a pair of adjoint functors, a left-adjoint functor from category of sets to a subcategory of commutative semigroups, see [Pad98a], Definition 3.1, or [Pad96], Definition 2.2.1
    Explanation: Domain of net markings
  – Flow structure (functor): A composite functor from a pair of adjoint functors, a left-adjoint functor from category of sets to a category of

flow structure, with a natural transformation to marking structure functor, see [Pad98b], Definition 2.1

Explanation: Domain of arc-weights

Frame Body

- Parametrized net classes (Abstract Petri nets), see [Pad98a], Definition 3.5, or [Pad96], Definition 3.3.1
- Firing rule, see [Pad98a], Definition 3.6 and [Pad96], Definition 3.3.2

Operations

Structuring

- Union, see [Pad96], Definition 4.5.6
- Fusion, see [Pad96], Definition 4.5.2
- Rule-Based Modification,
  see [Pad96], Definition 4.2.1 (Rules and Transformation)
- Rule-Based Refinement,
  see [Pad96], Section 4.3 (Q-Transformation)

Properties

Compatibilities

- Fusion and Union, see [Pad96,ML70]
- Rule-Based Modification and Rule-Based Refinement,
  see [Pad96], Fact 4.3.3 (Induced Q-Transformations and Pushouts in QCAT)
- Fusion and Rule-Based Modification,
  see [EHKP91] (for transformation)
- Union and Rule-Based Modification,
  see [EHKP91] (for transformation)
- Fusion and Rule-Based Refinement,
  see [Pad96], Fact 4.5.5 (for Q-Transformation)
- Union and Rule-Based Refinement,
  see [Pad96], Fact 4.5.10 (for Q-Transformation)
- Compatibility results for Parametrized net classes concerning parallel and sequential independent transformations, and horizontal structuring and transformations,
  see [Pad96], Theorem 4.6.7
  in detail:
- Local Church-Rosser I and II, see [EHKP91] (for transformation) and [Pad96] Theorems 4.4.5 and 4.4.7 (for Q-Transformation)
- Parallelism Theorem, see [EHKP91] (for transformation) and [Pad96], Theorem 4.4.11 (for Q-Transformation)

Properties

- Preservation of Behaviour by Morphisms,
  see [Pad96], Theorem 3.3.5

Further Theoretical Aspects

- Morphisms of Parametrized Net Class,
  see [Pad96], Definition 3.3.3
- Category of Parametrized Net Class,
  see [Pad96], Fact 3.3.4

    – Finite cocompleteness of Category of Parametrized Net Class, see [Pad96], Theorem 3.3.6

A.3   Actualization

The actualization of an abstract Petri net frame (as defined in Section A.2) is achieved by replacing the abstract mathematical entity described by the formal parameter by a concrete mathematical entity — the actual parameter. Operations and properties given on the level of the abstract Petri net frame are propagated via actualization leading to a corresponding formal Petri net technique. The core formalism of the formal Petri net technique can be enriched by further properties, aspects, or compatible operations to another formal Petri net technique such that the actualized formal Petri net technique is a subtechnique of the enriched one.

Summarizing, actualization of an abstract Petri net frame is given by replacement of the formal parameters by the actual parameters. The resulting instance may be a subtechnique of some other formal Petri net technique which — in this case — is called related formal technique.

The scheme for actualization below captures this description.

Actualization

Abstract Petri Net Frame
Formal Parameters replaced by Actual Parameters
Resulting Instance
Related Formal Technique

As example for the instantiation of the scheme for actualization, we present an actualization of the abstract Petri net frame of parameterized net classes leading to a variant of algebraic high-level nets, called algebraic high-level net schemes (see [EPR94]). Algebraic high-level net schemes do not include an explicit algebra as given in algebraic high-level nets. Nevertheless, all operations and properties derived from the abstract Petri net frame can be lifted to algebraic high-level nets as given in Section A.1.

Actualization "Algebraic High-Level Net Schemes"
Abstract Petri Net Frame
    Parametrized Net Classes, see [Pad96], Example 3.5.1
Formal Parameters replaced by Actual Parameters
    – Data structure $\mapsto$ Algebraic Specification
    – Marking structure $\mapsto$ free commutative monoid
    – Flow structure $\mapsto$ free commutative monid
Resulting Instance
    Algebraic high-level net schemes (without data model, see [EPR94])
Related Formal Technique
    Algebraic high-level nets (see Section A.1)

A.4   Transformations

A transformation is a construction which transforms Petri nets of one Petri net type into nets of another type. In the Expert View transformations are defined for formal Petri net techniques or even on the level of abstract Petri net frames, such that an actualization leads to a transformation of formal Petri net techniques. Transformations are essential for Petri net techniques, because they realize the transfer from one formal Petri net technique to another one within the process of system development.

An important aspect of a transformation concerns the transfer of operations on Petri nets from the source to the target type. Full compatibility with techniques would mean that every operation in the source technique is compatible with a corresponding one in the target technique. This cannot be expected in general, so that compatibility has to be stated explicitly.

Corresponding to the above description the following scheme captures transformations where the scheme also includes compatibility with other transformations and further properties.

> | Transformation |
>
> **Source**
> **Target**
> **Definition**
> **Compatibility with Techniques**
> **Compatibility with other Transformations**
> **Further Properties**

In the sequel we are going to discuss the transformation flattening and show how it can be obtained by instantiation of the transformation scheme.

Flattening of algebraic high-level nets is described by a functor $Flat : \textbf{AHL} \rightarrow \textbf{PT}$ from the category **AHL** of algebraic high-level nets to the category **PT** of place/transition nets, see [EPR94], Fact 3. Each algebraic high-level net is mapped to a place/transition net while preserving the behaviour. The data elements on places in the high-level net are coded into places of the low-level net. Analogously, the transitions together with consistent variable assignments in the high-level net are flattened to transitions in the low-level net. Note, that this transformation has also been studied in [Lil95], where it is called unfolding. Although the source of unfolding is based on slightly different algebraic high-level nets, results of [Lil95] clearly can be adapted to the (core formalism of the) formal Petri net technique presented in Section A.1. Performing first the folding construction and subsequently the inverse construction of unfolding leads to an isomorphic place/transition net[2].

---

[2] The corresponding Lemma 2.1.16. in [Lil95] obviously can be transferred to the notion of algebraic high-level nets as considered here.

Transformation "Flattening of Algebraic High-Level Nets"

Source

Algebraic High-Level Nets (see Section A.1)

Target

Place/Transition Nets

Definition

$Flat$: **AHL** $\rightarrow$ **PT** is a construction from the category **AHL** of algebraic high-level nets to the category **PT** of place/transition nets, as given in [EPR94], Fact 3

Compatibility with Techniques

Behavior equivalence, see [EPR94], Fact 3

Compatibility with other Transformations

$Flat(G(N)) \simeq N$, for a place/transition net N analogously to [Lil95], Lemma 2.1.16, where $G :$ **PT** $\rightarrow$ **AHL** is a folding functor

Further Properties

$Flat$ is a functor, see [EPR94], Remark to Fact 3

Improving the Usability of Petri Nets with the »Petri Net Baukasten«

Herbert Weber, Sabine Lembke, and Alexander Borusan*

Technical University of Berlin, Germany
{hweber,lembkes,aborusan}@cs.tu-berlin.de
http://www.cis.cs.tu-berlin.de

Abstract An enormously rich variety of Petri net concepts, techniques, and methods as well as various tools and tool environments has been developed to support the process-driven system development. But, system development under the use of Petri nets is still difficult since the one which best fits the application cannot be identified easily.

In this paper an application-oriented assistance approach is introduced which helps application developers to find the "right" Petri net variant, tool, technique, and development method for specific development tasks. It is based on the »Petri Net Baukasten«, more especially on its Application Developer View and its Common Base. Both are explained in detail.

1 Introduction

Petri nets are used in various application domains like workflow management and production automation to model, assess, and implement the behaviour of distributed systems. They originate in the Ph.D. thesis of Carl Adam Petri in 1962 [9]. Various additional concepts have been added to support the modeling of different system aspects and capabilities to assess and implement the modeled system behaviour have been developed. This has led to an enormously rich variety of Petri net concepts, techniques and development methods as well as to various tools and tool environments that still cannot be used effectively. Using Petri nets is still difficult since the Petri net variants, techniques, methods, and tools best fitting the respective development task cannot be identified easily. For this reason, improving the usability of Petri nets by assistance is needed strong by application developers to find the "right" one.

Tool surveys usually used by application developers to find a Petri net tool fulfilling specific requirements. The most user-friendly and current source of information on Petri net tools are the tool databases publicly accessible in the World Wide Web (WWW) at [4,12]. In this paper we present an assistance approach to additionally support application developers in finding suitable Petri

---

* This work is part of the joint research project "DFG-Forschergruppe PETRI NET TECHNOLOGY" between H. Weber (Coordinator), H. Ehrig (both from the Technical University Berlin), and W. Reisig (Humboldt University Berlin), supported by the Deutsche Forschungsgemeinschaft (DFG).

H. Ehrig et al. (Eds.): Unifying Petri Nets, LNCS 2128, pp. 54–78, 2001.

net variants, techniques, and methods. This approach is part of the »Petri Net Baukasten«, which has been developed by the research group PETRI NET TECHNOLOGY in Berlin, see further information at http://www.informatik.hu-berlin.de/PNT/.

The »Petri Net Baukasten« has been developed in order to enable a more straight forward understanding of Petri nets and its many variants. It takes net concepts and theories, tool support, and application of Petri nets into consideration. This is reflected in three different views upon Petri nets: the Expert View, the Tool Developer View, and the Application Developer View. A semi-formal Petri net classification is used to build the Common Base of these three views. It is a hierarchically structured representation of Petri net variants and their specialization/generalization relationships. The Expert View gives a precise description of Petri nets in formal and constructive terms while the Tool Developer View helps to develop Petri net tools, thus enabling the practical use of Petri nets (modeling, analyzing, etc.). The Application Developer View provides information which help practitioners of different application domains to understand the use of Petri nets and to find the "right" Petri net variants, tools, techniques, and development methods for their development tasks. A more detailed description of the »Petri Net Baukasten« approach and its Expert View can be found in [3] of this volume. A further paper of this volume discusses the Tool Developer View [2]. The Application Developer View and the Common Base are important to our assistance approach for application developers and are described in the corresponding sections of this paper.

This paper is organized as follows: in Section 2 we describe the Petri net classification in the Common Base of the »Petri Net Baukasten«. We explain its object-oriented description, discuss the principles used to construct it, and document its structural model. In Section 3 we present our understanding of assisted application development. We define notions like process model, method, and technique to explain our objectives and use cases of assisted application development. In Section 4 we describe the Application Developer View of the Petri net classification. We explain its structural model and its relationship to the Common Base which enables the assistance of application developers. In Section 5 we describe the system architecture and the repository of an assistance system supporting application developers according to the discussed assistance approach.

2   The Common Base of the »Petri Net Baukasten«

The central part of the »Petri Net Baukasten« and our assistance approach is the so-called Common Base. It encompass all Petri net variants and places them into a common scheme, thus defining the relationship between the Petri net variants. The three different views on Petri nets called Application Developer View, Expert View, and Tool Developer View are integrated by referring to this scheme.

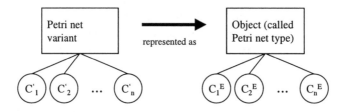

Fig. 1. Object-oriented description of Petri net variants

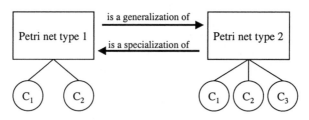

Fig. 2. Generalization/specialization of Petri net types

It is proposed that the Common Base may be structured as simple as possible and understandable for application developers, tool developers, and Petri net experts alike. The structuring concept considered to be simple enough and still powerful enough to capture the different properties of the many Petri net variants is borrowed from software engineering. The structuring scheme for the Common Base is object-oriented and described in Section 2.1. It is defined by relationships of specialization/generalization between the different Petri net variants which are represented by so-called Petri net types. The scheme represents the more simple Petri net types at the top and the more specific ones below.

The principle used to construct the hierarchy of Petri net types called Petri net classification are described in Section 2.2. The structural model of the Common Base comprising the structuring scheme introduced in Section 2.1 and taking the described principles into consideration is explained in Section 2.3.

2.1   Object-Oriented Description

The object-oriented semi formal description that is going to be used here introduces variants of Petri nets as objects and characteristics associated with Petri net variants as characteristics of objects, see Fig. 1. Such objects are called Petri net types in the following.

The classification of Petri net types and hence the classification of Petri net variants is based on the set of characteristics of Petri net types. Especially, Petri net types that have characteristics in common are considered as generalizations or specializations respectively, see Fig. 2.

A Petri net type T_1 is considered a generalization of a Petri net type T_2 if it carries a subset of the characteristics of T_2. Vice versa, T_2 is a specialization of T_1 if it carries all characteristics of T_1 and one or more further characteristics.

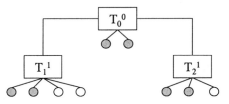

Fig. 3. Hierarchical order of Petri net types; T_0^0 represents the most simple Petri net type and $T_{j,k,\ldots}^i$ represents the specialization j,k of the degree i

Petri net types classified according to that classification scheme must hence have at least one common characteristic. This classification concept is considered to be appropriate for the classification of Petri net variants since all of them are expected to carry at least the characteristics of being a Petri net.

Classification based on sets of characteristics can be brought into a hierarchical order. The set of characteristics associated with every Petri net type on level i is a superset of the set of characteristics associated with the Petri net type on level $i - 1$, see Fig. 3.

In analogy to the common notions in object oriented techniques a hierarchy of that kind is called inheritance hierarchy since it depicts the inheritance of characteristics from top to bottom. Inheritance hierarchies do not need to be trees as depicted above, but may be acyclic directed graphs.

Petri net variants may be represented as inheritance hierarchies. For that purpose characteristics of Petri nets are ordered from "most basic" characteristics to "supplementary" characteristics of any degree. This ordering can be achieved through the selection of characteristics of interest for the respective classification level out of the larger set of possible characteristics. The selection of characteristics of interest happens in accordance to a separation of concerns. A first upgrading level of the hierarchy of Petri net types also called Petri net classification is introduced in [1].

2.2   Construction of the Petri Net Classification

The construction of the Petri net classification is based on similar concepts and follows similar principles as the construction of class hierarchies in UML [8]. None the less, the construction of the Petri net classification is more restricted. Therefore, we will explain the used concepts and principles more detailed in the following.

The following specialization principles have been used for the systematic construction of the Petri net classification:

- adding a new characteristic,
- limiting the type of a characteristic which has been introduced on a lower specialization level[1],

---

[1] A lower specialization level is placed at the top of the hierarchy. Accordingly, a higher specialization level is placed further down in the hierarchy.

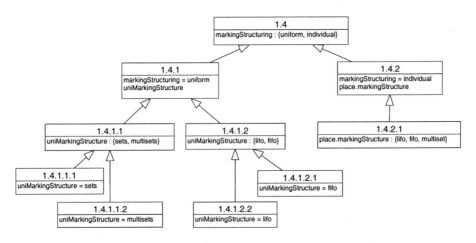

Fig. 4. Specialization example according the different marking structure of places of Petri nets

- assigning a value to a characteristic which has been introduced and typed on a lower specialization level (value assignment),
- defining the value of a characteristic more precisely by another characteristic (value nesting), and
- the specialization of more than one Petri net type by another Petri net type, possibly also using any of the other principles given above (multiple inheritance).

An example for the first three specialization principles is given in Figure 4. Here, we are focusing on the *uniMarkingStructure* characteristic introduced in the second level. Petri net type *1.4* is specialized by Petri net type *1.4.1*, which adds this new characteristic. Specialization by limiting the type and assigning values takes place by repeating the characteristic of the superior Petri net type. The type of the characteristic *uniMarkingStructure* is limited to an enumeration type with the elements *sets* and *multisets* by the Petri net type *1.4.1.1*, and, respectively, to an enumeration type with the elements *fifo* and *lifo* by the Petri net type *1.4.1.2*, see the third level of Fig. 4. An concrete value is assigned in the fourth level.

If a value of a characteristic has to be described more precisely in the adjacent level then this could be done by adding a new characteristic. But, in order to depict the dependencies between both characteristics without introducing new suitable characteristic names, the value name of the assigned value in the higher generalization level is used as the characteristic name in a further specialization level. Thus, this new characteristic can be specialized again on further levels. This principle is called value nesting and is illustrated in Fig. 5.

The characteristics of two or more Petri net types can be combined, if the last specialization principle is used. This is a useful principle for describing more specialized Petri net types again by reuniting the various specialization paths.

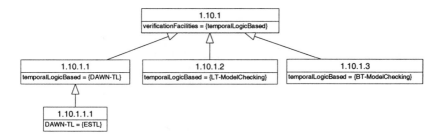

Fig. 5. Example of specialization using the principle of value nesting

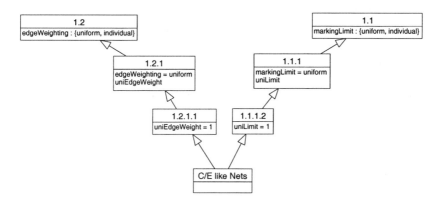

Fig. 6. Example of specialization using the principle of multiple inheritance

An example of multiple inheritance is depicted in Fig. 6. The Petri net type *C/E like Nets* inherits all characteristics of the Petri net types *1.2.1.1* and *1.1.1.2*.

2.3   The Structural Model of the Common Base

The structural model of the Common Base depicted as UML class diagram in Fig. 7 is described in the following.

Petri Net Classification. The foundation of the Common Base is the Petri net classification, which comprises the representation of various Petri net variants, called Petri net types. Only one Petri net classification with one root as an entry point is present in the Common Base. The entry point is a Petri net type which at least contains the characteristics of being a Petri net.

Petri Net Types. All Petri net types of the Petri net classification are identified by their classification name, a decimal name such as *1.4.1*, see Fig. 4. Additionally, a Petri net type can have a more understandable name such as *C/E like Nets*, see Fig. 6. A Petri net type can specialize other Petri net types

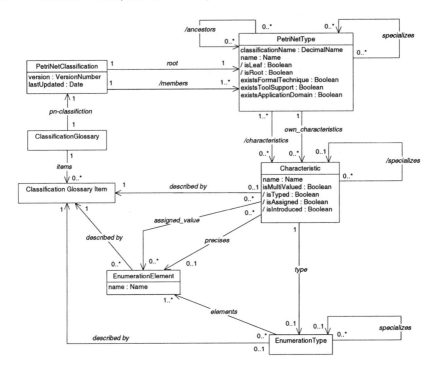

Fig. 7. Class diagram of the Common Base

and can be specialized itself by other Petri net types. This is described by the relationship *specializes* in Fig. 7. A Petri net type is a root if it does not specialize other Petri net types. It is a leaf if it is itself not specialized. Characteristics of a Petri net type are either inherited from its ancestors or owned. Characteristics of Petri net types themselves can either be newly introduced or specializations of their ancestors' characteristics. Every Petri net type holds indicators of references from the Application Developer View and/or the Tool Developer View and/or the Expert View, if relationships among these exist, see corresponding attributes of *PetriNetType* in Fig. 7. The relationships between the Application Developer View and the Common Base are explained in more detail in Section 4.2. The relationships among the other views and the Common Base are sketched in Section 5.2.

Characteristics. In UML, an attribute is the named property of a class that describes a range of values a property may hold. In our Petri net classification, an attribute is called a characteristic and is related to a Petri net type, since it represents a characteristic property of a Petri net variant, see class *Characteristic* in Fig. 7. A characteristic of a Petri net type can be typed and assigned a value. The types used to describe the range of a characteristic are user defined types consisting of several elements such as enumeration types. The possible values of a characteristic are elements of a certain type. A characteristic can be specialized by another characteristic of specialized Petri net types according to the principles described in Section 2.2. Furthermore, the value of a characteristic can be described more precisely by characteristics of more specialized Petri net types in accordance with value nesting as described in Section 2.2. We are not concerned with operations (like add_place, delete_transition,...) of the Petri net types, but rather with their ability to support Petri net operations such as verification, analysis, etc. These capabilities are characteristic properties of the Petri net type and therefore described as attributes or, respectively, as characteristics.

Classification Glossary. A further component of the Common Base is a classification glossary comprising various items giving informal explanations of terms used in the Petri net classification.

3   Assisted Application Development

The engineering-like development of application systems is always driven by a specific paradigm, e. g. a process-driven, an object-driven, or a function-driven paradigm. A whole variety of development techniques, methods, and tools exist for supporting the most diverse development paradigms. Hence, application developers require support in order to be able to carefully choose suitable ones for a particular application development.

Petri nets support process-driven application development, for example. None the less, the rich variety of Petri net variants makes it difficult for those who are not Petri net experts to find the suitable Petri net variant, Petri net tool, and development method for a particular development task. Therefore, we have to support the application of Petri nets more specifically, since there are no single Petri net variant, tool and method which are the *right* for the wide range of development tasks.

The outstanding feature of Petri net variants comprising a large number of language concepts is their higher and more compact expressive power. On the other hand, we often miss the benefit of using Petri nets such as the capability to be analyzed, if such Petri net variants could be applied. Using Petri net variants comprising fewer language concepts often leads to extensive Petri nets with reduced clarity and manageability. None the less, there are often possibilities to assess the behaviour of resulting Petri nets like place or transition invariant analysis. But, they are missed as soon as more complex Petri net variants are used. A similar situation arises when searching for a suitable Petri net tool. These differ within the supported Petri net variant, in terms of availability (free, commercial), functionality, etc. Moreover, application developers require support in methodical development with Petri nets. In particular, they require such methods which take the special characteristics of particular application domains into consideration. Finally, providing proposed solutions to particular application-oriented problems, such as Petri nets solving the mutual exclusion

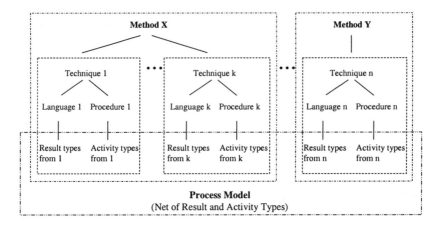

Fig. 8. Relationship between process model and development method

of two workflow activities, can reduce the amount of development work. The application of Petri nets would become easier and faster. In each case, application developers must weigh up the pros and cons as pertaining to the requirements of a particular development task before the correct decision can be made.

Assisted application development builds a bridge between the world of practice and the Petri net theory, making Petri nets more usable for application development. We intend to make it easier for practitioners of different application domains to understand the use of Petri nets. For this purpose, we will take objectives and aspects of several application domains into consideration. We want to provide information which allows application developers to make the best choice in different situations.

In Section 3.1, we will clarify some important notions which will be used in the following. Afterwards, we will explain the several objectives of assisted application development, see Section 3.2. Use cases of assisted application development in accordance with our opinion are described in Section 3.3.

3.1 Notions

Some of the notions in our paper, especially "method" and "technique", are used in different ways in the literature. We have tried to remain in accordance with the existing interpretations of these notions. Nevertheless, as much as possible, we consider it necessary to explain our understanding of these and other notions at this point. The relationships between these notions (process model, method, technique, language, and procedure) are illustrated in Fig. 8. This illustration is derived from an illustration in [6, p.140] and takes additionally the notion *technique* into consideration.

Process Model. A process model is an instrument for organizing of the development process. The particular activities (their order) can be organized in phases, on abstraction levels, in cycles (circular or spiral-shaped) or in stages of development. One can distinguish between result- (results are fixed) and activity-oriented (development activities are fixed) process models. A process model is described by a net of activity and result types as well as by conditions for transitions between activity and result types.

(Petri Net) Method. Each development process is a constructive and engineering-like process. This kind of creative activity cannot be prescribed by recipes and fix solutions. None the less, creative activities can be guided in the right direction by methodical rules. In other words, each method represents a more or less rough frame in order to formalize the development process or parts of it.[2] In [7], a method "is an orderly arrangement, development, or classification." A method is explained as a "plan" and "the habitual practice of orderliness and regular". It "implies an orderly logical effective arrangement usually in steps". We define methods as the application and problem-oriented orderliness to produce results within a development process. Methods can comprise several techniques and cover more then one phase within a process model.

In this paper we consider development methods based on the use of Petri nets. Therefore, we also call them Petri net methods (short for *Petri net-based development methods*).

(Petri Net) Technique. In [7], a technique is "the manner in which technical details are treated (as by a writer) or basic physical movements are used (as by a dancer)". It is also the "ability to treat such details or use such movements (good piano technique)." It is also explained as "a method of accomplishing a desired aim". We define techniques as a means to produce results within a development process. They comprise a language for representing development results as well as a (development) procedure prescribing the sequence of steps leading to a result. As compared to methods which support certain development phases, techniques are universal and may be used in several development phases. Thus, a technique can be an element of several methods. In this paper we consider techniques which use Petri nets as a language. Thus, they are called Petri net techniques. In the »Petri Net Baukasten« a Petri net technique is built up by a consistent set of Petri net types, a (development) procedure, a formal Petri net technique describing its formal foundation, and Petri net tools that enable its application. Further details are explained in paper [3] in this volume.

3.2 Objectives

Assisted application development pursues several objectives, which we will explain in the following.

---

[2] Compared with languages, methods are less prescriptive. There is lack of methods.

Application Orientation. A main objective of assisted application development is application orientation. Application orientation is realized by introducing an application domain specific interpretation of Petri net notions. In our opinion, application developers can relate to the world of Petri nets more easily and quickly using familiar technical terms from their own application domain. Therefore, technical terms of particular application domains are mapped to Petri net notions. For example, the technical terms in the area of business process modeling are business process, process activity and so-called roles which denote responsibilities.

Problem Orientation. Each application development of a certain application domain is characterized by its own problems which need to be examined. For example, business process modeling concentrates on the modeling of existing business processes within a certain enterprise. The typically modeled application aspects are business processes with particular process activities, the logical order between these process activities, responsibilities etc. Distribution and time consumption of business processes may be additional application aspects which have to be taken into consideration. The purpose of business process modeling may vary from pure documentation and analyzing to suggestions to improve existing processes. Assisted application development should therefore not only deal with the several application domains but also with the several typical problems in certain application domains.

Method Orientation. In the literature Petri nets are mainly described as a language for process modeling. With other words, the language concepts and their syntax, i. e. their graphical notation, are mainly discussed. Development methods explaining the usage of Petri nets in application development processes have not been examined and documented enough. The usage of Petri nets in application development can become more effective if development methods are made available since these allow application developers to orientate themselves within their own development processes and thus makes their work easier.

Solution Orientation. A good assisted application development should make standard and sample solutions available in order to reduce the time needed for development activities. It should be able to integrate such solutions into the developer's own work and to adapt them if necessary. Standard and sample solutions may be independent of applications , e. g. mutex algorithm – a protocol which realizes mutual exclusion. Moreover, standard and sample solutions may be application specific.

Tool Orientation. Modern application development doesn't work without suitable tool support during most phases of a development process. Therefore, assisted application development only makes sense if it provides suitable tools for application developers which support certain Petri net techniques, methods, etc.

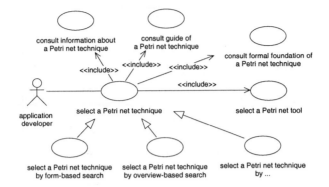

Fig. 9. Use case: select a Petri net technique

Need Orientation. Assisted application development shall be need-oriented in accordance with the concrete requirements of a certain application development process. Especially the assistance (selection of methods, techniques, tools, sample solutions) should follow the principle of *you only get what you really need*. Let us take the example of an application developer looking for a Petri net type which supports a certain set of language concepts. In this case, such Petri net types take up a prime position which only comprise the required language concepts or fewer more.

Multiple-way Orientation. Assisted application development should be provided by several assistance variants in order to provide the right development method, the right Petri net technique, the right Petri net type, the right tool or the right sample solution. All of these variants should be realized by several assistance methods, such as prescriptive, navigative or descriptive assistance methods. These allow every application developer to choose a suitable type of assistance.

3.3 Use Cases

In this section, we describe several use cases of assisted application development. The main use cases focus on selecting a suitable Petri net technique, Petri net method, Petri net type, Petri net tool, or sample solution.

Select a Petri Net Technique. The use case *select a Petri net technique* outlines the assistance variant for finding a suitable Petri net technique. A Petri net technique can be selected in different ways, for example by using a form-based or an overview-based search, compare Fig. 9. If application developers use a form-based search then they can specify required characteristics of a Petri net technique in predefined boxes. An example of a form-based search is depicted in Fig. 10. This form enables application developers to find a suitable Petri

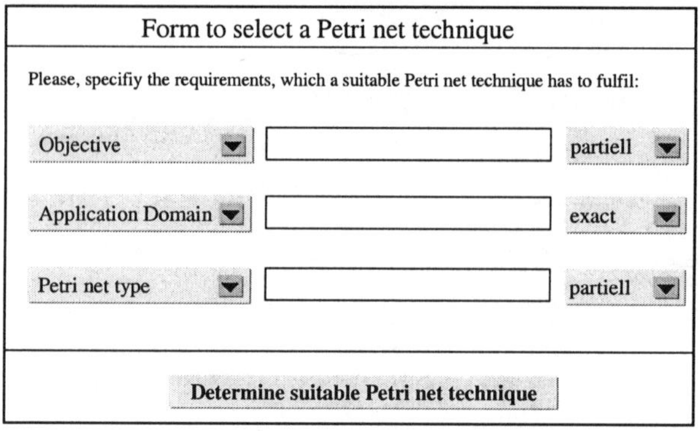

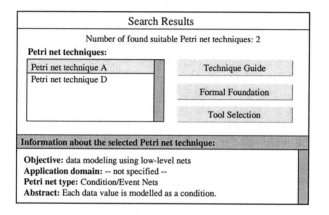

Fig. 10. A form to select a Petri net technique

Fig. 11. Presentation of the determined suitable Petri net techniques

net technique by keywords specifying its intended objective, application domain and its underlying Petri net type. Petri net techniques fulfilling the specified characteristics are determined automatically and offered to the application developers. For example, the found Petri net techniques are represented as depicted in Fig. 11. Before application developers select their suitable Petri net technique from the offered set, they may use several information about it, like information about the application domain, its objective, the underlying Petri net type, or the essentials of its procedure. Moreover, they may use its informal description, the so-called technique guide, or its formal foundation. Last but not least, a suitable Petri net tool supporting the preferred Petri net technique is selected.

Select a Petri Net Method. The use case *select a Petri net method* outlines the assistance variant for finding a suitable Petri net-based development method. Similarly as described above, a Petri net method can also be selected in different ways, compare Fig. 12. The form-based search is similar to the form-based search

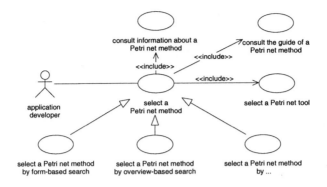

Fig. 12. Use case: select a Petri net method

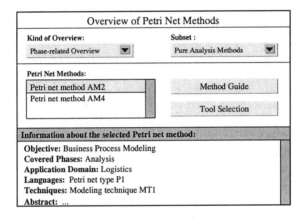

Fig. 13. An overview to select a Petri net method

described and depicted above, see Fig. 10. It allows search over objectives, application domains and covered development phases (analysis, design, etc.) of Petri net methods. An example of an overview-based search is depicted as a further possibility to find a suitable Petri net method in Fig. 13. Here, application developers can select a preferred kind of an overview, for example a phase-related overview of Petri net methods. In this overview the existing Petri net methods are sorted according to their covered phases, for example pure analysis methods, pure design methods, etc. A further overview may sort Petri net methods according to their intended application domain.

Application developers can use different information about each offered Petri net method. For example, its indented objective, its covered phases, its underlying languages, and techniques. Additionally, they may also consult the method guide before selecting their suitable Petri net method. Finally, application developers select tools which enables the application of the selected method.

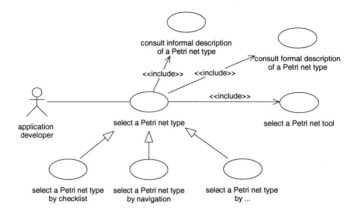

Fig. 14. Use case: select a Petri net type

Select a Petri Net Type. The use case *select a Petri net type* outlines the assistance variant for finding a suitable Petri net type which fulfills the requirements of a particular application development process. It is specialized by several use cases which reflect several assistance methods for finding a suitable Petri net type. These may be *select a Petri net type by checklist, select a Petri net type by navigation* etc., compare Fig. 14. The navigation based use case enables the selection of Petri net types by navigation through the Petri net classification described in Section 2. The checklist-based use case lists application-oriented aspects of a certain application domain as sketched in Fig. 15. We provide different checklists for different application domains. The application-oriented aspects are grouped in modeling-, assess-, and implementation-related aspects. For example, a checklist of business process modeling comprises modeling-related aspects like data/document specification, activity specification, role specification, time consumption, etc., assess-related aspects like critical path analysis, simulation, etc., and non implementation-related aspects. Thus, application developers can choose the most relevant aspects for their application purpose. Following this, Petri net types are determined automatically, supporting language concepts needed for dealing with the selected application-oriented aspects. The found Petri net types are offered to application developers like found Petri net techniques in Fig. 16.

Application developers can use different information about a certain Petri net type, for example its intended application domains, its informal and its formal description. Finally, application developers select tools which enable the application of the selected Petri net type.

Select a Petri Net Tool. The use case *select a Petri net tool* has been already mentioned in all use cases described above. It supports application developers to select suitable Petri net tools which enable the application of a Petri net technique, method, or type selected before. The selection of a Petri net tool to

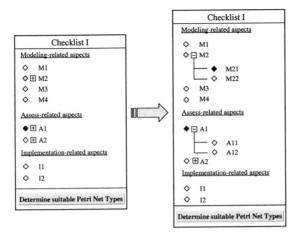

Fig. 15. Checklist to select a Petri net type

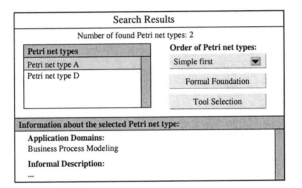

Fig. 16. Presentation of the determined suitable Petri net types

enable the application of a Petri net technique and the selection to enable the application of a Petri net method differ a little from the selection to enable the application of a Petri net type. But in all cases, application developers may use information about a Petri net tool and download it, compare Fig. 17.

Let us have a closer look to select a Petri net tool to enable the application of a Petri net method or Petri net technique. The application of a Petri net method or Petri net technique can be enabled by a certain set of Petri net tools which are integrated to a workbench or by a certain Petri net-based development environment. In both cases, underlying Petri net types and several modeling, assess, or implementation functionalities required by a certain Petri net method or Petri net technique have to be supported. Such tools or tool sets are recommended to application developers as depicted in Fig. 18.

Application developers may use information about a selected tool recommendation, for example about the required operating system (e.g. UNIX) or the

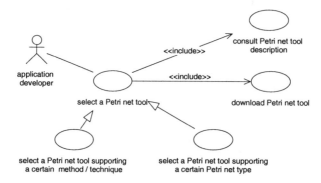

Fig. 17. Use case: select a Petri net tools

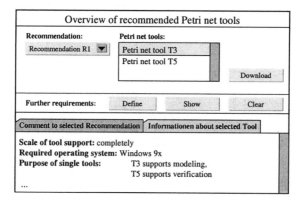

Fig. 18. Overview of recommended Petri net tools

purpose of a single tool (e.g. supports modeling), and about a selected single tool of a tool recommendation, for example about its usage costs or its usability. Such information explained in [11] as technical-functional criteria and socially-assessable criteria enable application developers to select the suitable tool or tool set for their development task. Application developers can also define such criteria as further requirements for suitable tool or tool sets. If application developers have identified a suitable tool or tool set they can download/order it from its supplier.

The selection of a Petri net tool to enable the application of a certain Petri net type additionally comprises the specification of the required functionality. In the cases described above, the required functionality is already determined by the Petri net technique or method. In this case, application developers determine the required modeling, assessment, and implementation functionality to enable the application of Petri net type selected before yourself.

Select a Sample Solution. The final use case described here is intended for finding the proposed solutions suitable for fulfilling the requirements of a certain

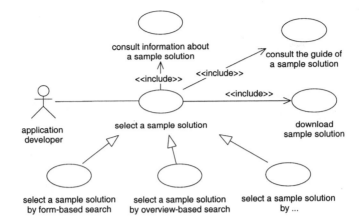

Fig. 19. Use case: select a sample solution

application development. It is called *select a sample solution* and is shown in Fig. 19. Basically, a sample solution can be selected in different ways, for example by a form-based or an overview-based search. If a form-based search is used then application developers can specify required characteristics of suitable sample solutions similar to the form-based search described for the selection of a Petri net technique or the selection of a Petri net method, compare Fig. 10. It enables application developers to find suitable sample solutions by their objectives, application domains, and their underlying Petri net types.

The overview-based search offers several kind of overviews of sample solutions, for example an application domain specific overview. Such an overview sorts sample solution according to their indented application domain. It is similar to the overview-based search for select a Petri net method described above, see also Fig. 13.

Before application developers select a certain sample solution to integrate it within their own application development they can use information about it, for example about the solved problem or the underlying Petri net type, and can consult the so-called solution guide for details.

4   Application Developer View

In this section, we will explain the structural model of the Application Developer View of our »Petri Net Baukasten« introduced in Section 1. This model comprises classes of objects and their required relationships for enabling an assisted application development as depicted in Section 3.

Accordingly, we will take this model as basis for explaining the relationship between the Application Developer View and the Common Base described in Section 2 as well as between the Application Developer View and the Tool Developer View.

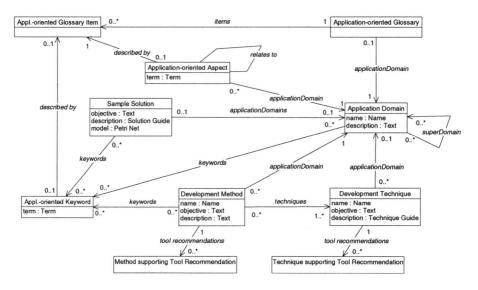

Fig. 20. Class diagram of the Application Developer View

4.1   The Structural Model

The structural model of the Application Developer View is depicted as a UML class diagram [8] in Fig. 20. It comprises classes specifying application domains, development methods, sample solutions, etc. Let's have a closer look at the model.

Application Domains. The core of our »Petri Net Baukasten« approach is application orientation. Therefore, application domains play a central role in our model. An application domain is identified by its name and explained by its description. Each application domain can be characterized by a number of application-oriented aspects reflecting all relevant entities and questions of a particular application domain. As an example, let us take the application domain *workflow management*, which can be characterized by application-oriented aspects according to the workflow process definition meta model of the Workflow Management Coalition (WfMC) documented in [13]. This includes workflow application, workflow participants, workflow process activity, workflow relevant data, etc. Each application-oriented aspect can be related to other application-oriented aspects as shown here by the *relates to*-relationship. For example, particular application aspects may be refined by another aspects if a number of subaspects stand in a subaspect/superaspect relationship to it. We will omit the description of further relationships at this point.

In principle, we would like to take different application domains into consideration. Therefore, one application domain may have a close relationship to another application domains, where one is the superdomain of the other. The superdomain of *logistics*, for example, could be *business process modeling*.

Development Methods. Furthermore, development methods play an important role in our model. A development method is specified by its name and its objective. It comprises a informal description, a so-called method guide, and one or more development techniques. Each and every development method is designated as a suitable method for supporting application development of a certain application domain, as depicted by the relationship *applicationDomain* between *Development Method* and *Application Domain*. There may be several tool recommendation for a development method.

Development Techniques. Development techniques are specified by their names and objectives. They comprise an informal description, a so-called technique guide. Development techniques are either universal or application domain-related. There may be several tool recommendation for a development technique.

Sample Solutions. Sample solutions represent several proposed Petri net based solutions for different problems which are or are not application specific. Accordingly, sample solutions are related to a application domain or not. Each sample solution comprise a description, a so-called solution guide, using the terminology of the corresponding application domain.

Keywords. Keywords can be assigned to several elements of our Application Developer View, e. g. development methods, sample solutions, etc. These keywords may be helpful in finding suitable Petri net methods, Petri net techniques, sample solutions, etc. later on.

Glossaries. Similarly to the Common Base described in Section 2, glossaries also form an important part of the Application Developer View. Our intention is to manage one glossary for each application domain. The usual terms of a particular application domain are explained within a so-called application-oriented glossary according to corresponding items, see *Application-oriented Glossary* and *Appl.-oriented Glossary Item* in Fig. 20.

4.2   Relation to the Common Base

Fig. 21 illustrates the close relation of the Application Developer View to the Common Base. It only depicts the elements of the Application Developer View which are directly related to elements of the Common Base. Compare Fig. 20 and Fig. 7. We will have a closer look at the several relations in the following.

Application Domains. A Petri net type may be related to an application domain, thus qualifying as a suitable one for a certain application domain, see relationship *usefulPNtypes* in Fig. 21. This relation may be differently used in several assistance methods. For example, before computing the suitable Petri

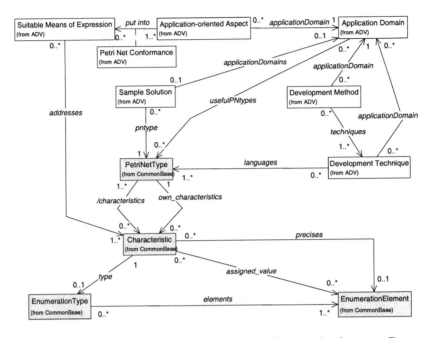

Fig. 21. Relation of the Application Developer View to the Common Base

net types fulfilling the predefined requirements, the search space of an assistance method may be reduced to the useful Petri net types of the application domain selected earlier.

Development Techniques. According to our definition of techniques, these are based on one or more languages. Petri net types are the only kinds of languages considered in the »Petri Net Baukasten«. Therefore, this relationship is defined by a corresponding relationship between development techniques and Petri net types, see relationship *languages* in Fig. 21.

Sample Solution. A sample solution is based on a certain Petri net type. This relation is fixed in a corresponding relationship, see *pntype* in Figure 21. As for all the relationships between the Application Developer View and the Common Base illustrated in Figure 21, this relationship plays an important role in assisting application developers. For example, only the sample solutions of a certain Petri net type selected earlier may be offered to application developers.

Application-Oriented Aspects. Application-oriented aspects stand in indirect relation to the Common Base. Let us assume that there may be more then one alternative possibility to express a single application-oriented aspect as regards the concepts of Petri nets. Each possibility – depicted as *Suitable Means of Expression* in Fig. 21 – may address more than one Petri net characteristic

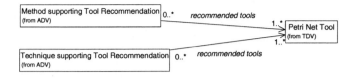

Fig. 22. Relation of the Application Developer View to the Tool Developer View

representing the required concepts supported by a Petri net type. Compare relationship *addresses* in Fig. 21. The Petri net conformity of application-oriented aspects is explicitly documented for each suitable means of expression.

4.3   Relation to the Tool Developer View

The relationship between the Application Developer View and the Tool Developer View is depicted in Fig. 22. We can see here, a recommendation of tools supporting a technique or a method comprises one or more Petri net tools.

5   An Assistance System for the Application Developer

This section describes a so-called assistance system which realizes the »Petri Net Baukasten« concept in order to support application developers in the Petri net-based development of applications. This system supports application developers as in the use cases described in Section 3.3. In particular, the assistance system provides application-oriented information about Petri nets as described in the Application Developer View in Section 4 in order to help application developers in finding suitable Petri net techniques.

Depending on the technical complexity, there are different possibilities for realizing such an assistance system. For example, it can be realized as an integrated development environment with integrated control during application development, meaning high technical complexity, or it can be realized as a system which only supplies information without any control, meaning a lower level of technical complexity. The realization described here corresponds to the second variant and is based on World Wide Web (WWW) technology.

5.1   Architecture of the Assistance System

The architecture of the assistance system is illustrated in Figure 23. It is based on middleware technologies in order to support distribution of the graphical user interface (GUI), functionality (Services), and data (Repository) on different servers.

Access to a repository is realized via an application programming interface (API). The API is a convenient interface, allowing abstraction from the repository realization. The services realize assistance methods corresponding to the use case described in Section 3.3 using the API. Therefore, the functionality of

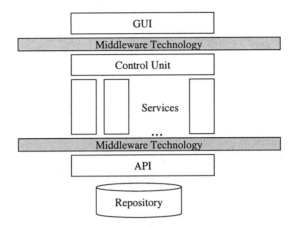

Fig. 23. Middleware-based architecture of the assistance system

the services is not influenced by the repository realization. The services are co-ordinated by a control unit. Interaction with application developers is realized via a graphical user interface (GUI).

As mentioned above, the assistance system is based on the Internet service World Wide Web (WWW). The WWW supports different techniques for dynamic information retrieval. For example, so-called JAVA applets implementing the graphical user interface can be integrated in HTML documents [5]. In this way, application developers can access the assistance system via the Internet using an Internet browser such as the Netscape® Communicator. A WWW-based system architecture is illustrated in Figure 24.

The functionality and the repository of the assistance system are located on servers connected via an intranet, which in turn is connected with the Internet. The Petri net tools may be located on download servers belonging to other intranets which are connected to the Internet. Maintenance of the assistance system is carried out via intranet access by »Petri Net Baukasten«-developers.

5.2   Structural Model of the Repository

The assistance system is based on a repository which manages all relevant data of the Application Developer View and the Common Base. As the assistance system is intended to support access to the formal descriptions described in the Expert View and existing Petri net tools managed in the Tool Developer View, the repository must store appropriate data of these views, too. Figure 25 illustrates the data model of the repository.

A more detailed illustration of the Application Developer View and the Common Base have already been described in Fig. 20 and Fig. 7.

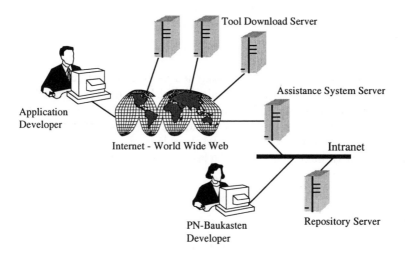

Fig. 24. World Wide Web-based architecture of the assistance system

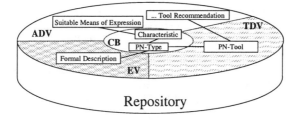

Fig. 25. The Common Base and the views upon Petri nets as part of the repository. Abbreviations are used as follows: *CB* – Common Base, *ADV* – Application Developer View, *EV* – Expert View, *TDV* – Tool Developer View

6   Conclusion

In this paper, we explained an assistance approach of the »Petri Net Baukasten« supporting application developers within a Petri net-based application development to improve the usability of Petri nets. This approach pursue objectives like application orientation, problem orientation, method orientation, tool orientation, and solution orientation. Several use cases are explained which describe the selection of suitable Petri net types, techniques, methods, and tools fulfilling the requirements of specific development tasks. Several assistance methods like overview-, form-, and checklist-based search are described for the different use cases. Information needed to realize the several use cases are described by structural models which represent the so-called Application Developer View and the so-called Common Base of the »Petri Net Baukasten« respectively. The Application Developer View comprises application-oriented information about Petri nets. These information are related to Petri net types which are hierarchically ordered in the Common Base. Moreover, an architecture of an WWW-based assistance system is proposed.

State of the Work. Parts of the sketched assistance system are realized on an experimental basis. The hierarchical order of Petri net types called Petri net classification was developed using Rational Rose [10]. The developed hierarchy representing the first upgrading level of our Petri net classification has been exported into an Oracle database. This database is used by a service implementing the navigation through the Petri net classification to enable the selection of a suitable Petri net type.

References

1. A. Battke, A. Borusan, J. Dehnert, H. Ehrig, C. Ermel, M. Gajewsky, K. Hoffmann, B. Hohberg, G. Juhás, S. Lembke, A. Martens, J. Padberg, W. Reisig, T. Vesper, H. Weber, and M. Weber. Initial realization of the »Petri Net Baukasten«. Informatik-Bericht 129, Humboldt-Universität zu Berlin, Berlin, Germany, 1999. DFG-Forschergruppe PETRI NET TECHNOLOGY.
2. C. Ermel and M. Weber. Implementation of Parameterized Net Classes with the Petri Net Kernel. In G. Rozenberg, H. Ehrig, J. Padberg, and G. Juhás, editors, *Unifying Petri Nets*, Advances in Petri Nets. Springer, 2001. To appear.
3. M. Gajewsky and H. Ehrig. The »Petri Net Baukasten«: An Overview. In G. Rozenberg, H. Ehrig, J. Padberg, and G. Juhás, editors, *Unifying Petri Nets*, Advances in Petri Nets. Springer, LNCS, 2001. To appear.
4. Wolf Garbe. Petri net tool survey, jun 1999.
   http://home.arcor-online.de/wolf.garbe/petrisoft.html; last visited: 26/11/2000.
5. Ivor Horton. *Beginning Java 2*. Wrox Press, 1999.
6. Stefan Jablonski, Markus Böhm, and Wolfgang Schulze, editors. *Workflow-Management – Entwicklung von Anwendungen und Systemen – Facetten einer neuen Technologie*. dpunkt Verlag, Heidelberg, 1. edition, 1997.
7. Merriam Webster's Collegiate Dictionary. Online-Version http://www.m-w.com/.
8. Object Management Group (OMG), http://www.omg.org/. *Unified Modeling Language Specification – Version 1.3*, March 2000.
9. Carl Adam Petri. *Kommunikation mit Automaten*. PhD thesis, Universität Bonn, 1962. Schriften des Instituts für Instrumentelle Mathematik.
10. Rational Software Corporation (http://www.rational.com). *Rational Rose*. Supports Visual Modeling with the Unified Modeling Language.
11. Dietmar Wikarski. Petri net tools—a comparative study. Forschungsberichte des Fachbereichs Informatik 97-4, Technische Universität Berlin, 1997.
12. World of Petri Net. World Wide Web, 2000. http://www.daimi.au.dk/PetriNets/; last visited: 01/10/2000.
13. Workflow Management Coalition, http://www.wfmc.org/. *The Workflow Reference Model, WFMC-TC-1003, Version 1.1*, Jan. 1995.

Implementation of Parameterized Net Classes with the Petri Net Kernel of the »Petri Net Baukasten« [*]

Claudia Ermel[1] and Michael Weber[2]

[1] Technische Universität Berlin, Fachbereich Informatik,
D-10587 Berlin, Germany
lieske@cs.tu-berlin.de
[2] Humboldt-Universität zu Berlin, Institut für Informatik,
D-10099 Berlin, Germany
mweber@informatik.hu-berlin.de

Abstract. We show in this paper how the formalism of Parameterized Net Classes is realized with the Petri Net Kernel. Parameterized Net Classes are an abstract notion of Petri nets using formal parameters to express Petri net type characteristics. This formalism allows the abstract formulation of formal concepts for a large variety of Petri net types. The Petri Net Kernel is a tool infrastructure supporting an easy implementation of Petri net algorithms. Moreover, the Petri Net Kernel is not restricted to a fixed Petri net type. Instead, only the net type has to be implemented as "net type specification". It is then used as basis for implemented application algorithms. In our paper we describe an implementation of the formal net type parameters via an interface such that the parameter implementation can be transformed to a net type specification for the Petri Net Kernel. This allows on the one hand a simple change of the net type by selecting a different combination of the actual net type parameters. On the other hand, applications (like simulation or analysis algorithms) can be developed generically, i.e. independently of the Petri net type, thus supporting rapid prototyping for Petri net tools. The implementation is embedded in the development of the »Petri Net Baukasten« and is therefore closely related to the contributions [3,8,24] in this volume.

1 Introduction

Unification Approaches Based on Parameterization. Since the introduction of Petri nets more than 30 years ago, many extensions and variants have been proposed in literature for different purposes and application areas. The fact that Petri nets are widely used and considered an important topic in research shows the usefulness and the power of this formalism. Nevertheless, the

---

[*] This work is part of the joint research project "DFG-Forschergruppe PETRINETZ-TECHNOLOGIE" between H. Weber (Coordinator), H. Ehrig (both from the Technical University of Berlin) and W. Reisig (Humboldt-Universität zu Berlin), supported by the German Research Council (DFG).

H. Ehrig et al. (Eds.): Unifying Petri Nets, LNCS 2128, pp. 79–102, 2001.
© Springer-Verlag Berlin Heidelberg 2001

situation in the field of Petri nets is unsatisfactory as the different notions, definitions and techniques, both in literature and in practice, make it difficult to find a common understanding and to provide good reasons for the practical use of the nets. Moreover, the unstructured variety of Petri net approaches causes reformulations and re-examinations of similar concepts for each Petri net variant. Most of the different concepts for Petri nets are defined explicitly for a single net type (e.g. place/transition nets) although many of these notions are essentially the same for different kinds of net types. Therefore, approaches to unification have been developed, e.g. [21,16,13] that employ the concept of parameterization. This permits the abstract formulation of formal concepts for a large variety of different net types.

Parameterized Net Classes and the Petri Net Kernel. The aim of this paper is to describe the implementation of an abstract notion of Petri net types, namely their formalization as Parameterized Net Classes, a uniform approach which is elaborated in another contribution to this volume [24]. The concept of Parameterized Net Classes is also known in the context of Abstract Petri Nets [21,20]. Here, the formal unification approach is combined with results concerning abstract Petri net morphisms that allow to express Petri net behavior and Petri net modifications. The formal framework of Abstract Petri Nets is based on category theory. Therefore, results obtained for Abstract Petri Nets and their morphisms hold as well in all instantiations, i.e. they are valid for all Petri net types that can be formalized as Abstract Petri Nets.

Our implementation of the concept of Parameterized Net Classes is realized via an interface to the Petri Net Kernel (PNK) [14]. The PNK provides an infrastructure offering methods for the administration and modification of Petri nets and allowing the user to define his own Petri net variant by implementing the specific characteristics of his net type. The design of the PNK was driven by the objective to support a quick implementation or integration of Petri net algorithms (e.g. for analysis, simulation, composition, ...) and to access basic net information via a simple interface.

The PNK already offers concepts for a modular, object oriented design of Petri net algorithms. Our implementation extends the PNK in order to reuse program code corresponding to formal parameters in Parameterized Net Classes. Tool support for Parameterized Net Classes on the one hand offers a validation of theoretical concepts, on the other hand, it allows the definition of a Petri net type by choosing an arbitrary combination of orthogonal parameters. Moreover, for the implementation of a new net type, it is possible to concentrate on the implementation of a new parameter and combine it with other already implemented parameters. Thus, Petri net algorithms implemented on the abstract level of Parameterized Net Classes are available without reimplementation for all kinds of Petri nets that can be defined by instantiating the formal parameters.

The »Petri Net Baukasten«. Both the formal concept of Abstract Petri Nets and the development of the Petri Net Kernel are part of the research group

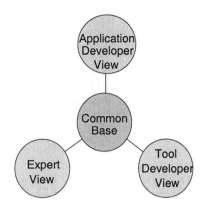

Fig. 1. The Views of the Petri Net Baukasten

project *Petri Net Technology* (see the footnote on page 79). The goal of the project is to point out and elaborate Petri net techniques appropriate for industrial sized applications. These techniques are comprised in a »Petri Net Baukasten« [23] to enable a more straightforward understanding of Petri net types, to support the use and development of Petri net tools, and the formal definition of Petri net semantics and theoretical results. The »Petri Net Baukasten« therefore establishes a classification concept for Petri nets as basis for these purposes. On the one hand, the »Petri Net Baukasten« introduces concepts and notations generally understandable to anybody with an engineering background, on the other hand it contains mathematical formal definitions to enable formal operation on nets like analysis, structuring, simulation and verification. Hence, the »Petri Net Baukasten« is designed to be presentable in different views serving particular purposes. Figure 1 shows these three views, related over the so-called *Common Base*, a classification of Petri net types. Here, we only name the main features of these views. For more details see the contributions [3,8,24] in this volume.

- The **Application Developer View** [3,18] is the basis of application-oriented support for project development based on Petri nets as specification technique. It supports developers to find the Petri net types, techniques, methodology, tools and process models most adequate for their applications. Therefore, the Application Developer View provides standard and example solutions which may be integrated into the developer's own work, descriptions of adequate methodologies, and application specific glossaries which relate technical terms from the application domain to Petri net notions used in the Common Base.
- The major task of the **Expert View** [8,7] is to provide the formal foundation of the »Petri Net Baukasten« in terms of mathematical presentation of the underlying notions and results. The Expert View provides a uniform structuring of the theory of Petri nets: The uniform definition of new notions is

supported, and the transfer of results is facilitated. For structuring the theory of Petri nets, the Expert View distinguishes essential notions which comprise abstract Petri net frames, formal Petri net techniques, instantiations, and transformations. Abstract Petri net frames have formal parameters which are integrated into an abstract description of Petri nets. Parameterized Net Classes as applied in this paper (introduced in Section 2) are one example for an abstract Petri net frame. The formal parameters are the data type parameter, formalized as specification frames and the net structure parameter, formalized as composite functor between adequate categories representing sets of places and structures on tokens [24,22].

- The **Tool Developer View** [32] supports Petri net tool developers in different ways: The tool developer who wants to announce an existing tool is supported by referring or distributing the tool. Developers who want to extend an existing Petri net tool or to build a new Petri net tool are helped by the Tool Developer View suggesting already implemented Petri net types and algorithms (administration of Petri net tools). Additionally, the Petri Net Kernel (PNK) as the main component of the Tool Developer View, offers an infrastructure for building new Petri net tools by providing standard operations on nets, a graphical Petri net editor and a graphical user interface. Moreover, the PNK is not restricted to a particular Petri net type. For more details see Sections 3 and 4. Furthermore, the PNK is extended by a repository of PNK application functions. These functions can be used by a developer of a PNK application.

The three views of the »Petri Net Baukasten« are linked over the **Common Base** that presents Petri nets in a semiformal classification hierarchy. Variants of Petri nets are informally described by the notions they comprise. The Application Developer View is linked to the Tool Developer View and provides an interface for the choice and use of Petri net tools and algorithms. Also, it is linked to the Expert View and provides an interface for the user to access the mathematical definition as well as theoretical foundation concepts concerning operations and semantics of a Petri net type. For a detailed description of the views and their relation via the Common Base see [4].

In our paper, we focus on the relation between the Tool Developer View and the Expert View. After a review of the theoretical concepts of Parameterized Net Classes and their algebraic formalization in Section 2, Section 3 sketches the concepts and the use of the PNK, a tool infrastructure for working with Petri nets of different net types. The main new result of our work consists of the design and implementation of interfaces that map the parameters of the theoretical concepts to attributes within the PNK implementation. These interfaces and their use for the sample implementation of Algebraic High-Level nets as an instance of Parameterized Net Classes are described in Section 4.

2   Introduction to Parameterized Net Classes

In order to allow a uniform approach to different kinds of Petri net classes, the concept of Parameterized Net Classes was introduced in [20]. By different actualizations of net type parameters, several well-known net variants can be formalized, like elementary nets, place-transition nets, coloured Petri nets, predicate transition nets and algebraic high-level nets, as well as several interesting new classes of low- and high-level nets. The basic idea of this uniform approach to Petri nets is to distinguish two parameters that describe the characteristics of a Petri net variant. These characteristics are the net structure defining the structure of the places and hence the markings [1], and (for high-level Petri nets) the data type specifying the internal structure of tokens for this net type. In the case of low-level Petri nets, the data type is trivial, as only black tokens are allowed on the places. The instantiation of these parameters leads to different Petri net types. In this section, we introduce the net structure parameter and the data type parameter. We sketch their instantiation with actual parameters in order to obtain actual net types. For a formal definition of Parameterized Net Classes and the instantiation of formal parameters with actual parameters, we refer the reader to the contribution [24] in this volume.

Parameters for Net Classes. The net structure parameter is sufficient to describe different low-level net types. In software industry, quite a large amount of low-level net variants have been developed over the last 30 years (see e.g. [27,29]) that are equipped with additional features and/or restrictions. We here review an abstraction of the net structure that can be instantiated to several low-level net types, including place/transition nets, elementary nets, variants of these and S-graphs.

For an abstract notion of high-level nets, we additionally need an abstraction of the data type used to describe the tokens, because several data type formalisms have been integrated with Petri nets leading to different notions of high-level nets. Typical examples are the following combinations: indexed sets with place/transition nets leading to Coloured Petri nets, predicate logic with elementary nets leading to predicate/transition nets [9], algebraic specifications with place/transition nets leading to algebraic high-level nets [25,28], ML with place/transition nets leading to Coloured Petri nets [11], OBJ2 with superposed automata nets leading to OBJSA-nets [1] and algebraic specifications with the Petri Box Calculus [2] leading to M-nets [17]. Object-oriented analysis and programming techniques are currently the de-facto standard of software development. This has lead to a variety of object-oriented Petri net formalisms combining Petri net structure with token objects or classes (see e.g. [30]).

Algebraic Presentation of the Net Structure Parameter. The formal basis for the definition of the parameters is the algebraic presentation of Petri nets using functions to relate a transition to its pre and post domain, introduced by Meseguer and Montanari in [19]. In this algebraic presentation, a place/transition net N is given by $N = (P, T, pre, post : T \rightarrow P^\oplus)$ with P and T being the

---

[1] As net and marking structure can differ for some net types, [16,22] introduces different parameters for these. The integration of a distinct flow parameter into our implementation of Parameterized Net Classes is subject to future work.

sets of places and transitions, and *pre* and *post* being functions from T to the free commutative monoid $P^\oplus$ over P. This construction corresponds to multisets over the set P of places. The pre domain of a transition t therefore can be written as a linear sum $pre(t) = \sum_{p_i \in P} n_i p_i$ for $n_i \in \mathbb{N}$, denoting how many black tokens have to be held by which places in the pre domain to enable transition t. For example, a pre domain function $pre(t) = 2p_1 \oplus 3p_2$ denotes that the arc inscription of the arc from p_1 to t is 2, and the inscription of the arc from p_2 to t is 3 and there are no other incoming arcs for t. The marking of a place/transition net then is given by some element $m \in P^\oplus$, and the operations for the computation of the firing behaviour are comparison, subtraction and addition based on linear sums, defined over the monoid operation.

Elementary nets consist of a set of places and a set of transitions, but the arc weight always equals one and the marking consists of at most one token on each place. A transition is enabled if there are enough tokens in the pre domain and no tokens in the post domain. The marking is a subset of the set of places, as there is only one token allowed on each place. The arc weight also can be expressed by a subset of the set of places. In the algebraic version of elementary nets, transitions are mapped to the powerset $\mathcal{P}(P)$ of P by a pre- and a post domain function $pre, post : T \to \mathcal{P}(P)$. Each element $m \in \mathcal{P}(P)$ can be considered as a marking of the elementary net. The firing behaviour makes use of the order on sets and the operations union and complement on sets.

These algebraic presentations [19] are equivalent to the classical presentations (see e.g. [27,6]), but have the advantage to be axiomatic, and thus simpler to generalize. The constructions $\mathcal{P}(P)$ and $P^\oplus$ for each set P of places then can be considered as functions from the class **Sets** of all sets via some class **Struct** of semigroups to the class **Sets**[2]. We consider $\mathcal{P}(P)$ and $P^\oplus$ as sets. The use of sets instead of semigroups allows the mapping from the transitions to these sets. This motivates that in general, an actual net structure parameter fo a net type can be considered as the composition of two functions: $Net : \textbf{Sets} \xrightarrow{F} \textbf{Struct} \xrightarrow{G} \textbf{Sets}$. Based on the function Net, we can describe Petri nets uniformly by $pre, post : T \to Net(P)$, where the specific net class depends on the choice of the function Net. Then, $F(P)$ denotes the markings and the pre and post domains of the transitions, and G relates the chosen construction (e.g. free monoids, power sets, ..) to sets.

Algebraic Presentation of the Data Type Parameter. For a generalization of the different data type representation used in combination with Petri nets, the notion of institutions [10]) is employed. Institutions provide a well-established abstract description of data type formalisms. The basic idea in generalizing different formalisms as algebraic specifications, predicate logic, ML, etc. is to assume axiomatically some data type specification $SPEC \in$ **SpecClass** and a class of models **Model(SpecClass)**. Based on this theory, an actual data type parameter consists of a class **SpecClass** of data type specifications and

---

[2] The constructions are in fact given by a pair of adjoint functors in [20,21,22] For simplicity is suffices to regard them as functions in this context.

for each $SPEC \in$ **SpecClass**, a class **Model(SpecClass)** of models satisfying the specification $SPEC$. Hence, the parameter can be represented by a function $Model :$ **SpecClass** $\rightarrow$ **ModelClasses** where **ModelClasses** is the (super)class of all model classes.

Example 2.1 (Algebraic High-Level Nets)
Algebraic High-Level Nets (short AHL nets) [25] combine place/transition nets and algebraic specifications in the sense of [5].

They are defined by the same actual net structure parameter as place/transition nets, namely $Net = (\_)^{\oplus}$, the construction of free commutative monoids over sets. The net structure is defined algebraically by the pre and post domain functions $pre, post : T \rightarrow (T_{OP}(X) \times P)^{\oplus}$ with $T_{OP}(X)$ being terms with variables over the signature. The respective marking structure is an element of the commutative monoid of $(A \times P)^{\oplus}$ with A being an adequate $SPEC$-algebra.

The data type is formalized as an algebraic specification $SPEC = (S, OP, E)$ with S being sorts, OP operation symbols and E equations over the signature $SIG = (S, OP)$. The arc inscriptions in AHL nets are given by terms with variables over the signature. The class of models for an AHL net data type is the class of SIG-algebras satisfying the equations E. One model, i.e. one of those SIG-algebras, defines the tokens (elements of the carrier sets for each sort) that are allowed on the places of the AHL net.

Thus, the actual data type parameter is given as follows: Let **SpecClass** be the class of all algebraic specifications $SPEC = (S, OP, E)$ and **Alg(SPEC)** be the class of all $SPEC$-algebras A (see [5]). Then we obtain the actual data type parameter as function $Model = Alg :$ **SpecClass** $\rightarrow$ **ModelClasses** that relates each specification $SPEC$ to the model class **Alg(SPEC)** of $SPEC$-algebras.
$\Diamond$

The following definition summarizes the notion of formal parameters of net classes:

Definition 2.2 (Formal Parameters of Net Classes)
The formal net structure parameter and the formal data type parameter are given by the functions $Net :$ **Sets** $\xrightarrow{F}$ **Struct** $\xrightarrow{G}$ **Sets** and $Model :$ **SpecClass** $\rightarrow$ **ModelClasses** as described above.
$\triangle$

Obviously, a uniform approach to Petri nets should comprise both low-level nets as well as high-level nets. We therefore consider low-level nets as a special case of high-level nets with a data type that yields only one data element, the black token. We call this the *trivial* actual data type parameter.

A survey of some actual net types defined by the actual parameters (i.e. instantiations of the formal parameters) is given in the Petri Net Square, shown in Figure 2. The parameters can be seen as the two dimensions of the square, whereas well-known Petri net types each correspond to one point in the Petri Net Square.

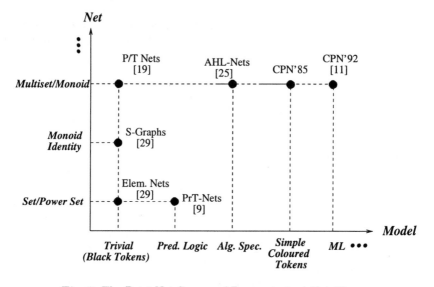

Fig. 2. The Petri Net Square of Parameterized Net Classes

3 The Petri Net Kernel

The development of the Petri Net Kernel (PNK) was initiated in 1996 as a vision of a universal Petri net tool [15]. Today, the Petri Net Kernel [14] is an infrastructure for an easy implementation of Petri net algorithms concerning simulation, analysis or verification of Petri nets. The PNK realizes standard operations on nets like loading and saving of nets, accessing and modifying net characteristics. Hence, the developer of a Petri net application is relieved from building a parser for loading nets from a file or from dealing with graphical user interfaces to represent nets graphically. Rather, one can concentrate on implementing the algorithmic idea. When the algorithm is implemented, the developer gets with the help of the PNK an executable Petri net tool prototype.

The PNK provides several interfaces. The *application interface* comprises some functions on Petri nets for programming applications. The *editor interface* describes the interaction between the PNK and an editor. It contains functions which are provided by the PNK. They must be provided by an editor for a proper interaction between a PNK based tool and its user. The PNK is equipped with a simple graphical net editor. This editor can be replaced by another editor, provided it conforms to the editor interface. Both, the application interface and the editor interface can be used by a *PNK user* for building prototypes of Petri net tools. The PNK itself as well as its applications are written in the programming language Python [26], an object oriented interpreted language. In the Python implementation, the PNK provides a textual storage format for net information which can also be used from external tools working with the PNK. Currently, a re-implementation of the PNK in Java is realized, together with the definition

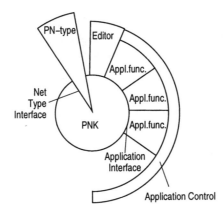

Fig. 3. The basic structure of a PNK based Petri net tool

of a generic interchange format for Petri nets based on XML [12]. The intended interchange format aims to support the transfer of net data between different tools. The PNK then can be seen as a tool integrator providing a universal interface for Petri net tools. On this basis, solutions of different tasks concerning Petri nets of different types could be exchanged between tools supporting this interchange format.

A very important feature of the PNK is that it is not restricted to a fixed Petri net type or a fixed set of Petri net types. It is possible to define new Petri net types. The *net type interface* describes how an application programmer may define his own Petri net type with specific extensions.

Figure 3 shows the basic structure of a PNK based Petri net tool consisting of several already implemented parts and several application functions which are implemented by a PNK user. The parts already implemented are the PNK itself, a graphical editor, the application control and some Petri net types. A Petri net type is passed as a parameter of the PNK via the net type interface. Whereas, an application function uses the PNK via the *application interface*.

In the following, we describe the application interface and the net type interface to the Petri Net Kernel before the implementation of an interface for Parameterized Net Classes as introduced in Section 2.

The Application Interface. To give a flavour of the usage of the PNK, we show the complete code of an executable application function in Listing 1. The execution of this code starts the PNK for P/T-nets and the editor of the PNK with a further menu button labelled with 'example_app'. Now, an application user may model or load a P/T-net. The example application function `example_app` starting after pressing the new menu button shows for each place in the preset of each transition an information string if the place is currently marked.

Listing 1. An application function of the PNK

```
1   from Build_Application import Build_Application
2   from Specification import PT_Specification
3
4   def example_app(net):
5       for transition in net.get_Transitions():
6           for place in transition.get_Preset():
7               if place.get_current_Mark().is_marked():
8                   net.show_information("The place " + place.get_Name() +
9                                        " in the preset of transition " +
10                                                  transition.get_Name() +
11                                                  " is marked.")
12
13  Build_Application(PT_Specification, example_app)
```

Line 1 of Listing 1 imports a function `Build_Application` from the PNK distribution package with the same name. The function is used as a link to the PNK and its editor. Line 2 imports the Petri net type definition (`PT_Specification`) specifying the net type for which the example application function should run—in our case P/T-nets—from the PNK package `Specification`. The definition of our application function starts with Line 4 and ends with Line 11. Finally, Line 13 links the application function to the PNK initialized for P/T-nets and the editor of the PNK in order to build an executable application.

An application function uses the several interfaces of the PNK allowing to access the net currently stored in the PNK. Our example, the application function in Listing 1 uses a few methods of the interfaces e.g. to get all transitions of the net (Line 5), to get all places in the preset of a certain transition (Line 6), to get the current marking of a certain place and an information whether a place is marked or not (Line 7). Lines 8 till 11 contain a user interface method to interact with a user. Of course, there are many methods (about 80) to get a comfortable access to different Petri net elements and to interact with an application user. For more details see [14].

The Net Type Interface. The PNK can be used with any kind of Petri net. The net type is given as an implementation of the net type interface, providing the following information:

- the representation of markings, as well as addition, subtraction and comparison operations on markings; (see for example the method `is_marked()` of Line 7 of Listing 1),
- a description of valid transition modes (e. g. assignments of variables) as well as a conversion operation from arc inscriptions into a corresponding marking for a given mode,
- the representation of arc inscriptions,

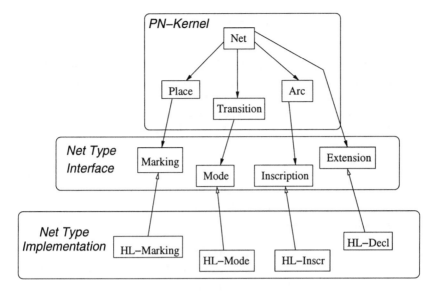

Fig. 4. The Petri Net Kernel and its Net Type Interface before the Implementation of
Parameterized Net Classes

- some further extensions concerning the net as a whole—such as definitions
  of variables and function symbols to be used in arc inscriptions of high-level
  Petri nets,
- extensions for certain elements of a Petri net, e. g. transition guards or time
  delay on places.

This information is provided by implementing derived classes of the classes
Marking, Mode, Inscription, and Extension. They are represented by attributes in the class Specification (of the net type) which is passed as the
net type parameter to the PNK.

The design idea of the net type interface follows intuitively from the PNK
architecture, where adequate and easy handling of net information was the main
goal. Figure 4 shows the different layers of a Petri net type specification using
the PNK. The basic PNK layer mainly consists of the four classes Net, Place,
Transition, and Arc defining modification and access methods for the net elements. The rest is defined by the net type implementation, which has to be
implemented for each new Petri net type that will be used by applications.
The net type implementation defines the net characteristics by implementing
attributes of the respective net classes: The class Place requires an implementation of Marking, the class Transition requires an implementation of Mode, and
the class Arc requires an implementation of Inscription. Additional information e. g. transition guards or variables are implemented in classes derived from
the class Extension. The classes Marking, Mode, Inscription and Extension

are realized as additional attributes of the classes `Place`, `Transition`, `Arc`, or `Net`, respectivly.

Example 3.1 (Implementing HL-Nets as Net Type for the PNK)
An example implementation of a simple high-level net type is sketched in Figure 4 by respective classes implementing the abstract net type interface classes `Marking`, `Mode`, `Inscription` and `Extension`. We here outline the implemented structures and methods. For a detailed documentation of this net type see [14].

- The implementation `HL-Marking` of the abstract class `Marking` uses Python dictionaries to store multisets of integers (tokens). For example, a marking of a place with the multiset of numbers $\{1, 1, 2, 3, 3, 3\}$ is represented by the dictionary $[1 : 2, 2 : 1, 3 : 3]$ where each pair denotes the token and how often it appears on the place. The class contains methods to check the validity of a marking, to get the current marking or set a new marking, and to add/subtract markings to/from the dictionary.
- The class `HL-Inscr` implements the abstract class `Inscription`. Here, we assume for simplicity that any string could be an arc inscription of our net type. Again, methods to set and get arc inscriptions are provided, as well as a method to check the syntactical correctness of a given arc inscription.
- The class `HL-Mode` is implementing the class `Mode` and defines the firing modes of transitions. A firing mode is an assignment of integers to all globally defined variables. Therefore, a mode is realized as dictionary in which each variable holds a value. Apart from get and set methods for modes, methods are implemented to compute a new assignment for the variables and to translate an arc inscription to a marking under the current variable assignment. The modes are connected tightly to the list of variables from the class `HL-Decl`.
- The class `HL-Decl` inherits from the abstract class `Extension`. In `HL-Decl`, methods are provided for two extensions, namely the declaration of variables (e.g. `x1`, `y1`, `z`) and the definition of functions (e.g. `def f(x): return x*x`). Variables and functions then can be used in arc inscriptions. Again, the get and set methods show or modify the current declarations of variables or functions, and a check method tests their syntactical correctness.

$\Diamond$

4   Implementing Parameterized Net Classes

This section deals with our main result, the implementation of Parameterized Net Classes with the Petri Net Kernel.

In the previous section we dealt with the net type interface of the original version of the Petri Net Kernel. Obviously, the method of reimplementing the net type specification makes it difficult to reuse parts of existing net type specifications (apart from copy & paste). A user could wish, for example, to have a high-level net with the behaviour of a condition/event net. Another user maybe wants to enhance a Petri net type by techniques to describe a data type for tokens. Based on the theory of formal parameters for net structure and data type, we therefore design a new interface to the PNK, called SquareInterface. This new interface enables the user to implement two corresponding actual parameters independently from each other. Moreover, it should be possible to combine two actual parameters arbitrarily to get different Petri net types as shown in the illustration of the Petri Net Square (Figure 2).

We describe a new net type specification layer that allows the definition of two parameters. The net type specification layer is called SquareSpecification and links the description of formal parameters to the original net type interface of the PNK. The parameters are accessed by the user via an interface called SquareInterface. The two dimensions of the Petri net square *net structure* and *data type* are required as parameters for SquareSpecification and correspond to the two formal parameters as described in Section 2. The main achievement of this implementation is that the formally described combination of orthogonal parameters now is supported by a tool. Thus, a new Petri net type can simply be defined by choosing a combination of already implemented parameters. The net type specification SquareSpecification generates from the actual parameters given via SquareInterface a Petri net type specification for the PNK. This section describes how the parameters are realized by SquareInterface.

Figure 5 shows the different layers of a Petri net type specification using the net type parameter interface SquareInterface and the net type specification SquareSpecification. The PNK and the original net type interface remain unchanged (see Figure 4) to allow a further use by applications that rely on net types already implemented without the SquareInterface parameters.

As sample implementation we again consider AHL nets (according to Example 2.1) and sketch how the net structure parameter and the data type parameter are implemented using the SquareInterface for Parameterized Net Classes.

Example 4.1 (Implementing AHL Nets as Parameterized Net Class)

In AHL nets, arcs connecting places and transitions are inscribed with terms with variables over an algebraic specification $SPEC$. These inscriptions symbolize the data taken from the pre domain or put to the post domain of a transition. A state of an AHL net is given by a marking $M \in (A \times P)^{\oplus}$, i.e. a distribution of data elements of a $SPEC$-algebra A on the places in P. The operational behaviour is realized by the firing of transitions under a consistent assignment asg of their variables. The firing effect is formalized by subtracting the data elements in the pre domain from the marking M and adding the data elements in the post domain to M according to the assignment asg: Let t be an enabled transition and M the current marking. The firing of t results in the follower marking M' which is computed by $M' = M \ominus pre_{asg}^{\oplus}(t) \oplus post_{asg}^{\oplus}(t)$.

Figure 6 shows a simple AHL net $N = (SPEC, P, T, pre, post, A)$ consisting of

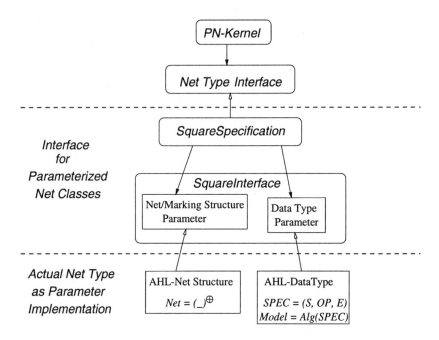

Fig. 5. The Petri Net Kernel and its Interface for Parameterized Net Classes

- an Algebraic Specification $SPEC = (S, OP, E)$ as shown in the Declaration Editor window of Figure 7. Here, the paragraphs *SORTS, OPNS* and *EQNS* correspond to the constituents *S, OP* and *E* of the specification;
- the set $P = \{Patients, Identity, Name, Sex\}$ of places;
- the set $T = \{Split\}$ of transitions;
- the pre domain function $pre : T \rightarrow (T_{OP}(X) \times P)^{\oplus}$ with $pre(Split) = (p, Patients)$;
- the post domain function $post : T \rightarrow (T_{OP}(X) \times P)^{\oplus}$ with $post(Split) = (getId(p), Identity) \oplus (getName(p), Name) \oplus (getSex(p), Sex) \oplus (p, Patients)$;
- the A-quotient term algebra realized over the algebra $A = (Sets, Functions)$ with the carrier sets *Sets* as given in the Declaration Editor window of Figure 7, paragraph *SETS* (where each set corresponds to one sort of the specification *SPEC*). The *Functions* are ground terms over the data elements from *Sets* with a special binding of constant symbols, again being defined in the Declaration Editor window of Figure 7, paragraph *CONST*. Elements of this algebra, e.g. $patient(Smith, masc, 1)$ are tokens and can be composed to a marking by the commutative monoid addition operation "$\oplus$"..

The example AHL net models a small part of the patient record administration of a hospital information management system. Here, we want to clarify the relation of the net inscriptions and the algebraic specification. The transition *Split* is enabled if there is a variable binding for the variable p (the only variable

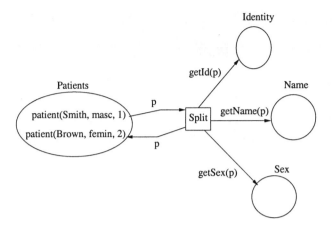

Fig. 6. AHL Net *Patient Record Administration*

in *Split's* pre domain). A variable in an arc expression must be bound to an element of the algebra, such that the evaluation of the arc expression then corresponds to a token on the place in the pre domain. In our example, we assume that the variable p is bound to the data element *patient(Brown, femin, 2)*. The firing of transition *Split* would produce the tokens 2 on place *Identity*, *Brown* on place *Name* and *femin* on place *Sex*. Let us consider the arc from the transition to the place *Identity* in its post domain. To compute the tokens produced on the place *Identity*, the term inscriptions of the arc have to be evaluated with the help of the equations of the algebraic specification. As variable p is bound to *patient(Brown, femin, 2)*, the left hand side of the equation $getId(patient(n, s, i)) = i$ is matching, with the variables of the equation bound by $n = Brown, s = femin$ and $i = 2$. The left hand side of the equation is replaced by its right hand side, namely i which evaluates to 2. Thus, token "2" is produced on place *Identity*.

Figure 7 shows the graphical user interface of the PNK. The AHL net *Patient Record Administration* has been drawn in the graphical net editor on the right, and the data type declaration has been edited with the declaration editor on the left-hand side.

The net structure parameter for AHL nets, $Net = (\_)^{\oplus}$, is implemented as list that may contain the same elements more than once. The implementation of the data type parameter for AHL nets requires a data type parser. This parser reads the declaration expression as given by the user in the declaration editor (see Figure 7). This declaration consists of the algebraic specification (the parts SORTS, VARS, OPNS and EQNS) and information to generate the model (the parts SETS and CONST), i.e. an A-qotient term algebra, see [5]). Note that both components, namely the abstract data type specification and the model are thus given by the user in form of one syntactical expression.

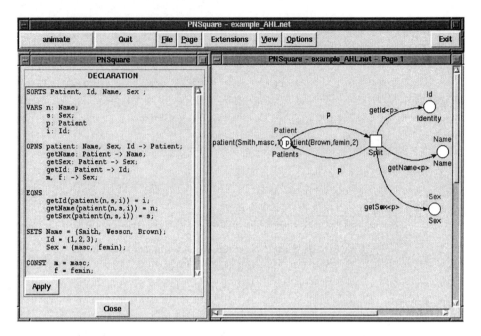

Fig. 7. The PNK GUI with the AHL Net *Patient Record Administration*

The parser then yields the syntactical components as attributes of the data type parameter's constructor. The formal actual data type $Model = Alg(SPEC)$ is realized by the data type parser in the following way: The algebraic specification $SPEC = (S, OP, E)$ is implemented by a list of sorts (attribute `sortlist`), the dictionary `opndict` relating operation symbols to their domain sorts and codomain sort and the list `eqnlist` of equations implemented as list of pairs of terms that correspond to the left and right hand sides of an equation. The algebra as model of the data type specification is the A-quotient term algebra realized over the signature and the algebra $A = (Sets, Functions)$ with $Sets$ given as the attribute `setdict` relating sorts from the signature to concrete sets of data elements. *Functions* are ground terms of the signature, i.e. terms without variables. But instead of constant symbols from the signature, these *Functions* contain data elements of *Sets*. The relation of constant symbols and data elements is specified by the user in the declaration editor and is implemented in the attribute `constdict`.

When generating the attribute structures from the declaration, the parser peforms several syntactical and static semantical (type) checks. Thus, it is guaranteed that the declaration corresponds to the syntax of an AHL data type, given implicitly by the parser being based on an EBNF-grammar. Type checks make sure, for example, that equations consist of terms of the signature, that the term sorts of the left and the right hand equation sides are matching and that variable sorts correspond to the sorts of the operation arguments they are contained in.

The generated attributes then are used for example in a simulator application to check whether arc inscriptions and place markings are syntactically correct or not, and to compute transition firing modes.

$\Diamond$

After having implemented both net structure and data type parameter for a net type (as shown for AHL nets in Example 4.1), generic Petri net algorithms can be applied that are designed for Parameterized Net Classes in general, thus independent of the actual Petri net type parameters. Currently, a generic simulation application is available for Petri net types implementing the interface for Parameterized Net Classes. The information to be provided by the data type parameter implementation concern the firing mode, i.e. a variable binding used to compute a possible marking from an arc inscription. The data type parameter interface ensures that there is a method eval implemented that generates a token from an arc inscription term in the current transition mode. In our example of AHL nets, the equations from the data type specification are used to reduce a generated token. The token is compared to the left-hand side of each equation. If a match is found, the equation's variables are bound to the respective data elements in the token, and the right-hand side of the equation is evaluated according to this variable binding and replaces the generated token. This process is repeated until no more matching equations are found. The data type specification equations therefore must not be cyclic.

In the following, we consider the parameter interface classes and their methods in more detail.

The Net Structure Parameter Interface. The actual net structure parameter is implemented as derived class of Collection. This abstract class describes operations working on a collection of elements, like adding an object to the collection, uniting two collections, deleting an element from a collection, subtracting a subcollection, and accessing elements. The detailed organization of the collection as set, multiset or FIFO list is realized in the implementation of the class of Collection. An implementation of Collection then realizes for example a structure that keeps track of the order when adding new elements (sequence, FIFO), or a structure that can contain the same elements with a multiplicity (multiset or commutative monoid), or a structure that contains each element at most once (set).

The Data Type Parameter Interface. The interface SquareInterface comprises quite a large amount of methods to enable the definition of an actual data type parameter (a declaration). Besides, firing modes of transitions can be related to arc inscriptions that are described as terms over the data type specification. Firing modes have to be able to generate tokens from terms, and tokens have to be able to be compared with other tokens. From the implementation of these features, SquareSpecification can generate structures that correspond to the net type interface of the PNK.

An actual data type parameter is implemented as a set of derived classes of the following abstract classes contained in the package SquareInterface:

- Token
  Here, the token structure is described (e.g. Integer, String, ...);
- Declaration
  This class contains the declaration of variables used in arc inscriptions, and, if necessary, signatures, data type specifications, algebras, ... (usually implemented as lists or dictionaries);
- Sort
  Here, the place sorts are described;
- Mode
  Contains methods for the computation of a new valid firing mode of transitions.
- Term
  This class describes the construction of terms over tokens. Usually, variables (found in the class Declaration) and simple tokens (as given in the class Token) are accepted as simple terms. Additionally, terms may be derived from a data type specification given in the class Declaration.

Defining Net Types by Parameter Selection. Once implemented, the parameters are accessible for a Petri Net Kernel user via a *PN-Square* graphical user interface (see Figure 8). Here, the user may choose between different combinations of actual parameters and gets a graphical editor for the concrete net type that results from the selected combination. This graphical user interface in combination with the realization of the SquareInterface is the tool corresponding to the formal Petri Net Square as given in Figure 2.

In the following, we describe some of the implemented parameters and their combinations according to Figure 8.

The following actual *Data Type Parameters* are implemented and made available via SquareInterface:

1. *Black Token*
   A token is represented as a '∗'-string in the graphical editor. The only trivial firing mode allows black tokens as arc inscriptions.
2. *Simple Coloured Tokens with finite Token Colours*
   Here simple high-level tokens are described, i.e. a finite number of integers.
3. *Simple Coloured Tokens with infinite Token Colours*
   Here simple high-level tokens are described again but potentially unbound according to the Python language's maximal integer.
4. *Algebraic Data Type Specification* This data type parameter describes an algebraic specification (sorts, operations, equations) and a model, i.e. an algebra to this specification. Tokens are data elements of the algebra. (see the sample implementation of the data type parameter for the AHL net in Example 2.1 above).

There are three actual *Net Structure Parameters* currently implemented and available to the PNK user via SquareInterface:

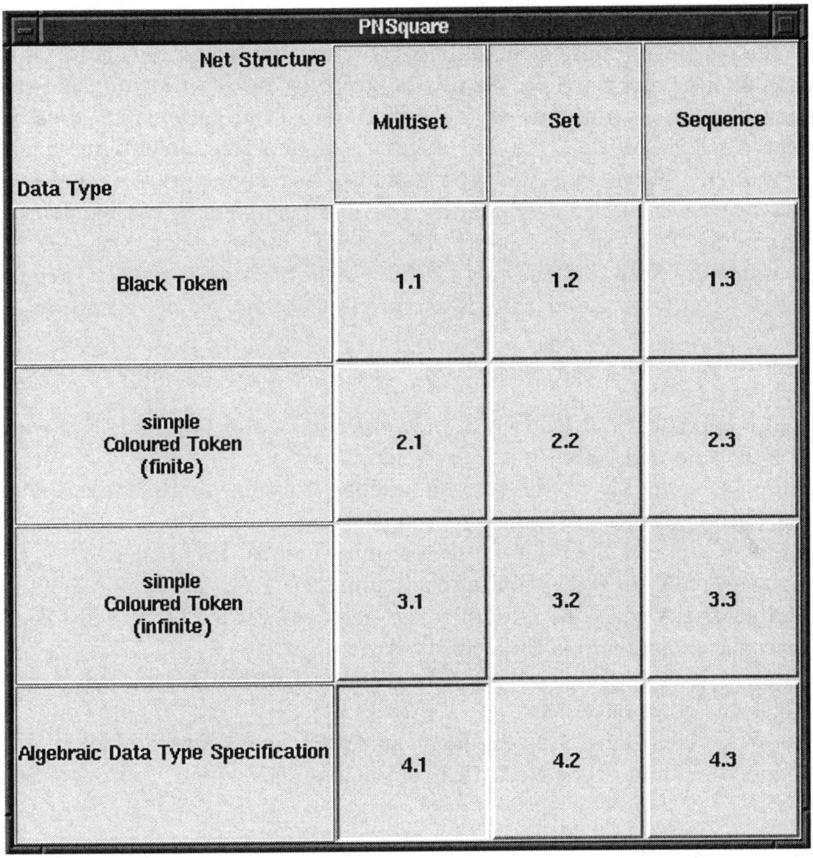

Net Structure			
	Multiset	Set	Sequence
Data Type			
Black Token	1.1	1.2	1.3
simple Coloured Token (finite)	2.1	2.2	2.3
simple Coloured Token (infinite)	3.1	3.2	3.3
Algebraic Data Type Specification	4.1	4.2	4.3

Fig. 8. The PN-Square Graphical User Interface of the PNK

1. *Multiset*

   A marking is implemented as a multiset over tokens, i.e. an element in a multiset may occur more than once. (see the sample implementation of the net structure parameter for the AHL net in Example 2.1 above).

2. *Set*

   A marking is implemented as a set over tokens. This means that an element in a set may occur only once, i.e. the insertion of an element into a set already containing this element fails.

3. *Sequence*

   A marking is implemented as a sequence over tokens. This means that the elements in a sequence are ordered by insertion. An element may occur more than once. Removing of an element is possible only at the head of a sequence, whereas insertion is realized at its tail.

Actual Parameter Combinations. An arbitrary combination of one actual data type parameter and one actual net structure parameter results in the specification of one Petri net type. We list the Petri net types whose implementation is realized by combining one of the actual data type parameters already implemented to one of the actual net structure parameters already implemented. The numbering of our Petri net type examples corresponds to the numbering in the PN-Square in Figure 8, where the first number refers to the number of the corresponding data type parameter, the second number denotes the net structure parameter. Note that the case "FIFO" (net structure parameter *Sequence*) makes sense only for the marking structure but not for the net structure.

1.1 Place/Transition Nets
1.2 Elementary Nets
1.3 Black-Token-FIFO Nets (which have the same behaviour as P/T Nets)
2.1 Simple Coloured Nets (finite token number)
2.2 Simple Coloured Nets (finite token number) without multiple tokens on each place, comparable to Predicate-Event Systems
2.3 Simple Coloured Nets (finite token number) with FIFO places
3.1 Simple Coloured Nets (infinite token number)
3.2 Simple Coloured Nets (infinite token number) without multiple tokens on each place, comparable to Predicate-Event Systems
3.3 Simple Coloured Nets (infinite token number) with FIFO places
4.1 Algebraic High-Level Nets
4.2 Algebraic High-Level Nets without multiple tokens on each place
4.3 Algebraic High-Level Nets with FIFO places.

A Generic Application for all Parameter Combinations. Using the example of a simulation algorithm animating the token game for one transition selected in the editor, we here sketch the implementation of a generic application which can be applied to all Petri net types that can be defined as combination of actual parameters, as represented in Figure 8.

Listing 2 sketches the relevant methods of our example application and shows that it is relatively easy to implement.

The simulator is initialized with one concrete Petri net of the chosen net type. The algorithm at first checks whether the selected transition is enabled or not. A method is_activated(transition) (see Listing 2, line 1 – line 13) uses methods from the data type parameter class Mode to compute a possible variable binding. Then, in this mode the arc inscriptions of incoming arcs are evaluated and compared to the tokens on places in the transition's pre domain using attributes from the data type parameter class Token. If a place is found which does not contain the required marking, the next possible firing mode is computed. Having found, at last, a firing mode under which the transition is enabled, the method fire(transition) (see Listing 2, line 15 – line 26) is executing one firing step: For all incoming arcs, the respective marking is removed from the places in the pre domain, and for all outgoing arcs, the respective marking is added to the places in the post domain.

Listing 2. The Generic Simulator Application

```
1    def is_activated(transition):
2        mode = transition.get_Mode()
3        while 1:
4            for arc in transition.get_Edges_in():
5                if not arc.get_Source().get_current_Mark().contains(
6                    mode.eval(arc.get_Inscription()))
7                    break # try next mode
8                else:
9                    return TRUE # found valid mode
10           if mode.exits_next():
11               mode.next()
12           else:
13               return FALSE # no valid mode found
14
15   def fire(transition):
16       for arc in transition.get_Edges_in():
17           place = arc.get_Source()
18           place.change_current_Mark(
19           place.get_current_Mark().subtract(
20               mode.eval(arc.get_Inscription())))
21
22       for arc in transition.get_Edges_out():
23           place = arc.get_Target()
24           place.change_current_Mark(
25           place.get_current_Mark().add(
26               mode.eval(arc.get_Inscription())))
```

The simulator algorithm is accessable in the Petri Net Kernel GUI (see Figure 7) via the animate button in the command line. The pressing of the animate button results in the call of the simulator method animate which computes all enabled transition in the net and presents them to the user. After the user has selected one transition from the list, animate calls the method fire(transition) and performs the firing step under the current firing mode.

This short example of a generic application gives an impression how the implementation of algorithms on Petri nets can be facilitated and be made usable for more than a single Petri net type.

5   Conclusion and Outlook

The basic idea of Parameterized Net Classes is to identify two parameters that allow a complete Petri net type description, namely the net structure and the data type formalism. In the tool infrastructure Petri Net Kernel, net types are described intuitively, without dissolving dependencies. Therefore, we have developed a new interface to the Petri Net Kernel that allows the integration of the two concepts. The Petri Net Kernel now offers a graphical user interface for the modelling with Petri nets whose types are defined in correspondence to the theory of Parameterized Net Classes. On the one hand, this work offers a validation of a theoretical/didactical concept as tool implementation. On the other hand, it allows the development of generic applications for Petri nets types that implement the SquareInterface, e.g. a generic algorithm simulating the token game.

Yet, the original PNK implementation containing the net type interface without distinguished parameters is preserved in the PNK tool infrastructure. On the one hand, this is for historical reasons: Existing applications relying on the "old" net type interface of course are still supported by the PNK. On the other hand, there are net variants that cannot yet be formalized in the theoretical framework of Parameterized Net Classes. For instance, parameter for organizational roles (e.g. for business process modeling) or for time have not yet been defined. Analogously, a formalization of the net structure parameter covering inhibitor arcs is not yet realized. Hence, future work is to define some important extensions of Petri nets in the terms of formal parameters which might be orthogonally combined with already existing parameters.

Related work concerning the practical implementation issues has been done recently by the integration of the Petri Net Cube [16] into the Petri Net Kernel. Here, the net and marking structure are treated as two separate parameters, a marking and a flow parameter, leading to the notion of a three-dimensional *Petri Net Cube* [31].

The main advantage of our approach is that the implementation of formal, orthogonal parameters as interfaces allows the definition of a new Petri net type by selecting a combination of previously implemented actual parameters. Thus, the gap between the theory of Parameterized Net Classes and their use in practice is reduced: A step towards rapid prototyping based on well-founded theoretical results has been done.

References

1. E. Battiston, F. De Cindio, and G. Mauri. OBJSA Nets: a Class of High-level Nets having Objects as Domains. In G. Rozenberg, editor, *Advances in Petri nets*, volume 340 of *LNCS*, pages 20–43. Springer Verlag Berlin, 1988.
2. E. Best, R. Devillers, and J. Hall. The Box Calculus: a new causal algebra with multi-label communication. In *Advances in Petri Nets*, pages 21–69. Springer, LNCS no. 609, 1992.
3. A. Borusan, S. Lembke, and H. Weber. Improving the Usability of Petri Nets with the »Petri Net Baukasten«. In G. Rozenberg, J. Padberg, H. Ehrig, and G. Juhás, editors, *Unifying Petri Nets*, Advances in Petri Nets. Springer, LNCS, 2001. (This Volume).
4. J. Dehnert. The Common Base of the »Petri Net Baukasten«. In H. Weber, H. Ehrig, and W. Reisig, editors, *Int. Colloquium on Petri Net Technologies for Modelling Communication Based Systems*, pages 211–229. Fraunhofer Gesellschaft ISST, October 1999.
5. H. Ehrig and B. Mahr. *Fundamentals of Algebraic Specification 1: Equations and Initial Semantics*, volume 6 of *EATCS Monographs on Theoretical Computer Science*. Springer Verlag, Berlin, 1985.

6. H. Ehrig and W. Reisig. Integration of Algebraic Specifications and Petri Nets. *Bulletin EATCS, Formal Specification Column*, (61):52–58, 1996.
7. M. Gajewsky. The Expert View of the »Petri Net Baukasten«. In H. Weber, H. Ehrig, and W. Reisig, editors, *Int. Colloquium on Petri Net Technologies for Modelling Communication Based Systems*, pages 243–265. Fraunhofer Gesellschaft ISST, October 1999.
8. M. Gajewsky and H. Ehrig. The "Petri Net Baukasten": An Overview. In G. Rozenberg, H. Ehrig, J. Padberg, and G. Juhás, editors, *Unifying Petri Nets*, Advances in Petri Nets. Springer, LNCS, 2001. (This Volume).
9. H.J. Genrich and K. Lautenbach. System modelling with high-level Petri nets. *Theoretical Computer Science*, 13:109–136, 1981.
10. J.A. Goguen and R.M. Burstall. Introducing institutions. *Proc. Logics of Programming Workshop, Carnegie-Mellon*, Springer LNCS 164:221–256, 1984.
11. K. Jensen. *Coloured Petri Nets. Basic Concepts, Analysis Methods and Practical Use*, volume 1. Springer, 1992.
12. M. Jngel, E. Kindler, and M. Weber. The Petri Net Markup Language. In *Proc. of Workshop on Algorithms and Tools for Petri Nets (AWPN 2000), Koblenz, Germany*, august 2000. http://www.informatik.hu-berlin.de/top/pnml/.
13. G. Juhás. *Algebraically generalised Petri nets*. PhD thesis, Institute of Control Theory and Robotics, Slovak Academy of Sciences, 1998.
14. E. Kindler and M. Weber. *The Petri Net Kernel. Documentation of the Application Interface. PNK Version 2.0*. Humboldt-Universität zu Berlin, Institut für Informatik, January 1999. http://www.informatik.hu-berlin.de/ kindler/PN-Kern/.
15. Ekkart Kindler and Jörg Desel. Der Traum von einem universellen Petrinetz-Werkzeug: Der Petrinetz-Kern. In Jörg Desel, Ekkart Kindler, and Andreas Oberweis, editors, *3. Workshop Algorithmen und Werkzeuge für Petrinetze*, Karlsruhe, October 1996.
16. Ekkart Kindler and Michael Weber. The dimensions of Petri nets: The Petri net cube. Informatik-Bericht, Humboldt-Universität zu Berlin, 1998.
17. H. Klaudel and E. Pelz. Communication as Unification in the Petri Box Calculus. Technical report, LRI, Universite de Paris Sud, 1995.
18. S. Lembke. The Application Developer View of the »Petri Net Baukasten«. In H. Weber, H. Ehrig, and W. Reisig, editors, *Int. Colloquium on Petri Net Technologies for Modelling Communication Based Systems*, pages 231–241. Fraunhofer Gesellschaft ISST, October 1999.
19. J. Meseguer and U. Montanari. Petri Nets are Monoids. *Information and Computation*, 88(2):105–155, 1990.
20. J. Padberg. *Abstract Petri Nets: A Uniform Approach and Rule-Based Refinement*. PhD thesis, Technical University Berlin, 1996. Shaker Verlag.
21. J. Padberg. Abstract Petri Nets as a Uniform Approach to High-Level Petri Net. In *Proc. WADT*, 1998.
22. J. Padberg. Classification of Petri Nets Using Adjoint Functors. *Bulletin of EACTS 66*, 1998.
23. J. Padberg. The »Petri Net Baukasten«: An Application-Oriented Petri Net Technology. In H. Weber, H. Ehrig, and W. Reisig, editors, *Int. Colloquium on Petri Net Technologies for Modelling Communication Based Systems*, pages 191–209. Fraunhofer Gesellschaft ISST, October 1999.
24. J. Padberg and H. Ehrig. Parameterized Net Classes: A Uniform Approach to Net Classes. In G. Rozenberg, H. Ehrig, J. Padberg, and G. Juhás, editors, *Unifying Petri Nets*, Advances in Petri Nets. Springer, LNCS, 2001. (This Volume).

25. J. Padberg, H. Ehrig, and L. Ribeiro. Algebraic high-level net transformation systems. *Mathematical Structures in Computer Science*, 5:217–256, 1995.
26. Python Consortium. *Python Language Homepage. http://www.python.org.*
27. W. Reisig. *Petri Nets*, volume 4 of *EATCS Monographs on Theoretical Computer Science*. Springer-Verlag, 1985.
28. W. Reisig. Petri Nets and Algebraic Specifications. *Theoretical Computer Science*, 80:1–34, 1991.
29. G. Rozenberg and P.S. Thiagarajan. Petri Nets: Basic Notions, Structure, Behaviour. In *Current Trends in Concurrency*, pages 585–668. 224, Springer, 1986.
30. R. Valk. Relating Different Semantics for Object Petri Nets . Technical Report FBI-HH-B-226/00, FB Informatik, Universität Hamburg, 2000.
31. M. Weber. Der Petrinetz-Würfel im Petrinetz-Kern. In Jörg Desel and Andreas Oberweis, editors, *6. Workshop Algorithmen und Werkzeuge für Petrinetze*, pages 69–74, J.W. Goethe-Universität Frankfurt/Main, Institut für Wirtschaftsinformatik, October 1999.
32. M. Weber. The Tool Developer View of the »Petri Net Baukasten«. In H. Weber, H. Ehrig, and W. Reisig, editors, *Int. Colloquium on Petri Net Technologies for Modelling Communication Based Systems*, pages 267–277. Fraunhofer Gesellschaft ISST, October 1999.

Process Landscaping: Modelling Distributed Processes and Proving Properties of Distributed Process Models

Volker Gruhn and Ursula Wellen

University of Dortmund, Department of Computer Science, Baroper Str. 301,
D-44221 Dortmund, Germany
{gruhn,wellen}@ls10.cs.uni-dortmund.de

1 Introduction

Traditional development and management of software processes is based on the idea of centralized real world processes carried out at one location. Reasons for modelling these processes, their chronological and hierarchical order, their interrelations and their deliverables at different levels of detail is to better understand their tasks and dependencies [Tul95]. We call a set of hierarchically structured process models, related via interfaces, a process landscape. Each activity of a process model belongs to the process landscape, but can also be refined by a landscape [GW00b] again. The predominant view onto process landscapes is a logical view, paying most attention to logical dependencies. We develop such a landscape by applying the Process Landscaping method [GW00a].

The globalization of companies leads to an increasing set of processes related within a process landscape [NKF95]. Different partners carrying out parts of the processes have varying degrees of autonomy and are distributed among different locations [GG95]. We can deduce, therefore, that globalization makes management of processes more and more difficult [LS99]. The logical view is no longer sufficient for management support. The distribution of processes to different locations requires a different view (called the locational view) onto a process landscape to analyze e.g. the distribution itself or the communication infrastructure between distributed processes. We call this view the locational view in order differenciate from the terms *local view* and *distributed view* which are also used in the context of distributed processes: A local view focuses on one part of a process landscape taking place at one single location, analogously to local views on distributed systems [FKT00]. The term *distributed view* is often used for locally distributed systems [TE00]. But in the context of Process Landscaping, we could also distribute parts of a process landscape for example with regard to different roles which are responsible for different sets of activities. Therefore, a distributed view does not express precisely enough that we talk about locational distributed activities.

Derniame et al. define a view as the particular approach to a software process conveyed by a (sub) model [DKW98]. They distinguish between models describing activities, organizational structures, products, resources or roles. This means different models representing different aspects of the same process landscape. Finkelstein and

H. Ehrig et al. (Eds.): Unifying Petri Nets, LNCS 2128, pp. 103–125, 2001.
© Springer-Verlag Berlin Heidelberg 2001

Sommerville define the term in a similar way. For them "the construction of a complex description .. involves many agents", each with "different perspectives or views of the artefact or system" where the views are "partial or incomplete descriptions .." [FS96]. The viewpoints framework has been used and documented e.g. in [FKN92, GMT99]. In the context of Process Landscaping the term *view* is used for describing a certain perspective of the entire process landscape, just by emphasizing different properties.

Irrespectively of the point of view or the level of detail, we need a suitable formal basis if we want to analyze properties of a process landscape. This basis should allow us:

- to model a process landscape by following the traditional way of representing process models and their logical interactions at different levels of refinement,
- to check properties of the given process landscape in the logical view,
- to check properties of the given process landscape in the locational view.

In this chapter we discuss a Petri net notation as a suitable formal basis fulfilling the requirements mentioned above. Other process modelling languages like event-driven process chains [KNS92], data flow diagrams or UML [omg99] do not support the explicit modelling of interfaces between process parts which are located at different sites. Additionally, verification and analysis techniques are not particularly capable due to the less formal basis.

For modelling the upper levels of a process landscape consisting of still complex activities and only key information objects, we developed PLL (Process Landscaping Language) as an abstract Petri net notation with a tree structure defining relations between activities. We denote PLL as abstract Petri net notation because we abstract from control flow information and do not deal with tokens or firing behaviour. The usage of the term *abstract* differs from the concept of abstract Petri nets described in [Pad98]. In this concept "the data type and the net structure can be considered as abstract parameters which can be instantiated to different concrete net classes" [Pad98].

PLL allows us to model static properties e.g. different locations, to describe requirements for communication infrastructure such as synchronous or encoded data interchange, to check some consistency conditions, and to analyze logical and locational aspects. With PLL as underlying formal notation, it is, therefore, already possible to analyze coarse-grained process landscapes. In order to analyze the more detailed levels we extend PLL to a Petri net notation (high level Petri nets with extended firing behaviour). In this extension we add some parameters useful for semantic analysis of a process landscape.

The process of extending a PLL landscape to a Petri net landscape can be compared to the idea of the "Petri Net Baukasten" of the research group on Petri Net Technology [PNK99]. Places, transitions and their relations serve as common components of all kinds of Petri net variants. Starting with PLL as rudimental kernel of a low level Petri net we stepwise extend it to different variants of high level Petri nets (e.g. hierarchically structured, with coloured token, with complex firing behaviour). Relations between the different variants are captured by transformations, in this paper illustrated by two functions *map* and *glue* (Section 4.1). They are essential for Petri net based modelling and analysis of process landscapes at different levels of abstraction. If we would declare the results of each step of the extension as special Petri net type, we could use it as unifying framework similar to the Common Base of the Petri Net Baukasten. There, a basic classification of Petri nets is provided with specialization/generalization relationships between the different Petri net types as structuring schema.

In this chapter we discuss an example process landscape representing processes and their relations at different levels of refinement, and discuss the logical and locational point of view. The example deals with the visual modelling and verification/validation of distribution and communication features in component-based software engineering. Related work can be found e.g. in [Stö00] where software architecture models in extended UML are mapped to Petri nets for automated analysis of communication properties and the result is mapped back to UML syntax. For Störrle, the architecture is a vital means of communication in the development process which should be understandable for all stakeholders. Therefore, he uses extended UML as graphical notation. In our approach, we abstract from information about the control flow in order to simplify the graphical representation the upper levels of a process landscape. This allows us to restrict ourselves on the formal basis of Petri nets instead of switching between different formal and informal notations.

In Section 2, we identify distribution properties of a software process landscape which are interesting to analyze. Section 3 discusses the key elements and the structure of PLL as underlying formal notation for the upper levels of a process landscape. We also introduce the graphical representation of a process landscape modelled in PLL (Section 3.2). Distribution and communication properties of the software process landscape are analyzed in Section 3.3. The mapping to a Petri net notation is discussed in Section 4.1. We analyze further properties of the resulting process landscape in Section 4.2. Section 5 summarizes the results of the analyses and gives a prospect to our future research.

2   Attributes Describing Distribution Properties of a Process Landscape

One purpose of modelling a process landscape is to analyze it with respect to certain properties. In the following, we identify the relevant properties.

To obtain a first overview about a process landscape, we first have to identify key processes and the most important information the processes need, produce or exchange. At the upper levels of the process landscape we deal with key activities like project management, software development, and quality management and with information objects like requirements documents, guidelines and source code. Simple static analysis is useful for checking these upper levels. Analysis becomes more complex when we consider the different types of information managed by different entities in different ways at different locations. By describing these distribution features of the different types of information as attributes of processes, information objects and interfaces, it is possible to analyze process landscapes with respect to properties which are considered important.

One aspect we want to analyze is the complexity and efficiency of distributed processes. For that purpose, we need information about the locations. We have to model where processes take place and where information objects are made persistent. The information whether or not data is stored locally by an activity sending or receiving it, allows to analyze the effort for keeping different representations of objects consistent. If this effort is "high", one should check whether a central database could be used to minimize the risk of inconsistencies due to redundancies.

We relate the property of autonomy to federated databases [SL90]. Correspondly, we distinguish two types of process autonomy:

- Those entities which control a database often share the data only if they retain the control. If, for example, the process "quality management" wants to retain the control of a tool recommendation document for software development environments, although the organizational entity that develops software knows other suitable tools, this entity is not allowed to change the document. We call this type of autonomy *data autonomy*. Only the controlling activity is allowed to change a document.
- *Operational autonomy* means, that an activity decides itself about the order in which its subactivities are performed and how they are carried out. It does not have to follow guidelines defined by a third party, but can define its own rules. Furthermore, it does not need to inform other processes about these rules. Therefore, activities with operational autonomy can be seen as a black box, where only the interfaces to this activity and the incoming and outcoming information objects are known.

Communication infrastructure can be described by communication channels which are associated to interfaces between processes. Communication channels describe how an activity sends or receives information objects, and whether it determines the communication infrastructure. The communication infrastructure can be based on:

- electronic data interchange like email or sms,
- synchronous communication infrastructure like telephone for oral information exchange or
- real document interchange like letter post.

Communication channels define whether the information exchange is

- persistent or not persistent
  (we call an information exchange persistent, if an activity stores the information it receives or sends)
- synchronous or asynchronous,
- private or not private
  (an information exchange is called private, if an information object is sent to exactly one recipient)

and whether information objects to be exchanged are encoded by senders and changeable by recipients. With attributes defining persistency, synchronity, privacy, coding and changeability we can analyze the communication

- of a process with its whole environment (all processes related),
- between exactly two processes,
- between refined processes (more detailed communication analysis between two processes) [Poh00].

In order to improve the process, we also want to analyze whether and how single activities of refined processes should be distributed in the process landscape. The so-called ping-pong communication is an example of such a situation. It means frequent exchange of information of the same type between processes with little modification of the exchanged information within the processes. This communication indicates that a refined process forwards information to another and that this information is only read or minimally modified before it is returned. This sort of communication may be the only option, but it can also indicate a situation where we should check the efficiency of distribution. Generally speaking, the ordering of activities of this type can be improved by internal checks before forwarding information to another activity via an interface.

Summarizing the discussion about properties of process landscapes, we have identified a set of attributes assigned to the landscape's key elements, namely activities, information objects and their relations to each other.

3   Formalization and Analysis of Process Landscapes in PLL

PLL (Process Landscaping Language) is a Petri net notation for the upper refinement levels of a process landscape. It is used to analyze interesting static properties like the communication infrastructure. The key elements and structure of PLL are explained in Section 3.1. In order to use PLL for modelling process landscapes, we also developed a graphical representation which is discussed in Section 3.2. In Section 3.3 we analyze an example process landscape depicting component-based software development which has been modelled in PLL.

3.1   PLL (Process Landscaping Language)

The main purpose of PLL is the identification of the most important information objects created or used by key activities. In PLL, they are described as document types. We consider the relations between activities and information objects as either being a reading access or a writing access. These considerations lead to the following definition:

A word $\omega \in$ PLL is defined as a triple $\omega = (V, D, Z)$ which is a Petri net. Elements of V are called activities, elements of D are called document types, and elements of Z are called (access) relations. A word $\omega \in$ PLL represents a process landscape, where

- $(v,d) \in Z$ means that activity $v \in V$ creates or writes a document of type $d \in D$,
- $(d,v) \in Z$ means that activity $v \in V$ reads a document of type $d \in D$.

For $\omega \in$ PLL, AB $\subseteq$ V×V describes the hierarchical composition of activities as a tree, more formally: $(v_1,v_2) \in$ AB means that v_2 is refining v_1. We call AB an activity tree. The root r of this activity tree does not denote an activity, but the process landscape itself.

With this definition it is possible to model a process landscape, to define activities and document types as key elements together with their relations to each other. The set $\{v \in V \mid \exists\, w \in V: (v,w) \in AB\}$ denotes refined activities. *leaves* = $\{v \in V \mid \nexists\, w \in V: (v,w) \in AB\}$ depicts activities which are not refined any further.

We define interfaces between activities by relating one document type to two activities, one reading and the other writing a document of that type. In PLL, an interface is defined as

interface: V×V → P(D) with
interface $((v_1,v_2)) := \{d \in D \mid \exists\, v_1, v_2 \in V: ((v_1,d) \in Z \wedge (d,v_2) \in Z) \vee$
$\qquad\qquad\qquad\qquad\qquad\qquad ((v_2,d) \in Z \wedge (d,v_1) \in Z)\}$

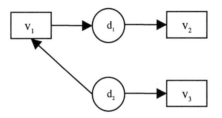

Fig. 1. Interfaces between activities

Figure 1 shows an example, where *interface* $((v_1,v_2)) = \{d_1\}$, *interface* $((v_1,v_3)) = \varnothing$ and *interface* $((v_2,v_3)) = \varnothing$. All document types belonging to interfaces are called interface document types. In figure 1, d_1 is an interface document type, d_2 is not. We refine an *interface* $((v_1,v_2))$ by adding further document types to *interface* $((v_1,v_2))$.

With the language core of PLL as defined above, we can express activities and document types as elements of a process landscape. We can model their relations to each other by defining read or write access and by defining interfaces. The refinements of activities and interfaces allow their hierarchical composition. In order to analyze properties of a given process landscape, we extend PLL by functions assigning different attributes to activities, document types and relations. Extensions relevant for process landscape analysis are discussed in the following:

- L is a set of locations and *loc*: V → L is a function assigning a location $l \in L$ to an activity $v \in V$. Based on this attribute it is possible to analyze the complexity of distribution for a given process landscape.

- *op-aut*: Z → {0, 1, undefined} is a function assigning either zero, one or undefined to each relation $z \in Z$.
  op-aut $((v_1,d_1)) = 1$, *op-aut* $((d_1,v_1)) = 1$ means that only activity v_1 defines rules how it creates, changes or uses a document of type d_1 and does not have to follow other guidelines.

op-aut $((v_1,d_1)) = 0$, *op-aut* $((d_1,v_1)) = 0$ means that activity v_1 may have to follow guidelines when it creates, changes or uses a document of type d_1.

op-aut $((v_1,d_1)) =$ undefined, *op-aut* $((d_1,v_1)) =$ undefined means that it is not yet defined, if activity v_1 has to follow guidelines when creating, changing or using a document of type d_1.

Function *op-aut* defines the operation autonomy of an activity concerning a single document type. If for all document types $d \in \{d \,|\, (v_1,d) \in Z \vee (d,v_1) \in Z\}$:

1. $(v_1,d) \in Z \Rightarrow$ *op-aut* $((v_1,d)) = 1 \; \forall \, d \in \{d \,|\, (v_1,d) \in Z\}$ and
2. $(d,v_1) \in Z \Rightarrow$ *op-aut* $((d,v_1)) = 1 \; \forall \, d \in \{d \,|\, (d,v_1) \in Z\}$

then activity v_1 is called operation autonomous.

- ***per*: $Z \to \{0, 1,$ undefined$\}$** is a function assigning either zero, one or undefined to each relation $z \in Z$.

  per $((v_1,d_1)) = 1$, *per* $((d_1,v_1)) = 1$ means that activity v_1 stores a document of type d_1 locally.

  per $((v_1,d_1)) = 0$, *per* $((d_1,v_1)) = 0$ means that activity v_1 does not store a document of type d_1 locally.

  per $((v_1,d_1)) =$ undefined, *per* $((d_1,v_1)) =$ undefined means that it is not yet defined, if activity v_1 stores a document of type d_1 locally.

  This attribute is important for the analysis of the effort for updates of redundant storages.

  If $|$ *per* $(z) = 1 \,|$ with $z = (v,d)$ or $z = (d,v)$ is "high" for a specific document type, one should consider about a central database where *per* $((v,d)) = 0$ and *per* $((d,v)) = 0$ for the affiliated relations.

- Let $Z' := \{(v,d) \,|\, (v,d) \in Z\} \subset Z$. ***d-aut*: $Z' \to \{0, 1,$ undefined$\}$** is a function assigning either zero, one or undefined to each relation $z \in Z$.

  d-aut $((v_1,d_1)) = 1$ means that only activity v_1 is allowed to change a document of type d_1.

  d-aut $((v_1,d_1)) = 0$ means that activity v_1 is not allowed to change a document of type d_1.

  d-aut $((v_1,d_1)) =$ undefined means that it is not defined, if activity v_1 is allowed to change a document of type d_1.

Function *d-aut* is restricted to relations representing write access because it does not make any sense to define an activity as data autonomous concerning a specific document when it only has read access to it. The following condition has to hold:

$$d\text{-}aut\,((v_1,d_1)) = 1 \Rightarrow \forall \, w \in V \colon d\text{-}aut\,((w, d_1)) = 0$$

We have to check this consistency condition, if we want to prove the data autonomy of an activity.

If for a given activity $v \in V$

1. *d-aut* $((v,d)) = 1 \ \forall \ d \in D$ with $(v,d) \in Z'$
2. *per* $((v,d)) = 1 \ \forall \ d \in D$ with $(v,d) \in Z'$

then activity v is called data autonomous.

- *synch*: $\mathbf{Z} \rightarrow \{\mathbf{0, 1, undefined}\}$ is a function assigning either zero, one or undefined to each relation $z \in Z$.
  synch $((v_1,d_1)) = 1$ means that activity v_1 sends a document type of d_1 to other activities synchronously.
  synch $((d_1,v_1)) = 1$ means that activity v_1 receives a document of type d_1 from other activities synchronously.
  synch $((v_1,d_1)) = 0$ means that activity v_1 sends a document of type d_1 to other activities synchronously.
  synch $((d_1,v_1)) = 0$ means that activity v_1 receives a document of type d_1 from other activities synchronously.
  synch $((v_1,d_1)) =$ undefined means that it is not yet defined, if activity v_1 sends a document of type d_1 to other activities synchronously.
  synch $((d_1,v_1)) =$ undefined means that it is not yet defined, if activity v_1 receives a document of type d_1 from other activities synchronously.

  This attribute has impact on the communication infrastructure between activities. Communication via letter post for example always has to be defined as asynchronous, whereas calling per telephone has to be defined as synchronous communication.

We define additional functions analogously to function *synch* in order to describe further communication attributes:

- Function *priv* defines whether information exchange between activities is private (*priv* $((v,d)) = 1$, *priv* $((d,v)) = 1$) or not (*priv* $((v,d)) = 0$, *priv* $((d,v)) = 0$). This attribute has impact on the way how documents can be distributed among several locations: If an activity wants to send information to others, e.g. via broadcasting, *priv* $((v,d))$ has to be zero.
- Function *coded* defines whether information exchange between activities is encoded (*coded* $((v,d)) = 1$, *coded* $((d,v)) = 1$) or not (*coded* $((v,d)) = 0$, *coded* $((d,v)) = 0$). Encoding documents before sending them indicates that no other but the recipients should read the content. It also requires that decoding mechanisms are available at the recipient's side. If one activity encodes information before sending it to another location, all other activities using and storing it have to encode it again after decoding and using the content. This is to ensure that no activity exchanges data to be encoded without encoding it before. Information is defined to be protected, as soon as it is encoded for the first time.
- Function *change* defines whether information to be exchanged is changeable by the receiving / sending activity (*change* $((v,d)) = 1$, *change* $((d,v)) = 1$) or not (*change* $((v,d)) = 0$, *change* $((d,v)) = 0$). This attribute corresponds with the data autonomy of an activity: If an activity wants to retain the control concerning a specific document not only *change* $((v,d))$ should be one, but also the corresponding function *d-aut* $((v,d))$.

- CC is a set of communication channels and *c-channel*: $(Z \times Z)' \rightarrow CC$ is a function assigning a communication channel $c \in CC$ to each tuple $(z_1, z_2) \in (Z \times Z)'$. $(Z \times Z)'$ is a subset of $Z \times Z$, where z_1 and z_2 are relations belonging to the same document type and at least z_1 or z_2 is defined as a write access. More formally: $(Z \times Z)' \subset Z \times Z$ and $\forall (z_1, z_2) \in (Z \times Z)': (z_1 = (v_1, d_1) \Rightarrow z_2 = (d_1, v_2)) \wedge (z_1 = (d_1, v_1) \Rightarrow z_2 = (v_2, d_1))$.

A communication channel is called congruent, if the attribute values of relations z_1 and z_2 belonging to this channel, assigned by functions *synch, priv, coded* and *change*, are equal in pairs: If, for example, persistency of z_1 is 1, it also has to be 1 for z_2 as precondition for a congruent communication channel. Each congruent communication channel defines how an activity v_1 receives or sends a document of type d_1 via an interface to another activity v_2. Non-congruent communication channels are not operative, which means they are unable to initiate an information exchange. They may occur, when the underlying process landscape is modelled by different modellers (at different locations) and can be identified by consistency checks.

With the set of functions *per, synch, priv, coded, change* and *c-channel* we are now able to analyze especially distribution and communication issues at the upper levels of a process landscape. Although PLL only describes static properties like locations, persistency and privacy, we are able to carry out some semantic analyses for distributed processes landscapes.

3.2   Graphical Notation of PLL Elements

In this section we explain how process landscapes are graphically represented. We developed several graphical views in order to optimize the consideration of different analysis aspects. For this purpose we model a process landscape which covers parts of a component-based software process landscape [AF98]. Key features of this process landscape are that we find activities like domain engineering and component engineering on the same level [BRS98]. Figure 2 sketches the activity tree of this landscape. Some activities are refined to more concrete levels than others, some details are omitted for the sake of a concise representation. Acitivities "Project Management", "Quality Management", "Application Management", "Domain Engineering" and "Component Engineering" form the top level of the software process landscape. Activity "Application Engineering", for example, is refined by a set of activities (the names of which all begin with "AE", thus, indicating their origin). Figure 3 illustrates this refinement in more detail.

The graphical view of an activity tree abstracts from the conventional view of a Petri net representation, where we always consider states, transitions and relations between them by a bipartite graph. An activity tree shows an overview of the hierarchical structure of the process landscape activities which means that only transitions of the underlying Petri net are depicted.

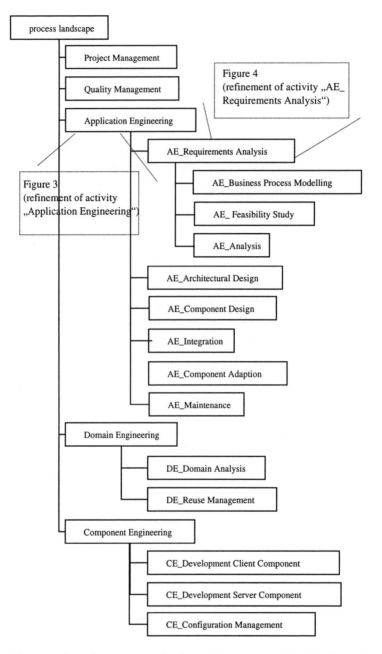

Fig. 2. Activity tree of a software process landscape for component-based software development

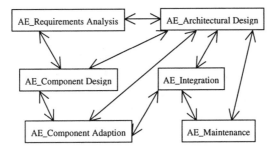

Fig. 3. Refinement of activity "Application Engineering" with interfaces indicated

As already mentioned, activities can be refined in terms of process landscapes. The view of an activity refinement illustrated in figure 3 for activity "Application Engineering" also does not show a conventional Petri net representation. It depicts six activities which are arranged on the same level of abstraction and belong to the same superordinated activity. States as second type of nodes in a conventional representation and their relations to transitions are indicated by bidirectional arrows. They do not express any details about the information objects to be exchanged, but they indicate that there is an information exchange between the related activities. Figure 3 shows, for example, that "AE_Requirements Analysis" and "AE_Architectural Design" are related via an interface.

Sometimes, further information about interfaces between activities is already known and has to be described. In this case we use a view of the refinement similar to a common Petri net representation. In the example, we assume that we already know some more details about activity "AE_Requirements Analysis". In its refinement (shown in figure 4) we recognize three activities and five document types. These document types specify the types of documents to be exchanged between the activities. For example, we can recognize that "system constraints" are exchanged between activities "AE_Feasibility Study" and "AE_Business Process Modelling". Moreover, we recognize two document types which appear as open ends in figure 4. They indicate where commitments with other activities in the process landscape are pending. In other words, documents of types "requirements document" and "process models" are provided for other activities, not covered by the refinement of "AE_Requirements Analysis". The indication of data flow by modelling an arrow without connecting it with the affiliated activity distinguishes this refinement view from common Petri net representations.

Despite the more activity-driven representations of PLL (as shown in figure 3 and 4) it may be useful, to immediately recognize which activities access documents of a certain document type. That is why PLL supports a document-driven view. Figure 5 shows this document view for the document type "architecture specification". It shows which activity access documents of this type without showing in detail where in the activity tree these activities are located. The naming conventions, however, allow a quick access to these activities in the activity tree.

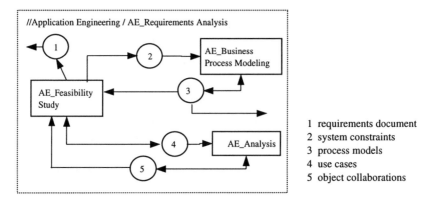

Fig. 4. Refinement of activity "AE_Requirements Analysis"

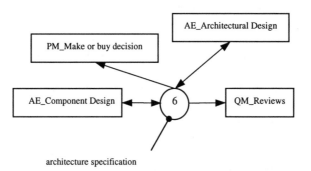

Fig. 5. Document view of document type "architecture specification"

In comparison to a common Petri net representation, we can identify the Petri net structure of states, transitions and relations in figure 5. The difference is the abstraction from control flow: we just have modelled the data flow but not the order in which the different activities access the architectural specification.

Summarizing the different views introduced, we distinguish four views for the graphical representation of PLL elements: the activity tree, two kinds of refinement views (with and without document types), and the document view. The first focuses on the hierarchical structure of the process landscape, the following two on relations between activities at the same refinement level, and the last focuses on relations to a specific document. All views represent the same process landscape or at least parts of it.

In the introduction, we also discussed the logical and the locational view of a process landscape. They are used for different analysis purposes. The logical view emphasizes logical dependencies, and the locational view is used for example for the analysis of communication features of locally distributed activities. These two analysis-driven views are represented in one or more of the "illustration-driven" views.

3.3 Analysis in PLL

We now analyze the example of a software process landscape for component-based software development. We focus on the properties of data autonomy, persistency and communication channels. For the sake of clarity, we restrict on simple examples.

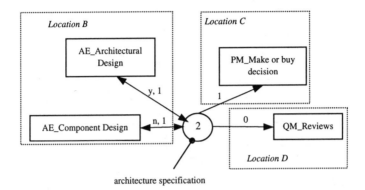

architecture specification

Fig. 6. Data autonomy and persistency within a software process landscape

The document view in figure 6 shows different activities associated with document "architecture specification". The activities take place at three different locations B, C and D. Some of them store the document locally, others do not. The value of the corresponding persistency attribute is indicated beside each relation arrow by either zero or one. The value of data autonomy is indicated by either y (if the value is one) or n (if the value is zero). If the value of this attribute is "undefined", it is not indicated. Activity "AE_Architectural Design" is the only one which keeps the document persistent and which is allowed to change the document simultaneously. For all activities at location C and D the value of data autonomy is undefined. Activity "AE_Component Design" has write access to the document, but the value of data autonomy is zero. This means, that the activity is only allowed to change the local copy of document "architecture specification", for example by adding remarks to the specification. Therefore, activity "AE_Architectural Design" retains the control of the document and is called data autonomous concerning "architecture specification".

In figure 6, $\lvert per\ (v,$"architcture specification"$) = 1 \wedge per\ ($"architecture specification"$,v) = 1 \rvert$ with $v \in \{$"AE_Architectural Design", "AE_Component Design", "PM_Make or buy decision", "QM_Reviews"$\}$ is "high" which means that nearly all activities store the document locally. It would be more efficient to implement a central database at location B with read access for activities "AE_Component Design", "PM_Make or buy decision" and "QM_Reviews" and write access for activity "AE_Architectural Design" as responsible activity for document "architecture specification".

Figure 7 shows a more complex overview of the process landscape for component-based software development. Activities at different levels of refinement are arranged according to their locations. This is indicated by dotted frames around sets of activities. The refinement of activity "AE_Requirements Analysis" (see also figure 4) is e.g. depicted as taking place at location A. Other activities concerning application engineering are located at B, domain engineering and project management take place at location C, and component engineering takes places at location D. The documents are explained in the legend below figure 7. For the sake of clarity, they are depicted multiply in the graphical representation, although they exist only once. This allows a better understanding of the different communication channels.

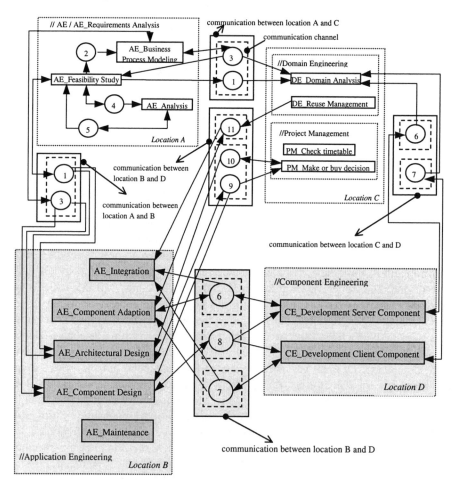

Fig. 7. Software process landscape with communication channels indicated

Legend: 1. requirements document   6 software code server component
           2 system constraints   7 software code client component
           3 (domain) process landscape   8 component specifications
           4 use cases   9 architectural specification
           5 object collaborations   10 make or buy decision document
                                           11 reusable components

Figure 7 shows some features of the communication infrastructure by indicating communication between different locations (full line frames), communication channels (dashed frames) and documents (circles) exchanged between different activities. We restrict our discussion of communication properties to the information exchange between locations B and D. The areas of the software process landscape participating at this communication are shaded grey. They consist of activities concerning application engineering (location B) and component engineering (location D). Documents "software code server component", "software code client component" and "component specifications" have to be exchanged via three different communication channels, indicated by three dotted frames around the documents.

We now have a closer look at the attributes assigned to relations between locations B and D. Tables 1 – 3 depict the attribute values for the example process landscape concerning each relation affiliated to either document "software code server component", "software code client component" or "component specifications".

Relation columns printed in bold/italic and belonging to the same document indicate a communication channel. In table 1, there is a communication channel between activities "AE_Component Adaption" and "CE_Development Server Component" exchanging document "software code server component" in both directions. Activity "CE_Development Server Component" also receives document "component specifications" from activity "AE_Component Design" (see table 2). Activity "CE_Development Client Component" sends document "software code client component" to activity "AE_Integration" (table 3).

Table 1. Relation attributes and communication channel concerning document "software code server component"

	software code server component, AE_Integration	*AE_Component Adaption, software code server component*	*software code server component, CE_Development Server Component*
persistent	1	*1*	*1*
synchronous	0	*0*	*0*
private	undefined	*1*	*1*
encoded	0	*1*	*1*
changeable	0	*1*	*1*

Table 1 describes a congruent communication channel between activities "AE_Component Adaption" and "CE_Development Server Component" because all attribute values fit together. Table 2 indicates a communication channel which is not congruent because the values of privacy and changeability are different. But "not congruent" does not always mean inconsistent. Privacy of the access relation from "CE_Development Server Component" to "component specification" is "undefined". Therefore, we just have to add a suitable value for this attribute. Concerning the changeability of this document the modeller has to decide if he can change the value for either the access relation of "CE_Development Server Component" or "AE_Component Adaption", or if he adds a transformation activity to the process landscape. This activity transforms document "component specifications" e.g. from a word-file (changeable) into a pdf-file (not changeable) and sends it to activity "CE_Development Server Component" afterwards.

Table 2. Relation attributes and communication channel concerning document "component specifications"

	AE_Component Design, component specifications	*component specifications, CE_Development Server Component*	component specifications, CE_Development Client Component
persistent	*1*	*1*	0
synchronous	*0*	*0*	0
private	*1*	*undefined*	undefined
encoded	*1*	*1*	1
changeable	*1*	*0*	0

Table 3. Relation attributes and communication channel concerning document "software code client component"

	CE_Development Client Component, software code client component	*software code client component, AE_Integration*	software code client component, AE_Component Adaption
persistent	*1*	*1*	1
synchronous	*undefined*	*undefined*	0
private	*1*	*1*	1
encoded	*1*	*0*	0
changeable	*1*	*1*	undefined

Table 3 also shows a communication channel which is not congruent because the values of the coding attribute are defined differently. In this case, the modeller has to decide whether document "software code client component" has to be encoded or not. If an activity creating / writing the document requires encoding he should perhaps extend this requirement to all other relations. If an activity reading the document requires encoding, it can perhaps encode the contents after reception.

There are many other interesting details in the tables to be analyzed. For example, almost all activities store the document, they are dealing with, locally. One should check if a central database can minimize redundancy and the effort for maintaining all existing copies. Not all relation combinations form a communication channel (see definition of function *c-channel* in Section 3.1). We identify eight possible communication channels for the three documents. They all have to be checked. We have to analyze if they are congruent and or not. In the latter case we have to consider whether to redesign the communication channel or to refine the communication infrastructure by adding a transformation activity. This fact indicates how complex the communication infrastructure of a process landscape is already on the upper levels of refinement.

But only the awareness of having a complex communication infrastructure does not justify the effort to define all communication attributes. The benefit of modelling communication channels and the related attributes is to express communication features of a (given or planned) communication infrastructure in terms of simple function values. They allow us to measure communication effort independently of concrete data interchange media like email, letter post or telephone. This avoids the discussion about a precise definition of what we mean by communication considering every possible communication aspect: some people focus on the hardware they need for communication, others on the underlying communication protocols or on some of the attributes considered in this chapter. The advantage to reduce communication to a small set of attributes affiliated to communication channels instead of talking about concrete media is first to avoid misunderstandings of what we focus on when we talk about communication. Secondly, it enables comparability of different media and allows therefore analysis already on the upper levels of a process landscape. Improvements can be suggested like it has been done for the example shown in figure 6 and inconsistencies like those in table 2 and 3 can be identified.

By now we have described how to check process landscapes at high levels of abstraction. But it is also useful to investigate more detailed levels. For doing so, we have to relate process landscapes represented in PLL to Petri nets giving more details about the control flow in processes. That is why we map PLL to a Petri net notation (described in the following section). This mapping is accompanied by adding information about control flow and dynamic behaviour of processes.

4   Formalization and Analysis of Process Landscapes in Petri Net Notation

"A software process is in fact the aggregation of numerous process fragments" [EB95]. This statement is a suitable description for the upper levels of our software process landscape modelled in PLL. The aggregation in PLL is coarse-grained, because the activity tree shows the hierarchical order of the activities, but not the control flow in a software process. Thus, we have depicted "numerous process fragments" consisting of core activities and core information objects. In order to analyze the behaviour of a software process we have to connect the fragments to a coherent net of activities and information objects. For that purpose we model additional information about process sequences and control flow.

With PLL we are not able to express and analyze dynamic features like process sequences and parallelism of activities. Modelling different types and different states of information objects is also not possible. This leads us to the requirement of mapping all process landscape elements expressed in PLL to a notation supporting the modelling of dynamic behaviour. In Section 4.1 we discuss the mapping of PLL elements to the Petri net notation as a suitable formalism. In Section 4.2 we analyze the example process landscape depicting component-based software development after it has been mapped to Petri net notation and refined.

4.1 Mapping PLL to Petri Net Notation

In order to model and analyze behavioural features of a process landscape we extend PLL to the notation of a Petri net, where a process landscape is defined as a tuple $PL = (C,S,F)$ with

- C is a set of activities
- S is a set of interfaces
- F is a set of flow relations

In PLL notation, each document type occurs only once. All activities reading/writing objects of this type are connected to it. This clearly shows, which activities depend on objects of which type. In order to insert information about the dynamic behaviour, states of objects have to be introduced. In doing so, we have to retain the data flow direction as defined in the process landscape. The mapping starts from the leaves of the PLL activity tree. Information about the hierarchical structure of a process landscape is abstracted away, the result is a flat Petri net. We need three functions for the mapping, each considering either activities, document types or relations. More formally:

Let $Z_{leaves} = \{(v,d) \in Z | v \in leaves\} \cup \{(d,v) \in Z | v \in leaves\}$.

- map_v: $leaves \rightarrow C$
  This function just maps the leaves of the PLL activity tree to elements of set C, representing activities in Petri net notation.

- map_d: $D \rightarrow P(S)$ with
  1. $|map\_d(d)| = |\{v \in leaves \,| \,(v,d) \in Z_{leaves} \vee (d,v) \in Z_{leaves}\}|$
  2. $\forall \, v \in leaves$ with $(v,d) \in Z_{leaves} \vee (d,v) \in Z_{leaves}$: $\exists \, s_v \in S$ with
     $(v,d) \in Z_{leaves} \Rightarrow (map\_v \,(v), \, s_v) \in F$
     $(d,v) \in Z_{leaves} \Rightarrow (s_v, \, map\_v \,(v)) \in F$

  If a document type is related to several activities, it has to be mapped to a power set of interfaces s_i. Condition 1 ensures that we map each document of type d to an interface s_i as often as there are relations (v,d), $(d,v) \in Z_{leaves}$ to it. Condition 2 requires for each activity v represented in PLL notation, that its target representation in Petri net notation is only related with an interface, which has a source representation related to the activity in PLL before.

- map_z: $Z_{leaves} \rightarrow F$ and
  $\forall \, (v,d) \in Z \, \exists! \, (c,s) \in F \wedge \forall \, (d,v) \in Z \, \exists! \, (s,c) \in F$ with $c \in C$, $s \in S$ such that
  $map\_z((v,d)) = (c,s) \wedge map\_z((d,v)) = (s,c)$ with $c = map\_v \,(v) \wedge s \in map\_d \,(d)$

  Function map_z maps each relation exactly once and such that the direction of the target relation remains the same as it has been in PLL notation. Additionally, each related activity and interface in Petri net notation has to have source representations in PLL notation.

Figure 8 denotes how activities, document types and affiliated relations have to be mapped. In our example, activities v_1, v_2 and v_3 out of set *leaves* are mapped with function map_v to activities c_1, c_2 and c_3. With function map_d, document type d_1 is mapped to interfaces s_1 and s_2, document type d_2 is mapped to s_3 and s_4 (indicated by dotted frames and arrows). The four access relations (v_1,d_1), (d_1,v_2), (d_2,v_1) and (d_2,v_3) are mapped to four data flow relations according to function map_z with relating interfaces s_1 and s_2 both to activity c_1. The result is a set of three isolated process fragments. Connecting them to the overall Petri net at the right places means adding information about the succession of the modelled activities. It is task of the modeller to add this information and cannot be done automatically.

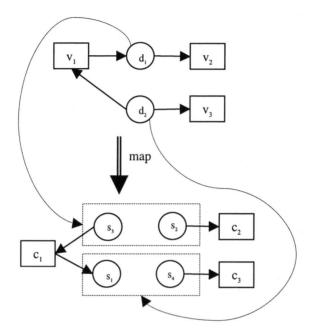

Fig. 8. Mapping an activity refinement in PLL to a Petri net notation

To merge Petri net fragment into to one Petri net is based on a function **_glue_**: $S^n \rightarrow$ S. This function glues together some of the interfaces which result from mapping one document type (on the PLL level) to the Petri net representation. More formally:

glue: $S^n \rightarrow S$ is applicable to $s_1, ..., s_n$ only if $\exists d_1 \in D : s_1, ..., s_n \in map\_d (d_1)$

Glued process fragments denote information exchange via interfaces and define the order in which activities are carried out. In figure 8, interfaces s_1 and s_2 may be glued together. Then they define in which order activities c_1 and c_2 have to be executed.

One might ask why a document type is mapped to a set of interfaces first which are glued together again, afterwards. The reason is the missing information about process sequences defining the order of different activities. The edges in PLL notation represent only document access and no temporal order of the execution of activities. This information is added after the mapping of the process landscape in Petri net notation. Thus, the edges represent control flow only after the mapping. Referring to the example in figure 8, interfaces s_1 and s_2 may also be connected by adding a sequence of new activities and interfaces between interfaces s_1 and s_2.

4.2 Analysis Example of a Process Landscape in Petri Net Representation

Obviously, attributes assigned to process landscape elements in PLL can be analyzed after the mapping. It is even possible to carry out some powerful analyses, because more information about the processes is available at the level of the Petri net representation. To analyze additional features of the process landscape concerning e.g. the efficiency of activity distribution within different refinements, we define attributes for process sequences with external interfaces involved. In this section, we only discuss one distribution aspect as an example for dynamic analysis of process landscapes in high-level Petri net representation. Currently, we are working on example process landscapes for the discussion of further analysis facilities on the Petri net representation concerning redundant storage of data at various locations and efficient use of telecommunication infrastructure.

Function *maxpath* $(c_1, s_1, s_2) \rightarrow \mathbb{Z}$ denotes the longest path within an activity refinement c_1 starting from one connected external interface s_1 and ending with another external interface s_2, where both interfaces connect c_1 with a second activity refinement c_2. If the length of the maximum path in a refinement connecting those two interfaces is "short", one could assume that there are activities which are isolated from others. A more process-oriented way of organising a process may be useful for avoiding this sort of ping-pong communication. Thereby it may help to reduce coordination and communication effort between different processes [GGK96].

Figure 9 shows another part of the process landscape for component-based software development, modelled with the process modelling tool LeuSmart [ade99]. The upper levels of the landscape have been mapped to Petri net notation, previously. Activities "AE_Maintenance" and "CE_Configuration Management" have been further refined and interfaces have been glued together. Moreover, we added information about locations of activities. This allows to come up with the process model representation shown in figure 9. Here we can analyze maxpath as explained above in the following way:

- *maxpath* ("AE_Maintenance", "released sw to maintenance", "release notes") = 9 denotes the longest path within activity "AE_Maintenance" starting at interface "released sw to maintenance" and ending at interface "release notes".
- *maxpath* ("CE_Configuration Mgmt", "release notes", "releases software to maintenance") = 5
- *maxpath* ("CE_Configuration Mgmt", "released sw from maintenance", "released sw to maintenance") = 5 denotes the longest path within activity "Configuration Mgmt", which starts with reading from "released sw from maintenance" and ends with writing to "released sw to maintenance". In figure 9 this path is marked as bold path. It consists of five nodes.

If the length of the maximum path connecting "CE_Configuration Mgmt", "released sw from maintenance" and "released sw to maintenance" is "short", one could assume that there are activities isolated from preparing data for release documents. Indeed, "released sw from maintenance" is only used in activity "CE_Configuration Mgmt" to extract data for release documents. When all release documents have been updated, the released software is sent back to activity "AE_Maintenance". It would be useful to check whether a closer integration with activities from activity "AE_Maintenance" is possible. For this example one could suggest to extract the data relevant for release documentation by activity "release software" within the refinement of activity "AE_Maintenance" because this activity also extracts data for the release notes.

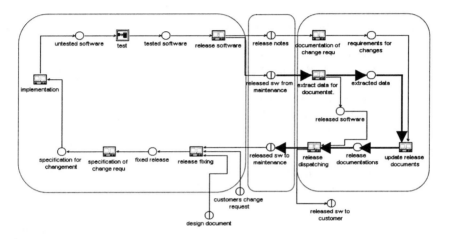

Fig. 9. ping-pong communication between activities "AE_Maintenance" and "CE_Configuration Mgmt"

5 Conclusion

In this chapter we introduced PLL as a formal basis for process landscapes. This notation allows to model distributed processes at different, but high level of abstractions. By incrementally adding further information about processes we reach the level of high level Petri nets, extended by some attributes defining behavioural properties of distributed processes. At this level we benefit from well-known Petri net analysis facilities and we propose further analysis mechanisms which are particularly relevant for distributed processes.

The focus of this chapter lies on different Petri net variants as formal basis of a process landscape. Therefore, we can compare it to the Expert View of the Petri Net Baukasten although it is not as detailed and sophisticated as in the Petri Net Baukasten itself. The Expert View presents formal descriptions of concepts described semiformally in the Common Base [Gaj99]. We are confident that the approach of the Process Landscaping method, respectively its underlying formal basis shown for some small examples in this paper, can be extended not only to further analyses of distribution properties but also to a unified and simplified construction kit for the modelling and analysis of process landscapes.

Our future research will be focused on tool support for PLL and on an automated mapping support from PLL to the Petri net representation. Furthermore, we are going to apply the Process Landscaping approach to further types of software processes, like, e.g., processes for the development of e-business applications. By doing so, we will be able to come up with improvement suggestions for the analysis facilities discussed above.

References

[ade99]   URL: http://www.adesso.de/leusmart
[AF98]   P. Allen, S. Frost, *Component-Based Development for Enterprise Systems – Applying The Select Perspective*, Cambridge University Press, 1998
[BRS98]   K. Bergner, A. Rausch, M. Sihling, A. Vilbig, *A Componentware Development Methodology based on Process Patterns*, In: PLOP'98, Proceedings of the 5th Annual Conference on the Pattern Language of Programs, August 1998
[DKW98]   J.-C. Derniame, B. A. Kaba, D. Wastell, *The Software Process: Modelling and Technology*, In: J.-C. Derniame, B. A. Kaba, D. Wastell (eds.), Software Process: Principles, Methodology and Technology, appeared as Lecture Notes in Computer Science No. 1500, Springer Verlag, pages 1-12, 1998
[EB95]   J. Estublier, N. Belkhatir, *A Generalized Multi-View Approach*, In: W. Schäfer (ed.), 4th European Workshop, EWSPT'95, Noordwijkerhout, The Netherlands, April 1995, appeared as Lecture Notes in Computer Science No. 913, Springer Verlag, pages 179-184, 1995
[FS96]   A. Finkelstein, I. Sommerville, *The Viewpoints FAQ*, Software Engineering Journal, IEE (Institution of Electronic Engineers), Vol. 11, No. 1, pages 2-4, 1996
[FKN92]   A. Finkelstein, J. Kramer, J. Nuseibeh, L. Finkelstein, M. Goedecke, *Viewpoints: A Framework for Integrating Multiple Perspectives in System Development*, Int. Journal of Software Engineering & Knowledge Engineering, World Scient. Publ., Vol. 2, No. 1, pages 31-57, 1992
[FKT00]   I. Fischer, M. Koch, G. Taentzer, *Local Views on Distributed Systems and their Communications*, In: H. Ehrig, G. Engels, H.-J. Kreowski, G. Rozenberg (eds.), Proceedings of the 6th International Workshop on Theory and Application of Graph Transformation (TAGT'98), November 1998, Paderborn, Germany, appeared as Lecture Notes in Computer Science No. 1764, Springer Verlag, 2000
[Gaj99]   M. Gajewski, *The Expert View of the "Petri Net Baukasten"*, In: H. Weber, H. Ehrig, W. Reisig (eds.), Proceedings of the Int. Colloquium on Petri Net Technologies for Modelling Communication Based Systems, Fraunhofer Gesellschaft ISST, pages 243-265, October 1999
[GE00]   G. Taentzer, H. Ehrig, *Semantics of Distributed System Specifications based on Graph Transformation*, presented at GI-Jahrestagung 2000, Workshop on "Rigorose Entwicklung software-intensiver Systeme", Berlin, 2000
[GG95]   G. Graw, V. Gruhn, *Process Management In-the-Many*, In: W. Schäfer (ed.), 4th European Workshop on Software Process Technology, EWSPT'95, Noordwijkerhout, The Netherlands, April 1995, appeared as Lecture Notes in Computer Science No. 913, Springer Verlag, pages 163-178, 1995
[GGK96]   G. Graw, V. Gruhn, H. Krumm, *Support for Cooperating and Distributed Business Processes*, In: Proceedings of the International Conference on Parallel and Distributed Systems (ICPADS'96), Los Alamitos, California, June 1996, IEEE Computer Society Press, pages 22-31, 1996

[GMT99]   M. Goedecke, T. Meyer, G. Taentzer, *ViewPoint-oriented Software Development by Distributed Graph Transformation: Towards a Basis for Living with Inconsistencies*, Proceedings of the 4th IEEE International Symposium on Requirements Engineering, Limerick, Ireland, 1999

[GW00a]   V. Gruhn, U. Wellen, *Structuring Complex Software Processes by "Process Landscaping"*, In: Reidar Conradi (ed.), 7th European Workshop on Software Process Technology, EWSPT 2000, Kaprun, Austria, February 2000, appeared as Lecture Notes in Computer Science No. 1780, Springer Verlag, pages 138-149, 2000

[GW00b]   V. Gruhn, U. Wellen, *Process Landscaping – Eine Methode zur Geschäftsprozessmodellierung*, In: Wirtschaftsinformatik, Vieweg Verlag, Vol. 4, pages 297-309, August 2000, in German

[KNS92]   G. Keller, M. Nüttgens, A.-W. Scheer, *Semantische Prozessmodellierung auf der Grundlage "Ereignisgesteuerter Prozessketten" (EPK)*, In: A.-W. Scheer (ed.), Publications of Institut für Wirtschaftsinformatik, Vol. 89, 1992, in German

[LS99]   E.C. Lupu, M. Sloman, *Conflicts in Policy-Based Distributed Systems Management*, In: IEEE Transactions on Software Engineering, IEEE Computer Society Press, Vol. 25, No. 6, pages 852-869, 1999

[NKF95]   B. Nuseibeh, J. Kramer, A. Finkelstein, U. Leonhardt, *Decentralized Process Modelling*, In: W. Schäfer (ed.), 4th European Workshop, EWSPT'95, Noordwijkerhout, The Netherlands, April 1995, appeared as Lecture Notes in Computer Science No. 913, Springer Verlag, pages 185-188, 1995

[omg99]   Object Management Group, *OMG Unified Modeling Language Specification*, Version 1.3, June 1999, URL: http://www.omg.org

[Pad98]   J. Padberg, *Abstract Petri Nets as a Uniform Approach to High-Level Petri Nets*, In: J. L. Fiadeiro (ed.), Proceedings of the 13th International Workshop on Algebraic Development Techniques, WADT 98, Lisbon, Portugal, April 1998, appeared as Lecture Notes in Computer Science No. 1589, Springer Verlag, pages 240-259, 1998

[Poh00]   A. Pohlmann, *Visualisierung und Simulation von Prozeßlandschaften – Die lokale Sichtweise*, master thesis at the University of Dortmund, Department of Computer Science, Software Technology, April 2000, in German

[PNK99]   A. Battke, A. Borusan, J. Dehnert, H. Ehrig, C. Ermel, M. Gajewski, K. Hoffmann, B. Hohberg, G. Juhas, S. Lemke, A. Martens, J. Padberg, W. Reisig, T. Vesper, H. Weber, M. Weber (research group on petri net technology, sponsored by the Deutsche Forschungsgemeinschaft DFG), *Initial Realization of the "Petri Net Baukasten"*, technical report at the Humboldt University Berlin, 1999, URL: http://www.informatik.hu-berlin.de/PNT/pnt-public.html

[Stö00]   H. Störrle, *Models of Software Architecture – Design and Analysis with UML and Petri-nets*, phd thesis, Ludwig-Maximilians-University of Munich, November 2000, URL: http://www.pst.informatik.uni-muenchen.de/personen/stoerrle/

[SL90]   S.P. Sheth, J.A. Larson, *Federated Database Systems for Managing Distributed, Heterogeneous, and Autonomous Databases*, ACM Computing Surveys, Vol. 22, No. 3, September 1990

[Tul95]   C. Tully, *The Software Process and the Modelling of Complex Systems*, In: W. Schäfer (ed.), 4th European Workshop on Software Process Technology, EWSPT'95, Noordwijkerhout, The Netherlands, April 1995, appeared as Lecture Notes in Computer Science No. 913, Springer Verlag, pages 138-143, 1995

Petri Nets over Partial Algebra

Jörg Desel, Gabriel Juhás, and Robert Lorenz*

Lehrstuhl für Angewandte Informatik
Katholische Universität Eichstätt, 85071 Eichstätt, Germany
{joerg.desel,gabriel.juhas,robert.lorenz}@ku-eichstaett.de

Abstract. Partial algebra is a suitable tool to define sequential semantics for arbitrary restrictions of the occurrence rule, such as capacity or context restrictions. This paper focuses on non-sequential process semantics of Petri nets over partial algebras. It is shown that the concept of partial algebra is suitable as a basis for process construction of different classes of Petri nets taking dependencies between processes that restrict concurrent composition into consideration.

Thus, Petri nets over partial algebra provide a unifying framework for Petri net classes in which some processes cannot be executed concurrently, such as elementary nets with context. We will illustrate this claim proving a one-to-one correspondence between processes constructed using partial algebra and processes based on partial orders for elementary nets with context. Furthermore, we provide compositional process term semantics using the presented framework for place/transition nets with (both weak and strong) capacities and place/transition nets with inhibitor arcs.

1 Introduction

Petri nets are applied in an increasing number of areas. As a consequence, numerous different variants of Petri nets have been developed, many of them based on the same behavioral principles but with slightly different occurrence rules. Examples include Petri nets extended by capacities, inhibitor arcs, read arcs or asymmetric synchronization of transitions.

The restrictions of the occurrence rule can be expressed by restricting the set of legal markings in the case of nets with capacities or by means of different kinds of arcs in the case of nets with inhibitor arcs, read arcs or asymmetric synchronization. Whereas the definition of sequential semantics for these variants can be obtained in a straightforward way from the occurrence rule, partial order semantics providing an explicit representation of concurrent transition occurrences is usually constructed in an ad-hoc way. The aim of this paper is to present a unifying concept for generalized Petri nets, i.e. for Petri nets with restricted occurrence rule, to obtain non-sequential semantics in a systematic way.

---

* supported by DFG: Project "SPECIMEN"

H. Ehrig et al. (Eds.): Unifying Petri Nets, LNCS 2128, pp. 126–172, 2001.

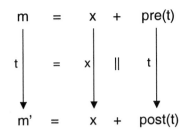

Fig. 1. Occurrence of a transition t from a marking m to a marking m' and its interpretation as a concurrent rewriting of the transition t and the marking x.

In [25, 26] and in [18] the authors realized that non-sequential semantics of elementary nets and place/transitions nets can be expressed in terms of concurrent rewriting using partial monoids and total monoids, respectively. In such an algebraic approach, a transition t is understood to be an elementary rewrite term allowing to replace the marking $pre(t)$ by the marking $post(t)$. Moreover, any marking m is understood to be an elementary term, rewriting m by m itself. A single occurrence of a transition t leading from a marking m to a marking m' (in symbols $m \xrightarrow{t} m'$) can be understood as a concurrent composition of the elementary term t and the elementary term corresponding to the marking x, satisfying $m = x + pre(t)$ and $m' = x + post(t)$, where $+$ denotes a suitable operation on markings (see Figure 1). For example, in [18] $+$ is the addition of multi-sets of places, and hence this approach describes place/transition nets. The non-sequential behaviour of a net is given by a set of process terms, constructed from elementary terms using operators for sequential and for concurrent composition, denoted by ; and $\|$, respectively.

Now, assume that for some class of Petri nets a suitable operation $+$ over the set of markings is given such that for each transition occurrence $m \xrightarrow{t} m'$ there exists a marking x satisfying $x + pre(t) = m$ and $x + post(t) = m'$. Then the occurrence of t at m is expressed by the term $x \| t$. Conversely, t cannot necessarily occur at any marking $x + pre(t)$ but its enabledness might be restricted. Such restrictions of the occurrence rule will be encoded by a restriction of concurrent composition, i.e. if $x + pre(t)$ does not enable t, then x and t are not allowed to be composed by $\|$. To describe such a restriction, we use an abstract set I of information elements together with a symmetric independence relation on I. Every marking x as well as every transition t has attached an information element. A marking x and a transition t can be composed concurrently if and only if their respective information elements are independent. For independent information elements we define an operation called concurrent composition with the intended meaning that the information of the composed term is the composition of the information elements of its components. Because the operation of concurrent composition between elementary terms and information elements is defined only partially, i.e. partial algebra is employed, such nets are called Petri nets over partial algebra [14, 15].

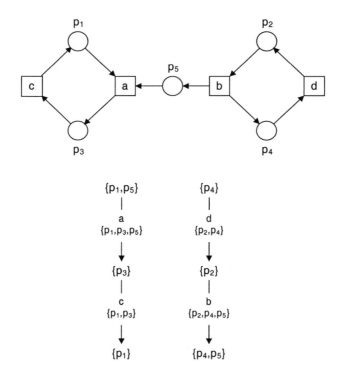

Fig. 2. An elementary net with places p_1, p_2, p_3, p_4, p_5 and the elementary terms corresponding to transitions. For example, transition a is enabled to occur if the places p_1 and p_5 are marked and the place p_3 is unmarked. Its occurrence removes a token from p_1 and p_5 and adds a token to p_3. In other words, transition a rewrites its preset $pre(a) = \{p_1, p_5\}$ by its post-set $post(a) = \{p_3\}$. It has attached the information element $\{p_1, p_3, p_5\}$, given by the union of its pre- and post-set.

For example, in the case of elementary nets, where markings are sets of places, we attach to a transition t as information element the union of $pre(t)$ and $post(t)$, while the information element for a marking m is the marking m itself. Two information elements are independent if they are disjoint. The concurrent composition of independent information elements is their union. For an illustrating example see Figure 2.

If a restriction of the occurrence rule is encoded by means of a partial algebra of information elements, one can build non-sequential semantics of nets over partial algebra. This semantics is given by process terms generated from the elementary terms (transitions and markings) using the partial operations sequential composition and concurrent composition.

Each process term has associated an initial marking, final marking and a set of information elements. For elementary process terms, the set of information elements is the one-element set containing the attached information element.

Initial and final markings are necessary for sequential composition: Two process terms can be composed sequentially only if the final marking of the first process term coincides with the initial marking of the second one. The set of information elements associated to the resulting process term is given by the union of the sets of information elements associated to the two composed terms.

Concurrent composition of two process terms is defined only if each information element associated to the first process term is independent from each information element associated to the second. Then the initial and final marking of the resulting term are given by concurrent composition of the initial markings and of the final markings of the two terms. The set of information elements of the resulting process term contains the concurrent composition of each information element associated to the first term with each information element associated to the second.

Thus, sets of information elements are employed for concurrent composition of terms. As already observed by Winkowski in [25, 26], for a process term of an elementary net (where information elements are markings, i.e. sets of places), instead of considering the set of information elements, it is sufficient to consider just those places which appear in at least one of the markings being information elements. In [6] we generalize this idea: Two sets of information elements A and B do not have to be distinguished, if for each set of information elements C either both A and B are independent from C^1 or both A and B are not independent from C. Therefore, we can use any equivalence $\cong \in 2^I \times 2^I$ that is a congruence with respect to the operations concurrent composition and union (for sequential composition) and satisfies: If $A \cong B$ and A is independent from C, then B is independent from C. That means, we can use any equivalence $\cong \in 2^I \times 2^I$ which is a *closed congruence* with respect to the operations concurrent composition and union. Equivalence classes of the greatest closed congruence represent the minimal information assigned to process terms necessary for concurrent composition. Thus, instead of sets of information elements we associate to process terms equivalence classes with respect to the greatest closed congruence.

There is a strong connection between the process term semantics described above and the usual partial order based semantics. Consider, for example, the process given in Figure 3. It determines that transition a occurs before b and c, and that transition d occurs before b. This process can be decomposed into the sequence ac occurring at the marking $\{p_1, p_4, p_5\}$ (described by the process term $(a; c) \parallel \{p_4\}$), followed by the sequence db occurring at the marking $\{p_1, p_4, \overline{p_5}\}$ (described by the process term $(d; b) \parallel \{p_1\}$). The resulting term is $((a; c) \parallel \{p_4\}); ((d; b) \parallel \{p_1\})$ (see Figure 4). Another interpretation of this process is the following: Transitions a and d occur concurrently at the marking $\{p_1, p_4, p_5\}$ replacing this marking by $\{p_2, p_3, \overline{p_5}\}$. At this marking transitions c and b occur concurrently. The corresponding term is $(a \parallel d); (c \parallel b)$ (see Figure 5). Each process term α defines a partially ordered set of events representing transition occurrences in an obvious way: an event e_2 *depends on* another event e_1 if the process term α contains a subterm $\alpha_1; \alpha_2$ such that e_1 occurs in α_1 and e_2 occurs in α_2. For example, the process term $\alpha = ((a; c) \parallel \{p_4\}); ((d; b) \parallel \{p_1\})$ generates

---

[1] Two sets of information elements X and Y are independent if and only if each information element of X is independent from each information element of Y.

130   Jörg Desel, Gabriel Juhás, and Robert Lorenz

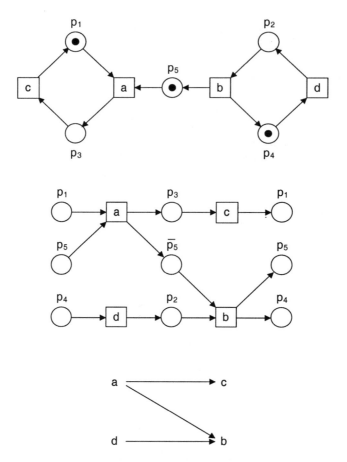

Fig. 3. The elementary net from Figure 2 with the initial marking $\{p_1, p_4, p_5\}$ together with a process and the corresponding partial order of the occurring transitions. The place annotated by $\overline{p_5}$ establishes an order between the occurrence of a and b, due to the contact situation at p_5 after the occurrence of d. For details how to construct processes of elementary nets with contacts see e.g. [23] or Subsection 8.1. The interpretation of $\overline{p_5}$ is that p_5 is not marked.

the partial order given in Figure 6, while the process term $\beta = (a \parallel d); (c \parallel b)$ generates the partial order given in Figure 7.

Unfortunately not all reasonable partial orders can be generated in this way. For example, consider the partial order shown in Figure 3, which is determined by the process from Figure 3. It is easy to show by induction on the structure of process terms that this partial order cannot be generated by any process term. However, this partial order can be constructed from the partial orders generated by process terms α and β, i.e. by two possible decompositions of the process from Figure 3, removing the contradicting connections between c and d. We will define an equivalence of process terms identifying exactly those process terms

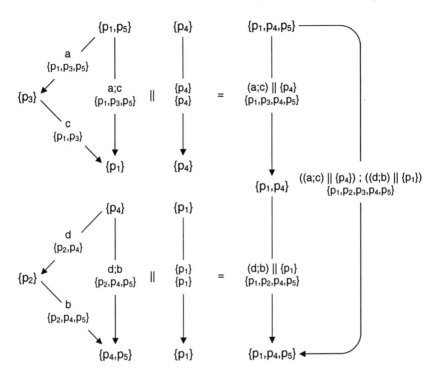

Fig. 4. Derivation of a process term of the elementary net from Figure 2. Instead of the whole set of information elements, each process term has attached only the set of all involved places, i.e. the set of places characterizing the greatest closed congruence class of the related set of information elements. For example, the process term $a; c$ has attached the information $\{p_1, p_3, p_5\}$ instead of the set of two information elements $\{\{p_1, p_3, p_5\}, \{p_1, p_3\}\}$.

representing the same run. Then each run is represented by an equivalence class of process terms.

The paper is organized as follows. Section 2 gives mathematical preliminaries. After introducing formally our concept in Section 3, we provide a couple of examples in Sections 4–9.

The first example given in Section 4 will re-formulate results achieved in [25, 26, 6] for elementary nets, showing that the information for concurrent composition used in [25, 26] is in fact (isomorphic to) the equivalence class of the greatest closed congruence of the related partial algebra and therefore is the minimal information necessary for concurrent composition.

Usually, if a transition depends on the state of a place, then this state is changed by the transition's occurrence. We call this a write operation and the place a write place. Extensions with context requirements release this property: a transition can only occur if in addition the context places are in a certain state but this state remains unchanged by the transition's occurrence (read operation). In [19] elementary nets with context are defined, generalizing the notions of

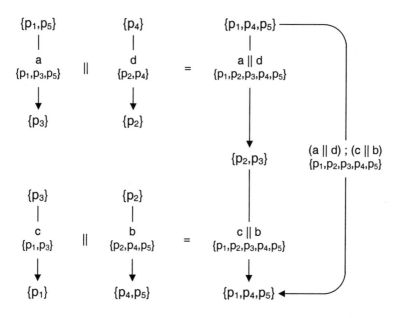

Fig. 5. Derivation of another process term of the elementary net from Figure 2.

Fig. 6. Partial order generated by the process term $\alpha = ((a;c) \parallel \{p_4\}); ((d;b) \parallel \{p_1\})$

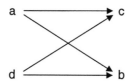

Fig. 7. Partial order generated by the process term $\beta = (a \parallel d); (c \parallel b)$

inhibitor arcs (negative context) and read arcs (positive context). In Sections 5–7 we apply our concept to elementary nets with context. For these nets, two enabled transitions using common places as (positive or negative) context can occur concurrently if their pre- and post-sets are disjoint. Accordingly, if two processes do not employ common places for the flow of tokens but partly use the same context, then the composition of these process terms should not be excluded. This means that read operations on a place can occur concurrently, whereas mixed read and write operations as well as two write operations are incompatible with respect to concurrent composition, just as concurrent access to a storage element is only possible for read access (this interpretation of context places was also chosen in [19] whereas [13, 17] allows concurrent read and write operations). Hence we need information about the nature (write or read) of the access to places for each process term. Therefore, the necessary information is more complex as the simple collection of markings associated to a process. In each case, i.e. in case of elementary nets with positive context (Section 5), negative context (Section 6), and mixed context (Section 7), the minimal information for concurrent composition is computed. Further, in Section 8 we prove that the non-sequential semantics given by process terms coincides with the partial order semantics given by process nets of elementary nets with context introduced in [19].

Section 9 illustrates the generality of our approach by applying it to two more net classes, namely place/transition nets with inhibitor arcs (negative context) and place/transition nets with capacities. In Subsection 9.1 we show that for place/transition nets with inhibitor arcs, concurrent composition of two processes should only be excluded if a place is a common context and write place, and therefore it is enough to store the set of context and the set of write places. Thus, the set of information is less complex (it is particularly a finite set) than the set of markings (which in this case is the infinite set of multi-sets over the set of places). We conclude by showing that our approach fits well for place/transition nets with strong and with weak capacities (Subsection 9.2).

2   Mathematical Preliminaries

We use $\mathbb{N}$ to denote the nonnegative integers and $\mathbb{N}^+$ to denote the positive integers. Given two arbitrary sets A and B, the symbol B^A denotes the set of all functions from A to B. Given a function f from A to B and a subset C of A we write $f|_C$ to denote the restriction of f to the set C. The symbol 2^A denotes the power set of a set A. Given a set A, the symbol $|A|$ denotes the cardinality of A and the symbol id_A the identity on the set A. We write id to denote id_A whenever A is clear from the context. The set of all multi-sets over a set A is denoted by $\mathbb{N}^A$. Given a binary relation $R \subseteq A \times A$ over a set A, the symbol R^+ denotes the transitive closure of R.

A partial groupoid is an ordered tuple $\mathcal{I} = (I, dom_+, +)$ where I is a set called the carrier of $\mathcal{I}$, $dom_+ \subseteq I \times I$ is the domain of $+$, and $+ : dom_+ \to I$ is the partial operation of $\mathcal{I}$. In the rest of the paper we will consider only partial groupoids $(I, dom_+, +)$ which fulfil the following conditions:

- If $a + b$ is defined then $b + a$ is defined and $a + b = b + a$.
- If $(a+b)+c$ is defined then $a+(b+c)$ is defined and $(a+b)+c = a+(b+c)$.

We use the symbol I for a set of information elements associated to elementary terms and the operation $+$ to express concurrent composition of information elements. Not each pair of process terms can be concurrently composed, hence $+$ is a partial operation. The relation dom_+ contains the pairs of elements which are independent and can be concurrently composed.

As explained in Introduction, generated terms have associated sets of information elements. So, the partial groupoid $(I, dom_+, +)$ is extended to the partial groupoid $(2^I, dom_{\{+\}}, \{+\})$, where

- $\quad dom_{\{+\}} = \{(X, Y) \in 2^I \times 2^I \mid X \times Y \subseteq dom_+\}.$

$-$ $X \{+\} Y = \{x + y \mid x \in X \wedge y \in Y\}$.

We will use more than one partial operation on the same carrier. A partial algebra is a set (called carrier) together with a couple of partial operations on this set (with possibly different arity). Given a partial algebra with carrier X, an equivalence $\sim$ on X satisfying the following conditions is a *congruence*: If op is an n-ary partial operation, $a_1 \sim b_1, \ldots, a_n \sim b_n$, $(a_1, \ldots, a_n) \in dom_{op}$ and $(b_1, \ldots, b_n) \in dom_{op}$, then $op(a_1, \ldots, a_n) \sim op(b_1, \ldots, b_n)$. If moreover $a_1 \sim b_1, \ldots, a_n \sim b_n$ and $(a_1, \ldots, a_n) \in dom_{op}$ imply $(b_1, \ldots, b_n) \in dom_{op}$ for each n-ary partial operation then the congruence $\sim$ is said to be *closed*. Thus, a congruence is an equivalence preserving all operations of a partial algebra, while a closed congruence moreover preserves the domains of the operations. For a given partial algebra there always exists a unique greatest closed congruence. The intersection of two congruences is again a congruence. Given a binary relation on X, there always exists a unique least congruence containing this relation. In general, the same does not hold for closed congruences. Given a partial algebra $\mathcal{X}$ with carrier X and a congruence $\sim$ on $\mathcal{X}$, we write $[x]_\sim = \{y \in X \mid x \sim y\}$ and $X/_\sim = \bigcup_{x \in X} [x]_\sim$. A closed congruence $\sim$ defines the partial algebra $\mathcal{X}/_\sim$ with carrier $X/_\sim$, and with n-ary partial operation $op/_\sim$ defined for each n-ary partial operation $op : dom_{op} \to X$ of $\mathcal{X}$ as follows: $dom_{op/_\sim} = \{([a_1]_\sim, \ldots, [a_n]_\sim) \mid (a_1, \ldots, a_n) \in dom_{op}\}$ and, for each $(a_1, \ldots, a_n) \in dom_{op}$, $op/_\sim([a_1]_\sim, \ldots, [a_n]_\sim) = [op(a_1, \ldots, a_n)]_\sim$. The partial algebra $\mathcal{X}/_\sim$ is called factor algebra of $\mathcal{X}$ with respect to the congruence $\sim$.

Let $\mathcal{X}$ be a partial algebra with k operations $op_i^{\mathcal{X}}, i \in \{1, \ldots, k\}$, and let $\mathcal{Y}$ be a partial algebra with k operations $op_i^{\mathcal{Y}}, i \in \{1, \ldots, k\}$ such that the arity $n_i^{\mathcal{X}}$ of $op_i^{\mathcal{X}}$ equals the arity $n_i^{\mathcal{Y}}$ of $op_i^{\mathcal{Y}}$ for every $i \in \{1, \ldots, k\}$. Denote by X the carrier of $\mathcal{X}$ and by Y the carrier of $\mathcal{Y}$. Then a function $f : X \to Y$ is called homomorphism if for every $i \in \{1, \ldots, k\}$ and $x_1, \ldots x_{n_i^{\mathcal{X}}} \in X$ we have: if $op_i^{\mathcal{X}}(x_1, \ldots, x_{n_i^{\mathcal{X}}})$ is defined then $op_i^{\mathcal{Y}}(f(x_1), \ldots, f(x_{n_i^{\mathcal{X}}}))$ is also defined and $f(op_i^{\mathcal{X}}(x_1, \ldots, x_{n_i^{\mathcal{X}}})) = op_i^{\mathcal{Y}}(f(x_1), \ldots, f(x_{n_i^{\mathcal{X}}}))$. A homomorphism $f : X \to Y$ is called closed if for every $i \in \{1, \ldots, k\}$ and $x_1, \ldots x_{n_i^{\mathcal{X}}} \in X$ we have: if $op_i^{\mathcal{Y}}(f(x_1), \ldots, f(x_{n_i^{\mathcal{X}}}))$ is defined then $op_i^{\mathcal{X}}(x_1, \ldots, x_{n_i^{\mathcal{X}}})$ is also defined. If f is a bijection, then it is called an isomorphism, and the partial algebras $\mathcal{X}$ and $\mathcal{Y}$ are called isomorphic.

There is a strong connection between the concepts of homomorphism and congruence in partial algebras: If f is a surjective (closed) homomorphism from $\mathcal{X}$ to $\mathcal{Y}$, then the relation $\sim \subseteq X \times X$ defined by $a \sim b \Longleftrightarrow f(a) = f(b)$ is a (closed) congruence and $\mathcal{Y}$ is isomorphic to $\mathcal{X}/_\sim$. Conversely, given a (closed) congruence $\sim$ of $\mathcal{X}$, the mapping $h : X \to X/_\sim$ given by $h(x) = [x]_\sim$ is a surjective (closed) homomorphism. This homomorphism is called the *natural homomorphism w.r.t.* $\sim$. For more details on partial algebras see e.g. [4].

3   The General Approach

An algebraic Petri net as introduced in [18] is based on a graph with vertices representing markings and edges labeled by transitions representing steps between markings. Moreover, an operator $+$ adds markings. The set of markings together with addition of markings denotes a commutative monoid $\mathcal{M} = (M, +)$ with neutral element e (the empty marking). To obtain the process term semantics of an algebraic Petri net, we assign to every marking and to every transition an information element used for concurrent composition. Two elementary process terms can be concurrently composed only if their associated information elements are independent. The set of all possible information elements is denoted by a partial groupoid $\mathcal{I} = (I, +, dom_+)$, where $+$ denotes the composition of independent information elements, and independence is given by the symmetric relation $dom_+ \subseteq I \times I$.

Since we will compose process terms concurrently and process terms have associated sets of information elements, we lift the partial groupoid $(I, +, dom_+)$ to the partial groupoid $(2^I, \{+\}, dom_{\{+\}})$.

A process term $\alpha \colon m_1 \to m_2$ represents a process transforming marking m_1 to marking m_2. Process terms $\alpha \colon m_1 \to m_2$ and $\beta \colon m_3 \to m_4$ can be sequentially composed, provided $m_2 = m_3$, resulting in $\alpha; \beta : m_1 \to m_4$. This notation illustrates the occurrence of β after the occurrence of α. The set of information elements of the sequentially composed process term is the union of the sets of information elements of the single process terms. The process terms can also be composed concurrently to $\alpha \parallel \beta : m_1 + m_3 \to m_2 + m_4$, provided the set of information elements of α is independent from the set of information elements of β. The set of information elements of $\alpha \parallel \beta$ contains the concurrent composition of each element of the set of information elements of α with each element of the set of information elements of β.

For sequential composition of process terms we need information about the start and the end of a process term, which are both single markings. For concurrent composition, we require that the associated sets of information elements are independent.

Two sets of information elements A and B do not have to be distinguished, if for each set of information elements C either both A and B are independent from C or both A and B are not independent from C. Therefore, we can use any equivalence $\cong \in 2^I \times 2^I$ that is a congruence with respect to the operations $\{+\}$ (concurrent composition) and $\cup$ (sequential composition) and satisfies $(A \cong B \wedge (A, C) \in dom_{\{+\}}) \implies (B, C) \in dom_{\{+\}}$, i.e. which is a *closed congruence* of the partial algebra $\mathcal{X} = (2^I, \{+\}, dom_{\{+\}}, \cup)$. The equivalence classes of the greatest (and hence coarsest) closed congruence represent the minimal information assigned to process terms necessary for concurrent composition. This congruence is unique ([4]).

Definition 1 (Algebraic $(\mathcal{M}, \mathcal{I})$-net and its process term semantics). *Let $\mathcal{M} = (M, +)$ be a commutative monoid and $\mathcal{I} = (I, dom_+, +)$ be a partial groupoid satisfying the properties defined in the previous section. Let $\cong \in 2^I \times 2^I$ be the greatest closed congruence of the partial algebra $\mathcal{X} = (2^I, \{+\}, dom_{\{+\}}, \cup)$.*

An algebraic $(\mathcal{M}, \mathcal{I})$-net is a quadruple

$$\mathcal{A} = (M, T, pre \colon T \to M, post \colon T \to M)$$

together with a mapping $\inf: M \cup T \to I$ *satisfying*

(a) $\forall x, y \in M : \quad (\inf(x), \inf(y)) \in dom_+ \implies \inf(x+y) = \inf(x) + \inf(y)$.

(b) $\forall t \in T : \{\inf(t)\} \cong \{\inf(t), \inf(pre(t)), \inf(post(t))\}$.

   Out of an algebraic net $\mathcal{A}$ *we can build process terms that represent all ab-stract concurrent computations of* $\mathcal{A}$. *Every process term* α *has associated an initial marking* $pre(\alpha) \in M$, *a final marking* $post(\alpha) \in M$, *and an information for concurrent composition* $Inf(\alpha) \in 2^I/_{\cong}$. *In the following, for a process term* α *we write* $\alpha : a \longrightarrow b$ *to denote that* a *is the initial marking and* b *is the final marking of* α.

   The elementary process terms are

$$id_a : a \longrightarrow a$$

with associated information $Inf(id_a) = [\{\inf(a)\}]_{\cong}$ *for each* $a \in M$, *and*

$$t : pre(t) \longrightarrow post(t)$$

with associated information $Inf(t) = [\{\inf(t)\}]_{\cong}$ *for each* $t \in T$.

   If $\alpha : a_1 \longrightarrow a_2$ *and* $\beta : b_1 \longrightarrow b_2$ *are process terms satisfying* $(Inf(\alpha), Inf(\beta)) \in dom_{\{+\}}/_{\cong}$, *their concurrent composition yields the process term*

$$\alpha \parallel \beta : a_1 + b_1 \longrightarrow a_2 + b_2$$

with associated information $Inf(\alpha \parallel \beta) = Inf(\alpha) \{+\}/_{\cong} Inf(\beta)$.

   If $\alpha : a_1 \longrightarrow a_2$ *and* $\beta : b_1 \longrightarrow b_2$ *are process terms satisfying* $a_2 = b_1$, *their sequential composition yields the process term*

$$\alpha; \beta : a_1 \longrightarrow b_2$$

with associated information $Inf(\alpha; \beta) = Inf(\alpha) \cup/_{\cong} Inf(\beta)$.

   The partial algebra of all process terms with the partial operations concurrent composition and sequential composition as defined above will be denoted by $\mathcal{P}(\mathcal{A})$.

We consider the used factor algebra $\mathcal{X}/_{\cong}$ up to isomorphism. Hence one can freely use any partial algebra isomorphic to $\mathcal{X}/_{\cong}$.

Requirement (a) in the previous definition means that the concurrent composition of information elements attached to markings respects the concurrent composition of the markings. Requirement (b) means that the information about the initial and the final marking of a transition is already included in the information associated to the transition.

As mentioned in Introduction, we now define an equivalence of process terms identifying exactly those process terms representing the same run. Then each run is represented by an equivalence class of process terms. We require this equivalence to preserve the concurrent composition and sequential composition of process terms, i.e. to be a congruence with respect to these operations.

Definition 2 (Congruence of process terms). *The congruence relation $\sim$ on the set of process terms of an algebraic $(\mathcal{M}, \mathcal{I})$-net is the least congruence on process terms with respect to the partial operations $\parallel$ and ; given by the following axioms for process terms $\alpha_1, \alpha_2, \alpha_3, \alpha_4$ and markings $x, y \in M$:*

1. $(\alpha_1 \parallel \alpha_2) \sim (\alpha_2 \parallel \alpha_1)$, *whenever $\parallel$ is defined for α_1 and α_2.*
2. $((\alpha_1 \parallel \alpha_2) \parallel \alpha_3) \sim (\alpha_1 \parallel (\alpha_2 \parallel \alpha_3))$, *whenever these terms are defined.*
3. $((\alpha_1; \alpha_2); \alpha_3) \sim (\alpha_1; (\alpha_2; \alpha_3))$, *whenever these terms are defined.*
4. $\alpha = ((\alpha_1 \parallel \alpha_2); (\alpha_3 \parallel \alpha_4)) \sim \beta = ((\alpha_1; \alpha_3) \parallel (\alpha_2; \alpha_4))$, *whenever these terms are defined and $Inf(\alpha) = Inf(\beta)$.*
5. $(\alpha; id_{post(\alpha)}) \sim \alpha \sim (id_{pre(\alpha)}; \alpha)$.
6. $id_{(x+y)} \sim (id_x \parallel id_y)$, *whenever these terms are defined.*
7. $\alpha \parallel id_x \sim \alpha$ *whenever the left term is defined, $pre(\alpha) + x = pre(\alpha)$ and $post(\alpha) + x = post(\alpha)$.*

In the sequel we will write x to denote the elementary term id_x.

Proposition 1. *By construction, $\alpha \sim \beta$ implies $pre(\alpha) = pre(\beta)$, $post(\alpha) = post(\beta)$ and $Inf(\alpha) = Inf(\beta)$.*

Axiom (1) represents commutativity of concurrent composition, axioms (2) and (3) associativity of concurrent and sequential composition. Axiom (4) states distributivity whenever both terms have the same information. It is also used in related approaches such as [18]. Notice that the partial order induced by β is a subset of the partial order induced by α. Therefore, the partial order induced by α can be understood as a partial sequentialization of the partial order induced by β, i.e. it is a partial sequentialization of the run represented by the corresponding equivalence class of process terms. Axiom (5) states that elementary terms corresponding to elements of M are partial neutral elements with respect to sequential composition. Axiom (6) expresses that composition of these neutral elements is congruent to the neutral element constructed from their composition. Finally, axiom (7) states that elements of M which are neutral to the initial and final marking of a term are neutral to the term itself.

For example, the process term $((a; c) \parallel \{p_4\}); ((d; b) \parallel \{p_1\})$ of the elementary net from Figure 2 generated in Figure 4 and the process term $(a \parallel d); (c \parallel b)$ of the elementary net from Figure 2 generated in Figure 5 are congruent:

$$((a; c) \parallel \{p_4\}); ((d; b) \parallel \{p_1\}) \quad \overset{(4),(5)}{\sim} \quad ((a \parallel \{p_4\}); (c \parallel \{p_4\})); ((d \parallel \{p_1\}); (b \parallel \{p_1\}))$$

$$\overset{(1),(3),(4)}{\sim} \quad (a \parallel \{p_4\}); ((c; \{p_1\}) \parallel (\{p_4\}; d)); (b \parallel \{p_1\})$$

$$\overset{(1),(5)}{\sim} \quad (a \parallel \{p_4\}); (d \parallel c); (b \parallel \{p_1\})$$

$$\overset{(5)}{\sim} \quad (a \parallel \{p_4\}); ((d; \{p_2\}) \parallel (\{p_3\}; c)); (b \parallel \{p_1\})$$

$$\overset{(4)}{\sim} \quad (a \parallel \{p_4\}); ((d \parallel \{p_3\}); (c \parallel \{p_2\})); (b \parallel \{p_1\})$$

$$\overset{(1),(3),(4),(5)}{\sim} \quad (a \parallel d); (c \parallel b).$$

Note that given a transition t of a $(\mathcal{M}, \mathcal{I})$-net, the elementary term t represents the single occurrence of the transition t leading from the marking $m = pre(t)$ to the marking $m' = post(t)$, and any term in the form $t \parallel x$, where $x \in M$, represents the single occurrence of the transition t leading from the marking $m = x + pre(t)$ to the marking $m' = x + post(t)$.

Despite the differences between different classes of Petri nets, there are some common features of almost all net classes, such as the notions of marking (state), transition, and occurrence rule (see our contribution [8]).

Thus, in the following definition we suppose a Petri net with a set of markings, a set of transitions and an occurrence rule characterizing whether a transition is enabled to occur at a given marking and if yes determining the follower marking. We suppose that the considered Petri net has no fixed initial marking.

Definition 3 (Corresponding algebraic $(\mathcal{M}, \mathcal{I})$-net). *Let N be a Petri net with a set of markings M_N, and a set of transitions T_N. Let $m \xrightarrow{t} m'$ denote that a transition t is enabled to occur in m and that its occurrence leads to the follower marking m'.*

Let $\mathcal{M} = (M, +)$ and $\mathcal{I} = (I, dom_+, +)$. Then an algebraic $(\mathcal{M}, \mathcal{I})$-net

$$\mathcal{A} = (M, T, pre \colon T \to M, post \colon T \to M)$$

together with a mapping $inf \colon M \cup T \to I$ is called a corresponding algebraic $(\mathcal{M}, \mathcal{I})$-net to the net N iff:

- *$\mathcal{A}$ has the same domain for markings as N, i.e. $M = M_N$*
- *transitions of $\mathcal{A}$ are those transitions of N which are enabled to occur in some marking, i.e. $T = \{t \in T_N \mid \exists m, m' \in M : m \xrightarrow{t} m'\}$, and*
- *the occurrence rule is preserved, i.e. $\forall m, m' \in M, t \in T : m \xrightarrow{t} m' \Longleftrightarrow ((m = pre(t) \wedge m' = post(t)) \vee (\exists x \in M : (inf(x), inf(t)) \in dom_+ \wedge x + pre(t) = m \wedge x + post(t) = m'))$.*

In the following sections we construct corresponding algebraic $(\mathcal{M}, \mathcal{I})$-nets for several classes of Petri nets using the following scenario:

- We give a classical definition of the considered net class including the occurrence rule.
- We identify $\mathcal{M}$ and construct $\mathcal{I}$ such that the requirements from Section 2 are satisfied.
- We construct functions $pre, post, inf$ in such a way that condition (a) from Definition 1 is valid and that dom_+, the independence relation of $\mathcal{I}$, encodes the restriction of the occurrence rule.
- We construct the greatest closed congruence $\cong$ of the partial algebra $(2^I, dom_{\{+\}}, \{+\}, \cup)$. Then, we construct a partial algebra isomorphic to $(2^I, dom_{\{+\}}, \{+\}, \cup)/_{\cong}$.
- We show that property (b) from Definition 1 is satisfied.

4   Elementary Nets

In this section we represent elementary nets as algebraic $(\mathcal{M}, \mathcal{I})$-nets.

An elementary net consists of a set of places P, a set of transitions T and relations between them. Places can be in different *states*. Transitions can occur, depending on the state of some places. The occurrence of a transition can change the state of some places.

Definition 4 (Elementary nets). *An elementary net is a triple $N = (P, T, F)$, where P (places) and T (transitions) are disjoint finite sets, and $F \subseteq (P \times T) \cup (T \times P)$ is a relation (flow relation). For a transition $t \in T$, ${}^\bullet t = \{p \in P \mid (p, t) \in F\}$ is the* pre-set *of t and $t^\bullet = \{p \in P \mid (t, p) \in F\}$ is the* post-set *of t. Throughout the paper we assume that each transition has nonempty pre- and post-sets.*

Each subset of P is called a marking. *A transition $t \in T$ is enabled to occur in a marking $m \subseteq P$ iff ${}^\bullet t \subseteq m \wedge (m \setminus {}^\bullet t) \cap t^\bullet = \emptyset$. In this case, its occurrence leads to the marking $m' = (m \setminus {}^\bullet t) \cup t^\bullet$.*

As usual, places are graphically expressed by circles, transitions by boxes and elements of the flow relation by directed arcs. A marking of the net is represented by tokens in places. For an example of an elementary net see Introduction.

The union of markings represents concurrent composition. Hence the approach of the previous section looks as follows: $\mathcal{M} = (M, +) = (2^P, \cup)$.

The information element associated to an elementary process term consists of the set of used places. An information element is independent from another information element, if they are disjoint. Hence we define the set of information elements $I = M = 2^P$ together with the independence relation $dom_+ = \{(w, w') \in M \times M \mid w \cap w' = \emptyset\}$ and the operation $w + w' = w \cup w'$. The partial groupoid $\mathcal{I} = (I, dom_+, +)$ respects the requirements of Section 2.

To find a $(\mathcal{M}, \mathcal{I})$-net corresponding to an elementary net $N = (P, T, F)$, we need to define mappings $pre, post : T \to M$ which assign an initial and final marking to every transition, and a function $inf : M \cup T \to I$ which assigns an information element to every marking $m \in M$ and every transition $t \in T$:

- For a transition $t \in T$, $pre(t) = {}^\bullet t$ and $post(t) = t^\bullet$.
- For a marking $m \in M$, $inf(m) = m$.
- For a transition $t \in T$, $inf(t) = {}^\bullet t \cup t^\bullet$.

It is easy to observe that the mapping inf satisfies the property (a) from Definition 1.

The following lemma shows that the occurrence rule is encoded by inf and dom_+, as described in Introduction.

Lemma 1. *A transition $t \in T$ is enabled to occur in a marking m and its occurrence leads to the marking m' iff there exists a marking x such that $(inf(x), inf(t)) \in dom_+$, $x + pre(t) = m$ and $x + post(t) = m'$.*

Proof. $\Rightarrow$: Choose $x = m \setminus {}^\bullet t$.

$\Leftarrow$: Assume an x with $x \cap ({}^\bullet t \cup t^\bullet) = \emptyset$. Obviously, ${}^\bullet t \subseteq (x \cup {}^\bullet t)$. Furthermore we have $x = (x \cup {}^\bullet t) \setminus {}^\bullet t$ and $x \cap t^\bullet = \emptyset$. Therefore t is enabled to occur in $x \cup {}^\bullet t = x + pre(t)$ and its occurrence leads to $x \cup t^\bullet = x + post(t)$. $\square$

To define a corresponding algebraic net and its process terms we have to find the greatest closed congruence on $(2^I, \{+\}, dom_{\{+\}}, \cup)$. Actually, instead of considering the set of all information elements associated with a process term, it will be enough to consider the information about all involved places of a process term. We define the mapping $supp : 2^I \to I$, $supp(A) = \bigcup_{w \in A} w$ and show that $supp$ is the natural homomorphism w.r.t. the greatest closed congruence $\cong$ on $(2^I, \{+\}, dom_{\{+\}}, \cup)$.

Lemma 2. *The relation* $\cong \subseteq 2^I \times 2^I$ *defined by* $A \cong B \iff supp(A) = supp(B)$ *is a closed congruence on* $(2^I, \{+\}, dom_{\{+\}}, \cup)$.

Proof. Straightforward observation.

Lemma 3. *The closed congruence* $\cong \subseteq 2^I \times 2^I$ *is the greatest closed congruence on* $(2^I, \{+\}, dom_{\{+\}}, \cup)$.

Proof. We show that any congruence $\approx$ such that $\cong$ is a proper subset of $\approx$ is not closed. Assume there are $A, A' \in 2^I$ such that $A \approx A'$ but $A \not\cong A'$. Then $supp(A) \neq supp(A')$.

We define a set $C \in 2^I$ such that $(A, C) \in dom_{\{+\}}$ but $(A', C) \notin dom_{\{+\}}$ or vice versa (which implies that $\approx$ is not closed). Denoting $supp(A) = \overline{w}$ and $supp(A') = \overline{w}'$ we have that $\overline{w} \neq \overline{w}'$.

Without loss of generality we assume $\overline{w}' \setminus \overline{w} \neq \emptyset$. Set $C = \{c\}$ with $c = \overline{w}' \setminus \overline{w}$. Then $c \cap \overline{w} = \emptyset$, but $c \cap \overline{w}' \neq \emptyset$, i.e. $(A, C) \in dom_{\{+\}}$, but $(A', C) \notin dom_{\{+\}}$. $\square$

Taking $pre, post, inf$ defined above, we have $supp(\{inf(t)\}) = {}^\bullet t \cup t^\bullet = ({}^\bullet t \cup t^\bullet) \cup {}^\bullet t \cup t^\bullet = supp(\{inf(t), inf(pre(t)), inf(post(t))\})$, and therefore the property (b) from Definition 1 is satisfied. Thus, we can formulate the following theorem.

Theorem 1. *Given an elementary net* $N = (P, T, F)$ *with* $\mathcal{M}, \mathcal{I}, pre, post, inf$ *as defined in this section, the quadruple* $\mathcal{A}_N = (2^P, T, pre, post)$ *together with the mapping* inf *is an algebraic* $(\mathcal{M}, \mathcal{I})$*-net. Moreover, it is a corresponding algebraic* $(\mathcal{M}, \mathcal{I})$*-net to the net* N.

Remark 1. In our definition of elementary nets we use an occurrence rule which slightly differs from the standard occurrence rule as given in [23]. Our main motivation of using the presented occurrence rule is to have a definition which is compatible with [19]. The only difference is that the occurrence of a transition with non-disjoint pre- and post-set is allowed in our definition, while using the standard occurrence rule for elementary nets such transitions are never enabled to occur and therefore, according to Definition 3, are irrelevant for a corresponding $(\mathcal{M}, \mathcal{I})$-net. In other words, the corresponding $(\mathcal{M}, \mathcal{I})$-net for the standard occurrence rule of an elementary net would differ from the one we presented in

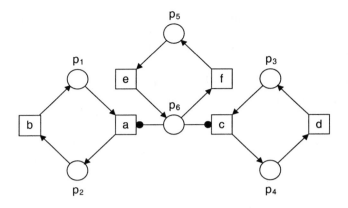

Fig. 8. An example of an elementary net with positive context. Observe that $^\bullet a = \{p_1\}$, $a^\bullet = \{p_2\}$ and $^+a = \{p_6\}$. Therefore, transition a is enabled to occur if the places p_1 and p_6 are marked and the place p_2 is unmarked. Its occurrence removes a token from p_1 and adds a token to p_2.

Theorem 1 only in the absence of transitions with non-disjoint pre- and post-set. In general, there is a more substantial difference between both occurrence rules. Namely, the occurrence rule used for elementary nets in this section corresponds in general to the occurrence rule of place/transition nets with weak capacity restrictions, while the standard occurrence rule for elementary nets corresponds in general to the occurrence rule of place/transition nets with the strong capacity restrictions. For more details on this difference we refer to the Section 9.2 and to [10, 9, 15].

5   Elementary Nets with Positive Context

In this section we represent elementary nets with positive context as algebraic $(\mathcal{M}, \mathcal{I})$-nets.

Definition 5 (Elementary nets with positive context). *An elementary net with positive context is a quadruple $N = (P, T, F, C_+)$, where (P, T, F) is an elementary net and $C_+ \subseteq P \times T$ is a positive context relation satisfying $(F \cup F^{-1}) \cap C_+ = \emptyset$. For a transition t, $^+t = \{p \in P \mid (p, t) \in C_+\}$ is the positive context of t.*

A transition t is enabled to occur in a marking m iff $(^\bullet t \cup {}^+t) \subseteq m \wedge (m \setminus {}^\bullet t) \cap t^\bullet = \emptyset$. Its occurrence leads to the marking $m' = (m \setminus {}^\bullet t) \cup t^\bullet$.

The positive context of a transition is the set of places which are tested on presence of a token as a neccessary condition for the possible occurrence of the transition. As usual, elements of the positive context relation are graphically expressed by arcs ending with a black bullet (so called read arcs). An elementary net with positive context is shown in Figure 8.

In comparison to elementary nets without context, an information element consists of two disjoint components: the set of write places and the set of positive context places. Information elements are independent, if each component of the first element is disjoint with each component of the second element except positive contexts, which may be overlapping. This reflects the fact that concurrent testing on presence of a token is allowed.

For the rest of this section, let $N = (P, T, F, C_+)$ be an elementary net with positive context.

Formally, we have $\mathcal{M} = (M, +) = (2^P, \cup)$. The set of information elements is given by $I = \{(w, p) \in 2^P \times 2^P \mid w \cap p = \emptyset\}$. The independence relation is defined by $dom_+ = \{((w, p), (w', p')) \mid w \cap w' = w \cap p' = w' \cap p = \emptyset\}$, and the operation $+$ by $(w, p) + (w', p') = (w \cup w', p \cup p')$.

$\mathcal{I} = (I, dom_+, +)$ satisfies the properties defined in Section 2.

To define a $(\mathcal{M}, \mathcal{I})$-net corresponding to the elementary net with positive context $N = (P, T, F, C_+)$, we need to define the mappings $pre, post : T \rightarrow \mathcal{M}$ which attach an initial and final marking to every transition, and the mapping $inf : M \cup T \rightarrow I$ which assigns an information element to every marking $m \in M$ and every transition $t \in T$:

- A transition t has the initial marking $pre(t) = {}^\bullet t \cup {}^+t$ and the final marking $post(t) = t^\bullet \cup {}^+t$.
- A marking m carries the information $inf(m) = (\emptyset, m)$.
- A transition t carries the information about the places which are contained in the pre- or post-set and about its positive context places, i.e. $inf(t) = ({}^\bullet t \cup t^\bullet, {}^+t)$.

For example, transition a from the net in Figure 8 has attached $pre(a) = \{p_1, p_6\}$, $post(a) = \{p_2, p_6\}$ and the information element $inf(a) = (w, p) = (\{p_1, p_2\}, \{p_6\})$. Transition c has attached the information element $inf(c) = (w', p') = (\{p_3, p_4\}, \{p_6\})$. These information elements are independent. They have the common positive context place p_6, but concurrent testing on presence of a token is allowed. On the other hand, transition e with information element $inf(e) = (w'', p'') = (\{p_5, p_6\}, \emptyset)$ is independent neither with a nor with c, because the write place p_6 of e is the positive context place of both a and c, i.e. $w'' \cap p \neq \emptyset$ as well as $w'' \cap p' \neq \emptyset$.

Property (a) from Definition 1 is valid for $(M, +)$, $\mathcal{I} = (I, dom_+, +)$ and inf defined above.

The following lemma shows that taking such mappings $pre, post, inf$, the partial groupoid $\mathcal{I}$ encodes the occurrence rule.

Lemma 4. *Given an elementary net with positive context, a transition t is enabled to occur in a marking m and its occurrence leads to the marking m' iff there exists a marking x such that $(inf(x), inf(t)) \in dom_+$, $x + pre(t) = m$ and $x + post(t) = m'$.*

Proof. $\Rightarrow$: Choosing $x = m \setminus ({}^\bullet t \cup {}^+t)$ we have that $(inf(x), inf(t)) \in dom_+$ and $m = x + pre(t) = x \cup ({}^\bullet t \cup {}^+t)$. We have to show that $x + post(t)$ equals m', i.e.

$x \cup (t^\bullet \cup {}^+t) = ((x \cup ({}^\bullet t \cup {}^+t)) \setminus {}^\bullet t) \cup t^\bullet$. This follows from the fact that by definition of elementary nets with positive context ${}^\bullet t \cap {}^+t = \emptyset$.

$\Leftarrow$: Taking any x such that $x \cap ({}^\bullet t \cup t^\bullet) = \emptyset$, we have $({}^\bullet t \cup {}^+t) \subseteq x \cup ({}^\bullet t \cup {}^+t)$. Furthermore (because of ${}^+t \cap {}^\bullet t = {}^+t \cap t^\bullet = \emptyset$) we have $x \cup {}^+t = (x \cup {}^+t \cup {}^\bullet t) \setminus {}^\bullet t$ and $(x \cup {}^+t) \cap t^\bullet = \emptyset$. Therefore t is enabled to occur in $x \cup ({}^\bullet t \cup {}^+t) = x + pre(t)$ and its occurence leads to $x \cup (t^\bullet \cup {}^+t) = x + post(t)$. $\qquad \Box$

Finally, we construct the greatest closed congruence $\cong$ of $(2^I, \{\!+\!\}, dom_{\{\!+\!\}}, \cup)$. Again we define a mapping $supp$ which turns out to be the natural homomorphism of this greatest closed congruence. Define two mappings $s_1, s_2 : 2^I \to 2^P$ by

$$s_1(A) = \bigcup_{(w,p) \in A} w \quad \text{and} \quad s_2(A) = \bigcup_{(w,p) \in A} p,$$

and $supp : 2^I \to I$ by $supp(A) = (s_1(A), s_2(A) \setminus s_1(A))$.

Lemma 5. *Let $\circ$ be the binary operation on I defined by $(w,p) \circ (w',p') = (w \cup w', (p \cup p') \setminus (w \cup w'))$. Then the mapping $supp : (2^I, \{\!+\!\}, dom_{\{\!+\!\}}, \cup) \to (I, +, dom_+, \circ)$ is a surjective closed homomorphism.*

Proof. The operation $\circ$ is well-defined because for any $x, y \in I$, we have $x \circ y \in I$.

(a) $supp$ is a homomorphism for the operations $\{\!+\!\}$ and $\cup$ on 2^I, because both equations $supp(A \{\!+\!\} A') = supp(A) + supp(A')$ (whenever both sides are defined) and $supp(A \cup A') = supp(A) \circ supp(A')$ follow directly from the properties of $\cup$.

(b) We show the closedness of $supp$, that is

$$(A, A') \in dom_{\{\!+\!\}} \iff (supp(A), supp(A')) \in dom_+$$

for any two $A, A' \subseteq I$. Denote $s_1(A) = \overline{w}$, $s_2(A) = \overline{p}$, $s_1(A') = \overline{w}'$ and $s_2(A') = \overline{p}'$. Then

$$\forall (w,p) \in A, \forall (w',p') \in A' : \quad w \cap w' = (w \cup w') \cap (p \cup p') = \emptyset$$
$$\iff \overline{w} \cap \overline{w}' = \emptyset \wedge (\overline{w} \cup \overline{w}') \cap ((\overline{p} \setminus \overline{w}) \cup (\overline{p}' \setminus \overline{w}')) = \emptyset.$$

(c) The mapping $supp$ is surjective, because, for any $(w,p) \in I$, we have $supp(\{(w,p)\}) = (w,p)$. $\qquad \Box$

Lemma 6. *The closed congruence $\cong \subseteq 2^I \times 2^I$ defined by*

$$A \cong B \iff supp(A) = supp(B)$$

is the greatest closed congruence on $(2^I, \{\!+\!\}, dom_{\{\!+\!\}}, \cup)$.

Proof. We will show that any congruence $\approx$ such that $\cong$ is a proper subset of $\approx$ is not closed. Assume there are $A, A' \in 2^I$ such that $A \approx A'$ but $A \not\cong A'$. Then $supp(A) \neq supp(A')$.

We define a set $C \in 2^I$ such that $(A, C) \in dom_{\{+\}}$ but $(A', C) \notin dom_{\{+\}}$ or vice versa (which implies that $\approx$ is not closed). If $supp(A) = (\overline{w}, \overline{p})$ and $supp(A') = (\overline{w}', \overline{p}')$, then $\overline{w} \cap \overline{p} = \overline{w}' \cap \overline{p}' = \emptyset$ (by definition of I) and $\overline{p} \neq \overline{p}' \vee \overline{w} \neq \overline{w}'$ (since $supp(A) \neq supp(A')$).

Let $\overline{w} \neq \overline{w}'$. Without loss of generality we assume $\overline{w}' \setminus \overline{w} \neq \emptyset$. Set $C = \{(c_w, c_p)\}$ with $c_w = \emptyset$ and $c_p = \overline{w}' \setminus \overline{w}$. Then $c_w \cap \overline{w} = c_w \cap \overline{p} = c_p \cap \overline{w} = \emptyset$, but $c_p \cap \overline{w}' \neq \emptyset$, i.e. $(A, C) \in dom_{\{+\}}$, but $(A', C) \notin dom_{\{+\}}$.

Now let $\overline{w} = \overline{w}'$ and $\overline{p} \neq \overline{p}'$. Without loss of generality we assume $\overline{p}' \setminus \overline{p} \neq \emptyset$. Set $C = \{(c_w, c_p)\}$ with $c_w = (\overline{p}' \setminus \overline{p})$ and $c_p = \emptyset$. Then $c_w \neq \emptyset$, $c_w \cap \overline{w} = c_w \cap \overline{p} = \overline{w} \cap c_p = \emptyset$ and $c_w \cap \overline{p}' \neq \emptyset$, and we are finished.   □

Easy computation, using $({}^\bullet t \cup t^\bullet) \cap {}^+ t = \emptyset$, proves condition (b) from Definition 1, i.e. $supp(\{inf(t)\}) = supp(\{inf(t), inf(pre(t)), inf(post(t))\})$.

Now we are able to represent an elementary net with positive context as an algebraic $(\mathcal{M}, \mathcal{I})$-net.

Theorem 2. *Let $N = (P, T, F, C_+)$ be an elementary net with positive context, together with $\mathcal{M}, \mathcal{I}, pre, post, inf$ defined throughout this section. Then the quadruple $\mathcal{A}_N = (2^P, T, pre, post)$ together with the mapping inf is an algebraic $(\mathcal{M}, \mathcal{I})$-net. Moreover, it is a corresponding algebraic $(\mathcal{M}, \mathcal{I})$-net to N.*

Remark 2. Taking an elementary net with empty positive context, Theorem 1 and Theorem 2 define algebraic nets $\mathcal{A}_1$ and $\mathcal{A}_2$ generating different sets of process terms: the set of the process terms $\mathcal{P}(\mathcal{A}_1)$ obtained using Theorem 1 is a subset of the set of process terms $\mathcal{P}(\mathcal{A}_2)$ obtained using Theorem 2. By Theorem 2 terms of the form $id_a \parallel id_a$ are allowed for any marking a. However, because we have $id_a \parallel id_a \sim id_a$, and id_a belongs to $\mathcal{P}(\mathcal{A}_1)$, the partial algebra $\mathcal{P}(\mathcal{A}_1)/_\sim$ according to Theorem 1 and the partial algebra $\mathcal{P}(\mathcal{A}_2)/_\sim$ according to Theorem 2 are isomorphic.

A possible process term of the net from Figure 8 is $\alpha = (e \parallel \{p_1, p_3\}); (a \parallel c); (b \parallel f \parallel d) : \{p_1, p_3, p_5\} \rightarrow \{p_1, p_3, p_5\}$ with the information $Inf(\alpha) = (\{p_1, p_2, p_3, p_4, p_5, p_6\}, \emptyset)$. Observe, that the place p_6, which is a write place of e and f but the positive context place of a and c appears as a write place of α.

In the case of elementary nets without context we have

$$\{inf(t)\} \cong \{inf(pre(t)), inf(post(t))\}.$$

That means that the information of a transition can be derived from the information of its initial and final marking. However, as it is illustrated by elementary nets with positive context, this is not the general case. For elementary nets with positive context $inf(t)$ contains more detailed information. This information about the nature of places distinguishes places whose state is changed by the occurrence of a transition and those places which are only tested.

6   Elementary Nets with Negative Context

In this section we represent elementary nets with negative context as algebraic $(\mathcal{M}, \mathcal{I})$-nets.

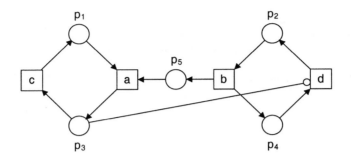

Fig. 9. An example of an elementary net with negative context. Observe that $^\bullet d = \{p_4\}$, $d^\bullet = \{p_2\}$ and $^-d = \{p_3\}$. Therefore, transition d is enabled to occur if p_4 is marked and both p_2 and p_3 are unmarked. Its occurrence removes a token from p_4 and adds a token to p_2.

Definition 6 (Elementary net with negative context). *An elementary net with negative context is a quadruple* $N = (P, T, F, C_-)$, *where* (P, T, F) *is an elementary net and* $C_- \subseteq P \times T$ *is a negative context relation satisfying* $(F \cup F^{-1}) \cap C_- = \emptyset$. *For a transition* t, $^-t = \{p \in P \mid (p, t) \in C_-\}$ *is the* negative context *of* t.

A transition t *is enabled in a marking* m *iff* $^\bullet t \subseteq m \wedge (m \setminus {}^\bullet t) \cap ({}^- t \cup t^\bullet) = \emptyset^2$. *Its occurrence leads to the marking* $m' = (m \setminus {}^\bullet t) \cup t^\bullet$.

The negative context of a transition t is the set of places which are tested on absence of a token for the possible occurrence of a transition. Elements of the negative context relation are graphically expressed by arcs ending with a circle (so called inhibitor arcs). Figure 9 shows an elementary net with negative context.

Similarly to elementary nets with positive context, we need information elements which consist of two disjoint components: the set of write places, and the set of negative context places. Concurrent composition of information elements is allowed if each component of the first element is disjoint with each component of the second element except negative contexts, which may be overlapping. This reflects the fact that concurrent testing on absence of a token is allowed.

Formally, we have the same algebra $\mathcal{M}$ for markings and the same partial algebra $\mathcal{I}$ for information elements as for elementary nets with positive context, and therefore requirements from Section 2 are fulfilled.

We define $pre, post$ by $pre(t) = {}^\bullet t$, $post(t) = t^\bullet$ for each $t \in T$ and inf by $inf(m) = (m, \emptyset)$ for each $m \in 2^P$ and $inf(t) = ({}^\bullet t \cup t^\bullet, {}^- t)$ for each $t \in T$.

For example, transition d from the net in Figure 9 has attached $pre(d) = \{p_4\}$, $post(d) = \{p_2\}$ and the information element $inf(d) = (w, p) = (\{p_2, p_4\}, \{p_3\})$.

---

^{2} Remember that $^- t \cap {}^\bullet t = {}^- t \cap t^\bullet = \emptyset$ but $^\bullet t \cap t^\bullet$ can be nonempty.

Property (a) from Definition 1 is fulfilled and $\mathcal{I}$ encodes the occurrence rule of the net with negative context. Moreover, property (b) from Definition 1 is preserved. So, we can formulate the following theorem.

Theorem 3. *Given an elementary net with negative context $N = (P, T, F, C_-)$ together with $\mathcal{M}, \mathcal{I}, pre, post, inf$ defined in this section, the quadruple $\mathcal{A}_N = (2^P, T, pre, post)$ together with the mapping inf is a $(\mathcal{M}, \mathcal{I})$-net. Moreover, it is a corresponding $(\mathcal{M}, \mathcal{I})$-net to N.*

Remark 3. For nets with positive context, idle tokens generated by an elementary process term m can be concurrently composed with each other. Hence the respective places belong to the second component representing the context. However, for nets with negative context, an additional token can spoil the enabledness of a transition. So, for this class places carrying tokens generated by elementary process terms m belong to the first component representing write places. This way, a concurrent composition of a process term using a place for inhibition with a process term using the same place for an (idle or moving) token is prevented.

A possible process term of the net from Figure 9 is

$$(d \parallel \{p_1, p_5\}); (a \parallel \{p_2\}); (b \parallel c) : \{p_1, p_4, p_5\} \rightarrow \{p_1, p_4, p_5\}$$

with information $(\{p_1, p_2, p_3, p_4, p_5\}, \emptyset)$.

7   Elementary Nets with Mixed Context

In this section we associate to an elementary net with (mixed) context an algebraic $(\mathcal{M}, \mathcal{I})$-net.

Definition 7 (Elementary net with (mixed) context). *An elementary net with (mixed) context is a five-tuple $N = (P, T, F, C_+, C_-)$, where (P, T, F) is an elementary net, and $C_+, C_- \subseteq P \times T$ are positive and negative context relations satisfying $(F \cup F^{-1}) \cap (C_+ \cup C_-) = C_+ \cap C_- = \emptyset$. For a transition t, $^\bullet t, t^\bullet$, ^+t and ^-t are defined as in the previous sections.*
A transition t is enabled to occur in a marking m iff $(^\bullet t \cup {}^+t) \subseteq m \wedge (m \setminus {}^\bullet t) \cap (^-t \cup t^\bullet) = \emptyset$. Its occurrence leads to the marking $m' = (m \setminus {}^\bullet t) \cup t^\bullet$, in symbols $m \xrightarrow{t} m'$.

Figure 10 shows an elementary net with (mixed) context.

Again we have $\mathcal{M} = (M, +) = (2^P, \cup)$. An information element consists of three disjoint components: the set of write places, the set of positive context places and the set of negative context places. Information elements are independent if each component of the first element is disjoint from each component of the second element, except positive contexts (the second components) and negative contexts (the third components). This reflects the fact that concurrent testing on presence of a token as well as concurrent testing on absence of a token is allowed.

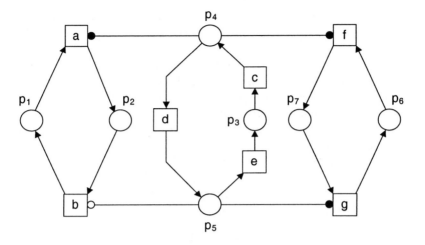

Fig. 10. An elementary net with (mixed) context.

Formally, we define the set of information elements

$$I = \{(w, p, n) \in 2^P \times 2^P \times 2^P \mid w \cap p = w \cap n = p \cap n = \emptyset\},$$

together with the independence relation

$$dom_+ = \{((w, p, n), (w', p', n')) \mid w \cap w' = w \cap (p' \cup n')$$
$$= w' \cap (p \cup n) = p \cap n' = p' \cap n = \emptyset\},$$

and the operation

$$(w, p, n) \dotplus (w', p', n') = (w \cup w', p \cup p', n \cup n').$$

For $\mathcal{I} = (I, dom_+, \dotplus)$ the requirements from Section 2 are fulfilled.

To define a $(\mathcal{M}, \mathcal{I})$-net corresponding to an elementary net with context $N = (P, T, F, C_+, C_-)$, we need to define the mappings $pre, post : T \to M$ attaching an initial and final marking to every transition t, and the function $inf : M \cup T \to I$ assigning an information to every marking m and every transition t:

- A transition t has the initial marking $pre(t) = {}^{\bullet}t \cup {}^+t$ and the final marking $post(t) = t^{\bullet} \cup {}^+t$.
- A marking m carries the information $inf(m) = (\emptyset, m, \emptyset)$.
- A transition t carries information about write places and extra information about positive and negative context places, i.e. $inf(t) = ({}^{\bullet}t \cup t^{\bullet}, {}^+t, {}^-t)$.

For example, transition b from the net in Figure 10 has attached $pre(b) = \{p_2\}$, $post(b) = \{p_1\}$ and the information element $inf(b) = (w, p, n) = (\{p_1, p_2\}, \emptyset, \{p_5\})$, while transition g has attached the information element $inf(g) = (w', p',$

$n') = (\{p_6, p_7\}, \{p_5\}, \emptyset)$. These information elements are not independent, because the negative context place p_5 of b is the positive context place of g, i.e. $n \cap p' \neq \emptyset$.

The mapping inf satisfies property (a) from Definition 1.

Similarly to Lemma 4 one can show for the functions $pre, post, inf$ that the partial algebra $\mathcal{I}$ encodes the occurrence rule of the net with mixed context.

Again, we have to find the greatest closed congruence $\cong$ of $(2^I, \{+\}, dom_{\{+\}}, \cup)$. We define a mapping $supp$ which turns out to be the natural homomorphism of this greatest closed congruence.

Define three mappings $s_1, s_2, s_3 : 2^I \to 2^P$ by

$$s_1(A) = \bigcup_{(w,p,n) \in A} w, \quad s_2(A) = \bigcup_{(w,p,n) \in A} p \quad \text{and} \quad s_3(A) = \bigcup_{(w,p,n) \in A} n.$$

Define $s : 2^I \to 2^P$ by $s(A) = s_1(A) \cup (s_2(A) \cap s_3(A))$.

Finally, define $supp : 2^I \to I$ by $supp(A) = (s(A), s_2(A) \backslash s(A), s_3(A) \backslash s(A))$.

Lemma 7. *Let $\circ$ be the binary operation on I defined by*

$$(w, p, n) \circ (w', p', n') = supp(\{(w, p, n), (w', p', n')\}).$$

Then the mapping $supp : (2^I, dom_{\{+\}}, \{+\}, \cup) \to (I, dom_+, +, \circ)$ is a surjective closed homomorphism.

Proof. First we show the closedness of $supp$, i.e.

$$(A, A') \in dom_{\{+\}} \iff (supp(A), supp(A')) \in dom_+.$$

We write shortly s_1, s_2, s_3 and s to denote $s_1(A), s_2(A), s_3(A)$ and $s(A)$ resp. s_1', s_2', s_3' and s' to denote $s_1(A'), s_2(A'), s_3(A')$ and $s(A')$.

$\Rightarrow$: Suppose that $(A, A') \in dom_{\{+\}}$ but $(supp(A), supp(A')) \notin dom_+$.

Case 1: $s \cap s' \neq \emptyset$, i.e. $(s_1 \cup (s_2 \cap s_3)) \cap (s_1' \cup (s_2' \cap s_3')) \neq \emptyset$.

 – $s_1 \cap s_1' \neq \emptyset$ contradicts $\forall (w, p, n) \in A, (w', p', n') \in A' : w \cap w' = \emptyset$,
 – $s_1 \cap (s_2' \cap s_3') \neq \emptyset$ contradicts $\forall (w, p, n) \in A, (w', p', n') \in A' : w \cap (p \cup n) = \emptyset$,
 – $(s_2 \cap s_3) \cap (s_2' \cap s_3') \neq \emptyset$ contradicts $\forall (w, p, n) \in A, (w', p', n') \in A' : p \cap n' = \emptyset$.

Case 2: $(s_2 \backslash s) \cap s' \neq \emptyset$, i.e. $(s_2 \backslash (s_1 \cup (s_2 \cap s_3))) \cap (s_1' \cup (s_2' \cap s_3')) \neq \emptyset$.

 – $(s_2 \backslash (s_1 \cup (s_2 \cap s_3))) \cap s_1' \neq \emptyset$ contradicts $\forall (w, p, n) \in A, (w', p', n') \in A' : p \cap w' = \emptyset$.
 – $(s_2 \backslash (s_1 \cup (s_2 \cap s_3))) \cap (s_2' \cap s_3') \neq \emptyset$ contradicts $\forall (w, p, n) \in A, (w', p', n') \in A' : p \cap n' = \emptyset$.

All remaining cases are similar.

$\Leftarrow$: Suppose that $(A, A') \notin dom_{\{+\}}$ but $(supp(A), supp(A')) \in dom_+$.

Case 1: $\exists (w, p, n) \in A, (w', p', n') \in A' : w \cap w' \neq \emptyset$ contradicts $s \cap s' = \emptyset$.

Case 2: $\exists (w, p, n) \in A, (w', p', n') \in A' : p \cap w' \neq \emptyset$:

$$-(p \cap w') \cap ((\bigcup_{(x,y,z)\in A} x) \cup (\bigcup_{(x,y,z)\in A} z)) \neq \emptyset \text{ contradicts } s \cap s' = \emptyset,$$

$$-(p \cap w') \cap ((\bigcup_{(x,y,z)\in A} x) \cup (\bigcup_{(x,y,z)\in A} z)) = \emptyset \text{ contradicts } (s_2 \setminus s) \cap s' = \emptyset.$$

All remaining cases are similar.

Now we show that $supp(A \{+\} A') = supp(A) + supp(A')$, whenever defined. Let $supp(A \{+\} A') = (w, p, n)$, where $w = s_1 \cup s_1' \cup ((s_2 \cup s_2') \cap (s_3 \cup s_3'))$, $p = (s_2 \cup s_2') \setminus w$ and $n = (s_3 \cup s_3') \setminus w$. Since $(supp(A), supp(A')) \in dom_+$, we have

$$(s_2 \setminus s) \cap (s_3' \setminus s') = s \cap s' = (s_2' \setminus s') \cap (s_3 \setminus s) = \emptyset, \tag{1}$$
$$(s_2 \setminus s) \cap s' = s \cap s' = (s_2' \setminus s') \cap s = \emptyset. \tag{2}$$

Equations (1) and (2) imply $(s_2 \cap s_3') = (s_2' \cap s_3) = \emptyset$. This gives $w = s_1 \cup (s_2 \cap s_3) \cup s_1' \cup (s_2' \cap s_3') = s \cup s'$. Together with equation (2) this gives $s_2 \cap s' = s_2' \cap s = \emptyset$. Then $p = (s_2 \setminus s) \cup (s_2' \setminus s')$. Similarly, $n = (s_3 \setminus s) \cup (s_3' \setminus s')$.

Finally, we have to show that

$$supp(A \cup A') = supp(A) \circ supp(A') = supp(\{supp(A), supp(A')\}).$$

We have $s = s_1 \cup (s_2 \cap s_3)$ and $s' = s_1' \cup (s_2' \cap s_3')$, and therefore

$$s_1 \cup s_1' \subseteq s \cup s' \subseteq s_1 \cup s_1' \cup ((s_2 \cup s_2') \cap (s_3 \cup s_3')) = s(A \cup A').$$

Since $s(\{supp(A), supp(A')\}) = s \cup s' \cup (((s_2 \setminus s) \cup (s_2' \setminus s')) \cap ((s_3 \setminus s) \cup (s_3' \setminus s')))$, we have $s(A \cup A') = s(\{supp(A), supp(A')\})$. Similarly

$$s_2(A \cup A') \setminus s(A \cup A') = s_2(\{supp(A), supp(A')\}) \setminus s(\{supp(A), supp(A')\})$$

and

$$s_3(A \cup A') \setminus s(A \cup A') = s_3(\{supp(A), supp(A')\}) \setminus s(\{supp(A), supp(A')\}).$$

To show surjectivity, let $(w, p, n) \in I$. Then $supp(\{(w, p, n)\}) = (w, p, n)$. $\quad \square$

Lemma 8. *The closed congruence* $\cong \subseteq 2^I \times 2^I$ *defined by*

$$A \cong B \Longleftrightarrow supp(A) = supp(B)$$

is the greatest closed congruence on the partial algebra $\mathcal{X} = (2^I, dom_{\{+\}}, \{+\}, \cup)$.

Proof. Assume there is a closed congruence $\approx$ on $\mathcal{X}$ with $\cong \subsetneq \approx$. Let $A, A' \in 2^I$ with $A \approx A'$ but $A \ncong A'$. This means $supp(A) \neq supp(A')$. We will define a set $C \in 2^I$ with $(A, C) \in dom_{\{+\}}$ and $(A', C) \notin dom_{\{+\}}$ or vice versa, what contradicts the closedness of $\approx$.

Let $supp(A) = (\overline{w}, \overline{p}, \overline{n})$ and $supp(A') = (\overline{w}', \overline{p}', \overline{n}')$. Then $\overline{w} \neq \overline{w}'$ or $\overline{p} \neq \overline{p}'$ or $\overline{n} \neq \overline{n}'$.

Assume first that $\overline{w}' \setminus \overline{w} \neq \emptyset$. Set $C = \{(\emptyset, \overline{w}' \setminus (\overline{w} \cup \overline{n}), \overline{n})\}$. Clearly, $(A, C) \in dom_{\{+\}}$. If $\overline{w}' \setminus \overline{w} \subseteq \overline{n}$ then $\overline{w}' \cap \overline{n} \neq \emptyset$ and therefore $(A', C) \notin dom_{\{+\}}$. If $\overline{w}' \setminus \overline{w} \nsubseteq \overline{n}$ then $\overline{w}' \cap (\overline{w}' \setminus (\overline{w} \cup \overline{n})) \neq \emptyset$ and therefore $(A', C) \notin dom_{\{+\}}$.

Now assume $\overline{w} = \overline{w}'$ and $\overline{p}' \setminus \overline{p} \neq \emptyset$. Set $C = \{(\emptyset, \emptyset, \overline{p}' \setminus \overline{p})\}$. Assume finally $\overline{w} = \overline{w}'$ and $\overline{n}' \setminus \overline{n} \neq \emptyset$. Set $C = \{(\emptyset, \overline{n}' \setminus \overline{n}, \emptyset)\}$. In both previous cases $(A, C) \in dom_{\{+\}}$ but $(A', C) \notin dom_{\{+\}}$. $\qquad\square$

The partial algebra $(2^I, dom_{\{+\}}, \{+\}, \cup)/_\cong$ is isomorphic to the partial algebra $(I, dom_+, +, \circ)$. For elementary nets with context we only have to use one element of the set I as the information of a process term. This element consists of three sets of places - the set of write places, the set of positive context places which are not write places, and the set of negative context places which are not write places.

For example, the process term $\alpha = a; (b \parallel \{p_4\}) : \{p_1, p_4\} \rightarrow \{p_1, p_4\}$ of the net in Figure 10 has the information $Inf(\alpha) = (\{p_1, p_2\}, \{p_4\}, \{p_5\})$ and the process term $\beta = f : \{p_4, p_6\} \rightarrow \{p_4, p_7\}$ has the information $Inf(\beta) = (\{p_6, p_7\}, \{p_4\}, \emptyset)$. Observe that they can be concurrently composed yielding the process term $\gamma = \alpha \parallel \beta = (a; (b \parallel \{p_4\})) \parallel f : \{p_1, p_4, p_6\} \rightarrow \{p_1, p_4, p_7\}$ with $Inf(\gamma) = (\{p_1, p_2, p_6, p_7\}, \{p_4\}, \{p_5\})$.

Property (b) from Definition 1 is valid, and therefore we can give the theorem:

Theorem 4. *Given an elementary net with (mixed) context $N = (P, T, F, C_+, C_-)$ together with $\mathcal{M}, \mathcal{I}, pre, post, inf$ defined in this section, the quadruple $\mathcal{A}_N = (2^P, T, pre, post)$ together with the mapping inf is an algebraic $(\mathcal{M}, \mathcal{I})$-net. Moreover, it is a corresponding algebraic $(\mathcal{M}, \mathcal{I})$-net to the net N.*

Remark 4. Similarly to Remark 2, given an elementary net with negative context, the equivalence classes of process terms obtained using Theorems 3 and 4 are isomorphic.

8   Relationship between Process Terms and Processes of Elementary Nets with Context

In this section we prove for elementary nets with mixed context a one-to-one correspondence between the obtained non-sequential semantics and the partial-order based semantics obtained in the usual way using process nets. Analogous results hold for elementary nets without context, for elementary nets with (only) positive context, and for elementary nets with (only) negative context.

8.1   Process Semantics of Elementary Nets with Context

In this subsection we give the definition of partial-order based process semantics of elementary nets with context as introduced in [19].

We say that a marking m' is reachable from a marking m, if $m = m'$ or if there is a finite sequence of transitions $t_1, \ldots, t_n$ such that

$$m \xrightarrow{t_1} m_1 \ldots m_{n-1} \xrightarrow{t_n} m'.$$

An elementary net with positive context is said to be contact-free w.r.t. an initial marking m_0, if for each marking m reachable from m_0 and each transition $t : ({}^\bullet t \cup {}^+t) \subseteq m \Rightarrow t^\bullet \cap m = \emptyset$.

As it is shown in [19], an elementary net with mixed context can be transformed via complementation into a contact-free elementary net with positive context exhibiting the same behaviour. For technical reasons we assign complement-places (co-places) to every place. The complementation is defined as follows:

Definition 8 (Complementation). *Given an elementary net with context* $N = (P, T, F, C_+, C_-)$, *let* P' *be a set satisfying* $|P'| = |P|$ *and* $P' \cap (P \cup T) = \emptyset$, *and let* $c : P \to P'$ *be a bijection.*

The complementation $\overline{N} = (\overline{P}, \overline{T}, \overline{F}, \overline{C_+})$ *of* N *is defined by*

$$
\begin{aligned}
\overline{P} &= P \cup P', \\
\overline{T} &= T, \\
\overline{F} &= F \cup \{(t, c(p)) \mid (p, t) \in F \wedge (t, p) \notin F\} \\
&\quad \cup \{(c(p), t) \mid (t, p) \in F \wedge (p, t) \notin F\}, \\
\overline{C_+} &= C_+ \cup \{(c(p), t) \mid (p, t) \in C_-\}.
\end{aligned}
$$

Given an initial marking m_0 *of* N, *its complementation* $\overline{m_0}$ *is defined by*

$$\overline{m_0} = m_0 \cup \{c(p) \mid p \in P, \, p \notin m_0\}.$$

Given an elementary net with context N, the construction of $\overline{N}$ is unique up to isomorphism.

Proposition 2 ([19]). *Given an elementary net with context* N *and an initial marking* m_0 *of* N, *its complementation* $\overline{N}$ *is contact-free w.r.t.* $\overline{m_0}$.

Figure 11 shows a complementation of the net from Figure 10 w.r.t. the initial marking $\{p_1, p_3, p_6\}$. We only draw the co-places, which are necessary to express negative context places using positive context places and to obtain a contact-free net according to the given initial marking. In Figure 11 the only co-place we need to draw is $\overline{p_5}$.

Definition 9 (The causality relation $\leqslant$ of an elementary net with positive context). *Let* $N = (P, T, F, C_+)$ *be a net with positive context. Then* $\leqslant_N$ *denotes the minimal transitive and reflexive binary relation on* $P \cup T$ *satisfying the following conditions:*

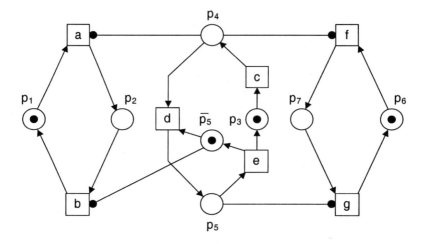

Fig. 11. The complementation of the net from Figure 10 w.r.t. the initial marking $\{p_1, p_3, p_6\}$.

(a) $(x, y) \in F$ implies $x \leqslant_N y$.
(b) $(t, p) \in F$ and $(p, s) \in C_+$ implies $t \leqslant_N s$.
(c) $(p, t) \in C_+$ and $(p, s) \in F$ implies $t \leqslant_N s$.

Furthermore we define $<_N = \leqslant_N \backslash \{(x, x) \mid x \in P \cup T\}$. Whenever the net N is clear from the context we simply write $\leqslant$ instead of $\leqslant_N$ and $<$ instead of $<_N$.

The intuition behind the definition of the causality relation is that the flow relation defines causality between transitions in the usual way, i.e.:

- If a place of the post-set of a transition t belongs to the pre-set of a transition s than t causally precedes s,

while the positive context relation defines causality in the following two ways:

- If an occurrence of a transition t produces a token in a place p and a transition s tests the place p on presence of a token, then transition t causally precedes transition s.
- If a transition t tests a place p on the presence of a token and an occurrence of a transition s removes a token from the place p then transition t causally precedes transition s.

Definition 10 (Contextual occurrence net). *A contextual occurrence net is an elementary net with positive context $K = (B_K, E_K, F_K, C_K)$ such that*

(a) $\leqslant_K$ is a partial order,
(b) $|{}^\bullet b|, |b^\bullet| \leqslant 1$ for all $b \in B_K$[3] (places are unbranched).

---

[3] where ${}^\bullet b = \{e \in E_K \mid (e, b) \in F_K\}$ is the *pre-set* of b and $b^\bullet = \{e \in E_K \mid (b, e) \in F_K\}$ is the *post-set* of b

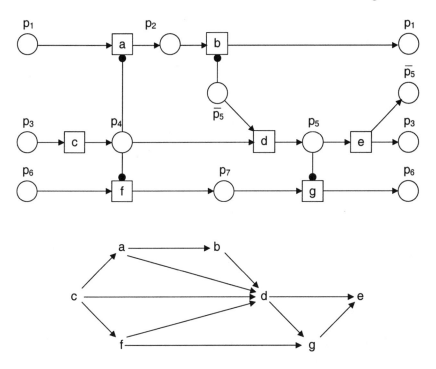

Fig. 12. An example of a contextual occurrence net and the underlying partial order. The contextual occurrence net together with the identity function is a process of the elementary net with context from Figure 10.

Graphically, a contextual occurrence net might have read arcs (arcs for positive context), but each place has at most one ingoing and one outgoing proper arc. Two ordered transitions are connected by a sequence of directed proper arcs (at least one) and undirected read arcs.

An example of a contextual occurrence net and the underlying partial order is shown in Figure 12.

Definition 11 (Co-set, slice). *A co-set of a contextual occurrence net K is a subset $S \subseteq B_K$ such that for no $a, b \in S$: $a <_K b$. A slice is a maximal co-set.*

Denote $^\bullet K = \{b \in B_K \mid |^\bullet b| = 0\}$ and $K^\bullet = \{b \in B_K \mid |b^\bullet| = 0\}$.

Definition 12 (Process of a contact-free elementary net with positive context). *Let $N = (P, T, F, C_+)$ be an elementary net with positive context and let $m_0 \subseteq P$ be an initial marking of N, such that N is contact-free w.r.t. m_0. A process K of N w.r.t. m_0 is a five-tuple $K = (B_K, E_K, F_K, C_K, \rho_K)$, where (B_K, E_K, F_K, C_K) is a contextual occurrence net and $\rho_K : B_K \cup E_K \to P \cup T$ is a mapping satisfying*

(a) $\rho_K(^\bullet K) = m_0$,
(b) $\rho_K|_D$ is injective for every slice D of K,

(c) $\rho_K(D)$ *is reachable from* m_0 *for every slice* D *of* K,
(d) For each $e \in E_K$: $\rho(\,^\bullet e) = \,^\bullet(\rho(e))$, $\rho(e^\bullet) = (\rho(e))^\bullet$ *and* $\rho(\,^+ e) = \,^+(\rho(e))$.

Given a process $K = (B_K, E_K, F_K, C_K, \rho_K)$ of a contact-free elementary net with positive context N (w.r.t. an initial marking m_0) and a set A of isolated places of K (i.e. $\forall b \in A, e \in E_K$: $b \notin \,^\bullet e \cup e^\bullet \cup \,^+ e$) we have that $(B_K \setminus A, E_K, F_K, C_K, \rho_K|_{(B_K \setminus A) \cup E_K})$ is a process of N (w.r.t. the initial marking $m_0' = m \setminus \rho_K(A)$). In other words, after removing isolated places from a process of N we still have a process of N.

For technical reason, we assume that processes contain no isolated places which are mapped to co-places.

Definition 13 (Process of an elementary net with (mixed) context).
Let $N = (P, T, F, C_+, C_-)$ *be an elementary net with context,* m_0 *be an initial marking of* N, $\overline{N} = (\overline{P}, \overline{T}, \overline{F}, \overline{C_+})$ *be the complementation of* N *and* $K = (B_K, E_K, F_K, C_K, \rho_K)$ *be a process of* $\overline{N}$ *w.r.t. the initial marking* $\overline{m_0}$. *Denote by* $B_{ICO}^K = \{b \in B_K \mid \rho_K(b) \notin P \wedge (\forall e \in E_K : b \notin \,^\bullet e \cup e^\bullet \cup \,^+ e)\}$ *the set of all isolated places of* K *which are mapped to co-places of* N. *Then* $(B_K \setminus B_{ICO}^K, E_K, F_K, C_K, \rho_K|_{(B_K \setminus B_{ICO}^K) \cup E_K})$ *is called a process of* N *w.r.t. an initial marking* m_0.

Let $\mathcal{P}(N, m)$ *be the set of all processes of* N *w.r.t. an initial marking* m. *By* $\mathcal{P}(N) = \bigcup_{m \subseteq P} \mathcal{P}(N, m)$ *we denote the set of all processes of* N.

The contextual occurrence net in Figure 12 together with the identity function is a process of the elementary net with context from Figure 10.

Processes $K_1 = (B_1, E_1, F_1, C_1, \rho_1)$ and $K_2 = (B_2, E_2, F_2, C_2, \rho_2)$ are isomorphic (in symbols $K_1 \simeq K_2$) iff there exist bijections $\gamma : B_1 \to B_2, \delta : E_1 \to E_2$ such that $\forall b \in B_1, e \in E_1$:

$$(b, e) \in F_1 \Longleftrightarrow (\gamma(b), \delta(e)) \in F_2,$$
$$(e, b) \in F_1 \Longleftrightarrow (\delta(e), \gamma(b)) \in F_2,$$
$$(b, e) \in C_1 \Longleftrightarrow (\gamma(b), \delta(e)) \in C_2,$$
$$\rho_1(b) = \rho_2(\gamma(b)), \rho_1(e) = \rho_2(\delta(e)).$$

8.2   Compositionality of Processes

In this section we show how processes of elementary nets with context can be concurrently and sequentially composed. The results are similar to those given for elementary nets without context in [25, 26] (sequential composition is due to [19]).

Let $N = (P, T, F, C_+, C_-)$ be an elementary net with context, let $\overline{N} = (\overline{P}, \overline{T}, \overline{F}, \overline{C_+})$ be its complementation, and let c denote the bijection associating co-places to places from P.

For a process $K = (B_K, E_K, F_K, C_K, \rho_K)$ of N, we define:

- $^\circ K$ as the set of write places of K mapped by ρ to places of N, formally

$$^\circ K = \{b \in B_K \mid \rho_K(b) \in P \wedge (\exists e \in E_K : b \in {}^\bullet e \cup e^\bullet)\},$$

- ^+K as the set of positive context places of K, which do not correspond to negative context places of N and are not write places of K, formally

$$^+K = \{b \in B_K \mid \rho_K(b) \in P \wedge (\forall e \in E_K : b \notin {}^\bullet e \cup e^\bullet)\},$$

- and ^-K as the set of places, which correspond to negative context places of N and are not write places of K, formally

$$^-K = \{b \in B_K \mid \rho_K(b) \notin P \wedge (\forall e \in E_K : b \notin {}^\bullet e \cup e^\bullet)\}.$$

We now define *elementary processes* w.r.t. markings and transitions of N.

Definition 14 (Elementary process associated to a marking). *Let* $m \subseteq P$ *be a marking of* N. *Then the process*

$$K(m) = (m, \emptyset, \emptyset, \emptyset, id_m)$$

of N *is called elementary process associated to* m.

Definition 15 (Elementary process associated to a transition). *Let* $t \in T$ *be a transition of* N. *Then the process* $K(t)$ *of net* N *defined by*

$$K(t) = ({}^\bullet t \cup t^\bullet \cup {}^+t, \{t\}, ({}^\bullet t \times \{t\}) \cup (\{t\} \times t^\bullet), {}^+t \times \{t\}, id_{{}^\bullet t \cup t^\bullet \cup {}^+t \cup \{t\}}),$$

where $^\bullet t$, $t^\bullet$ *and* ^+t *are defined w.r.t.* $\overline{N}$, *is called elementary process associated to* t.

Processes can be composed concurrently and sequentially:

Proposition 3. *Let* c *be the bijection associating co-places to places of* N *and* c^{-1} *its inverse. Let* $K_1 = (B_1, E_1, F_1, C_1, \rho_1)$ *and* $K_2 = (B_2, E_2, F_2, C_2, \rho_2)$ *be two processes of* N *w.r.t. initial markings* m_1 *and* m_2 *with disjoint sets of transitions such that* $\forall b_1 \in B_1, b_2 \in B_2$:

$$b_1 = b_2 \iff (b_1 \in {}^+K_1 \cup {}^-K_1 \wedge b_2 \in {}^+K_2 \cup {}^-K_2 \wedge \rho_1(b_1) = \rho_2(b_2)), \quad (3)$$

and

$$\emptyset = \rho_1({}^\circ K_1) \cap \rho_2({}^\circ K_2), \tag{4}$$

$$\emptyset = \rho_1({}^\circ K_1) \cap (\rho_2({}^+K_2) \cup c^{-1}(\rho_2({}^-K_2))), \tag{5}$$

$$\emptyset = \rho_2({}^\circ K_2) \cap (\rho_1({}^+K_1) \cup c^{-1}(\rho_1({}^-K_1))), \tag{6}$$

$$\emptyset = \rho_1({}^+K_1) \cap c^{-1}(\rho_2({}^-K_2)), \tag{7}$$

$$\emptyset = \rho_2({}^+K_2) \cap c^{-1}(\rho_1({}^-K_1)). \tag{8}$$

Then $K = (B_K, E_K, F_K, C_K, \rho_K)$, *where* $B = B_1 \cup B_2, E = E_1 \cup E_2, F = F_1 \cup F_2, C = C_1 \cup C_2, \rho = \rho_1 \cup \rho_2$, *is a process of* N.

Definition 16 (Concurrent composition of processes). *With notions of Proposition 3, the process K is called the concurrent composition of the processes K_1 and K_2. It is denoted by $K_1 \parallel K_2$.*

Proof of Proposition 3. No element of F_i $(i = 1, 2)$ contains glued places, i.e. places in $B_1 \cap B_2$. Therefore, (B, E, F, C) is an occurrence net.

Since $b_1 = b_2 \Longrightarrow \rho_1(b_1) = \rho_2(b_2)$ for any $b_1 \in B_1, b_2 \in B_2$, ρ is well defined. Every slice D of $K_1 \parallel K_2$ can be written in the form

$$D = D_1 \cup D_2$$

with slices D_1 of K_1 and D_2 of K_2. We show that $\rho|D$ is injective. Suppose that this is not true, i.e. that there exists $b_1 \in D_1$ and $b_2 \in D_2$ satisfying $b_1 \neq b_2$ and $\rho(b_1) = \rho(b_2)$. It is enough to consider the following four situations:

a) $\rho(b_1) \notin P, b_1 \in {}^\circ K_1, b_2 \in {}^\circ K_2$: by construction of complement places, there exists $b_1' \in {}^\circ K_1$ and $b_2' \in {}^\circ K_2$ such that $\rho_1(b_1') = \rho_2(b_2') \in P$, contradicting (4),
b) $\rho(b_1) \in P, b_1 \in {}^\circ K_1, b_2 \in {}^\circ K_2$, contradicting (4),
c) $\rho(b_1) \notin P, b_1 \in {}^\circ K_1, b_2 \in {}^- K_2$: from properties of complementation there exists $b_1' \in {}^\circ K_1$ such that $\rho_1(b_1') = c^{-1}(\rho_2(b_2))$, contradicting (5),
d) $\rho(b_1) \in P, b_1 \in {}^\circ K_1, b_2 \in {}^+ K_2$, contradicting (5).

Take a marking m reachable from $m_0 = \rho({}^\bullet K_1)$ and let

$$m_0 \xrightarrow{t_1} m_1 \dots m_{n-1} \xrightarrow{t_n} m_n = m.$$

Since we replaced negative context by positive context, for any marking m' with $m' \cap \bigcup_{0 \leqslant i \leqslant n} m_i = \emptyset$, the marking $m \cup m'$ is reachable from $m_0 \cup m'$, firing the same sequence of transitions. Using the fact that

$$\rho_1({}^\bullet K_1) \cap \rho_2({}^\bullet K_2) = \rho_1(D_1) \cap \rho_2(D_2),$$

it is easy to see that $\rho(D)$ is reachable from $\rho({}^\bullet(K_1 \parallel K_2)) = \rho_1({}^\bullet K_1) \cup \rho_2({}^\bullet K_2)$.

Since F_i contains no glued places $(i = 1, 2)$, ρ preserves pre- and post-sets of transitions. The preservation of the positive contexts of transitions follows directly from the construction of C.

Thus, $K = K_1 \parallel K_2$ is a process of $\overline{N}$ w.r.t. the initial marking $\rho({}^\bullet K)$.

It remains to show that K is also a process of N w.r.t. the initial marking $\rho({}^\bullet K) \cap P$, i.e. there is a process $\overline{K} = (\overline{B}, \overline{E}, \overline{F}, \overline{C}, \overline{\rho})$ of $\overline{N}$ w.r.t. the initial marking $\overline{\rho({}^\bullet K) \cap P}$ such that $K = (\overline{B} \setminus B_{ICO}^K, \overline{E}, \overline{F}, \overline{C}, \overline{\rho}|_{\overline{B} \setminus B_{ICO}^K \cup \overline{E}})$.

Without loss of generality, suppose that $B \cap \overline{P} = \emptyset$. Set

$$\overline{K} = (B \cup (\{c(p) \mid p \notin \rho({}^\bullet K) \cap P\} \setminus \rho(B)), E, F, C, \overline{\rho}),$$

where $\overline{\rho} = \rho$ on $B \cup E$ and $\overline{\rho} = id$ on $\{c(p) \mid p \notin \rho({}^\bullet K) \cap P\} \setminus \rho(B)$. Clearly $\overline{K}$ is a process of $\overline{N}$ with respect to the initial marking

$$\rho({}^\bullet K) \cup (\{c(p) \mid p \notin \rho({}^\bullet K) \cap P\} \setminus \rho(B)).$$

Because K_1 and K_2 have no isolated places which are copies of co-places, also $K_1 \parallel K_2$ contains no isolated places which are mapped to co-places, i.e. $B = \overline{B} \setminus B_{ICO}^{K}$.

To prove that $\overline{K}$ is a process of $\overline{N}$ w.r.t. the initial marking

$$\overline{\rho(^\bullet K) \cap P} = (\rho(^\bullet K) \cap P) \cup \{c(p) \mid p \notin \rho(^\bullet K) \cap P\}$$

it suffices to show that

$$(\rho(^\bullet K) \cap P) \cup \{c(p) \mid p \notin \rho(^\bullet K) \cap P\} = \rho(^\bullet K) \cup (\{c(p) \mid p \notin \rho(^\bullet K) \cap P\} \setminus \rho(B)).$$

To see that the first set is a subset of the second set, observe that all co-places removed from the set $\{c(p) \mid p \notin \rho(^\bullet K) \cap P\}$ belong to the set $\rho(^\bullet K)$: Because K has no isolated places which are mapped to co-places, for every place $b \in B$ with $\rho(b) \in P'$ and $c^{-1}(\rho(b)) \notin \rho(^\bullet K) \cap P$, either $b \in {}^-K$ or $b \in {}^\circ K$. In the first case, $b \in {}^\bullet K$. In the second case, either $b \in {}^\bullet K$ or there exists $e_1 \in E$ such that $b \in e_1^\bullet$. By construction of the complementation, there exists $b_1 \in {}^\bullet e_1$ such that $\rho(b_1) = c^{-1}(\rho(b))$. By the assumption $c^{-1}(\rho(b)) \notin \rho(^\bullet K) \cap P$ there exists $e_2 \in E$ such that $b_1 \in e_2^\bullet$. By induction, there exists $b_n \in {}^\bullet K$ such that $\rho(b_n) = \rho(b)$.

To prove that the second set is a subset of the first set, it is enough to show that

$$p \in \rho(^\bullet K) \implies c(p) \notin \rho(^\bullet K). \tag{9}$$

Assume that this is not true. K_1 and K_2 are processes of N and therefore (9) holds for K_1 and K_2. Without loss of generality, let $b_1 \in {}^\bullet K_1$ and $b_2 \in {}^\bullet K_2$ such that $c(\rho(b_1)) = \rho(b_2)$. We have either $b_1 \in {}^\circ K_1$ or $b_1 \in {}^+K_1$. Because K_2 has no isolated placed which are mapped to co-places, there exists $e_2 \in E_2$ such that either $b_2 \in {}^\bullet e_2$ or $b_2 \in {}^+e_2$. If $b_2 \in {}^\bullet e_2$, by definition of the complementation, there exists $b_2' \in e_2^\bullet$ such that $c(\rho_2(b_2')) = \rho_2(b_2)$ which contradicts $\emptyset = \rho_1({}^\circ K_1) \cap \rho_2({}^\circ K_2)$ if $b_1 \in {}^\circ K_1$, and contradicts $\emptyset = \rho({}^+K_1) \cap \rho_2({}^\circ K_2)$ if $b_1 \in {}^+K_1$. If $b_2 \in {}^+e_2$, then $b_1 \in {}^\circ K_1$ contradicts $\emptyset = \rho_1({}^\circ K_1) \cap c^{-1}(\rho_2({}^-K_2))$, and $b_1 \in {}^+K_1$ contradicts $\emptyset = \rho_1({}^+K_1) \cap c^{-1}(\rho_2({}^-K_2))$. $\qquad \square$

Given processes K_1, K_2, K_3, K_4 such that $K_1 \parallel K_2, K_3 \parallel K_4$ are defined and $K_1 \simeq K_3, K_2 \simeq K_4$, we have $K_1 \parallel K_2 \simeq K_3 \parallel K_4$, i.e. we have that isomorphism between processes is a congruence w.r.t. the partial operation of concurrent composition defined in the previous proposition.

Proposition 4. *Let $K_1 = (B_1, E_1, F_1, C_1, \rho_1)$ and $K_2 = (B_2, E_2, F_2, C_2, \rho_2)$ be two processes of N with disjoint sets of transitions such that $\forall b_1 \in B_1, b_2 \in B_2$:*

$$b_1 = b_2 \iff (b_1 \in K_1^\bullet \wedge b_2 \in {}^\bullet K_2 \wedge \rho_1(b_1) = \rho_2(b_2)), \quad and \tag{10}$$

$$\rho_1(K_1^\bullet) \cap P = \rho_2(^\bullet K_2) \cap P. \tag{11}$$

Then $K = (B, E, F, C, \rho)$, where $B = B_1 \cup B_2, E = E_1 \cup E_2, F = F_1 \cup F_2, C = C_1 \cup C_2, \rho = \rho_1 \cup \rho_2$, is a process of N.

Proof. See [19].

Definition 17 (Sequential composition of processes). *With notions of Proposition 4, K is called the sequential composition of the processes K_1 and K_2. It is denoted by $K_1; K_2$.*

Isomorphism between processes is a congruence also w.r.t. the partial operation of sequential composition defined in the previous proposition.

Furthermore, given two isomorphic processes $K_1 \simeq K_2$, we have:

$$\rho_1({}^\bullet K_1) = \rho_2({}^\bullet K_2), \quad \rho_1(K_1^\bullet) = \rho_2(K_2^\bullet),$$

and

$$\rho_1({}^\circ K_1) = \rho_2({}^\circ K_2), \quad \rho_1({}^+ K_1) = \rho_2({}^+ K_2), \quad \rho_1({}^- K_1) = \rho_2({}^- K_2).$$

8.3 Relationship between Process Terms and Processes of Elementary Nets with Context

For the most general case of an elementary net with mixed context we prove a one-to-one correspondence between isomorphism classes of its processes and equivalence classes of process terms of the corresponding $(\mathcal{M}, \mathcal{I})$-net from Section 7 with respect to $\sim$. As a consequence, the partial order constructed in a canonical way from an equivalence class of process terms by considering the ordering of transitions of all process terms in the equivalence class coincides with the partial order derived from the corresponding process net.

In the sequel, let $\mathcal{A}_N$ together with *inf* be the $(\mathcal{M}, \mathcal{I})$-net corresponding to an elementary net with context $N = (P, T, F, C_+, C_-)$, as defined in Section 7. With the help of the above definitions and propositions we will inductively construct isomorphism classes A_α of processes of N associated to process terms $\alpha : a \to b \in \mathcal{P}(\mathcal{A}_N)$ with information $Inf(\alpha)$ according to the four construction rules of process terms. We will also show that processes $K_\alpha \in A_\alpha$ enjoy the following properties:

$$\rho_\alpha({}^\bullet K_\alpha) \cap P = a \text{ and } \rho_\alpha(K_\alpha^\bullet) \cap P = b, \tag{12}$$
$$(\rho_\alpha({}^\circ K_\alpha), \rho_\alpha({}^+ K_\alpha), c^{-1}(\rho_\alpha({}^- K_\alpha))) = Inf(\alpha). \tag{13}$$

Proposition 5. *Let $m : m \to m$ be the reflexive process term of a marking m of N with associated information $Inf(m) = (\emptyset, m, \emptyset)$. According to Definition 14, $K(m)$ is a process of N. Clearly the properties (12) and (13) hold for $K(m)$.*

Definition 18 (Isomorphism class of processes associated to markings). *With notions of Proposition 5 define $A_m = [K(m)]_\sim$ to be the isomorphism class of processes associated with the elementary term m.*

Proposition 6. *Let $t : pre(t) \to post(t)$ be the process term generated by a transition t with associated information $Inf(t) = ({}^\bullet t \cup t^\bullet, {}^+ t, {}^- t)$. The process $K(t)$ of N satisfies properties (12) and (13).*

Definition 19 (Isomorphism class of processes associated to transitions). *With notions of Proposition 6 define* $A_t = [K(t)]_\simeq$ *to be the isomorphism class associated with the elementary term* t.

Proof of Proposition 6. According to Definition 15, $K(t)$ is a process of N. Property (12) follows from $^\bullet t \cup {}^+t = pre(t)$ and $t^\bullet \cup {}^+t = post(t)$, where $^\bullet t$, $t^\bullet$ and ^+t are taken w.r.t. N, and $^\bullet K(t) = {}^\bullet t \cup {}^+t \cup \{c(p) \mid p \in {}^-t\}$, $K(t)^\bullet = t^\bullet \cup {}^+t \cup \{c(p) \mid p \in {}^-t\}$. Property (13) follows from:

$$^-t = c^{-1}({}^-K(t)).$$

$\square$

Proposition 7. *Let* α_1, α_2 *be process terms of* $\mathcal{A}_N$, *such that* $\alpha = \alpha_1 \parallel \alpha_2$ *is a defined process term. Then there exist processes* $K_1 = (B_1, E_1, F_1, C_1, \rho_1) \in A_{\alpha_1}$ *and* $K_2 = (B_2, E_2, F_2, C_2, \rho_2) \in A_{\alpha_2}$, *such that the preconditions for concurrent composition of* K_1 *and* K_2 *are fulfilled. Moreover, the process*

$$K_\alpha = K_1 \parallel K_2 = (B_\alpha, E_\alpha, F_\alpha, C_\alpha, \rho_\alpha),$$

satisfies the properties (12) and (13).

Definition 20 (Isomorphism class of processes associated to concurrent composed process terms). *With notions of Proposition 7 define* $A_\alpha = [K_\alpha]_\simeq$ *to be the isomorphism class associated with the term* α.

Proof of Proposition 7. Take processes $K_1 \in A_{\alpha_1}, K_2 \in A_{\alpha_2}$, such that the sets $B_1 \setminus ({}^+K_1 \cup {}^-K_1), B_2 \setminus ({}^+K_2 \cup {}^-K_2), \overline{P}$ are disjoint, and $^+K_i \cup {}^-K_i \subseteq \overline{P} \wedge \rho_i|_{{}^+K_i \cup {}^-K_i} = id$ for $i = 1, 2$ (what can be achieved by an appropriate renaming). Then the precondition (3) formulated in Proposition 3 is fulfilled.

Denoting $Inf(\alpha_1) = (w_1, p_1, n_1)$ and $Inf(\alpha_2) = (w_2, p_2, n_2)$ we have by the definition of dom_+:

$$w_1 \cap w_2 = w_1 \cap (p_2 \cup n_2) = w_2 \cap (p_1 \cup n_1) = p_1 \cap n_2 = p_2 \cap n_1 = \emptyset.$$

From property (13) of K_1 and K_2 we have for $i = 1, 2$:

$$w_i = \rho_i(^\circ K_i), \quad p_i = \rho_i(^+K_i), \quad n_i = c^{-1}(\rho_i(^-K_i)).$$

Therefore the remaining preconditions for concurrent composition of K_1 and K_2 formulated in Proposition 3 are fulfilled.

We have that

$$^\bullet(K_1 \parallel K_2) = {}^\bullet K_1 \cup {}^\bullet K_2, (K_1 \parallel K_2)^\bullet = K_1^\bullet \cup K_2^\bullet,$$

and, because joined places are neither in $^\circ K_1$ nor in $^\circ K_2$,

$$^\circ(K_1 \parallel K_2) = {}^\circ K_1 \cup {}^\circ K_2,$$
$$^+(K_1 \parallel K_2) = {}^+K_1 \cup {}^+K_2,$$
$$^-(K_1 \parallel K_2) = {}^-K_1 \cup {}^-K_2,$$

which easily implies properties (12) and (13).

$\square$

Proposition 8. *Let α_1 and α_2 be two process terms such that $\alpha = \alpha_1 ; \alpha_2$ is a defined process term. Then there exist $K_1 = (B_1, E_1, F_1, C_1, \rho_1) \in A_{\alpha_1}$ and $K_2 = (B_2, E_2, F_2, C_2, \rho_2) \in A_{\alpha_2}$ such that $K_\alpha = K_1 ; K_2$ is a defined process, which fulfills properties (12) and (13).*

Definition 21 (Isomorphism class of processes associated to sequential composed process terms). *With notions of Proposition 8 define $A_\alpha = [K_\alpha]_\simeq$ to be the isomorphism class associated with the term α.*

Proof of Proposition 8. Take processes $K_1 \in A_{\alpha_1}, K_2 \in A_{\alpha_2}$ such that the sets $B_1 \backslash K_1^\bullet, B_2 \backslash {}^\bullet K_2, \overline{P}$ are disjoint, $K_1^\bullet \subseteq \overline{P} \wedge \rho_1|_{K_1^\bullet} = id$ and ${}^\bullet K_2 \subseteq \overline{P} \wedge \rho_2|_{{}^\bullet K_2} = id$ (what can be achieved by an appropriate renaming). Then the precondition (10) formulated in Proposition 4 is fulfilled.

From property (12) of processes K_1, K_2 and from $post(\alpha_1) = pre(\alpha_2)$ we have $\rho_1(K_1^\bullet) \cap P = \rho_2({}^\bullet K_2) \cap P$ and therefore precondition (11) formulated in Proposition 4 is fulfilled.

The new process $K_\alpha = K = (B, E, F, C, \rho)$ obviously satisfies property (12). We have ${}^\circ K = {}^\circ K_1 \cup {}^\circ K_2$ and therefore

$$\rho({}^\circ K) = \rho_1({}^\circ K_1) \cup \rho_2({}^\circ K_2)$$

Moreover, ${}^+K = ({}^+K_1 \cup {}^+K_2) \backslash {}^\circ K$. Since ρ is injective on ${}^+K_1 \cup {}^+K_2$, we have

$$\rho({}^+K) = (\rho_1({}^+K_1) \cup \rho_2({}^+K_2)) \backslash \rho({}^\circ K).$$

Let ${}^\circ K' = \{b \in B \mid \rho(b) \notin P \wedge (\exists e \in E : b \in {}^\bullet e \cup e^\bullet)\}$.
Then ${}^-K = ({}^-K_1 \cup {}^-K_2) \backslash {}^\circ K'$.
By injectivity of ρ on ${}^-K_1 \cup {}^-K_2$ we have $\rho({}^-K) = (\rho_1({}^-K_1) \cup \rho_2({}^-K_2)) \backslash \rho({}^\circ K')$. By construction of complementation we have $c^{-1}(\rho({}^\circ K') \subseteq \rho({}^\circ K)$. Since $p \in \rho({}^\bullet K) \Rightarrow c(p) \notin \rho({}^\bullet K)$, by induction we have

$$p \in \rho(D) \Rightarrow c(p) \notin \rho(D) \qquad (14)$$

for each slice D of K. Since each slice of K contains ${}^-K = ({}^-K_1 \cup {}^-K_2) \backslash {}^\circ K'$, we have $(\rho({}^\circ K) \backslash c^{-1}(\rho({}^\circ K'))) \cap \rho({}^-K) = \emptyset$. Thus, we have

$$c^{-1}(\rho({}^-K)) = (c^{-1}(\rho_1({}^-K_1)) \cup c^{-1}(\rho_1({}^-K_1))) \backslash \rho({}^\circ K).$$

Since each slice of K contains ${}^+K \cup {}^-K$, by (14) we have

$$\rho({}^+K) \cap c^{-1}(\rho({}^-K)) = \emptyset.$$

Thus, process K_α enjoys property (13).   □

Definition 22. *Given an elementary net with mixed context N, let $\tau : \mathcal{P}(\mathcal{A}_N) \to (\mathcal{P}(N))/_\simeq$ be the mapping defined by $\tau(\alpha) = A_\alpha$.*

Lemma 9. *Let $K = (B_K, E_K, F_K, C_K, \rho_K)$ be a process of N and $e_1, e_2 \in E_K$ with $e_1 \not< e_2 \wedge e_2 \not< e_1$. Then $\rho_K(e_1) \parallel \rho_K(e_2)$ is a defined process term.*

Proof. It suffices to show that:

(a) $(\rho_K(\,^\bullet e_1 \cup e_1^\bullet) \cap P) \cap (\rho_K(\,^\bullet e_2 \cup e_2^\bullet) \cap P) = \emptyset$,

(b) $((\rho_K(\,^+e_1) \cap P) \cup c^{-1}(\rho(\,^+e_1) \cap P')) \cap (\rho_K(\,^\bullet e_2 \cup e_2^\bullet) \cap P) = \emptyset$, and

(c) $(\rho_K(\,^+e_1) \cap P) \cap c^{-1}(\rho_K(\,^+e_2) \cap P') = \emptyset$.

(a) follows from: $e_1 \not\leqslant e_2 \wedge e_2 \not\leqslant e_1$ implies that the sets $^\bullet e_1 \cup \,^\bullet e_2 \cup \,^+K$ and $e_1^\bullet \cup e_2^\bullet \cup \,^+K$ are subsets of slices of K. Since ρ_K is injective on slices, $\rho_K(\,^\bullet e_1) \cap \rho_K(\,^\bullet e_2) = \rho_K(e_1^\bullet) \cap \rho_K(e_2^\bullet) = \emptyset$. Assume there is a place $p \in \rho_K(\,^\bullet e_1) \cap \rho_K(e_2^\bullet)$ or $p \in \rho_K(\,^\bullet e_2) \cap \rho_K(e_1^\bullet)$. Without loss of generality let $p \in \rho_K(\,^\bullet e_1) \cap \rho_K(e_2^\bullet)$. There are places $b_1 \in \,^\bullet e_1$ and $b_2 \in e_2^\bullet$ such that $\rho_K(b_1) = \rho_K(b_2)$. Then either $b_1 \not\leqslant b_2 \wedge b_2 \not\leqslant b_1$ (which would be a contradiction to the injectivity of ρ_K on slices) or $b_2 \leqslant b_1$ (which would be a contradiction to $e_2 \not\leqslant e_1 \wedge e_1 \not\leqslant e_2$ by the transitivity of $\leqslant$) or finally $b_1 \leqslant b_2$, which would imply $e_1 \leqslant e_2$ because places are unbranching, what is again a contradiction.

To show (b), assume there is a place $p \in (\rho_K(\,^+e_1) \cap P) \cup c^{-1}(\rho_K(\,^+e_1) \cap P')) \cap (\rho(\,^\bullet e_2 \cup e_2^\bullet) \cap P)$. Then there are places $b_1 \in \,^+e_1$ and $b_2 \in (\,^\bullet e_2 \cup^\bullet e_2) \cap P$ such that $c(\rho_K(b_1)) = \rho_K(b_2)$ or $\rho_K(b_1) = \rho_K(b_2)$. In the first case we observe:

– Assume $b_2 \in \,^\bullet e_2$. By construction of the complementation $\overline{N}$ of N there is a place $b_2' \in e_2^\bullet$ such that $\rho_K(b_1) = \rho_K(b_2')$. We can distinguish 4 situations:
  $b_1 \not\leqslant b_2' \wedge b_2' \not\leqslant b_1$ leads to a contradiction similar as in case (a).
  $b_1 = b_2'$ implies $e_2 \leqslant e_1$.
  $b_1 < b_2'$ implies the existence of a transition $e' \in E$ such that $b_1 \in \,^\bullet e' \wedge e_1 < e'$. Because places are unbranched, this implies $e' < e_2$ and therefore $e_1 < e_2$.
  $b_2' < b_1$ implies the existence of a transition e' such that $b_2 < e'$ and $b_1 \in (e')^\bullet$. It follows $e' < e_1$ and therefore $e_2 < e_1$.
– The proof for $b_2 \in e_2^\bullet$ is similar.

The second case obviously reduces to the situations considered in the first case.

Finally we obtain (c) by assuming that there is a place $p \in (\rho_K(\,^+e_1) \cap P) \cap c^{-1}(\rho_K(\,^+e_2) \cap P')$. Then there are places $b_1 \in \,^+e_1$ and $b_2 \in \,^+e_2$ with $c(\rho_K(b_1)) = \rho_K(b_2)$. Since $p \in \rho_K(\,^\bullet K) \Rightarrow c(p) \notin \rho_K(\,^\bullet K)$, by induction we have $p \in \rho_K(D) \Rightarrow c(p) \notin \rho_K(D)$ for each slice D of K. This implies either $b_1 < b_2$ or $b_2 < b_1$ which again gives a contradiction to $e_1 \not\leqslant e_2 \wedge e_2 \not\leqslant e_1$.   □

Remark 5. (a) Given process terms α_i, $i = 1, \ldots, 4$ of $\mathcal{A}_N$ such that the terms $\alpha = ((\alpha_1 \parallel \alpha_2); (\alpha_3 \parallel \alpha_4))$ and $\beta = ((\alpha_1; \alpha_3) \parallel (\alpha_2; \alpha_4))$ are defined. Then $Inf(\alpha) = Inf(\beta)$.

(b) For any two process terms α_1 and α_2 such that $\alpha_1 \parallel \alpha_2$ is defined, we have $\alpha_1 \parallel \alpha_2 \sim (\alpha_1; post(\alpha_1)) \parallel (pre(\alpha_2); \alpha_2) \sim (\alpha_1 \parallel pre(\alpha_2)); (\alpha_2 \parallel post(\alpha_1))$ and analogously $\alpha_1 \parallel \alpha_2 \sim (\alpha_2 \parallel pre(\alpha_1)); (\alpha_1 \parallel post(\alpha_2))$

(c) If $(\alpha_1; \alpha_2) \parallel m$, m being a marking, is defined, then we have $(\alpha_1; \alpha_2) \parallel m \sim (\alpha_1 \parallel m); (\alpha_2 \parallel m)$.

Theorem 5. *The mapping $\tau : \mathcal{P}(\mathcal{A}_N) \to (\mathcal{P}(N))/_\simeq$ is surjective.*

Proof. Let $K = (B, E, F, C, \rho)$ be a process of N. We inductively construct a process term α with $K_\alpha = K$ by the method of maximal steps analogously to the proof of the similar theorem in [6, Theorem 1]: Beginning with the slice $D = {}^\bullet K$, we take all transitions $\{e_1, \ldots, e_m\} \in E$ with ${}^\bullet e_i \subset D$ such that there is no transition $e \in E$ with $e < e_i$ ($1 \leqslant i \leqslant m$). Then the transitions $\rho(e_1), \ldots, \rho(e_m)$ can be composed by $\|$ as process terms. The resulting process term then is sequentially composed with the next one, which we derive by the same procedure now starting with the follower slice of D after firing $e_1, \ldots, e_m$. This is repeated until the follower slice equals $K^\bullet$. $\square$

Theorem 6. *For two process terms $\alpha, \beta \in \mathcal{P}(\mathcal{A}_N)$, $\alpha \sim \beta$ implies $\tau(\alpha) = \tau(\beta)$.*

Proof. It is sufficient to show the proposition for every (of the seven) construction rules of $\sim$ (Definition 2).

(1) The proof for the rule (1) is obvious.

(2) Given $\alpha_1, \alpha_2, \alpha_3$ such that terms $(\alpha_1 \| \alpha_2) \| \alpha_3$ and $\alpha_1 \| (\alpha_2 \| \alpha_3)$ are defined, take processes

$$K_1 \in A_{\alpha_1},\ K_2 \in A_{\alpha_2},\ K_3 \in A_{\alpha_3},$$

such that sets

$$B_1 \setminus ({}^+K_1 \cup {}^-K_1),\ B_2 \setminus ({}^+K_2 \cup {}^-K_2),\ B_3 \setminus ({}^+K_3 \cup {}^-K_3),\ \overline{P}$$

are disjoint and

$$ {}^+K_i \cup {}^-K_i \subseteq \overline{P} \wedge \rho_i|_{{}^+K_i \cup {}^-K_i} = id, \quad i \in \{1, 2, 3\}$$

(what can be achieved by an appropriate renaming). Then processes $(K_1 \| K_2) \| K_3, K_1 \| (K_2 \| K_3)$ are defined and equal.

(3) Given $\alpha_1, \alpha_2, \alpha_3$ such that terms $(\alpha_1; \alpha_2); \alpha_3$ and $\alpha_1; (\alpha_2; \alpha_3)$ are defined, let G be a set satisfying $|G| = |\overline{P}|$ and $G \cap \overline{P} = \emptyset$, and let $g : G \to \overline{P}$ be a bijection. Take processes

$$K_1 \in A_{\alpha_1},\ K_2 \in A_{\alpha_2},\ K_3 \in A_{\alpha_3},$$

such that sets

$$B_1 \setminus K_1^\bullet,\ B_2 \setminus ({}^\bullet K_2 \cup K_2^\bullet),\ B_3 \setminus {}^\bullet K_3,\ G,\ \overline{P}$$

are disjoint and

$$K_1^\bullet \subseteq \overline{P} \wedge \rho_1|_{K_1^\bullet} = id,\ {}^\bullet K_2 \subseteq \overline{P} \wedge \rho_2|_{{}^\bullet K_2} = id,$$

$$K_2^\bullet \setminus ({}^+K_2 \cup {}^-K_2) \subseteq G \wedge \rho_2|_{K_2^\bullet \setminus ({}^+K_2 \cup {}^-K_2)} = g|_{K_2^\bullet \setminus ({}^+K_2 \cup {}^-K_2)},$$

$$ {}^\bullet K_3 \subseteq G \wedge \rho_3|_{{}^\bullet K_3} = g|_{{}^\bullet K_3}$$

(what can be achieved by an appropriate renaming).

Set $Q = {}^\bullet K_3 \cap g^{-1}({}^+K_2 \cup {}^-K_2)$. Now, take the process $K_3' \in A_{\alpha_3}$ obtained from the process K_3 by renaming every place $b \in Q$ by the place $g(b) \in {}^+K_2 \cup {}^-K_2$. Then processes $(K_1; K_2); K_3', K_1; (K_2; K_3)$ are defined and equal.

(4) Given $\alpha_1, \alpha_2, \alpha_3, \alpha_4$ such that terms $(\alpha_1 \parallel \alpha_2); (\alpha_3 \parallel \alpha_4)$ and $(\alpha_1; \alpha_3) \parallel (\alpha_2; \alpha_4)$ are defined, take processes

$$K_1 \in A_{\alpha_1}, \ K_2 \in A_{\alpha_2}, \ K_3 \in A_{\alpha_3}, \ K_4 \in A_{\alpha_4},$$

such that sets

$$B_1 \setminus K_1^\bullet, \ B_2 \setminus K_2^\bullet, \ B_3 \setminus {}^\bullet K_3, \ B_4 \setminus {}^\bullet K_4, \ \overline{P}$$

are disjoint and

$$K_1^\bullet \subseteq \overline{P} \wedge \rho_1|_{K_1^\bullet} = id, \ K_2^\bullet \subseteq \overline{P} \wedge \rho_2|_{K_2^\bullet} = id,$$

$${}^\bullet K_3 \subseteq \overline{P} \wedge \rho_3|_{{}^\bullet K_3} = id, \ {}^\bullet K_4 \subseteq \overline{P} \wedge \rho_4|_{{}^\bullet K_4} = id$$

(what can be achieved by an appropriate renaming).

Then processes $(K_1 \parallel K_2); (K_3 \parallel K_4), (K_1; K_3) \parallel (K_2; K_4)$ are defined and equal.

(5-7) The proof for rules (5-7) is similar.

□

Theorem 7. *For two process terms* $\alpha, \beta \in \mathcal{P}(\mathcal{A}_N)$, $\tau(\alpha) = \tau(\beta)$ *implies* $\alpha \sim \beta$.

Proof. Without loss of generality let α and β be process terms with $K(\alpha) = K(\beta) = K = (B, E, F, C, \rho)$ and $\gamma = \gamma_1; \ldots; \gamma_m$ be the process term constructed from the process K in the proof of Theorem 5 by considering maximal steps. Then γ_i is of the form

$$\gamma_i = \rho(e_1^i) \parallel \ldots \parallel \rho(e_{n_i}^i) \parallel \rho(a^i),$$

$e_1^i, \ldots, e_{n_i}^i \in E$ and $a^i \subseteq B$, $i = 1, \ldots, m$. We show that α is equivalent to γ. By symmetry, the same holds for β, and we are done.

According to Remark 5, we assume without loss of generality that α is of the form

$$\alpha = \rho(e_1) \parallel (\rho(a_1) \cap P); \ldots; \rho(e_k) \parallel (\rho(a_k) \cap P)$$

with transitions $e_1, \ldots, e_k \in E$ and subsets $a_1, \ldots, a_k \subseteq B$. We will use shorthands $\alpha = e_1; \ldots; e_k$, and ignore the sets $a_1, \ldots, a_k$, because they are determined by the definition of the sequential composition of process terms. Clearly, α and γ 'contain' the same transitions, i.e.

$$\{e_1, \ldots, e_k\} = \{e_1^1, \ldots, e_{n_1}^1, \ldots, e_1^m, \ldots, e_{n_m}^m\}.$$

Assume $e_i = e_1^1$ for an $i \geq 2$. It suffices to prove

$$e_1; \ldots; e_i \sim e_1; \ldots; e_i; e_{i-1} \sim \ldots \sim e_i; e_1; \ldots; e_{i-1},$$

because firstly the same procedure applied to $e_2^1, \ldots, e_{n_1}^1$ provides $e_1; \ldots; e_n \sim \gamma_1; \delta$ (where δ is the rest of the term α after removing transitions of γ_1), and secondly this procedure applied to $\gamma_2, \ldots, \gamma_m$ finishes the proof. In fact, it is enough to show that we can exchange e_i and e_{i-1} in α. A sufficient condition is that $\rho(e_i) \parallel \rho(e_{i-1})$ is a defined process term.

We have to distinguish two cases: If $e_{i-1} = e_j^1$ for some $j \in \{2, \ldots, n_1\}$, $\rho(e_i) \parallel \rho(e_{i-1})$ is defined according to the process term γ. The other possibility is $e_{i-1} = e_j^l$ for an $l \in \{2, \ldots, m\}$ and $j \in \{1, \ldots, k_l\}$. By construction of the process K_α from α follows $e_i \not\lessgtr e_{i-1}$. On the other hand, by construction of γ follows $e_{i-1} \not\lessgtr e_i$. By Lemma 9, $\rho(e_i) \parallel \rho(e_{i-1})$ is defined. □

Remark 6. The set of all processes of an elementary net with mixed context w.r.t. an initial marking m_0 corresponds to the set of all equivalence classes of process terms containing process terms of the form $\alpha = m_0; \beta$ (i.e. process terms starting with m_0).

Finally, looking at the definition of τ, we can state the main result for elementary nets with mixed context, which now follows easily from the previous theorems.

Theorem 8. *Given any elementary net N, there exists a one-to-one correspondence between the isomorphism classes of processes $\mathcal{P}(N)$ of N and the $\sim$-congruence classes of the process terms $\mathcal{P}(\mathcal{A}_N)$ of the corresponding algebraic $(\mathcal{M}, \mathcal{I})$-net defined in Section 7. This correspondence preserves the initial marking, final marking and the information about write places, positive context places and negative context places of processes and process terms, as well as concurrent composition and sequential composition of processes (resp. congruence classes of process terms).*

Remark 7. Clearly, according to Remarks 2 and 4 the previous theorem holds also for elementary nets without context and elementary nets with negative context, although in these examples we considered a slightly different process term semantics.

Using terminology from partial algebra [4] we can rephrase Theorem 8 as follows: Given an elementary net with context N and a process term $\alpha \in \mathcal{P}(\mathcal{A}_N)$ of the corresponding net over partial algebra, the congruence class $[\alpha]_\sim \in (\mathcal{P}(\mathcal{A}_N))/\sim$ corresponds to the isomorphism class $\tau(\alpha) = [K]_\simeq$ of a process $K \in \mathcal{P}(N)$ such that the initial and final marking are preserved, i.e. $\rho({}^\bullet K) \cap P = pre(\alpha)$, $\rho(K^\bullet) \cap P = post(\alpha)$, and information for concurrent composition is preserved, i.e. $(\rho({}^\circ K), \rho({}^+ K), c^{-1}(\rho({}^- K))) = Inf(\alpha)$. The factor algebra $(\mathcal{P}(\mathcal{A}_N))/\sim$ is isomorphic to the factor algebra $(\mathcal{P}(N))/\simeq$, (i.e. τ is a surjective closed homomorphism between $\mathcal{P}(\mathcal{A}_N)$ and $(\mathcal{P}(N))/\simeq$).

9   Place/Transition Nets

In this section we give algebraic definitions of place/transition Petri nets with inhibitor arcs and place/transition Petri nets with capacities.

Here we provide semantics corresponding to collective token philosophy [5]. In this case an equivalence class of process terms corresponds to an equivalence class of partial orders, according to collective token semantics of place/transition nets without capacity restriction (see [1] and [5]). In the case of individual token philosophy, where the single partial orders are of interest, one can use more sophisticated algebras, such as for example concatenated processes [24].

Let us briefly mention another possibility how to deal with individual token semantics without using different algebra from those used for the collective token semantics. As it was discussed in Introduction, any process term defines naturally a partial order of events labeled by transitions. Thus, an equivalence class of process terms defines a set of partial orders. As we have illustrated in the example from Introduction, one can modify these partial orders comparing each other and removing causalities which are not defined by the net itself. The idea for further research is to generalize this modification procedure in order to obtain the set of partial orders containing only those causalities which are given by the net itself. Such set of partial orders would correspond to collective token semantics, while obtained single partial orders would correspond to individual process semantics.

Clearly, one can also combine restrictions given by inhibitor arcs and capacities and extend them further, or combine them with other approaches such as positive context to get a more complicated enabling rule. In such cases one could use more complicated algebras, see e.g. [11, 3].

Definition 23 (Place/transition nets). *A place/transition Petri net (shortly a p/t net) is a quadruple $N = (P, T, F, W)$, where P, T and F are defined as for elementary nets, and $W : F \to \mathbb{N}^+$ is the weight function. Given a transition t, define $^\bullet t, t^\bullet \in \mathbb{N}^P$ as follows:*

$$^\bullet t(p) = \begin{cases} W((p, t)) & \text{if } (p, t) \in F, \\ 0 & \text{otherwise,} \end{cases}$$

$$t^\bullet(p) = \begin{cases} W((t, p)) & \text{if } (t, p) \in F, \\ 0 & \text{otherwise.} \end{cases}$$

9.1   Place/Transition Nets with Inhibitor Arcs

Definition 24 (Place/transition nets with inhibitor arcs). *A p/t net with inhibitor arcs is a five-tuple $N = (P, T, F, W, C_-)$, where (P, T, F, W) is a p/t net, and $C_- \subseteq P \times T$ is an inhibitor relation (set of inhibitor arcs) satisfying $(F \cup F^{-1}) \cap C_- = \emptyset$. As usual, $^-t = \{p \mid (p, t) \in C_-\}$ for each $t \in T$. A marking of N is a multi-set $m \in \mathbb{N}^P$. A transition t is enabled to occur at m iff $\forall p \in P : m(p) \geqslant {}^\bullet t(p) \wedge ((p, t) \in C_- \Rightarrow m(p) = 0)$. Its occurrence leads to the marking $m' = m - {}^\bullet t + t^\bullet$.*

For p/t nets with inhibitor arcs the cardinality of the information set I is smaller than the cardinality of the marking set of the net: $\mathcal{M} = (M, +) = (\mathbb{N}^P, +)$, where $+$ is multi-set addition. For concurrent composition it is obviously enough to check that one process does not use negative context places of the other process as write places. Therefore, the necessary information for concurrent composition consists of the set of those places which appear in a marking of the process term and the set of negative context places. For a marking m over the set P of places we denote $m_s = \{p \mid m(p) \neq 0\}$. It follows $\mathcal{I} = (I, +, dom_+)$ with $I = 2^P \times 2^P$, $dom_+ = \{((w, n), (w', n')) | w \cap n' = w' \cap n = \emptyset\}$ and $\forall ((w, n)(w', n')) \in dom_+ : (w, n) + (w', n') = (w \cup w', n \cup n')$.

The partial groupoid $\mathcal{I}$ satisfies the requirements given in Section 2.

For a transition t and a marking m define

$$pre(t) = {}^{\bullet}t, post(t) = t^{\bullet},$$
$$inf(m) = (m_s, \emptyset), inf(t) = ((pre(t))_s \cup post(t))_s, {}^{-}t).$$

The function inf preserves property (a) from Definition 1. One can also easily prove that the independence relation of $\mathcal{I}$ encodes the restriction of the occurrence rule by restriction of concurrent occurrences of a transition and a marking.

Lemma 10. *Let* $supp : 2^I \to I$ *be defined by*

$$supp(A) = \left(\bigcup_{(w,n) \in A} w, \bigcup_{(w,n) \in A} n \right).$$

Then relation $\cong$ *defined by* $x \cong y \Leftrightarrow supp(x) = supp(y)$ *is the greatest closed congruence on the partial algebra* $(2^I, dom_{\{+\}}, \{+\}, \cup)$.

Proof. It is a straightforward observation that $supp$ is a surjective closed homomorphism from $(2^I, dom_{\{+\}}, \{+\}, \cup)$ to $(I, dom_+, +, \circ)$, where $\forall (w, n), (w', n') \in I : (w, n) \circ (w', n') = (w \cup w', n \cup n')$. Hence $\cong$ is a closed congruence.

To prove that $\cong$ is the greatest closed congruence it suffices to show that any congruence $\approx$ satisfying $\cong \subsetneq \approx$ is not closed. The proof is similar to the proof of Lemma 6. Assume there are $A, A' \in 2^I$ such that $A \approx A'$ but $A \not\cong A'$. Then $supp(A) \neq supp(A')$.

We construct a set $C \in 2^I$ such that $(A, C) \in dom_{\{+\}}$ but $(A', C) \notin dom_{\{+\}}$ or vice versa (which implies that $\approx$ is not closed). If $supp(A) = (\overline{w}, \overline{n})$ and $supp(A') = (\overline{w}', \overline{n}')$ then $\overline{n} \neq \overline{n}' \vee \overline{w} \neq \overline{w}'$ (since $supp(A) \neq supp(A')$).

Let $\overline{w} \neq \overline{w}'$. Without loss of generality we can assume $\overline{w}' \setminus \overline{w} \neq \emptyset$. Set $C = \{(c_w, c_n)\}$ with $c_w = \emptyset$ and $c_n = \overline{w}' \setminus \overline{w}$. Therefore $c_w \cap \overline{n} = c_n \cap \overline{w} = \emptyset$, but $c_n \cap \overline{w}' \neq \emptyset$, i.e. $(A, C) \in dom_{\{+\}}$, but $(A', C) \notin dom_{\{+\}}$.

Now let $\overline{n} \neq \overline{n}'$. Without loss of generality we have $\overline{n}' \setminus \overline{n} \neq \emptyset$. Set $C = \{(c_w, c_n)\}$ with $c_w = (\overline{n}' \setminus \overline{n})$ and $c_n = \emptyset$. Then $c_w \neq \emptyset$, $c_w \cap \overline{n} = \overline{w} \cap c_n = \emptyset$ and $c_w \cap \overline{n}' \neq \emptyset$, and we are finished. □

Because also property (b) from Definition 1 is preserved, we can formulate the following theorem.

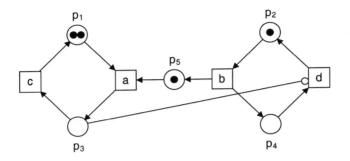

Fig. 13. An example of a p/t net with inhibitor arcs. A possible process term is $(a \parallel b \parallel p_1); (c \parallel (p_1 + p_4 + p_5); d \parallel (2p_1 + p_5))$.

Theorem 9. *Given a p/t net with inhibitor arcs $N = (P, T, F, W, C_-)$ with $\mathcal{M}, \mathcal{I}, pre, post, inf$ as defined in this subsection, the quadruple $\mathcal{A}_N = (2^P, T, pre, post)$ together with the mapping inf is an algebraic $(\mathcal{M}, \mathcal{I})$-net. Moreover, it is a corresponding algebraic $(\mathcal{M}, \mathcal{I})$-net to the net N.*

Figure 13 shows an example of a p/t net with inhibitor arcs.

9.2   Nets with Capacities

There are two different interpretations of consuming and producing tokens for Petri nets with capacities (for more details see e.g. [9, 10, 15]). According to the order of consuming and producing tokens one can distinguish the following situations:

- A transition t first consumes the tokens given by $pre(t)$ yielding an intermediate marking 0 (empty multiset) and then produces tokens $post(t)$. This interpretation corresponds to classical rewriting and such capacities are said to be weak [9].
- A transition t first produces tokens (given by $post(t)$), yielding an intermediate marking $pre(t) + post(t)$ and then consumes tokens (given by $pre(t)$) yielding the marking $post(t)$. Such capacities are said to be strong [9].

Definition 25 (Place/transition nets with capacities). *A place/transition net with capacities is a p/t net together with a partial function $K : P \to \mathbb{N}^+$ with a domain $P_K \subseteq P$.*
A marking of a net with capacities is a multi-set $m \in \mathbb{N}^P$ such that $\forall p \in P_K : m(p) \leqslant K(p)$.
A transition t is said to be weakly enabled at a marking m iff $\forall p \in P : m(p) \geqslant {}^\bullet t(p)$ and $\forall p \in P_K : K(p) \geqslant m(p) - {}^\bullet t(p) + t^\bullet(p)$.
A transition t is said to be strongly enabled at a marking m iff $\forall p \in P : m(p) \geqslant {}^\bullet t(p)$ and $\forall p \in P_K : K(p) \geqslant m(p) + t^\bullet(p)$.
The occurrence of an enabled transition t at a marking m leads to the marking $m' = m - {}^\bullet t + t^\bullet$.

The concurrent occurrence of transitions, and more general concurrent composition of processes, have to respect capacities. In the case of strong capacities the information about the intermediate marking $pre(t) + post(t)$ is attached to transition t.

Thus, as the set of markings we set $\mathcal{M} = (\{a \in \mathbb{N}^P \mid \forall p \in P_K : a(p) \leqslant K(p)\}, \tilde{+})$, where the operation $\tilde{+}$ is defined by $a(p)\tilde{+}b(p) = min(a(p) + b(p), K(p))$ for all $p \in P_K$ and $a(p)\tilde{+}b(p) = a(p) + b(p)$ for all $p \in P \setminus P_K$.

The partial groupoid of information $\mathcal{I} = (I, +, dom_+)$ is defined by

$$I = (\{w \in \mathbb{N}^{P_K} \mid \forall p \in P_K : w(p) \leqslant K(p)\},$$
$$dom_+ = \{(w, w') \in I \times I \mid \forall p \in P_K : w(p) + w'(p) \leqslant K(p)\}$$
$$+ = +|_{dom_+}.$$

This partial groupoid satisfies the requirements from Section 2.

Define $pre(t) = {}^\bullet t, post(t) = t^\bullet$ for every transition t. Moreover, for weak capacities define a mapping $inf_w : M \cup T \to I$ by:

- For a marking m, $inf_w(m) = m|_{P_K}$.
- For a transition t and a place $p \in P_K$, $inf_w(t)(p) = max(pre(t)(p), post(t)(p))$.

For strong capacities define a mapping $inf_s : M \cup T \to I$ by:

- For a marking m, $inf_s(m) = m|_{P_K}$.
- For a transition t and a place $p \in P_K$, $inf_s(t)(p) = (pre(t)(p) + post(t)(p))$.[4]

Again, property (a) from Definition 1 is satisfied. The considered independence relation encodes the restriction of the occurrence rule.

In the sequel, we define a mapping $supp : 2^I \to I$ and prove that $supp$ is the natural homomorphism of the greatest closed congruence $\cong$ of the partial algebra $(2^I, dom_{\{+\}}, \{+\}, \cup)$.

Lemma 11. *Given $\mathcal{I}$ as above, let $supp : 2^I \to I$ be defined for all $p \in P_K$ by*

$$supp(A)(p) = max_{a \in A}a(p).$$

Then the relation $\cong$ defined by $A \cong A' \iff supp(A) = supp(A')$ is the greatest closed congruence on the partial algebra $(2^I, dom_{\{+\}}, \{+\}, \cup)$.

Proof. By the properties of maximum and the definition of the mapping $supp$, $supp$ is a surjective closed homomorphism from $(2^I, dom_{\{+\}}, \{+\}, \cup)$ to $(I, dom_+, +, \circ)$, where $\forall a, a' \in I : a \circ a' = supp(\{a, a'\})$, and therefore $\cong$ is a closed congruence. To prove that $\cong$ is the greatest closed congruence we show that any congruence $\approx$ satisfying $\cong \subsetneq \approx$ is not closed. We construct a set $C \in 2^I$ such that $(A, C) \in dom_{\{+\}}$ but $(A', C) \notin dom_{\{+\}}$ or vice versa. Assume there are

---

[4] In the case of strong capacities we implicitly suppose for each transition t and each place $p \in P_K$ that $pre(t)(p) + post(t)(p) \leqslant K(p)$. Otherwise transition t is never enabled to occur and therefore according to the Definition 3 it is irrelevant for the corresponding net

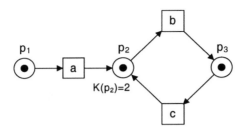

Fig. 14. An example of a p/t net with capacity

$A, A' \in 2^I$ such that $A \approx A'$ but $A \ncong A'$. Then there is a place $p \in P$ such that $max_{a \in A}a(p) \neq max_{a' \in A'}a'(p)$. Without loss of generality let $max_{a' \in A'}a'(p) > max_{a \in A}a(p)$. It suffices to take, for example, $C = \{a\}$ for the multi-set $a(p) = K(p) - max_{a \in A}a(p)$ and $a(p') = 0$ for all $p' \in P_K$ such that $p' \neq p$.   □

The property (b) from Definition 1 is satisfied both for inf_w and inf_s. Thus, we have the following theorem for place/transition nets with capacities.

Theorem 10. *Given a p/t net with capacity $N = (P, T, F, W, K)$ with $\mathcal{M}, \mathcal{I}, pre,$ post, inf_w, inf_s as defined in this subsection, the quadruple $\mathcal{A}_N = (M, T, pre, post)$ together with inf_w for weak capacities and inf_s for strong capacities is a corresponding $(\mathcal{M}, \mathcal{I})$-net to the net N.*

Notice that in the case that there are no self-loops in the net, as it is in Figure 14, weak and strong capacities coincide. Nets with capacities represent a class of $(\mathcal{M}, \mathcal{I})$-nets where information can violate the distributive law (see Definition 2, (4)). For example, we have the following process terms of the net from Figure 14: $\alpha = (b \parallel p_1); (p_3 \parallel a)$ with $Inf(\alpha) = p_2$ and $\beta = (b; p_3) \parallel (p_1; a)$ with $Inf(\beta) = 2p_2$. The information of the term α corresponds to the fact that during the execution of α there is at most one token in place p_2, while the information of β expresses the fact that during the execution of β place p_2 can obtain two tokens. Because terms α and β have different information, they are not equivalent. As a consequence of the difference of information, α can run concurrently with c, but β cannot. If the place p_2 had no capacity restriction, then α and β would be equivalent according to the distributive law and α and β would represent the same run.

10   Conclusion

There are several approaches to unifying Petri nets (see e.g. [22, 20, 21, 16]). They enable to unify different classes of Petri nets which use different underlying algebras and different treatment of data-type part, defining them as formal parameters which can be actualized by choosing an appropriate structure. However, in these approaches enabling condition of the occurrence rule is not a parameter, but it is fixed. Both definitions in [20, 16] capture elementary nets but they let open more complicated restrictions of enabling condition in occurrence rule, such as inhibitor arcs or even capacities.

In our paper we have focused on unified description of Petri nets with modified occurrence rule. Namely, we have described a unifying approach to non-sequential semantics of Petri nets with modified occurrence rule. We have demonstrated that methods of partial algebra, such as greatest closed congruence, represent a suitable mathematical tool for such an approach. By restricted domains of operations we were able to generate precisely just those runs of the net which are allowed. In comparison with methods based on partial order where concurrency is defined implicitly if there is no causal connection between runs, we define explicitly when runs can be composed concurrently. Thus, in our approach causality is defined using two partial operations to generate runs, namely concurrent and sequential composition.

On the other hand, we did not discuss unifying of data type part. So, we did not discuss high-level Petri nets in this paper. There are also other restrictions of the occurrence rules in various high-level nets (e.g. transition guards, time intervals, roles etc.) which are of different characters and were not discussed in the paper. It would be interesting to discuss those kinds of restrictions in order to see the implication of the unifying approach for high-level nets. Namely, it would be interesting to combine the approach presented in [22] and the approach presented in this paper.

The presented approach opens many interesting questions. We can further distinguish between synchronous and concurrent occurrences of transitions. In such an extension of our approach one first needs to generate steps from transitions using a partial operation of synchronous composition and then to use this steps to generate process terms using partial operations of concurrent and sequential composition. In terms of causal relationships, such an extension corresponds to the approach described in [13, 17], where two kinds of causalities are defined, first saying (as usual) which transitions cannot occur earlier than others, while the second indicating which transitions cannot occur later than others. In [13, 17] the principle is illustrated for a variant of nets with inhibitor arcs, where testing for zero precedes the execution of a transition. Thus, if a transition t tests a place for zero, which is in a post-set of another transition t', this means that t cannot occur later than t' and therefore they cannot occur concurrently - but still can occur synchronously. There are also other net extensions employing steps of transitions (distinguishing between synchronous and concurrent composition), such as nets with asymmetric synchronization [12]. We are currently working on the extension of our approach using a partial operation for synchronous composition to cover such cases.

Another area of further research is to investigate whether the presented framework would lead to a unifying and mathematically elegant way of producing the causal semantics for nets with restricted occurrence rule. Namely, as it was discussed in Introduction, any process term defines naturally a partial order of events labeled by transitions. Thus, an equivalence class of process terms defines a set of partial orders. As we have illustrated in the example from Introduction, one can modify these partial orders comparing each other and removing causalities which are not defined by the net itself. The idea for further research is to generalize this modification procedure in order to obtain the set of partial orders containing only those causalities which are given by the net itself.

References

1. E. Best and R. Devillers. Sequential and Concurrent Behaviour in Petri Net Theory. *Theoretical Computer Science*, 55, pp. 87–136, 1987.
2. R. Bruni, J. Meseguer, U. Montanari and V. Sassone A Comparison of Petri Net Semantics under the Collective Token Philosophy. *Proc. of ASIAN 1998*, Springer, LNCS 1538, pp. 225–244, 1998.
3. R. Bruni and V. Sassone Algebraic Models for Contextual Nets. *Proc. of ICALP 2000*, Springer, LNCS 1853, pp. 175–186, 2000.
4. P. Burmeister. Lecture Notes on Universal Algebra – Many Sorted Partial Algebras. Technical Report, TU Darmstadt, 1998.
5. E. Degano, J. Meseguer and U. Montanari. Axiomatizing the Algebra of Net Computations and Processes. *Acta Informatica*, 33(7), pp. 641–667, 1996.
6. J. Desel, G. Juhás and R. Lorenz. Process Semantics of Petri Nets over Partial Algebra. In M. Nielsen and D. Simpson (Eds.) *Proc. of 21th International Conference on Application and Theory of Petri Nets*, Springer, LNCS 1825, pp. 146–165,2000.
7. J. Desel, G. Juhás and R. Lorenz. Petri Nets over Partial Algebra. To appear in H. Ehrig, G. Juhás, J. Padberg and G. Rozenberg (Eds.) *Unifying Petri Net,* Advances in Petri Nets, Springer, LNCS, 2001.
8. J. Desel and G. Juhás. What is a Petri Net? To appear in H. Ehrig, G. Juhás, J. Padberg and G. Rozenberg (Eds.) *Unifying Petri Net,* Advances in Petri Nets, Springer, LNCS, 2001.
9. J. Desel and W.Reisig. Place/Transition Petri Nets. In *Lectures on Petri nets I: Basic Models*, LNCS 1491, pp. 123–174,1998.
10. R. Devillers. The Semantics of Capacities in P/T Nets. In *Advances in Petri Nets 1989*, LNCS 424, pp. 128–150,1990.
11. F. Gadducci and U. Montanari. Axioms for Contextual Net Processes. In *Proc of. ICALP'98*, Springer, LNCS 1443, pp. 296-308, 1998.
12. H-M. Hanisch and A. Lüder A Signal Extension for Petri nets and its Use in Controller Design. To appear in Fundamenta informaticae, 2000.
13. R. Janicki and M. Koutny. Semantics of Inhibitor Nets. *Information and Computations*, 123, pp. 1–16, 1995.
14. G. Juhás. Reasoning about algebraic generalisation of Petri nets. In S. Donatelli and J. Klein (Eds.) *Proc. of 20th International Conference on Application and Theory of Petri Nets*, Springer, LNCS 1639, pp. 324-343, 1999.
15. G. Juhás. Petri nets over partial algebra. In J. Pavelka, G. Tel and M. Bartosek (Eds.) *Proc. of 26th Seminar on Current Trends in Theory and Practice of Informatics SOFSEM'99*, Springer, LNCS 1725, pp. 408-415, 1999.
16. E. Kindler and M. Weber. The Dimensions of Petri Nets: The Petri Net Cube. *EATCS Bulletin*, No. 66, pp. 155-166, 1998.
17. H.C.M. Kleijn, M. Koutny. Process Semantics of P/T-Nets with Inhibitor Arcs. In M. Nielsen and D. Simpson (Eds.) *Proc. of 21th International Conference on Application and Theory of Petri Nets*, Springer, LNCS 1825, pp. 261–281,2000.

18. J. Meseguer and U. Montanari. Petri nets are monoids. *Information and Computation*, 88(2):105–155, October 1990.
19. U. Montanari and F. Rossi. Contextual Nets. *Acta Informatica*, 32(6), pp. 545–596, 1995.
20. J. Padberg. *Abstract Petri Nets: Uniform Approach and Rule-Based Refinement*, Ph.D. Thesis, TU Berlin, Germany, 1996.
21. J. Padberg. Classification of Petri Nets Using Adjoint Functors *Bulletin of EACTS* No. 66, 1998.
22. J. Padberg and H. Ehrig. Parametrized Net Classes: A uniform approach to net classes. To appear in H. Ehrig, G. Juhás, J. Padberg and G. Rozenberg (Eds.) *Unifying Petri Net*, Advances in Petri Nets, Springer, LNCS, 2001.
23. G. Rozenberg, and J. Engelfriet. Elementary Net Systems. In W. Reisig and G. Rozenberg (Eds.) *Lectures on Petri Nets I: Basic Models*, Springer, LNCS 1491, pp. 12-121, 1998.
24. V. Sassone An Axiomatization of the Category of Petri Net Computations. *Mathematical Structures in Computer Science*, vol. 8, pp. 117–151, 1998.
25. J. Winkowski. Behaviours of Concurrent Systems. *Theoretical Computer Science*, 12, pp. 39–60, 1980.
26. J. Winkowski. An Algebraic Description of System Behaviours. *Theoretical Computer Science*, 21, pp. 315–340, 1982.

Parameterized Net Classes:
A Uniform Approach to Petri Net Classes

Julia Padberg and Hartmut Ehrig*

Technische Universität Berlin,
{padberg,ehrig}@cs.tu-berlin.de

Abstract The concept of *parameterized net classes* is introduced in order to allow a uniform approach to different kinds of Petri net classes. By different actualizations of the *net structure parameter* and the *data type formalism parameter* we obtain several well-known net classes, like elementary nets, place-transition nets, colored nets, predicate transition nets, and algebraic high-level nets, as well as several interesting new classes of low- and high-level nets. First the concept of *parameterized net classes* is defined on a purely set theoretical level, subsequently we give the concepts taking into account also morphisms and universal properties in the sense of category theory. We explain the underlying notions in an intuitive way. Moreover we give extracts from two of our case studies, where the application of these notions are illustrated in specific net classes, i.e. in instantiations of the parameterized net class.

The formal foundation of parameterized net classes this the uniform theory of abstract Petri nets. Low-level abstract Petri nets are a special case of high-level abstract Petri nets, but for better understanding they are presented separately. The theory of abstract Petri nets yields sufficient concepts and results for a specification technique of parameterized net classes. Operational behavior of nets is so presented in a uniform way. Different notions of horizontal structuring, rule-based refinement and their compatibility become available. The horizontal structuring techniques comprise union and fusion of nets. Last but not least we present some examples from our case studies using the notions and results introduced in this paper.

Keywords: Petri Nets, high-level nets, actual and formal parameter, uniform approach, union, fusion, rule-based refinement

1 Introduction

Petri nets have been used successfully for more than three decades to model concurrent processes and distributed systems. Various kinds of Petri net classes with numerous features and analysis methods have been proposed in literature

---

* This work is part of the joint research project "DFG-Forschergruppe PETRI NET TECHNOLOGY" between H. Weber (Coordinator), H. Ehrig (both from the Technical University Berlin), and W. Reisig (Humboldt University Berlin), supported by the Deutsche Forschungsgemeinschaft (DFG).

H. Ehrig et al. (Eds.): Unifying Petri Nets, LNCS 2128, pp. 173–229, 2001.
© Springer-Verlag Berlin Heidelberg 2001

(see e.g. the following surveys [PR91, JR91, RR98b, RR98a, WER99]) for different purposes and application areas. The fact that Petri nets are widely used and are still considered to be an important topic in research, shows the usefulness and the power of this formalism. Nevertheless, the situation in the field of Petri nets is far from satisfactory, partly due to the enormous interest in Petri nets, that has been leading to a vast accumulation of dissimilar approaches. The different notions, definitions and techniques, both in literature and practice, make it hard to find a common understanding and to provide good reasons for the practical use of the nets. Moreover, the unstructured variety of Petri net approaches causes the new formulation and examination of similar concepts. The relation of these concepts requires complicated, but boring conversions. Most of the different concepts for Petri nets are defined explicitly for a single net class, although many of these notions are essentially the same for different kinds of net classes. Since the mid-nineties abstract notions of Petri net have become an increasingly important issue in order to permit an abstract formulation of such notions for a large variety of different net classes. This volume itself is the first comprehensive collection of such approaches. This uniform approach to Petri net classes captures the common components of different kinds of Petri nets like places, transitions, net structure, and – in case of high-level nets – a data type part. Moreover, this approach treats low- and high-level nets in the same way, considering a trivial data type for low-level nets. General notions, like firing behavior being essential for all kinds of Petri nets are formulated in the frame of abstract Petri nets independently of their specific definition within a fixed net class. We do not consider this uniform approach as a net formalism for application purposes. Nevertheless, such an approach allows the easy transfer of results between different Petri net formalisms and thus has an impact on Petri nets used in practice. Hence, this concept comprises many known and several new net classes as special cases and allows their mutual comparison. Results achieved within this frame can be generalized to several different net classes. This means notions and results are achieved without further effort in each of these net classes, provided the general assumptions have been verified for the specific instance.

This paper is organized as follows. The first part is *Part I: Introduction to Parameterized Net Classes*. In Section 2 we present a purely set theoretic description of this uniform approach first introduced in [EP97]. This includes the definition of the net structure parameter and the data type formalisms parameter. Then in Section 3 we extend these parameters into a categorical frame that allows the precise definition of the formal and actual parameters. We explain the used categorical concepts in a way that is easy to understand. Moreover we list our results and give an intuitive explanation. These notions and results are then illustrated in Section 4 with a few examples from our case studies (first presented in [Erm96, PGH99]). We show how the notions given for parameterized net classes are applied in concrete net classes, that are obtained by actualization of the formal net structure and data type formalism parameter. The second part presents *Part II: Theory of Parameterized Net Classes Based on Abstract*

Petri Nets. In Section 5 we give first the formal basis for parameterized net classes in terms of low-level abstract Petri nets, first introduced in [Pad96]. In this section we introduce the formal parameter for the net structure only. The formal parameter for the data type part is investigated subsequently in Section 6. High-level abstract Petri nets are presented in Section 7 together with the corresponding categorical results. We investigate horizontal and vertical structuring techniques and their compatibility results in Section 8. The conclusion summarizes the achieved results.

In Part I of the paper we do not assume any previous knowledge of category theory. On the contrary we carefully explain the use of the categorical notions in the area of Petri nets and present the main results on a conceptual level. In Part II we present the mathematical precise version of the results. These, the review of high-level replacement systems in Appendix A, and the proofs of the main results in Appendix B assume some basic knowledge in category theory.

Related Work

In [MM90] a basis for various approaches to a uniform description of Petri nets has been provided. There the marking graph of a place/transition net (understood as directed graph over a commutative monoid) is constructed by symmetric and additive closure of the category of place/transition nets. In [EPR94b, Pad96, EP97] the above idea has been developed further using adjoint functors between category of sets and a (sub)category of commutative semigroups. Moreover, a suitable treatment based on institutions is given for the data type part of high-level nets. This paper gives a comprehensive survey over this approach.

The transfer of this parameterization concept to nets closely related to FUN-SOFT [DG91, DG94, DG98], a net class well-known in industrial practice, has been suggested in [EGLP97, GL98, GL99]. An extension to partial algebras has been extensively investigated in [Juh98b, Juh99]. This approach allows the treatment of relaxed enabling rules. In [DM93] positive cones of Abelian groups are used to relate Petri nets in a uniform way to automata. In [BD95, BD98] non-algebraic parameterizations of Petri nets have been achieved by a characterization of automata determined by different Petri net types. The use of rewriting logic for the unification of Petri nets has been investigated in [Mes92]. Another line of research follows the idea to present abstract description of Petri nets and their semantics, for example [GR83, Win87, MM90, Sas98]. The classification of Petri nets and their extensions as investigated in [BDC92, DA94, KW98] is an important basis for the identification of further parameters. In [KW98] an approach has been introduced that focuses on describing orthogonal dimensions of Petri net notions. These dimensions can be considered as parameters as well.

Since the sole topic of this book concerns unifying Petri nets, related work concerning parameterization of Petri nets is also presented in this volume. A first step towards this work has been presented in [EPR94b] (a revised version is [PER01]). The extension to partial algebras is presented in [DJL01]. [DS01] is based on parameters as well but that contribution concentrates on low-level nets and transition systems. In [BBD01] a parameterization is based on different automata and their representations as Petri nets.

Part I:
Introduction to Parameterized Net Classes

In Part I we give a set theoretical and also a categorical introduction to parameterized net classes, summarize the main results at a conceptual level, and discuss two relevant applications of parameterized net classes.

2   Set-Theoretical Approach
to Parameterized Net Classes

The basic idea of this uniform approach to Petri nets is to identify two parameters, that describe each of the net classes entirely. In case of the usual Petri nets this is the net structure and in case of high-level nets it is the net structure and the data type formalism. We call these parameters net structure parameter and data type (formalism) parameter. For convenience we often use the expression data type parameter instead of data type formalism parameter. Nevertheless we convey the formalism, not a specific data type. The instantiation of these parameters leads to different types of Petri nets, or more precisely Petri net classes. In this section we introduce parameters for net classes, parameterized net classes and their instances leading to several well-known and some interesting new net classes. In more detail this work is presented in [PER01].

2.1   Relevance of Parameters for Net Classes

The net structure parameter describes different low-level net classes. Several different net classes have been proposed over the last 40 years. Moreover, the developments in software industry have yielded quite a large amount of variants that are equipped with additional features and/or restrictions. We propose an abstraction of net structure that can be instantiated to several net classes, including place/transition nets, elementary nets, variants of these and S-graphs. We have shown in [Pad96] that the underlying construction is general enough to comprise several different kinds of low-level nets and their net structure. The data type parameter is necessary, because several data type formalisms have been integrated with Petri nets leading to different notions of high-level nets. Typical examples are: predicate logic with elementary nets leading to predicate/transition nets [GL81], algebraic specifications with place/transition nets leading to different versions of algebraic high-level nets [Vau86, Rei91], ML with place/transition nets leading to colored nets [Jen92], OBJ2 with superposed automata nets leading to OBJSA-nets [BCM88], and algebraic specifications with the Petri Box Calculus [BDH92] leading to A-nets [KP95]. In practice, there

are also other data type formalisms like entity/relationship diagrams, SADT
and many other semi-formal techniques that are combined with Petri nets in an
informal way.

2.2   Algebraic Presentation of Place/Transition Nets

We use the algebraic presentation of Petri nets, that uses functions to relate
a transition with its pre- and post-domain. This approach relates a transition
on the one hand with all those places in its pre-domain using the function *pre*
and on the other hand with its post-domain using the function *post*. In this
algebraic presentation a place/transition net N is simply given by a 4-tuple
$N = (P, T, pre, post)$ where P and T are the sets for places and transitions respectively and *pre* and *post* are functions $pre, post : T \longrightarrow P^\oplus$ from T to the
free commutative monoid $P^\oplus$ over P. This construction is similar to the construction of words over some given alphabet. Due to the axiom of commutativity
the elements of the free commutative monoid are considered to be linear sums
over P, that is for each $t \in T$ we have $pre(t) = \sum_{p \in P} n_p * p$ for $n_p \in \mathbb{N}$. Note,
that this is just the same as multisets. The marking is given by some element
$m \in P^\oplus$ and the operations for the computation of the firing behavior are comparison, subtraction and addition based on linear sums, defined over the monoid
operation. This algebraic presentation [MM90] is equivalent to the classical presentation (see e.g. [Rei85,ER96]), but has the advantage of a clear and axiomatic
presentation, thus it is much simpler to generalize.

2.3   Algebraic Presentation of Elementary Nets

In the case of elementary nets[1] the algebraic presentation is given by the power
set construction $\mathcal{P}(P)$, that is $pre, post : T \longrightarrow \mathcal{P}(P)$, because $^\bullet t = pre(t)$
and $t^\bullet = post(t)$ are given by subsets of P. Moreover, each element $m \in \mathcal{P}(P)$
can be considered as a marking of the elementary net. The firing behavior makes
use of the order on sets and the operations union and complement of sets.

2.4   Variants of Net Classes

Note, that there are several variants of place/transition nets, and similar for
other types of nets, where nets are considered with initial marking or with labels
on places, transitions, and/or arcs. However, in this paper we only consider a
basic variant without initial markings and without labels. The neglect of the
initial marking is due to our focus on structural composition techniques. The
composition of nets without initial marking yields techniques as union and fusion.
These techniques are independent from the behavior of the net. The composition
of nets with initial marking yields techniques as substitution and invocation,
where the composition is dependent from the behavior of the net. In the case

---

[1] We talk about elementary nets as elementary (net) systems without an initial marking.

of high-level nets our basic variant means in addition that a net includes one explicit model with total operations (resp. predicates) and that there are no firing conditions for the transitions (unguarded case). In [ER96] several kinds of variants of algebraic high-level nets are discussed that can be considered as well.

2.5   Mathematical Notation of Parameters for Net Classes

In the following we distinguish between formal and actual parameters for the net structure and data type formalisms. The actual net structure parameter for a net class is based on the algebraic presentation of nets, more precisely the codomain of the pre- and post-domain functions. The algebraic presentation of different kinds of Petri nets as the actual parameter allows the generalization in an axiomatic way, that is the formal parameter. Hence, it is the basic construction in order to express the difference of an actual and a formal parameter in an uniform approach to Petri nets.

For place/transition nets (in Subsection 2.2) the codomain uses the construction of the free commutative monoid $P^\oplus$ over the set of places P. For elementary nets (in Subsection 2.3) the power set construction is used. The calculation with markings is based on the operations union of $\mathcal{P}(P)$ and addition of $P^\oplus$ respectively. In order to generalize this computation a semigroup structure is employed in both classes. Hence, the constructions $\mathcal{P}(P)$ and $P^\oplus$ for each set P can be considered as functions from the class **Sets** of all sets via some class **Struct** of semigroups to the class **Sets**. These constructions are used as the actual parameter Net for the parameterized net classes. We consider $\mathcal{P}(P)$ and $P^\oplus$ as sets. The use of sets instead of semigroups allows the mapping from the transitions to these sets. Moreover, this has the advantage to consider nets, where the structure of the marking is different from the structure of the pre- and post-domain of the transitions, as for example in S-graphs, where markings contain multiple tokens, but the arc weight always equals one (see Example 1.4).

This motivates that in general an actual net structure parameter for a net class can be considered as the composition of two functions: $Net : $ **Sets** $\xrightarrow{\ F\ }$ **Struct** $\xrightarrow{\ G\ }$ **Sets**.

Based on the function Net we can describe Petri nets uniformly by $pre, post : T \longrightarrow Net(P)$, where the specific net class depends on the choice of the function Net. Then $F(P)$ denotes the markings and the pre- and post-domain of the transitions, $F(T)$ yields transition vectors, and G relates the used construction (i.e. free monoids, power sets) with sets.

For high-level net classes we use the notion of institutions (see [GB84,ST84]), which is well-established in the area of abstract data types. Institutions are an abstract description of data type formalisms and generalize different formalisms, as algebraic specifications, predicate logic, functional programming languages, and so on. The basic idea is to assume axiomaticly some specification $SPEC$ and a class of models **Mod(SPEC)**. Based on this theory an actual data type parameter for a net class consists of a class **SPEC** of data type specifications and for each $SPEC \in$ **SPEC** a class **Mod(SPEC)** of models satisfying the specification $SPEC$. Hence, it can be represented by a function $Mod : \textbf{SPEC} \longrightarrow \textbf{ModelClasses}$ where $\textbf{ModelClasses}$ is the (super)class of all model classes.

Concept: Formal Parameters of Net Classes   *The formal net structure parameter, respectively formal data type formalism parameter, of a parameterized net class is given a pair of functions* $Net : \textbf{Sets} \xrightarrow{F} \textbf{Struct} \xrightarrow{G} \textbf{Sets}$ *and a function* $Mod : \textbf{SPEC} \longrightarrow \textbf{ModelClasses}$ *respectively, as motivated above.*

Example 1. Actual Parameters of Net Classes

1. The free commutative monoid $P^{\oplus}$ over a set P defines the two functions $(\_)^{\oplus} : \textbf{Sets} \longrightarrow \textbf{Struct} \longrightarrow \textbf{Sets}$, where $P^{\oplus}$ together with the addition operation on linear sums is a semigroup (without this operation $P^{\oplus}$ is a set, thus the function G simply "forgets" the addition), which is the actual net structure parameter for the class of place/transition nets. The free commutative monoid $P^{\oplus}$ can be represented by formal sums $P^{\oplus} = \{\sum_{p \in P} n_p * p \mid n_p \in \mathbb{N}\}$ with component-wise addition.

2. Analogously the free Abelian groups P^* over a set P determine an actual net parameter where we obtain a specific class of place/transition net, that allows negative tokens.

3. The powerset construction $\mathcal{P}(P)$ over a set P defines two functions $\mathcal{P} : \textbf{Sets} \longrightarrow \textbf{Struct} \longrightarrow \textbf{Sets}$, where $\mathcal{P}(P)$ with union operation is semigroup (without the union operation $\mathcal{P}(P)$ is a set), which is the actual net structure parameter for the class of elementary nets.

4. The actual net structure parameter for the subclass of place/transition nets, called S-Graphs (see [RT86]), where each transition has exactly one place in its pre- and postdomain respectively makes use of the compositionality of the function Net. The corresponding net can be considered as graph, where the nodes are the places and the edges are the transitions, that is transitions are mapped to places by $pre, post : T \longrightarrow P$. Nevertheless, markings are elements $m \in P^{\oplus}$ (as usual in place/transition nets) rather than $m \in P$, which would allow only one token at all in the net, thus the intermediate construction has to be the free commutative monoid $P^{\oplus}$. This is expressed by the pair of functions $SG : \textbf{Sets} \longrightarrow \textbf{Struct} \longrightarrow \textbf{Sets}$, defined by $SG : X \mapsto X^{\oplus} \mapsto X$ for each set X.

5. Let $\textbf{SPEC}$ be the class of all algebraic specifications $SPEC = (S, OP, E)$ with signature (S, OP) and equations E and $\textbf{Alg}(\textbf{SPEC})$ the class of all $SPEC$-algebras A (see [EM85]). Then we obtain a function $Alg : \textbf{SPEC} \longrightarrow \textbf{ModelClasses}$ which is the actual data type parameter for the class of algebraic high-level nets ([PER95]).

6. Let $\textbf{FOSPEC}$ be the class of all first order predicate logic specifications $FOSPEC = (\Omega, \Pi, AXIOMS)$ with the signature (Ω, Π) and $AXIOMS$ being a set of closed formulas, and $\textbf{FOMod}(\textbf{FOSPEC})$ the class of all non-empty models satisfying the formulas in $AXIOMS$. Then we obtain a

function $FOMod$: **FOSPEC** $\longrightarrow$ **ModelClasses**, which is the actual data type parameter for the class of predicate/transition nets ([GL81]).

7. As a uniform approach to Petri nets should comprise low- as well as high-level nets, we consider low-level nets as a special case of high-level nets with a data type that yields only one data element, the usual black token. This is merely a technical extension, because in [Pad96] it has been shown, that these high-level nets with a trivial data type correspond one-to-one to the usual low-level nets. The great advantage is that this conception allows to consider low- and high-level nets within one uniform approach. In order to be able to define low-level nets as special case of high-level nets we define the following trivial actual data type parameter $Triv$: **TRIV** $\longrightarrow$ **ModelClasses**, where the class **TRIV** consists only of one element called trivial specification $TRIV$, and **Triv(TRIV)** is the model class, consisting only of one model, the one-element set $\{\bullet\}$, representing a single token.

8. There are also actual data type parameters for the class of colored nets in the sense [Jen92], which is based the functional language ML (see [Pad96]).

Note, that the theory of institutions allows to treat different data type descriptions in the same way, but does not neglect the differences. This is due to the abstract formulation of this theory. Our uniform approach to Petri nets is motivated by this abstraction.

Now we are able to define parameterized net classes and their instantiations mentioned above.

Though parameterized net classes cannot yield concrete nets, as the formal parameters are not yet actualized, they give rise to abstract Petri nets. Abstract Petri nets constitute a pattern for Petri nets consisting of places, transitions, pre- and post-domain, specification and a data type model. Nevertheless, neither the structure of pre- and post-domain is fixed – due to the net parameter – nor the kind of the specification and its model – due to the data type parameter.

Parameterized Net Classes
*A parameterized net class is defined by a formal net structure parameter (Net, Mod) with $Net = G \circ F$ and Mod : **SPEC** $\longrightarrow$ **ModelClasses** consists of all abstract Petri nets $N = (P, T, SPEC, A, pre, post)$ satisfying the following conditions:*

- P *and* T *are sets of places and transitions respectively,*
- $SPEC \in$ **SPEC** *is the data type specification*
- $A \in$ **Mod(SPEC)***, called data type model*
- $pre, post : T \longrightarrow Net(T_{SPEC} \times P)$ *are the pre- and post-domain functions,*
  where T_{SPEC} is a distinguished model with respect to the specification SPEC (e.g. where the elements are congruence classes of terms over the specification SPEC).

In the case of low-level nets we have $Mod = Triv$ (see Example 1.7) and hence **SPEC** $=$ **TRIV** *and* $A = \{\bullet\}$ *which are omitted as components of a net.*

Since T_{SPEC} consists of a single element, we obtain $N = (P, T, pre, post)$ with $pre, post : T \longrightarrow Net(P)$

Other parameterized and actual net classes can be defined by other variants of net classes discussed in example 1. In our notion above we only consider the basic variant without initial markings, without labels, with only one explicit (total) model and without firing conditions for transitions.

The behavior can be given already for the abstract Petri nets, although the abstract Petri net is no "real net" unless the formal parameters are actualized. The behavior of abstract Petri nets yields a uniform description of the behavior in different Petri net classes: The firing of a transition vector can be defined axiomaticly using the pair of functions $Net :$ **Sets** $\xrightarrow{F}$ **Struct** $\xrightarrow{G}$ **Sets**, the operation $+$, given by the semigroup structure, and the extensions $\overline{pre}$ (resp. $\overline{post}$) of pre (resp. $post$) to the semigroup structure. The marking is given by $m \in F(P)$. A transition vector $v \in F(T)$ is enabled under $m \in F(P)$ if there exists $\hat{m} \in F(P)$ so that $m = \hat{m} \oplus \overline{pre}(v)$. The follower marking $m' \in F(P)$ obtained by firing v under m is given by $m' = \hat{m} \oplus \overline{post}(v)$

The firing of high-level nets involves additionally the assignment of values to variables (see Section 7).

Example 2. Actual Net Classes A survey of actual net classes defined by the actual parameter given in example 1 is given in the table below, where well-known net classes are shown as explicit entries in the table while each blank entry corresponds to a new or less well-known net class.

Mod / *Net*	Triv	Alg. Spec.	Pred. Logic	ML	Indexed Sets
powerset	Elem. Nets [RT86]	*NEW*	PrT-Nets [GL81]	*NEW*	*NEW*
free c. monoid	P/T-Nets [Rei85]	AHL-Nets [PER95]	*NEW*	CPN '92 [Jen92]	CPN '81 [Jen81]
free A. group	neg. P/T-net	*NEW*	*NEW*	*NEW*	*NEW*
monoid identity	S-Graph [RT86]	*NEW*	*NEW*	*NEW*	*NEW*

In more detail we have:

- Place/Transition nets defined by $Net = (\_)^{\oplus}$ (see Example 1.1) are given by $N = (P, T, pre, post)$ with $pre, post : T \longrightarrow P^{\oplus}$ (see Subsections 2.2 and 2.5 as well as [PER01]).
- Elementary nets defined by $Net = \mathcal{P}$ (see Example 1.3) are given by $N = (P, T, pre, post)$ with $pre, post : T \longrightarrow \mathcal{P}(P)$ (see Subsection 2.5 as well as [PER01]).
- S-graphs defined by $Net = SG$ (see Example 1.4) are given by $N = (P, T, pre, post)$ with $pre, post : T \longrightarrow P$.

- AHL-nets defined by $Net = (\_)^\oplus$ and $Mod = Alg$ (see Example 1.5) are given by $N = (P, T, SPEC, A, pre, post)$ with $pre, post : T \longrightarrow (T_{OP}(X) \times P)^\oplus$. Here the distinguished model is the quotient term algebra $T_{OP}(X)$, that is $T_{OP}(X)$ are the terms with variables X over $SPEC$.
- For a presentation of other high-level nets see Section 7.

3 Categorical Approach to Parameterized Net Classes

The main concepts of parameterized net classes is now expressed in terms of category. We extend the notion of parameterized net classes studied in the previous section by morphisms on different levels. Hence, the classes become categories and the corresponding functions become functors in the sense of category theory [AHS90]. First we motivate the benefits of morphisms for Petri nets. For this purpose we first review net morphisms for place/transition nets in algebraic presentation (see Subsection 2.2). Then we discuss the benefit of morphisms in this case which is also valid for other types of net classes. Finally, we discuss general constructions and results which have been obtained in this framework.

3.1 Introduction to Categorical Concepts

Category theory is a universal formalism which is successfully used in several fields of mathematics and theoretical computer science. It has been developed for about 50 years and its influence can be found in most branches of structural mathematics and, for about 25 years in several areas of theoretical computer science. In the survey [EGW96] it has been shown that the following areas in computer science have been influenced by category theory: Automata and system theory, flow charts and recursion, λ-calculus and functional languages, algebraic specifications, logical systems and type theory, graph transformation, Petri nets and replacement systems. The aim of category theory is to present structural dependencies and universal notions that can be found in many (mathematical) areas and to give a uniform frame independently of internal structures. This uniform frame and the universality of the concepts distinguish category theory as a common language for the modeling and analysis of complex structures, as well as for a unified view of the development of theories, and for the integration of different theories within computer science. The main purpose of category theory is to have a uniform frame for different kinds of mathematical structures, mappings between structures, and constructions of structures. The most fundamental notions in category theory are on the one hand *categories* consisting of *objects* and *morphisms* and on the other hand *functors* defining structure compatible mappings between categories. Another important concept of category theory is that of limits and colimits, especially product, equalizers, and pullbacks and the dual concepts of coproducts, coequalizers and pushouts. In [Pad96] we especially need pushouts and coequalizers corresponding to a union of objects with shared subobjects and the fusion of subobjects, respectively. Universal properties express that these constructs are generated. These universal properties imply that the corresponding construction is essentially unique. They are strongly exploited for the results we obtain in [Pad96]. Special colimits as pushouts and coequalizers are the basis for the just mentioned structuring techniques. Injective pullbacks correspond to the intersection of nets. Furthermore, the notions and results for rule-based modification depend on these constructions. The marking graph construction and the realization construction are derived from the net structure, and the preservation of colimits by free constructions yields the compatibility of the marking graph construction with the structuring techniques (see [Pad96]).

3.2 Morphisms of Place/Transition Nets

Given two place/transition nets N_1 and N_2 with $N_i = (P_i, T_i, pre_i, post_i)$ for $(i = 1, 2)$ (see algebraic presentation in 2.2) a morphism $f : N_1 \longrightarrow N_2$ of place/transition nets is a pair $f = (f_P : P_1 \longrightarrow P_2, f_T : T_1 \longrightarrow T_2)$ of functions, such that the diagram aside commutes for pre- and post-domain functions respectively.

$$
\begin{array}{ccc}
T_1 & \overset{pre_1}{\underset{post_1}{\rightrightarrows}} & P_1^{\oplus} \\
f_T \downarrow & & \downarrow f_P^{\oplus} \\
T_2 & \overset{pre_2}{\underset{post_2}{\rightrightarrows}} & P_2^{\oplus}
\end{array}
$$

This means that we have $f_P^{\oplus} \circ pre_1 = pre_2 \circ f_T$ and $f_P^{\oplus} \circ post_1 = post_2 \circ f_T$ where $f_P^{\oplus} : P_1^{\oplus} \longrightarrow P_2^{\oplus}$ is defined by $f_P^{\oplus}(\sum_{p \in P_1} n_p * p) = \sum_{p \in P_1} n_p f_P(p)$.

Example 3. Example of Union in Low-Level Nets

This example illustrates the union of place/transition nets of N_1, and N_2 with the interface N_0 resulting in N_3 (that is $N_3 = N_1 +_{N_0} N_2$). The net N_3 consists of the subnet N_1 (the grey and lighter grey colored part) and the subnet N_2 (the grey and darker grey colored part) sharing the common subnet N_0 (the grey colored part).

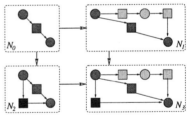

3.3 Benefits of Morphisms for Petri Nets

Similar to Subsection 3.2 for place/transition nets morphisms can also be defined for all other kinds of Petri nets, including low-level and high-level nets. The main benefits – illustrated using place/transition nets – are the following:

1. A morphism $f : N_1 \longrightarrow N_2$ of nets allows to express the structural relationship between nets N_1 and N_2. If f is injective (in all components) then N_1 can be considered as a subset of N_2. In general f may map different places p_1 and p_1' or transitions t_1 and t_1' of N_1 to only one place p_2 or one transition t_2 of N_2. Then only a homomorphism image of N_1 is a subnet of N_2. In fact, there may be different morphisms $f, g : N_1 \longrightarrow N_2$ which

corresponds to different occurrences of the net N_1 in the net N_2. In Example 3 all morphisms are injective such that N_0 can be considered as subnet of N_1 via f_1 and of N_2 via f_2. Moreover, N_1 and N_2 can be considered as subnet of N_3 via g_1 and g_2 respectively.

2. A bijective morphism $f : N_1 \xrightarrow{\sim} N_2$ is called isomorphism. In this case the nets N_1 and N_2 are called isomorphic, written $N_1 \cong N_2$, which means that they are equal up to renaming of places and transitions.

3. The composition of net morphisms $f_1 : N_1 \longrightarrow N_2$ and $f_2 : N_2 \longrightarrow N_3$ is again a net morphism $f_2 \circ f_1 : N_1 \longrightarrow N_3$. Moreover, this composition is associative and for each net N there is an identity morphism $id_N : N \longrightarrow N$ such that we have $f_1 \circ id_{N_1} = f_1$ and $id_{N_2} \circ f_1 = f_1$ for all $f_1 : N_1 \longrightarrow N_2$. This means that the class of all nets together with all net morphisms constitutes a category. This allows to apply constructions and results from category theory to different types of nets and net classes. Note, that each pair (Net, Mod) of actual parameters defines an actual net class (see Example 2) which is the object class of the corresponding category.

4. Morphisms can be used to define the horizontal structuring of nets, for example the net N_3 in Example 3 as union of N_1 and N_2 via the common subnet N_0. Vice versa, the nets N_1 and N_2 with subnet N_0 (distinguished by morphisms $f_1 : N_0 \longrightarrow N_1$ and $f_2 : N_0 \longrightarrow N_2$) can be composed leading to net $N_3 = N_1 +_{N_0} N_2$. In fact, this union of nets is also a pushout construction in the corresponding category. This allows to apply general results of category theory like composition and decomposition properties of pushouts to the union construction of nets, for example associativity and commutativity of union up to isomorphism.

5. Morphisms can also be used to define refinement of nets. In several cases more general morphisms than those in Subsection 3.2 should be considered for this purpose. One simple generalization is to replace $f_P^\oplus : P_1^\oplus \longrightarrow P_2^\oplus$ generated by $f_P : P_1 \longrightarrow P_2$ by an arbitrary monoid homomorphism $\hat{f}_P : P_1^\oplus \longrightarrow P_2^\oplus$. This allows to map one place p_1 in P_1 to a sum of places, that is $p_1 \mapsto \sum_{p \in P_2} n_p * p$ for $p_{2_i} \in P_2$, which is important for refinement. In [PGH99] we introduce various morphisms that preserve safety properties in the sense of [MP92]. These are illustrated in the examples in Section 4.

6. A morphism $f : N_1 \longrightarrow N_2$ of place/transition nets preserves the firing behavior: If transition vector $v \in F(T)$ is enabled under marking m in net N_1 leading to marking m', that is $m[v > m'$ then also the transition $f_T(v)$ is enabled under marking $f_P^\oplus(m)$ in net N_2 leading to marking $f_P^\oplus(m')$, that is $f_P^\oplus(m)[f_T(v) > f_P^\oplus(m')$. In a similar way morphisms preserve the firing behavior also for other types of nets. Specific kinds of net morphisms can be considered to preserve other kinds of Petri net properties, for example deadlock-freeness. Especially, isomorphisms preserve all kinds of net properties which do not depend on a specific notation.

3.4   Parameters of Net Classes Based on Functors

The parameters of net classes considered in Section 2 are expressed in terms of categories and functors instead of classes and functions. In more detail the classes of sets, structures, specifications, and model classes are extended by suitable morphisms leading to categories **Sets** of sets, **Struct** of structures, **SPEC** of specifications, and **ModelClasses** of model-classes, and for each $SPEC \in$ **SPEC** a category **Mod(SPEC)** of $SPEC$-models. Moreover, the functions are extended to become functors: $Net : Sets \xrightarrow{F} Struct \xrightarrow{G} Sets$, the net structure parameter and $Mod :$ **SPEC**$^{op} \longrightarrow$ **ModelClasses**, the data type (formalism) parameter. Note, that we use an overloaded notation, where **Sets**, **Struct**, **SPEC**, **ModelClasses**, and **Mod(SPEC)** denote classes and Net, F, G, and Mod denote functions in Section 2, while they denote categories and functors respectively in this section. In fact, all the examples of actual parameters of net classes given in Example 1 can be extended to universal parameters of net classes with well-known categories and functors (see Sections 5 and 7).

Parameterized classes have the actual parameters (Net, Mod) that are functors. In this case we have in addition to the abstract Petri nets $N = (P, T, SPEC, A, pre, post)$ of the corresponding parameterized net class also abstract Petri net morphisms $f : N_1 \longrightarrow N_2$ leading to the corresponding category. In the case of low-level net patterns $N_i = (P_i, T_i, pre_i, post_i)$ for $(i = 1, 2)$ of type (Net, Mod) an abstract Petri net morphism $f : N_1 \longrightarrow N_2$ is a pair of functions $f = (f_P : P_1 \longrightarrow P_2, f_T : T_1 \longrightarrow T_2)$ such that $pre_2 \circ f_T = Net(f_P) \circ pre_1$ and $post_2 \circ f_T = Net(f_P) \circ post_1$. In the special case $Net = (\_)^{\oplus}$ we obtain the notion of morphisms for place/transition nets (see Subsection 3.2). Moreover, these morphisms preserve the firing behavior of Petri nets.

The Formal Net Structure Parameter

The net structure parameter is given by the functor $Net : Sets \xrightarrow{F}$ $Struct \xrightarrow{G} Sets$ where $Net = G \circ F$ and the functor F is a *free functor* with respect to the *forgetful functor* G. We only consider the net structure parameter of place/transition nets in more detail (see Example 1.1):

Let **Struct** $=$ **CMon** *be the category of commutative monoids,* $F :$ **Sets** $\longrightarrow$ **CMon** *the free commutative monoid construction, that is* $F(P) = (P^{\oplus}, 0, \oplus)$, $G :$ **CMon** $\longrightarrow$ **Sets** *the forgetful functor, defined by* $G(M, \epsilon, \circ) = M$, *forgetting only about the neutral element* ϵ *and the monoid operator* $\circ$. *Then* $Net :$ **Sets** $\xrightarrow{F}$ **CMon** $\xrightarrow{G}$ **Sets** *with* $Net(P) = P^{\oplus}$ *is the universal net structure functor for the class of place/transition nets.*

In fact, F *is a free functor with respect to* G, *because for each set* P *the free construction* $F(P) = (P^{\oplus}, 0, \oplus)$ *together with the inclusion* $u_P : P \longrightarrow G \circ F(P) = P^{\oplus}$ *satisfies the following universal property: For each commutative monoid* $(M, \epsilon, \circ)$ *and each function* $f : P \longrightarrow G(M, \epsilon, \circ) = M$ *there is a unique monoid homomorphism* $\overline{f} : F(P) = (P^{\oplus}, 0, \oplus) \longrightarrow (M, \epsilon, \circ)$ *such that* $G(\overline{f}) \circ u_P = f$:

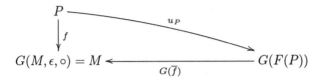

In fact, $\overline{f} : (P^{\oplus}, 0, \oplus) \longrightarrow (M, \epsilon, \circ)$ is uniquely defined by $\overline{f}(0) = \epsilon$ and $\overline{f}(\sum_{p \in P} n_p * p) = \sum_{p \in P} n_p * f(p)$. This universal property allows to extend the pre- and postdomain functions of place/transition nets $pre, post : T \longrightarrow P^{\oplus}$ – and similar for other types of nets – to monoid homomorphisms $\overline{pre}, \overline{post} : T^{\oplus} \longrightarrow P^{\oplus}$ and hence to parallel firing of transitions.

The Formal Data Type Formalism Parameter

The data type (formalism) parameter is given by $Mod :$ **SPEC**$^{op} \longrightarrow$ **ModelClasses** where Mod is a contravariant functor from **SPEC** to **ModelClasses** in the sense of category theory [AHS90]. We use the idea of specification frames [EBCO91, CBEO99] for the representation of the data type formalism parameter. According to the main concepts of algebraic specifications, for each signature there is a category **Cat**(Σ) of models. Also signatures and signature morphisms constitute a category. The categorical formulation of these concepts is given by a functor Cat that provides for each signature its category of models. Moreover, each signature morphism implies a forgetful functor in the opposite direction. More formally, for each signature morphism, $f_{\Sigma} : \Sigma_1 \longrightarrow \Sigma_2$ we can map models of Σ_2 to models of Σ_1, that is $V_{f_{\Sigma}} :$ **Cat**$(\Sigma_2) \longrightarrow$ **Cat**(Σ_1). For each model $M_2 \in$ **Cat**(Σ_2) there is $V_{f_{\Sigma}}(M_2) \in$ **Cat**(Σ_1), the model M_2 restricted to the syntax of Σ_2. This construction is generalized by a suitable (contravariant) functor $Cat :$ **ASIG**$^{op} \longrightarrow$ **CATCAT**. In fact **CATCAT** is not a proper category, but only a quasi-category in the sense of [AHS90] (see Definition 3.49 there). This basic idea is extended to obtain suitable data type formalisms parameter for high-level abstract Petri nets. This extension involves a natural transformation, which can be regarded as mapping of functors. This mapping is given by a family of morphisms, relating the target objects of both functors. We use this concept to relate the model of the data type signature with the set of places.

3.5 Uniform Constructions and Results

In Section 2 we have shown how to obtain several well-known and new net classes in a uniform way by the notion of parameterized and actual net classes. Now we raise the question, how far it is possible to obtain well-known results for each of these net classes in a uniform way. At first sight this seems hopeless, because each type of Petri net has its own notation and own kind of problems, although the general idea of most constructions and results is quite similar. However, the presentation of net classes as instances of parameterized net classes opens the way to study the theory on the level of parameterized net classes rather than for specific actual ones. In the following we summarize some main constructions and results for abstract Petri nets in the terminology of this paper. In this way we obtain uniform constructions and results for all the actual net classes (see Example 2) which are instantiations of parameterized net classes:

1. There is a uniform notion of abstract Petri nets, their marking, enabling and firing of transitions (see Definitions 12 and 13).
2. There is a uniform notion of morphisms for abstract Petri nets leading to the corresponding category. This category is cocomplete, which includes as special cases existence and construction of pushouts and coequalizers corresponding to union and fusion of nets (see Definition 15, Theorem 3).
3. Morphisms preserve the firing of transitions (see Theorems 1 and 2).
4. Firing of transitions can be extended to parallel and concurrent firing in a uniform way (see Subsection 3.3 and Theorem 2).
5. In the case of low level nets there is a uniform construction of the marking graph of a net in terms of F-graphs and a characterization of all those F-graphs, which are realizable by nets in the net class defined by $Net = G \circ F$ (see [Pad96]).
6. There is a uniform construction of the operations union and fusion for nets in the sense of [Jen92], which are most important for horizontal structuring of nets (see Definitions 21 and 23).
7. Important results concerning independence and parallelism of rule-based refinement – developed first in the theory of graph grammars – have been extended to parameterized net classes. Under certain independence conditions rule-based refinement is shown be locally confluent. Moreover, the parallel application of rules is possible. These parallel rules can be sequentialized in arbitrary order, provided they are independent (see Theorem 4, Theorem 5, and Theorem 6)
8. Refinement is an essential technique for vertical structuring of the software development process. Several refinement notions are known in the Petri net literature (see for example [BGV90]). Rule-based refinement can comprise these, provided they are based on morphisms. Examples are transition-gluing and place-preserving morphisms that refine nets so that safety properties are preserved.
9. Horizontal structuring of nets based on union and fusion is compatible with rule-based modification of nets, provided that certain independence conditions are satisfied (see Theorems 7 and 8).
10. There is a uniform construction of flattening from high-level abstract Petri nets of to low-level abstract Petri nets (see [Pad96]).

4   Applications of Parameterized Net Classes

We now give two applications of our results in specific net classes that are instantiations of parameterized net classes. A detailed version of these applications has been already presented in [PGH99,Erm96,EPE96]. The first application is given in terms of algebraic high-level nets (see Example 7). The second application in Subsection 4.2 is given in terms of place/transition nets. This category is obtained by actualization of parameterized net classes using the actual parameter given in Example 1.

Note that we have examples and applications at different levels of abstraction. Examples at the level of net classes are given as examples for instantiations of parameterized net classes (e.g. Examples 1 or 4). These are instantiations of the general notion of abstract Petri nets. Moreover, these examples constitute a specific Petri net class together with a corresponding theory for this class. In contrast to these example at a high level of abstraction the applications we present subsequently are examples at a lower level of abstraction. This means we illustrate our notions within a specific net class that is an example itself for the instantiation of parameterized net classes.

4.1   Requirements Engineering for a Medical Information [EPE96]

We now justify our concepts of horizontal structuring and rule-based refinement by sketching their role in the case study (see [Erm96, EPE96]).

The medical information system, called Heterogeneous Distributed Information Management System (HDMS), has been a large project, that included the whole reorganization of the medical and management data of the German Cardiac Center Berlin, Deutsches Herz-Zentrum Berlin (DHZB). This project has been developed by the Projektgruppe Medizin/Informatik (PMI) at the DHZB and the Technical University Berlin. The DHZB is a clinical center which is dedicated to the treatment of all kinds of cardiac diseases. It is a specialized hospital which lacks many of the typical features normal hospitals have, for example there are no emergency admission or general clinical laboratories. The high grade of necessary machine support inherent in most of the medical treatments concerning the human heart motivate the need of an integrated and complete computer support. In fact, many computers and other electronic devices have been already used and have been necessary in many of the medical areas. Most surgeries or intensive care urgently require very fast and 'intelligent' machines. Even an x-ray device is a sort of a computer and the recording, the saving and the diagnostic radiology of x-ray films or angiographic films is rather impossible without computer systems.

The aim of the project has been the *development of a support and information system for all activities of the medical and the non-medical personnel at the DHZB, which is able to digitally record and store all medical data which are produced during the treatment of DHZB patients, which is able to communicate these data within the whole system and to present these in a unique form at the user interface for further human processing* ([FHMO91, CHL95]).

An adequate, formal requirements engineering has been the aim of this case study. We first introduce the actual state analysis of the core of the German Heart Center Berlin (DHZB). The integration of routines and documents is achieved by using algebraic high-level nets. Hence, in the case study [Erm96, EPE96] the actual state description and its development towards the functional essence is shown as an algebraic high-level net. The presentation of the actual state involves several algebraic high-level nets and uses the structuring techniques union and fusion (as introduced in Section 8). The case study concerns the development of the actual state towards the functional essence. This involves mainly abstraction from irrelevant routines and documents. The transition from actual state to functional essence is realized using concepts of rule-based refinement. The algebraic high-level nets modeling the actual state contain about 130 places and 50 transitions. The transition from actual state to functional essence comprises about 100 rules. Different strategies for the development of software systems demand independence of different refinement steps and independence from structuring. The possibility of local refinement that is valid for the global system is crucial for the practical relevance of such an formalism for requirements engineering. That means refinement cannot be achieved for the whole system, as this requires abolishing the structuring. Hence, refinement of the whole net has to be derived from the local refinement. The compatibility of structuring and rule-based refinement meets this demand. In [Erm96] this compatibility has been shown on the basis of compatibility between fusion and union and rule-based refinement according to Theorems 7 and 8.

4.2   Developing a Model of an Elevator [PGH99]

In this subsection we stepwise develop a Petri net model of an elevator. The development of the model goes along with the development of safety properties for the model. These safety properties have to be proven only when introducing them, because they are preserved by all further modifications of the model. So this examples makes use of the results mentioned in Subsection 3.5, namely items 1, 8, and 7.

Some basic notions of the models (place/transition nets), temporal logic formulas, and refinement of models by transformations are given on an intuitive level in order to explain the example. We distinguish two major steps in the modeling of the elevator. First we derive a simple elevator which can arbitrarily move up and down. This model is equipped in a second step with a simple control mechanism to call the elevator. The first floor of the elevator is model-led by the net given in Figure 1. Analogously to graph grammars we call this initial model *start net*. There is a floor denoted by **f** and two states of the door. The places **dc** and **do** denote a closed, respectively opened door. The state of doors can be changed by the transitions **o** and **c** meaning opening and closing. The elevator can either go up, model-led by transition **u**, or come down, by transition **d**. From the viewpoint of the first floor, the elevator vanishes by going up. Analogously, it appears by coming down in an unpredictable way, that is the pre domain of **d** is empty. The initial marking $M_0 = \mathbf{f} \oplus \mathbf{dc}$ denoted by black dots expresses that there is an elevator and the door is closed. Together with the start net E_0 there is given a safety property. For security reasons it should always be guaranteed that if the door stands open the elevator is on the floor.

This is expressed by the temporal logic formula at the bottom line of Figure 1. Intuitively, a temporal logic formula states facts
about the markings and is given in terms of
numbers of tokens on places. That is, the
static formula 5**a** ∧ 2**b** is true for a marking M where at least 5 tokens are on place
a and at least 2 tokens are on place **b**.
The always operator in an invariant formula
□(5**a** ∧ 2**b**) states that this is true for all
reachable markings from M.

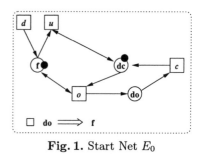

Fig. 1. Start Net E_0

In our case the safety property □**do** ⟹ **f**, meaning that "At any time, if the
door is open then the elevator is on the floor" is satisfied. In fact, we can argue
as follows: The formula **do** ⟹ **f** is satisfied in the initial state. Moreover, **u** is
the only transition which deletes the token on **f** and therefore may violate the
formula. After its firing, **o** — the only transition to change the state of the door
— is not enabled. Therefore, the door stays closed. Summarizing, the formula
do ⟹ **f** is always satisfied. We are now going to enhance the model with
further floors and requests bottoms. This will be done by adding floors to the
start net, i. e. the application of the rules r_{int} and r_{fin} given in Figures 2 and 3.

Application of a rule to
a place/transition net informally means replacing a
subnet specified by the left-
hand side of the rule with
the net specified by the
right-hand side. As the left-
hand sides of r_{int} and r_{fin}
are empty, we simply add
the right-hand side to the
(start) net. The property
which should hold for each
floor separately is again that
the doors must be closed if
the elevator is not in that
floor.

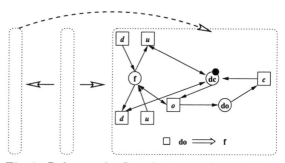

Fig. 2. Rule r_{int} for Introducing an Intermediate
Floor

Correspondingly, the rules are introducing new safety properties depicted at the
bottom of the net in the right-hand side. The formula □**do** ⟹ **f** is satisfied for
the net in the right-hand side of the rule, which can be seen analogously to the
start net E_0. Applying these rules, of course, we do not want to lose the safety
property, which we already proved for the start net. Moreover, the introduced
safety properties should be propagated to the resulting net. The preservation of
old safety properties and the satisfaction of the newly introduced safety properties is stated in Theorem 3 in [PGH99]. This means that the resulting net
satisfies all the safety properties introduced by the rules and also all originally
stated safety properties in the start net.

The application of the rule r_{fin} depicted in Figure 3 and iterated application of the intermediate rule r_{int} yields an elevator with many (disconnected) floors and corresponding safety conditions. In order to connect the floors we have to identify the up-going of the elevator from one floor with the coming-from-below from the next floor. his is achieved by the rule r_{glu} in Figure 4 that *glues the corresponding transitions.* It is compatible with the safety properties. Hence in the derived net still all safety properties hold.

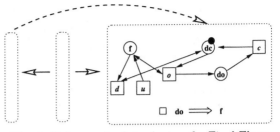

Fig. 3. Rule r_{fin} for Introducing the Final Floor

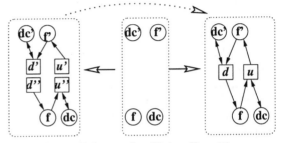

Fig. 4. Rule r_{glu} for Gluing Transitions

A sample transformation sequence $E_0 \xrightarrow{r_{int}+r_{fin}} E_1 \xrightarrow{r_{glu}+r_{glu}} E_2$ yields the model E_2 of a simple elevator depicted in Figure 5, where $r_i + r_j$ designates the parallel application of the rules r_i and r_j.

The set of safety properties satisfied by E_2 is given by $\{\Box do1 \implies f1, \Box do2 \implies f2, \Box do3 \implies f3\}$. This means that for all floors the safety property "At any time, if the door is open then the elevator is on the floor" holds. For enhancing this simple model, we want to add a simple control mechanism for calling the elevator. Three rules r_{rq_fin}, r_{rq_int}, and r_{rq_exc} introduce exclusive requests to the elevator E_2. If there is a request at a floor, the elevator may not leave that floor, unless the door has been opened and subsequently closed.

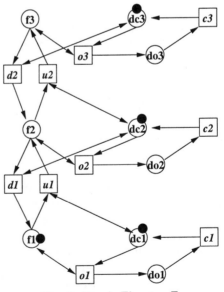

Fig. 5. Simple Elevator E_2

The insertion of requests to floors where only one direction of movement is possible is described by rule $r_{rq\text{-}fin}$. The marked place **nrq** designating no request and transition r (requesting) are added. Furthermore, the elevator may only move if there is no request on this floor which is captured by an additional arc to the transition m. By closing of the door the request is cleared which is model-led by the additional arc from c to **nrq**. Similarly, rule $r_{rq\text{-}int}$ – not depicted in this paper – describes the insertion of requests to intermediate floors. The last rule $r_{rq\text{-}exc}$ describes the mutual exclusion of the requests. All the rules $r_{rq...}$ do not change the environment of places, i.e. they also preserve safety properties.

Applying them to our simple elevator E_2 via $E_2 \xrightarrow{r_{rq\text{-}fin}+r_{rq\text{-}fin}+r_{rq\text{-}int}} E_3 \xrightarrow{r_{rq\text{-}exc}} E_4$ results in an elevator E_4 with a request mechanism. For the definition of these parallel application we use the results stated in Subsection 3.5, namely item 8 and 7. The main advantage of our approach is that we do not have to prove the safety property in the net E_4 but just for the start net and for the rule introducing new safety properties. By these we could add further safety properties, which were preserved by transition gluing as well as place preserving rules. For software development this significantly decreases the cost of proving safety properties.

Part II:
Theory of Parameterized Net Classes
Based on Abstract Petri Nets

In Part II we introduce the formal foundation of parameterized net classes based on abstract Petri Nets and present the main results for parameterized net classes. The net structure and data type parameter of high-level abstract Petri nets are integrated orthogonally, so that different combinations are possible. The general theory of high-level abstract Petri nets allows us to define the operational behavior on an abstract level, to show its compatibility with net morphisms and to prove that the corresponding category of high-level abstract Petri nets is finitely cocomplete. The existence of specific colimit constructions is essential in order to apply general results concerning structuring and rule-based refinement to abstract Petri nets. In Section 5 we define low-level abstract Petri nets based on a net structure functor. the dara rtype formalism is introduced in Section 6 leading to high-level abstract Petri nets in Section 7. In Section 8 we study rule-based refinement and horizontal structuring. Rule-based refinement is studied within the frame of high-level replacement systems. Results concerning independence, and parallelism of derivations, as given in [EHKP91b], are extended to a new type of rules, which allow us to consider different kinds of refinement as special cases [Pad99b]. Horizontal structuring is given by the notions of union and fusion, motivated by the constructions in [Jen92] but they are defined in a categorical way independently of Petri nets. These results are the formal basis for the notions and results presented in the Examples in Section 3.

5   Low-Level Abstract Petri Nets

In this section we introduce a categorical version of Petri nets, called low-level abstract Petri nets. The constructions used in Section 2, free monoids for place/transition nets, the powerset construction for elementary nets and the construction for S-graphs are special cases of a categorical construction, that states that the constructions are generated or free and that a universal property with respect to all other objects exists. Low-level abstract Petri nets are based on a functor $Net = G \circ F$, which is the composition of such a left-adjoint functor F and the corresponding right-adjoint functor G. Instantiations of low-level abstract Petri nets to various low-level nets have been given in Section 2.

Definition 1 (Net Structure Functor).

1. *We assume to have two categories* **Sets** *and* **Struct**, *called category of structure, and functors* $F \dashv G : \mathbf{Struct} \longrightarrow \mathbf{Sets}$, *where* F *is left adjoint to* G *with universal morphisms* $\gamma^S : S \longrightarrow G \circ F(S)$ *for all sets* S *in* **Sets***. The composition*

$$Net = G \circ F : \mathbf{Sets} \longrightarrow \mathbf{Sets}$$

   is called net structure functor.
2. *Furthermore let* **Struct** *be some category with commutative semigroups as objects.*

The basis for low-level abstract Petri nets are sets, which are used to generate some structure. This structure, given by the category **Struct** defines the structure of the places and hence the markings. The second condition provides the addition, that is needed for the definition of firing and has not been included in [PER01]. This condition ensures that each object in **Struct** is supplied with an associative operation $+$.

Definition 2 (Low-Level Abstract Petri Nets). *A low-level abstract Petri net* $N = (T, P, pre, post)$ *is given by sets* T *and* P, *called transitions and places, and functions* $T \overset{pre}{\underset{post}{\rightrightarrows}} Net(P)$ *called pre- and postcondition of* T, *where* $Net = G \circ F : \mathbf{Sets} \longrightarrow \mathbf{Sets}$ *with* $F \dashv G : \mathbf{Struct} \longrightarrow \mathbf{Sets}$ *is a net structure functor (see Definition 1).*

The characterization of the operational behavior of low-level abstract Petri nets uses the adjunction, that is given by the net structure functor. The unique extensions allow the definition of enabling and the computation of the follower marking using the addition given in the category **Struct**.

Definition 3 (Marking, Enabling, Firing). *Given a low-level abstract Petri net with* $N = (T, P, pre, post)$ *and the unique extensions of* pre *and* $post$, *namely* $\overline{pre}$ *and* $\overline{post}$:

1. *The marking of a low-level abstract Petri net is given by* $m \in F(P)$.
2. *A transition vector is defined by* $v \in F(\{t\})$.

3. $v \in F(\{t\})$ *is enabled under* $m \in F(P)$ *if there exists* $\overline{m} \in F(P)$ *so that*
$m = \overline{m} + \overline{pre}(v)$

4. *The* **follower marking** $m' \in F(P)$ *obtained by firing* v *under* m *is given by:* $m' = \overline{m} + \overline{post}(v)$

In case of a unique $+$-complement $\overline{m}$ with $m = pre(t) + \overline{m}$, we obtain a unique follower marking as well. This is the case for place/transition nets, because the $+$-complement for free commutative monoids is unique (see Example 1.1).

If the $+$-complement is not unique, it has to be specified by additional condition in the specific instance. This is the case for elementary nets, where the $\cup$-complement is not unique, but can be easily specified more precisely, demanding the usual complement on sets (see Example 1.3).

Next we define the category **LLAPN** of low-level abstract Petri nets, where we use a generated homomorphism for the mapping of the places. These morphisms are more restricted than the monoid-homomorphisms in [MM90], as our kind of morphisms do not allow to map a place to a sum of places. But their advantage are the structuring techniques presented in Section 8, that cannot be obtained with usual monoid-homomorphisms (see [MM90] p. 115: The category **Petri** is not cocomplete).

Definition 4 (Category LLAPN of Low-Level Abstract Petri Nets).

Given low-level abstract Petri nets $N_i = (T_i, P_i, pre_i, post_i)$ *with* $i = 1, 2$ *a low-level abstract Petri net morphism* $f : N_1 \longrightarrow N_2$ *is given by a pair*
$f = (f_T, f_P)$ *of functions* $f_T : T_1 \longrightarrow T_2$,
$f_P : P_1 \longrightarrow P_2$ *such that we have compatibility of the pre- and postdomain, that is the diagram to the right commutes separately for pre- and postconditions.*

$$
\begin{array}{ccc}
T_1 & \underset{post_1}{\overset{pre_1}{\rightrightarrows}} & Net(P_1) \\
f_T \downarrow & & \downarrow Net(f_P) \\
T_2 & \underset{post_2}{\overset{pre_2}{\rightrightarrows}} & Net(P_2)
\end{array}
$$

Low-level abstract Petri nets together with low-level abstract Petri net morphisms yield the category **LLAPN** *of low-level abstract Petri nets.*

The **LLAPN**-morphisms defined above preserve firing, that is if a transition of the source net is enabled, then the image of the transition in the target net is enabled as well. Moreover the follower marking of the source net is mapped to the follower marking of the target net.

Theorem 1 (LLAPN-Morphisms Preserve Firing).

Given an **LLAPN**-*morphism* $f : N_1 \longrightarrow N_2$ *(as in Definition 4) and let* $v \in F(\{t\})$ *with* $t \in T_1$ *be enabled under* m,

(that is $\exists \overline{m} \in F(P_1) : m = \overline{m} + \overline{pre_1}(v)$ *and the follower marking* $m' \in F(P_1)$ *is given by* $m' = \overline{m} + \overline{pre_1}(v)$ *)*

then:

1. $F(f_T)(v)$ *is enabled under* $F(f_P)(m)$:

$$F(f_P)(m) = F(f_P)(\overline{m}) + \overline{pre_2} \circ F(f_T)(v)$$

2. *the follower marking after firing $F(f_T)(v)$ is preserved:*

$$F(f_P)(m') = F(f_P)(\overline{m}) + \overline{post_2} \circ F(f_T)(v)$$

This theorem is a special case of Theorem 2, so we refer to that proof in Appendix B.1.

Examples of instances of low-level Petri nets have been briefly discussed in Subsections 2.2 to 2.4. In some more detail you find examples in another contribution to this book, namely in [PER01].

6   The Data Type Formalism Parameter

High-level abstract Petri nets, that are used as a frame for the uniform treatment of high-level nets are introduced in Section 7. High-level abstract Petri nets can be considered as an extension of low-level abstract Petri nets where the tokens are structured according to a given data type formalism. In order to allow different data type formalisms for different kinds of high-level nets we use the concept of specification frames. Motivated by [Mah89], specification frames have been introduced in [EBO91, EG94] as a categorical framework that allows to cover different kinds of algebraic specifications and other formalisms for the description of data types. It is a special advantage that different kinds of logics can be treated within this frame.

6.1   Specification Frames as the Data Type Formalism

First we need some signature part for high-level abstract Petri nets in order to obtain terms for the decoration of the net structure. The subsequent definitions are closely related to the concepts of specification frames [EBO91, EG94, Wol95] and institutions [GB84]. The main idea is to give the signature formalism in an abstract way, that only relates signatures to models. The following definitions lead to the data type parameter for high-level abstract Petri nets, presented in Definition 11, step by step. As examples we consider algebraic specifications, predicate logic, and the functional programming language ML.

Definition 5 (Signature Part for High-Level Abstract Petri Nets). *The signature part for high-level abstract Petri nets is given by a specification frame, that is a category* **ASIG** *of (abstract) signatures Σ and a contravariant functor $Cat : \mathbf{ASIG}^{op} \longrightarrow \mathbf{CATCAT}$, where* **CATCAT** *is the category of all categories. This means, that for each $f_\Sigma \in MOR_{\mathbf{ASIG}}$ with $\Sigma 1 \xrightarrow{f_\Sigma} \Sigma_2$ there is the forgetful functor $V_{f_\Sigma} : \mathbf{Cat}(\Sigma_2) \longrightarrow \mathbf{Cat}(\Sigma_1)$ with $V_{f_\Sigma} := Cat(f_\Sigma)$.*

Next we define the specification of the data type parameter. For this purpose we introduce sentences for signatures in the sense of institutions [GB84]. We want the possibility of restricting the data type without restricting the signature for the arc inscriptions. Due to the aim of developing the data type parameter for high-level Petri nets we have to define sentences for the data type as as well as for the firing conditions. The data type is equipped with axioms, that have to be satisfied by the models. This is the usual treatment of specifications as in equational algebraic specifications, conditional algebraic specifications, predicate logic and others.

Definition 6 (Specifications for High-Level Abstract Petri Nets). *Given the signature part* $Cat : \textbf{ASIG}^{op} \longrightarrow \textbf{CATCAT}$, *a functor* $Sen : \textbf{ASIG} \longrightarrow \textbf{Sets}$ *and a family of sentence satisfaction relations* $\models^{Sen}_{\Sigma} \subseteq |Cat(\Sigma)| \times Sen(\Sigma)$, *such that the following sentence satisfaction condition is satisfied for* $f_\Sigma : \Sigma_1 \longrightarrow \Sigma_2$, $M_2 \in |\textbf{Cat}(\Sigma_2)|$ *and* $\varphi \in Sen(\Sigma_1)$

$$V_{f_\Sigma}(M_2) \models^{Sen}_{\Sigma_1} \varphi \Leftrightarrow M_2 \models^{Sen}_{\Sigma_2} Sen(f_\Sigma)(\varphi)$$

then we have specifications $SPEC = (\Sigma, AXIOMS)$ *where* $AXIOMS \subseteq Sen(\Sigma)$. *Specifications* $SPEC$ *together with specification morphisms (that are signature morphisms* $f_\Sigma : \Sigma_1 \longrightarrow \Sigma_2$, *such that* $Sen(f_\Sigma)(AXIOMS_1)$ *is derivable from* $AXIOMS_2$) *yield* **ASPEC**, *called the category of (abstract) specifications.*

Note, we usually omit the index for the family of sentence satisfaction relations and write $\models^{Sen}$ instead of $\models^{Sen}_{\Sigma}$. Moreover, we have a model functor for specifications $SPEC = (\Sigma, AXIOMS)$, that yields the category of models satisfying the axioms. This category **Mod(SPEC)** is a subcategory of **Cat(Σ)**. These constructions, including the following two facts, are well-known for institutions [GB84]. The relationship between institutions and specification logics and frames is discussed in [EBO91] and [EG94].

Lemma 1 (Model Functor for Specification [EBO91]). *Given* **ASIG**, *Cat, Sen, and* $\models^{Sen}$ *as in Definition 6 then there is the category* **ASPEC** *of (abstract) specifications* $SPEC$ *and a contravariant model functor* $Mod : \textbf{ASPEC}^{op} \longrightarrow \textbf{CATCAT}$, *where* Mod *is a restriction of the functor* Cat.

This means for each $f_\Sigma \in MOR_{\textbf{ASPEC}}$ *with* $SPEC1 \xrightarrow{f_\Sigma} SPEC2$ *there is the forgetful functor* $V_{f_\Sigma} : Mod(SPEC2) \longrightarrow Mod(SPEC1)$ *with* $V_{f_\Sigma} := Mod(f_\Sigma)$.

Lemma 2 (Cocompleteness of ASPEC [GB84]). *The category* **ASPEC** *is cocomplete if the category of signatures* **ASIG** *is cocomplete.*

Definition 7 (Amalgamation [EG94]). *A specification frame has amalgamations, if for every pushout* $SPEC_1 \xrightarrow{g_1} SPEC_3 \xleftarrow{g_2} SPEC_2$ *of* $SPEC_2 \xleftarrow{f_2} SPEC_0 \xrightarrow{f_1} SPEC_1$ *in* **ASPEC** *we have*

1. *For every* $A_i \in |Mod(SPEC_i)|$ *for* $i = 0, 1, 2$ *such that* $V_{f_1}(A_1) = A_0 = V_{f_2}(A_2)$ *there is a unique* $A_3 \in |Mod(SPEC_3)|$, *called the amalgamation of* A_1 *and* A_2 *via* A_0, *written* $A_1 +_{A_0} A_2$, *such that we have* $V_{g_1}(A_3) = A_1$ *and* $V_{g_2}(A_3) = A_2$.

2. *Conversely, every $A_3 \in |Mod(SPEC_3)|$ has a unique decomposition $A_3 = V_{g_1}(A_3) +_{V_{g_1 \circ f_1}(A_3)} V_{g_2}(A_3)$.*

3. *Similar properties to Items (1) and (2) above are required if we replace the objects A_i by morphisms h_i in $Mod(SPEC_i)$ for $i = 0, 1, 2, 3$ leading to a unique amalgamated sum of morphisms $h_3 = h_1 +_{h_0} h_2$ with $V_{g_1}(h_3) = h_1$ and $V_{g_2}(h_3) = h_2$.*

As the variables are essential for high-level nets we have to express them at this abstract level. The usual treatment of variables as indexed set assumes a certain knowledge about the sorts available in the signature. Such an approach restricts the possibility of instantiation. To express variables in the context of specification frames we treat them as additional constants. Thus we introduce a set of morphisms V in **ASIG** that expresses variables as an inclusion of signatures.

Definition 8 (Signatures with Variables). *Variables are given by a class V of morphisms, that are preserved by pushouts (meaning: given a pushout*

$$\Sigma_1 \xrightarrow{g_1} \Sigma_3 \xleftarrow{g_2} \Sigma_2 \text{ of } \Sigma_2 \xleftarrow{f_2} \Sigma_0 \xrightarrow{f_1} \Sigma_1 \text{ in } \textbf{ASIG} \text{ then we have}$$

$f_1 \in V$ implies $g_2 \in V$).
A signature with variables (Σ', ϕ) is given by $\phi : \Sigma \longrightarrow \Sigma'$ and $\phi \in V$.
The expansion $Exp_{\Sigma'}(M)$ of a model $M \in |\textbf{Cat}(\Sigma)|$ denotes all models $M' \in |\textbf{Cat}(\Sigma')|$ such that $V_\phi(M') = M$. Each of the models of this expansion denotes an assignment.
Furthermore, we demand, that the class V is compatible with the given set-based specification frame, that means for each $\phi \in V$ we have $\pi_\phi = id_{U_\Sigma}$ the natural identity.

We also have to express firing conditions. Note, there is no reason that sentences for the description of the data type and for the firing conditions are of the same kind. These two kinds of sentences are not necessarily related, although they are similar in most kinds of high-level nets. We choose for the sake of generality two kinds and thus have two kinds of satisfaction relations, the sentence satisfaction defined above and the condition satisfaction defined below.

Definition 9 (Conditions for the Firing of Transition). *Conditions are given by a functor $Cond : \textbf{ASIG} \longrightarrow \textbf{Sets}$ and the following condition satisfaction relation $\models_{\Sigma}^{Cond} \subseteq |Cat(\Sigma)| \times Cond(\Sigma)$, such that the satisfaction condition is satisfied for $f_\Sigma : \Sigma_1 \longrightarrow \Sigma_2$, $M_2 \in |Cat(\Sigma_2)|$ and $\varphi \in Cond(\Sigma_1)$*

$$V_{f_\Sigma}(M_2) \models_{\Sigma_1}^{Cond} \varphi \Leftrightarrow M_2 \models_{\Sigma_2}^{Cond} Cond(f_\Sigma)(\varphi)$$

Conditions with free variables for a signature Σ are given by conditions of the signature with variables (Σ', ϕ) with $\phi : \Sigma \longrightarrow \Sigma'$, that is by $Cond(\Sigma')$.

Note, we usually omit the index for the family of condition satisfaction relations and write $\models^{Cond}$ instead of $\models_\Sigma^{Cond}$.

The following lemma is crucial for a uniform treatment of arc inscriptions with variables. Hence, it is fundamental for the definition of abstract variable assignments (see Definition 13) and the proof of Theorem 2.

Lemma 3 (Translation of Signature with Variables). *The translation of the signature with variables (Σ_1', ϕ_1) with $\phi : \Sigma_1 \longrightarrow \Sigma_1'$ along the morphism $f_\Sigma : \Sigma_1 \longrightarrow \Sigma_2$ is given by the pushout construction in **ASIG** and leads to the translated signature with variables $(\Sigma_2', \phi 2)$.*

Proof. The translation is given by the following pushout in **ASIG**, where f_Σ is the signature morphism and $\phi_1 \in \mathcal{V}$ denotes the variables. Because $\mathcal{V}$ is preserved by pushouts, we have $\phi_2 \in \mathcal{V}$. Thus (Σ_2', ϕ_2) is signature with variables.

$$
\begin{array}{ccc}
\Sigma_1 & \xrightarrow{\ \phi_1\ } & \Sigma_1' \\
{\scriptstyle f_\Sigma}\big\downarrow & & \big\downarrow{\scriptstyle f_\Sigma'} \\
\Sigma_2 & \xrightarrow{\ \phi_2\ } & \Sigma_2'
\end{array}
$$

Lemma 4 (Translation of Expansions). *Given a specification frame with amalgamation as in Definition 7, then for each signature with variables (Σ_1', ϕ_1), each signature morphisms $f_\Sigma : \Sigma_1 \longrightarrow \Sigma_2$, the translated signature with variables $(\Sigma_2', \phi 2)$, and models $M_1 \in \mathbf{Cat}(\Sigma_1)$, $M_2 \in \mathbf{Cat}(\Sigma_2)$ each expansion $Exp_{\Sigma_1'}(M_1) \in \mathbf{Cat}(\Sigma_1')$ of M1 can be translated to an expansion $Exp_{\Sigma_2'}(M_2) \in \mathbf{Cat}(\Sigma_2')$ of M_2.*

Proof. Given a model $M_1 \in \mathbf{Cat}(\Sigma_1)$, an expansion $Exp_{\Sigma_1'}(M1) \in \mathbf{Cat}(\Sigma_1')$, and a model $M_2 \in \mathbf{Cat}(\Sigma_2)$ with $V_{f_\Sigma}(M_2) = M_1$, then we obtain the translated expansion $Exp_{\Sigma_2'}(M_2) \in \mathbf{Cat}(\Sigma_2')$ by amalgamation $Exp(M_2)_{\Sigma_2'} = Exp_{\Sigma_1'}(M_1) +_{M_1} M_2$.

Due to the condition satisfaction we furthermore have for each expansion :
$Exp_{\Sigma_1'}(M_1) \in \mathbf{Cat}(\Sigma_1')$:
$$Exp_{\Sigma_1'}(M_1) \models^{Cond} \varphi \ \Leftrightarrow \ Exp_{\Sigma_2'}(M_2) \models^{Cond} Cond(f_{\Sigma_,})(\varphi).$$

This translation of expansions provides the possibility to define as many variables as wanted and to give arbitrary names. The translation due to the pushout construction and amalgamation inhibits the identification of variables. But it permits the definition of new variables and renaming of variables. Renaming is possible, because pushouts are unique only up to isomorphism that is unique up to renaming. Identification of variables has to be avoided, because then morphisms cannot preserve firing behavior.

Due to the fact, that Petri nets are based on sets of transitions and places we have to provide the compatibility of specification frames with sets. This means we have to relate the models of the data type with the underlying sets.

Definition 10 (Set-Based Specification Frames). *A set-based specification frame (Cat, U) is given by a specification frame $Cat : \mathbf{ASIG}^{op} \longrightarrow$ **CATCAT** and a family of functors $U = (U_\Sigma)_{\Sigma \in |\mathbf{ASIG}|}$ with $U_\Sigma : \mathbf{Cat}(\Sigma) \longrightarrow$ **Sets** and for each $f_\Sigma \in MOR_{\mathbf{ASIG}}$ with $f_\Sigma : \Sigma_1 \longrightarrow \Sigma_2$ and $U_{\Sigma_i} : \mathbf{Cat}(\Sigma_i) \longrightarrow$ **Sets** for $i = 1, 2$ there is a natural transformation $\pi_{f_\Sigma} : U_{\Sigma_1} \circ V_{f_\Sigma} \longrightarrow U_{\Sigma_2}$, so that:*

1. *π is compatible with the identity of signatures, that is for each $\Sigma \in \mathbf{ASIG}$ we have $\pi_{id_\Sigma} = id_{U_\Sigma}$ the natural identity.*

2. *π is compatible with the composition of signature morphisms, that is for* $\Sigma_1 \xrightarrow{f_\Sigma} \Sigma_2$ *and* $\Sigma_2 \xrightarrow{g_\Sigma} \Sigma3$ *we have* $\pi_{g_\Sigma \circ f_\Sigma} = \pi_{g_\Sigma} \circ \pi_{f_\Sigma}$. *So the following diagram commutes:*

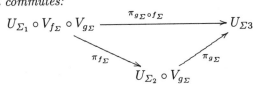

Summarizing, a data specification technique, called data type parameter for high-level abstract Petri nets, consists of a set-based specification frame, sentences, variables, and conditions.

Definition 11 (Data Type Formalism Parameter). *The data type formalism parameter for high-level abstract Petri nets* $DT = (Cat, U, Sen, \mathcal{V}, Cond)$ *consists of*

1. *a set-based specification frame* (Cat, U) *with amalgamation (Definition 7), where we have* $Cat : \mathbf{ASIG}^{op} \longrightarrow \mathbf{CATCAT}$ *(see Definition 5) and* $U = (U_\Sigma)_{\Sigma \in |\mathbf{ASIG}|}$ *(see Definition 10),*
2. *sentences* $Sen : \mathbf{ASIG} \longrightarrow \mathbf{Sets}$ *(see Definition 6) that allow the description of data type specifications, leading to the category* $\mathbf{ASPEC}$,
3. *variables denoted by a class of morphisms* $\mathcal{V}$ *(see Definition 8), and*
4. *conditions for the firing of transitions* $Cond : \mathbf{ASIG} \longrightarrow \mathbf{Sets}$ *(see Definition 9).*

Furthermore we demand that

5. $\mathbf{ASIG}$ *is cocomplete, and*
6. $Cat(\Sigma)$ *has an initial object* T_Σ *for all* $\Sigma \in \mathbf{ASIG}$.

Note that (5) implies cocompleteness of $\mathbf{ASPEC}$ due to Lemma 2.

In the remaining part of this section we provide different examples of specification techniques, which can be used as data type parameters for high-level abstract Petri nets.

Example 4 (Algebraic Specifications). Algebraic specifications are a data type parameter for high-level abstract Petri nets. This data type parameter gives rise to algebraic high-level nets as in Example 7.

1. The set-based specification frame $Cat : \mathbf{ASIG}^{op} \longrightarrow \mathbf{CATCAT}$ is given by the category of algebraic signatures $\mathbf{SIG}$ in the sense of [EM85] and by the model functor $Alg : \mathbf{SIG} \longrightarrow \mathbf{CATCAT}$.

   Algebraic signatures are set-based: $U_\Sigma : \mathbf{Alg}(\Sigma) \longrightarrow \mathbf{Sets}$ is given by $U_\Sigma(A) = \biguplus_{s \in S} A_s$ the disjoint union of the carrier sets. For each signature morphism $f = (f_S, f_{OP}) : \Sigma_1 \longrightarrow \Sigma_2$ the natural transformation $\pi : U_{\Sigma_1} \circ V_f \longrightarrow U_{\Sigma_2}$ is given for all $A_2 \in |\mathbf{Alg}(\Sigma_2)|$ by inclusion of data elements: $\pi_f^{A_2} : \biguplus_{s \in S_1}(V_f(A_2))_s \longrightarrow \biguplus_{s \in S_2} A_{2s}$. $\pi_f^{A_2}$ is well-defined due to the definition of the forgetful functor, that is for each $s \in S1$ we have $(V_f(A_2))_s = A_{2f_S(s)}$. Amalgamation is given in [EM85] (see Definition 8.10).

2. The sentence functor $Sen :$ **SIG** $\longrightarrow$ **Sets** yields for each signature the set of all equations over some set X, that is $Sen(SPEC) = \{(X, L, R) | L, R \in T_{OP}(X)\}$. The satisfaction relation is defined by the extended assignment with respect to an assignment of variables onto the same data element, that is $A \models^{Sen} (X, L, R) \iff \overline{asg}_A(L) = \overline{asg}_A(R)$. Satisfaction is preserved due to $V_f(A) \models^{Sen} (X, L, R) \iff A \models^{Sen} (X^\#, f^\#(L), f^\#(R))$ see Fact 8.3 in [EM85]. This yields the category of algebraic specifications **SPEC** in the sense of [EM85] with $SPEC = (\Sigma, AXIOMS)$. Mod is given by the model functor Alg.

3. $\mathcal{V}$, the class of morphisms denoting variables is given by those inclusions, that given $\phi : \Sigma \longrightarrow \Sigma'$ with $\phi \in \mathcal{V}$ the signature Σ' only has additional constants. Given $\Sigma = (S, OP)$, $\Sigma' = (S, OP')$, and $\phi : \Sigma \longrightarrow \Sigma'$ with $\phi \in \mathcal{V}$ then $\phi = (id_S, \phi_{OP})$ is given by the identity on the sorts id_S and an inclusion on operations $\phi_{OP} : OP \longrightarrow OP'$, such that all operation not included in the image of ϕ_{OP} are constants.
   Formally: $\forall N \in OP' : N \notin \phi_{OP}(OP) \Longrightarrow N \in OP'_{\lambda, s}$
   $\mathcal{V}$ is preserved under pushouts:

   Given $\Sigma_i = (S_i, OP_i)$ and $\Sigma'_i = (S'_i, OP'_i)$ for $i = 1, 2$ and the pushout aside in **SIG** then $S'_1 = S_1$ due to the component-wise construction of pushouts and due to $\phi = (id_{S_1}, \phi_{OP})$.

   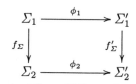

   Let $N \in OP'_2$ but $N \notin \phi_{2OP}(OP_2)$ then there is $N' \in OP'_1$ so that $f_{\Sigma'_{OP}}(N') = N$ and $N' \notin \phi_{1OP}(OP_1)$ due to the pushout construction in **Sets**. Thus N' is constant, that is $N' \in OP'_{1\lambda, s}$ and as signature morphisms preserve the arity of operations we conclude N is constant as well, that is $N \in OP_{2\lambda, f_s(s)}$

4. The condition functor $Cond :$ **SPEC** $\longrightarrow$ **Sets** yields for each signature the set of all ground terms, that is $Cond(SPEC) = \{(L, R) | L, R \in T_{OP}\}$. Here, we only use ground terms, because the definition for variables as they are used for the decoration of the net, is given already on the abstract level using the class $\mathcal{V}$ (see Item 3). The idea is to distinguish between variables used for the description of the specification and the variables used for the decoration of the net. The latter ones are given due to signatures with variables in Definition 8, thus need not to be expressed for the conditions explicitly. They are used implicitly as the conditions are given for a signature with variables. The satisfaction relation is defined by the evaluation onto the same data element, that is $A \models^{Cond} (L, R) \iff eval_A(L) = eval_A(R)$ and satisfaction is preserved due to $V_f(A) \models^{Cond} (L, R) \iff A \models^{Cond} (f^\#(L), f^\#(R))$ see Fact 8.3 in [EM85].

5. **SIG** is cocomplete (see [GB84]).

6. For each signature Σ there is the initial object, that is the termalgebra T_Σ in **Alg(Σ)** ([EM85] Theorem 3.7).

Example 5 (Predicate Logic). The predicate logic data type part for high-level abstract Petri nets is based on the formulation of first order predicate logic in the frame of institutions [GB84] and is adapted to the predicate logic used in [GL81, Gen91]. This gives rise to predicate/transition nets as in Example 8.

1. The set-based specification frame is given by one-sorted first order signatures $\Sigma = (\Omega, \Pi)$ where $\Omega = (\Omega^n)_{n\in\mathbb{N}}$ denotes the set of n-ary operations and $\Pi = (\Pi^n)_{n\in\mathbb{N}}$ the set of n-ary predicates. Together with morphisms preserving the arity, we have the category **FOSIG**. A model for $\Sigma = (\Omega, \Pi)$ consists of a **possibly empty** domain R and functions and predicates (R, Ω_R, Π_R) with respect to the signature. Thus we have the contravariant functor $Fosig : \textbf{FOSIG}^{op} \longrightarrow \textbf{CATCAT}$ where **Fosig(Σ)** denotes the category of models.

   First order signatures are set-based: For each signature Σ there is a functor $U_\Sigma : \textbf{Fosig}(\Sigma) \longrightarrow \textbf{Sets}$ that is given by the domain, $U_\Sigma(R, \Omega_R, \Pi_R) = R$. Thus π is the identity.

   Due to the fact that there is only one sort, amalgamation is given by identity. Hence, it is a special case of amalgamation of algebraic specifications.

2. Sentences are given by closed formulas. Abstract specifications are given by abstract signatures and closed formulas. Their models are models of the signature that satisfy these formulas.

   Given a set of variables X we use the usual definition of terms:
   - $x \in X$ is a term.
   - $f^n \in \Omega^n$ and $v_1, \ldots, v_n$ are terms, then $f^n(v_1, \ldots v_n)$ is a term.
   - No other expression is term.

   and formulas:
   - v_1, v_2 are terms, then $v_1 = v_2$ is a formula.
   - $p^n \in \Pi^n$ and $v_1, \ldots, v_n$ are terms, then $p^n(v_1, \ldots v_n)$ is a formula.
   - p is a formula then $\neg p$ is a formula.
   - p_1, p_2 are formulas then $p_1 \vee p_2$ is a formula.
   - $x \in X$ and p is a formula then $\exists x : p$ is a formula.
   - No other expression is a formula.

   The functor $FoSen : \textbf{FOSIG} \longrightarrow \textbf{Sets}$ yields for each abstract signature $FoSen(\Omega, \Pi)$ the set of all closed formulas. First order signatures together with a set of closed formulas $AXIOMS \subseteq FoSen(\Omega, \Pi)$ denote the first order specifications $FOSPEC = (\Omega, \Pi, AXIOMS)$. The model functor $FoMod : \textbf{FOSPEC}^{op} \longrightarrow \textbf{CATCAT}$ yields for each specification $FOSPEC = (\Omega, \Pi, AXIOMS)$ the category **Fomod($\Omega, \Pi, $AXIOMS)** with the **nonempty** models that satisfy the formulas in $AXIOMS$. The satisfaction relation $\models^{FoSen}$ is the usual one for predicate logic.

3. $\mathcal{V}$, the class of morphisms denoting variables is given by those inclusions, that given $\phi : \Sigma \longrightarrow \Sigma'$ with $\phi \in \mathcal{V}$ the signature Σ' only has additional constants similar to Example 4, Item 3. Variables are given by:

   $\Sigma \xrightarrow{\phi} \Sigma' \in \mathcal{V}$ if $\phi = (\phi_\Omega, id_\Pi)$ and ϕ_Ω is inclusion so that $\forall N \in \Omega' :$ $N \notin \phi_\Omega(\Omega) \Longrightarrow N \in \Omega'^0$. $\mathcal{V}$ is preserved under pushouts due to the same argumentation as in Example 4, Item 3.

4. The firing conditions are defined in the same way as the sentences (see Item 2). Note, we have closed formulas over some signature (Σ', ϕ) with $\phi :$

$\Sigma \longrightarrow \Sigma' \in \mathcal{V}$. This means, the formulas in $FoCond(\Sigma')$ with $FoCond$: **FOSIG** $\longrightarrow$ **Sets** are closed with respect to the signature Σ', but in view of the arc inscriptions of the net we have free variables as they are given by $\phi \in \mathcal{V}$ using additional constants. These are assigned by models of the expansion. Due to this construction there is no confusion between variables bound by some quantifier and variables that belong to the net.

5. **FOSIG** is cocomplete ([GB84]).
6. **Fosig(Σ)** has an initial object, that is T_Σ for all $\Sigma \in$ **FOSIG** ([GB84]).

The empty carrier set problem is based on the subsequent facts:

- **Nonempty** carrier sets may yield model categories without initial object.
- **Empty** carrier sets may yield an unsound logic.

This has some impact on high-level abstract Petri nets. In the first case we could not necessarily use the term algebra T_Σ and would loose the cocompleteness of the category of high-level abstract Petri nets. But this cocompleteness is the basis for the structuring techniques presented in the following subsection. The second case has to be prevented anyhow.

We have avoided the problem by allowing empty carrier sets for the model of the first order signature (thus we can assume the initial object) and by demanding nonempty carrier sets for the models of the specification (thus we obtain a sound logic).

Example 6 (ML). Including ML into this frame yields a variant of colored nets in the sense of [Jen92]. The presupposition is to express ML and its semantics within institutions. This task has been solved in principle, but the details are not yet finished. Hence, we claim that also ML is a suitable data type part for high-level abstract Petri nets.

7   High-Level Abstract Petri Nets

In this section we first introduce the basic notions of high-level abstract Petri nets and discuss in Subsection 7.2 interesting instantiations.

7.1   Basic Notions of High-Level Abstract Petri Nets

We now introduce high-level abstract Petri nets, based on a data type parameter as defined above and a net structure functor as given in Subsection 5, that are fixed in this section. We define pre- and post-functions, that map each transition to a linear sum consisting of pairs of terms and places, where terms are data elements of the term algebra T_Σ and places are elements of P. These terms represent the arc inscriptions. These inscriptions and the firing conditions have to include variables, which are given as a family of variables for each transition.

Definition 12 (High-Level Abstract Petri Nets). *Given a data type parameter $DT = (Cat, U, Sen, \mathcal{V}, Cond)$ for high-level nets (Definition 11) with*

a category **ASPEC** *of abstract specifications (Definition 6) and a net structure functor* $Net = G \circ F$ *(Definition 1) then a high-level abstract Petri net is given by*
$$N = (P, T, SPEC, Var, pre, post, cond)$$
with

- P : *the set of places,*
- T : *the set of transitions,*
- $SPEC \in |\mathbf{ASPEC}|$: *some specification with* $SPEC = (\Sigma, AXIOMS)$
- $Var = (Var(t))_{t \in T} = (\Sigma^t, \phi_t)_{t \in T}$ *the signature with variables with* ϕ_t : $\Sigma \longrightarrow \Sigma^t \in \mathcal{V}$ *for each transition* $t \in T$
- $pre, post : T \longrightarrow Net(U_{\Sigma^t}(T_{\Sigma^t}) \times P)$
  the pre- and postcondition functions of T, *defining for each transition with adjacent arcs the arc inscriptions and the weight, such that* $pre(t), post(t) \in Net(U_{\Sigma^t}(T_{\Sigma^t}) \times P)$.
- $cond : T \longrightarrow \mathcal{P}_{fin}(Cond(\Sigma^t))$
  the function that maps each transition to a finite set of conditions over the signature with variables representing the firing conditions such that $cond(t) \in \mathcal{P}_{fin}(Cond(\Sigma^t))$.

Definition 13 (Marking, AAS^t-Enabling, Firing). *Given a high-level abstract Petri net* $N = (P, T, SPEC, Var, pre, post, cond)$ *with the model functor* Mod *(see Definition 1) then we have for each data type model* $M \in Mod(SPEC)$:

- *A marking of* N *is given by* $m \in F(U_\Sigma(M) \times P)$, *where* F *is the left adjoint functor of the net structure functor* $Net = G \circ F$ *(see Definition 1).*
- *An abstract assignment for a transition* $t \in T$ *is given by* $AAS^t \in Exp_{\Sigma^t}(M)$, *where* $Exp_{\Sigma^t}(M)$ *is the expansion of* M *(see Definition 8) with respect to the signature with variables* (Σ^t, ϕ^t) *for* $t \in T$.
  The abstract assignment AAS^t *defines an abstract assignment function* aas^t : $F(U_{\Sigma^t}(T_{\Sigma^t}) \times P) \longrightarrow F(U_\Sigma(M) \times P)$ *with* $aas^t = F(U_{\Sigma^t}(eval^t) \times id_P)$,
  where $eval^t$ *is the unique morphism* $T_{\Sigma^t} \xrightarrow{eval^t} AAS^t$ *due to initiality of* T_{Σ^t}. aas^t *is well-defined due to the fact that* $U_\Sigma(M) = U_\Sigma \circ V_{\phi^t}(AAS^t) = U_{\Sigma^t}(AAS^t)$ *and* $U_{\Sigma^t}(eval^t) : U_{\Sigma^t}(T_{\Sigma^t}) \longrightarrow U_{\Sigma^t}(AAS^t)$.
- *A transition vector is defined by* $v \in F(\{t\})$.
- AAS^t *satisfies the firing condition* $cond(t)$ *if and only if* $AAS^t \models^{Cond} \varphi$ *for all* $\varphi \in cond(t)$.
- *A transition vector* $v \in F(\{t\})$ *is* AAS^t*-enabled under* $m \in F(U_\Sigma(M) \times P)$ *if there exists* $\overline{m} \in F(U_\Sigma(M) \times P)$ *so that* $m = \overline{m} + aas^t(\overline{pre}(v))$, *where* $\overline{pre} : F(T) \longrightarrow F(U_{\Sigma^t}(T_{\Sigma^t}) \times P)$ *is the unique extension of* pre.
  Then t *is* AAS^t*-enabled, if* v *is* AAS^t*-enabled.*
- *The firing of a transition vector* v *under a marking* m *and an abstract assignment* $AAS^t \in Exp_{\Sigma^t}(M)$ – *provided that* v *is* AAS^t*-enabled under* m – *is defined by the follower marking* $m' \in F(U_\Sigma(M) \times P)$, *given by:*
$$m' = \overline{m} + aas^t(\overline{post}(v))$$
where $\overline{post}$ *is the unique extension of* $post$.

The uniqueness of the follower marking still depends on the category **Struct** and the uniqueness of the +-complement.

Morphisms for high-level abstract Petri nets are composed using functions between the sets of transitions, between the sets of places and between the specifications. The other components of the nets have to be preserved in order to obtain further results, especially concerning structuring. High-level abstract Petri nets together with the morphisms yield the category **HLAPN**.

Definition 14 (High-Level Abstract Petri Net Morphisms). *Given high-level abstract Petri nets $Ni = (P_i, T_i, SPEC_i, Var_i, pre_i, post_i, cond_i)$ for $i = 1, 2$ then a high-level abstract Petri net morphism $f : N_1 \longrightarrow N_2$ is given by $f = (f_P, f_T, f_\Sigma)$ with*

- $f_P : P_1 \longrightarrow P_2$ *maps places to places in* **Sets,**
- $f_T : T_1 \longrightarrow T_2$ *maps transitions to transitions in* **Sets,**
- $f_\Sigma : SPEC_1 \longrightarrow SPEC_2$ *maps specifications to specifications in* **ASPEC,**

such that for all $t_1 \in T_1$ and $f_T(t_1) = t_2 \in T_2$ the subsequent conditions hold:

1. *Preservation of variables :*
   The translation of the signature with variables $(\Sigma_1^{t_1}, \phi_1^{t_1})$ with $\phi_1^{t_1} : \Sigma_1 \longrightarrow \Sigma_1^{t_1}$ along the morphism $f_\Sigma : \Sigma_1 \longrightarrow \Sigma_2$ to the signature with variables $(\Sigma_2^{t_2}, \phi_2^{t_2})$ is given by the pushout $\Sigma_2^{t_2}$ and the corresponding pushout morphisms f_Φ and $\phi_2^{t_2}$ in **ASIG** *(see Lemma 3):*

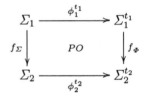

2. *Compatibility of pre- and postcondition function:*
   The following diagram

$$T_1 \xrightarrow[\;post_1\;]{\;pre_1\;} Net(U_{\Sigma_1^{t_1}}(T_{\Sigma_1^{t_1}}) \times P_1)$$

$$\downarrow f_T \qquad\qquad\qquad \downarrow Net(f_{ins})$$

$$T_2 \xrightarrow[\;post_2\;]{\;pre_2\;} Net(U_{\Sigma_2^{t_2}}(T_{\Sigma_2^{t_2}}) \times P_2)$$

commutes componentwise, for
$f_{ins} = (\pi_{f_\Phi} \circ U_{\Sigma_1^{t_1}}(eval)) \times f_P : U_{\Sigma_1^{t_1}}(T_{\Sigma_1^{t_1}}) \times P_1 \longrightarrow U_{\Sigma_2^{t_2}}(T_{\Sigma_2^{t_2}}) \times P_2$
with the natural transformation
$\pi_{f_\Phi} : U_{\Sigma_1^{t_1}} \circ V_{f_\Phi}(T_{\Sigma_2^{t_2}}) \longrightarrow U_{\Sigma_2^{t_2}}(T_{\Sigma_2^{t_2}})$ *see Definition 10*
where eval $: T_{\Sigma_1^{t_1}} \longrightarrow V_{f_\Phi}(T_{\Sigma_2^{t_2}})$ is the unique morphism defined by $AAS_1^{t_1}$ due to initiality of $T_{\Sigma_1^{t_1}}$.

3. *Compatibility of firing conditions:*
   The following diagram commutes

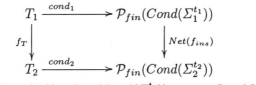

$$\text{for } f_{cond} = \mathcal{P}_{fin}(Cond(f_\Phi)) \; : \mathcal{P}_{fin}(Cond(\Sigma_1^{t_1})) \longrightarrow \mathcal{P}_{fin}(Cond(\Sigma_2^{t_2}))$$

Definition 15 (Category HLAPN). *High-level abstract Petri nets (Definition 12) and high-level abstract Petri net morphisms (Definition 14) are defining a category* **HLAPN**, *called category of high-level abstract Petri nets.*

Due to the component-wise definition of morphisms we obtain composition, identities, as well as associativity of composition, and hence a category of high-level abstract Petri nets.

Next we show that morphisms preserve the operational behavior, if the models of the specification are compatible. We allow new sorts and operations in the specification. Thus new data elements may occur in the model. But the reduct of the model in the target net has to be identical to the model of the source net. Clearly, more changes of the data elements would change the firing behavior.

Theorem 2 (HLAPN-Morphism Preserve Firing). *Given an* **HLAPN**-*morphism* $f : N_1 \longrightarrow N_2$ *(as in Definition 14) and compatible data type models (that is* $M_1 \in Mod(SPEC_1)$ *and* $M_2 \in Mod(SPEC_2)$ *with* $V_{f_\Sigma}(M_2) = M_1$*) and let* $v \in F(\{t\})$ *with* $t \in T_1$ *be* AAS_1-*enabled under* m, *(that is* $\exists \overline{m} \in F(U_{\Sigma_1}(M_1) \times P_1) : m = \overline{m} + aas_1(\overline{pre_1}(v)))$ *then there is an abstract assignment* AAS_2 *so that for* $f_m = \pi_{f_\Sigma} \times f_P$:

1. $F(f_T)(v)$ *is* AAS_2-*enabled :*

   $$F(f_m)(m) = F(f_m)(\overline{m}) + aas_2(\overline{pre_2}(F(f_T)(v)))$$

2. *the follower marking after firing* $F(f_T)(v)$ *is preserved :*

   $$F(f_m)(m') = F(f_m)(\overline{m}) + aas_2(\overline{post_2}(F(f_T)(v)))$$

The proof is given in Appendix B.1.

Theorem 3. HLAPN *is finitely cocomplete*

The proof is given in Appendix B.2.

It is likely, that the category **HLAPN** of high-level abstract Petri nets has also arbitrary coproducts, that means it is even infinitely cocomplete. But we have not treated this proposition, because this result is not relevant for practical issues, it would only imply some kind of infinite composition.

Corollary 1 (Decomposition of Pushouts). *Given a pushout in* **HLAPN** *then the components yield the corresponding pushouts in* **ASPEC** *and* **Sets**.

Direct consequence from the constructions in proof of Theorem 3.

7.2   Instances of High-Level Abstract Petri Nets

We present some interesting instantiations of high-level abstract Petri nets in this subsection. We give the definitions as they result from the net structure functors given in Section 5 and the data type parts exemplified in Examples 4, 5 and 6. These instantiations do not correspond exactly to the definitions found in literature, they are closely related and the remaining differences are discussed in the subsequent examples.

Example 7 (Algebraic High-Level Nets). There are different variants of algebraic high-level nets in literature (see [Rei91, EPR94a, PER95, Hum89, Lil95]), here we have chosen the definition given in [PER95]. An algebraic high-level net is given by $AHL = (SPEC, P, T, pre, post, cond, A)$, where $SPEC = (S, OP, E)$ is an algebraic specification and A is a $SPEC$-algebra in the sense of [EM85],

$$pre, post : T \longrightarrow (T_{OP}(X) \times P)^{\oplus}$$
$$\text{and } cond : T \longrightarrow \mathcal{P}_{fin}((T_{OP}(X) \times (T_{OP}(X)))$$

with the set of variables $X \cong X_{fix} \times S$ so that $(x, s), (x, s') \in Var(t) \implies s = s'$ and $Var(t)$ is the set of all variables occurring in $pre(t)$, $post(t)$ or $cond(t)$.

The instantiation of high-level abstract Petri nets with algebraic specifications as the data type formalism parameter (see Example 4) and the net structure functor of place/transition nets (see Example 1.1) yields algebraic high-level nets without algebras (also called algebraic-high-level net schemes in [EPR94a]), that is $N = (P, T, SPEC, Var, pre, post, cond)$
with $pre, post : T \longrightarrow (T_{OP_t} \times P)^{\oplus}$. The differences to the above defined algebraic high-level nets are:

- The instantiation lacks a $SPEC$-algebra A, which however, is available in the corresponding instantiation of abstract Petri nets with models.
- The variables of the instantiation are defined depending on the transitions. This dependency is given in algebraic high level nets implicitly, by the additional condition for the set of variables X. Nevertheless both formulations denote the same net, if the signature with variables is given for each $t \in T$ by $\Sigma_t = \Sigma +$

  opns: $x : \underline{\qquad} \longrightarrow s$ for all $(x, s) \in X$

  This means, that each variable x of sort s is taken as an additional constant of sort s.

Other variants of algebraic high-level nets can be obtained by slight changes of the data type parameter. Let the condition functor $Cond : \textbf{ASIG} \longrightarrow \textbf{Sets}$ be the constant functor, that yields for each signature the empty set $Cond(\Sigma) = \emptyset$, then the corresponding instantiation of high-level abstract Petri nets is closer to the definition of algebraic high-level nets as defined in [Rei91]. Another example, if we choose order sorted algebraic specifications, which are shown to be an institution in [GB84], we obtain an instantiation that is closely related to the order sorted algebraic high-level nets in [Lil95].

Example 8 (Predicate/Transition Nets). Predicate/transition nets as defined in [Gen91] are given subsequently:

Definition
Let **L** be a first-order language and let $\mathbf{L}_s$ designate the
sublanguage using only Π_s, the predicate denoting static
relations. The class PRT_L consists of marked annotated net,
$MN = (N, A, M^0)$ where N is the underlying directed net, A is its
annotation in L and M^0 is its representative marking.
1. N is a directed net, $N = (S, T, F)$.
2. A is annotation of N, $A = (A_N, A_S, A_T, A_F)$ where
    (a) $A_N = \mathcal{R}$ is a first-order structure for $\mathbf{L}_s$, called the
        support of MN (it is the kind of legend that annotates
        the whole net rather than a particular element);
    (b) A_S is a bijection between the set of places, S, and the
        set of variable predicates, Π_s;
    (c) A_T is a mapping of the set of transitions, T, into the
        set of formulae (called transition selectors) that only use
        operators and static predicates (i.e. are in $\mathbf{L}_s$);
    (d) A_F is a mapping of the set of arcs, F, into the set of
        symbolic sums of tuples of terms of L, LC, such that for
        an arc $(x, y) \in F$ leading into or out of a place (i.e. $x = s$
        or $y = s$) and n being the index of the predicate annotating
        s, $A_F(x, y)$ is in $LC^{(n)}$.
3. M^0 is a consistent marking of places [...]

The corresponding instantiation of high-level abstract Petri nets consist of
the net structure parameter similar to place/transition nets (Example 1.3) and
the data type parameter of predicate logic (Example 5). This means that the
instantiation of high-level abstract Petri nets in this case is given by $PRT = (P, T, \Sigma, var, pre, post, cond)$. The differences to the above definition are:

1. Our instantiation is not supplied with a first order structure, but this is the
   case for the corresponding instantiation of high-level abstract Petri nets with
   models.
2. We have no annotation for the places.
3. We allow a set of firing conditions, where in [Gen91] there is only one formula,
   the transition selector.

Example 9 (Colored Petri Nets (81)). Colored Petri nets [Jen81] are based on
indexed sets [TBG87], where the colors denote the index and the color sets the
indexed sets.

Definition
A colored Petri net $R = (P, T, C, I^+, I^-)$ is defined by[2] :
  - P the set of places
  - T the set of transitions with $P \cup T \neq \emptyset$ and $P \cap T = \emptyset$
  - C the color function from $P \cup T \longrightarrow W$ where W is some
    finite set of finite and nonempty sets. An item of $C(s)$ is
    called a color of s and $C(s)$ is called the color set of s.

---

[2] We omit the initial marking.

- I^+ (I^-, respectively) is the forward (backward) incidence
  matrix of $P \times T$, where I^+ (p,t) is a function from $C(p) \times C(t)$
  to $\mathbb{N}$.

This class of high-level nets treats the data type on a purely semantical level.
Thus there are different ways to present this class in the context of high-level
abstract Petri nets.

- One instantiation is achieved by the supplementation of the data type description. The class of colored Petri nets [Jen81] provides distinguishable tokens, but no explicit functions, terms, variables, or assignment. Thus – seen
  strictly – it is not possible to give arc inscriptions or firing conditions. The
  data type model is merely given implicitly, due to the forward and backward
  incidence matrices, describing linear functions for each arc between places
  and transitions.

  To make explicit the specification describing the model, results in some other
  kind of high-level net, depending on the chosen formalism. One possibility
  is to use algebraic signatures. Then we use for each colored net above the
  following signature:

  cn-sig
  <u>   </u>

        <u>sorts:</u> p for all $p \in P$

        <u>opns:</u> $I(p,t) : t \longrightarrow p$ for all $(p,t) \in P \times T$

  In this case colored Petri (81) nets can be considered as a special case of
  algebraic high-level nets.
- Another instantiation, obtained by using the net structure functor of place/
  transition nets (Example 1.1) and indexed, pointed sets as the data type
  parameter yields nets without (real) inscriptions, but with a sufficient set of
  colors namely $(C_p)_{p \in P}$ and $(C_t)_{t \in T}$. The drawback is that the firing behavior
  needs to be redefined in order to use the forward and backward incidence
  matrices.

In both cases we do not supply an initial marking and the places are not
sorted, that is each color is allowed on any place.

Example 10 (Colored Petri Nets (92)). An instantiation of high-level abstract
Petri nets similar to colored Petri nets (92), in the sense of [Jen92] requires
a data type parameter considering ML as an institution. The corresponding
instantiation of high-level abstract Petri nets would be a close variant of colored
Petri nets.

8   Structuring Results for Abstract Petri Nets

We now show the structuring techniques and the compatibility results for (high-
level) abstract Petri nets. First we introduce rule-based refinement in Subsection
8.1 with results concerning local confluence and parallelismin in Subsection 8.2
and horizontal structuring techniques in Subsection 8.3.

8.1 Rule-Based Refinement

The concept of refinement is a well-known technique within software engineering in general and for stepwise development of Petri nets in particular. In fact, several different concepts for the refinement of nets have been proposed in literature, above all the refinement of one place, one transition or even a subnet by some other subnet. Based on the idea of formal grammars we propose rule-based refinement, to present rules denoting the replacement of a subnet by another one, without changing the remaining part of the whole net. This has the advantage of a simple local presentation of the refinement even if the whole system is large and complex. We consider to have a rule (or production) p with a left-hand side net L that is replaced by a right-hand side net R. This rule can be applied to some net N, yielding the new net M. This application of a rule, called transformation, is denoted by $N \overset{p}{\Longrightarrow} M$. Rule-based refinement is based on a construction consisting of two pushouts, called double-pushout diagram. The proofs of the different compatibility results make use of this close relation between the categorical concepts of rule-based refinement and horizontal structuring. In fact, all of them are based on specific colimits.

Since the general theory of rule-based refinement is presented in the frame of high-level replacement systems (see [EHKP91b, Pad99b]), the proofs in Appendix B make use of specific conditions for high-level replacement systems. The underlying theory of high-level replacements is given purely categorical, so we formulate the following notions and results in these terms. The application to abstract Petri nets is due to the satisfaction of the HLR- and Q-conditions under the following assumptions.

Definition 16 (HLR-Assumption for Abstract Petri Nets). *The assumptions for being a HLR-category are for abstract Petri nets the following:*

1. *There is a class of $\mathcal{M}$-morphisms for high-level replacement system in the category* **HLAPN** *given by the class of* **HLAPN***-morphisms, that are injective functions and a suitable class of $\mathcal{M}_{\textbf{ASPEC}}$-morphisms for the data type parameter.*
2. *The high-level replacement system $(\textbf{ASPEC}, \mathcal{M}_{\textbf{ASPEC}})$ of abstract specifications satisfies the HLR-conditions (see Definition 26).*
3. *There are pullbacks of $\mathcal{M}$-morphisms in* **HLAPN** *(see Definition 6)*
4. *There is some category* **QHLAPN** *and an inclusion functor*
   $I : \textbf{HLAPN} \longrightarrow \textbf{QHLAPN}$ *that satisfies the Q-conditions (see Definition 27).*

In the case of abstract Petri nets – and similar in the general case – the rule is split into a deleting part L, an adding part R and an interface K which is presented, such that the rule p is given by $p = (L \overset{l}{\longleftarrow} K \overset{r}{\longrightarrow} R)$ where l and r are mappings of Petri nets, called Petri net morphisms. Deleted are those parts of the net L that are not in the image of the morphism $l : K \longrightarrow L$. In general terms, the 'difference' between L and K is deleted. Adding works symmetrically, all those parts of R are added, that are not in the image of the morphism $r : K \longrightarrow R$. The transformation $G \overset{p}{\Longrightarrow} H$ is defined using two pushouts **(1)** and **(2)**. Since the general notion of refinement cannot be expected to be compatible with all different kinds of analysis techniques and behavior, we have introduced the notion of Q-morphisms and Q-rules in [Pad99b], which can be adapted to different kinds of refinement for nets found in literature. In the following the concept of rule-based refinement is formalized using the notions of transformations and Q-transformations.

Definition 17 (Rules and Transformations).

1. *A rule in* **HLAPN** $p = (L \overset{l}{\longleftarrow} K \overset{r}{\longrightarrow} R)$ *consists of the abstract Petri nets L, K and R, called left-hand side, interface (or gluing net), and right-hand side, respectively,*

   and two morphisms $K \overset{l}{\longrightarrow} L$ and $K \overset{r}{\longrightarrow} R$ with both morphisms $l, r \in \mathcal{M}$, the distinguished class of morphisms in **HLAPN**. *Given a rule $p = (L \overset{u}{\longleftarrow} K \overset{v}{\longrightarrow} R)$ a direct transformation $G \overset{p}{\Longrightarrow} H$, from an abstract Petri net G to an abstract Petri net H is given by two pushout diagrams*

   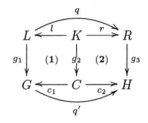

   (1) *and* **(2)** *in the category* **HLAPN**. *The morphisms $L \overset{g_1}{\longrightarrow} G$ and $R \overset{g_3}{\longrightarrow} H$ are called occurrences of L in G and R in H, respectively. By an occurrence of rule $p = (L \overset{l}{\longleftarrow} K \overset{r}{\longrightarrow} R)$ in a a net G we mean an occurrence of the left-hand side L in G.*

   A transformation sequence $G \overset{*}{\Longrightarrow} H$, short transformation, between nets G and H means G is isomorphic to H or there is a sequence of $n \geq 1$ direct transformations:
   $$G = G_0 \overset{p_1}{\Longrightarrow} G_1 \overset{p_2}{\Longrightarrow} \ldots \overset{p_n}{\Longrightarrow} G_n = H$$

2. *A Q-rule (p, q) is given by a rule $p = (L \overset{l}{\longleftarrow} K \overset{r}{\longrightarrow} R)$ in* **HLAPN** *(see above) and a Q-morphism $q : L \longrightarrow R$, so that $q \circ l = r$ in* **QHLAPN**. *Moreover there is a unique Q-morphism $q' : G \longrightarrow H$, such that $q' \circ c_1 = c_2$. Moreover, we have the pushout $R \overset{g_3}{\longrightarrow} H \overset{q'}{\longleftarrow} G$ of $G \overset{g_1}{\longleftarrow} L \overset{q}{\longrightarrow} R$ in* **QHLAPN**. *The transformation $(G \overset{p}{\Longrightarrow} H, q' : G \longrightarrow H)$, or short $G \overset{(p,q')}{\Longrightarrow} H$, is called Q-transformation in* **QHLAPN**.

8.2   Local Confluence and Parallelism for Rule-Based Refinement

This subsection deals with the independence of transformations and Q-transformations. Independence intuitively means that the changes of subsequent or parallel transformations do not infer with each other. Then three of our main theorems are given concerning local confluence and parallelism of transformations.

General Assumption:
The assumptions as in Definition 16 hold.
This means speaking in the following of transformations and Q-transformations we mean those in the categories **HLAPN** and **QHLAPN**.

Definition 18 (Sequential Independence [EHKP91b]). *Given two direct transformations $G \overset{p_1}{\Longrightarrow} H$ and $H \overset{p_2}{\Longrightarrow} X$ as in the subsequent diagram:*

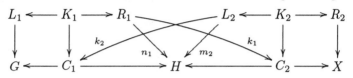

Then $G \overset{p_1}{\Longrightarrow} H$ and $H \overset{p_2}{\Longrightarrow} X$ are called sequentially independent if and only if there are the morphisms $k_1 : R_1 \longrightarrow C_2$ and $k_2 : L_2 \longrightarrow C_1$, so that $n_1 = g_2 \circ k_1$ and $m_2 = h_1 \circ k_2$.

Definition 19 (Parallel Independence [EHKP91b]). *Given two direct transformations $G \overset{p_1}{\Longrightarrow} H_1$ and $G \overset{p_2}{\Longrightarrow} H_2$ as in the subsequent diagram*

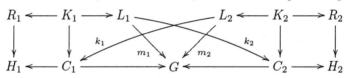

Then $G \overset{p_1}{\Longrightarrow} H_1$ and $G \overset{p_2}{\Longrightarrow} H_2$ are called parallel independent if and only if there are the morphisms $k_1 : L_1 \longrightarrow C_2$ and $k_2 : L_2 \longrightarrow C_1$, so that $m_1 = g_2 \circ k_2$ and $m_2 = g_1 \circ k_1$.

The following results, called local Church-Rosser Theorem I and II, show that independent direct transformations and Q-transformations commute, where independence of Q-transformations means independence of the underlying transformations.

Theorem 4 (Local Church-Rosser I [EHKP91b, Pad99b]).

1. *Given parallel independent direct transformations $G \overset{p_1}{\Longrightarrow} H_1$ and $G \overset{p_2}{\Longrightarrow} H_2$ there is a net X and direct transformations $H_1 \overset{p_2}{\Longrightarrow} X$ and $H_2 \overset{p_1}{\Longrightarrow} X$, so that the transformations $G \overset{p_1}{\Longrightarrow} H_1 \overset{p_2}{\Longrightarrow} X$ and $G \overset{p_2}{\Longrightarrow} H_2 \overset{p_1}{\Longrightarrow} X$ are sequentially independent.*

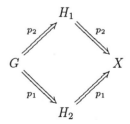

2. *Given two parallel independent Q-transformations $(G \overset{p_1}{\Longrightarrow} H_1, q_1 : G \longrightarrow H_1)$ and $(G \overset{p_2}{\Longrightarrow} H_2, q_2 : G \longrightarrow H_2)$ then there are two Q-transformations $G \overset{(p_1,q_1)}{\Longrightarrow} H_1 \overset{(p_2,r_2)}{\Longrightarrow} X$ and $G \overset{(p_2,q_2)}{\Longrightarrow} H_2 \overset{(p_1,r_1)}{\Longrightarrow} X$ so that $r_2 \circ q_1 = r_1 \circ q_2$.*

For the proof see Appendix B.3.

Theorem 5 (Local Church-Rosser II [EHKP91b, Pad99b]).

1. *Given a sequentially independent transformations $G \overset{p_1}{\Longrightarrow} H_1 \overset{p_2}{\Longrightarrow} X$ there also exists a sequentially independent transformation $G \overset{p_2}{\Longrightarrow} H_2 \overset{p_1}{\Longrightarrow} X$. Moreover, the transformations $G \overset{p_1}{\Longrightarrow} H_1$ and $G \overset{p_2}{\Longrightarrow} H_2$ are parallel independent.*

2. *Given two sequentially independent Q-transformations $(G \overset{p_1}{\Longrightarrow} H_1, q_1 : G \longrightarrow H_1)$ and $(H_1 \overset{p_2}{\Longrightarrow} X, r_2 : H_1 \longrightarrow X)$ then there is a Q-transformation $G \overset{(p_2,q_2)}{\Longrightarrow} H_2 \overset{(p_1,r_1)}{\Longrightarrow} X$ so that $r_2 \circ q_1 = r_1 \circ q_2$. Moreover, the transformations $(G \overset{p_1}{\Longrightarrow} H_1, q_1 : G \longrightarrow H_1)$ and $(G \overset{p_2}{\Longrightarrow} H_2, r_1 : G \longrightarrow H_2)$ are parallel independent.*

For the proof see Appendix B.3.

Definition 20 (Parallel Rules [EHKP91b, Pad99b]).

1. *Given rules $p_1 = (L_1 \overset{l_1}{\longleftarrow} K_1 \overset{r_1}{\longrightarrow} R_1)$ and $p_2 = (L_2 \overset{l_2}{\longleftarrow} K_2 \overset{r_2}{\longrightarrow} R_2)$ the rule $p_1 + p_2 = (L_1 + L_2 \overset{l_1+l_2}{\longleftarrow} K_1 + K_2 \overset{r_1+r_2}{\longrightarrow} R_1 + R_2)$ defined by binary coproducts in **HLAPN** is called parallel rule of p_1 and p_2.*

   Transformations $G \overset{p_1+p_2}{\Longrightarrow} X$ defined by parallel rules are called parallel transformations.

2. *Given two Q-rules $p_i = L_i \overset{q_i}{\longleftarrow K_i \longrightarrow} R_i$ for $i = 1, 2$ then there is a parallel $Q-$rule*

$$p_1 + p_2 = L_1 + L_2 \overset{q_1+q_2}{\longleftarrow K_1 + K2 \longrightarrow} R_1 + R_2 \text{ , where } + \text{ denotes the cor-}$$

*responding coproduct-constructions, provided that the corresponding coproduct exists in **QHLAPN**. In this case we have $p_1 + p_2 \in Q$, because Q is closed under coproducts (Definition 27).*

Theorem 6 (Parallelism [EHKP91b]). *Let p_1 and p_2 be rules and $p_1 + p_2$ the corresponding parallel rule as defined in Definition 20, then we have:*

1. **Synthesis:**

For a sequentially independent transformation $s_1 : G \overset{p_1}{\Longrightarrow} H_1 \overset{p_2}{\Longrightarrow} X$ as in Definition 18 there is a parallel transformation $t : G \overset{p_1+p_2}{\Longrightarrow} X$.

2. **Analysis:**

Given a parallel transformation $t : G \overset{p_1+p_2}{\Longrightarrow} X$ as in Definition 20 there are two sequentially independent transformations $s_1 : G \overset{p_1}{\Longrightarrow} H_1 \overset{p_2}{\Longrightarrow} X$ and $s_2 : G \overset{p_2}{\Longrightarrow} H_2 \overset{p_1}{\Longrightarrow} X$.

3. **Bijective correspondence:**

There is a bijective correspondence between sequentially independent and parallel transformations.

That means, given the sequentially independent transformation s_1 the "synthesis" construction leads to the parallel transformation t and the "analysis" construction leads back to the same sequentially independent transformation s_1 (up to isomorphism), and vice versa as shown in the diagram to the right.

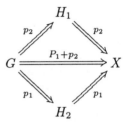

Given a parallel Q-transformation $G \overset{(p_1+p_2,q)}{\Longrightarrow} X$ and its sequentializations $G \overset{p_1,q_1}{\Longrightarrow} H_1 \overset{p_2,r_2}{\Longrightarrow} X$ and $G \overset{p_2,q_2}{\Longrightarrow} H_2 \overset{p_1,r_1}{\Longrightarrow} X$ then we have additionally:

$$r_1 \circ q_2 = q = r_2 \circ q_1$$

For the proof see Appendix B.3.

8.3   Horizontal Structuring and Its Compatibility with Rule-Based Refinement

We now introduce two basic constructions for high-level structures. The first one, allowing the construction of larger structures from smaller ones with shared subpart is called union. The second one is called fusion, a construction which allows to identify distinguished items. This is a generalization of the notions introduced by [Jen92] for coloured Petri nets. We adopt this notion and reformulate it in the frame of high-level replacement systems, thus we can apply it to other application areas. These constructions yield a horizontal structuring for high-level structures, while transformations in high-level replacement systems are regarded as rule-based refinement. Compatibility results under suitable assumptions between horizontal and vertical structuring are shown as the main results of this section.

General Assumption:

The assumptions as in Definition 16 hold.

This means speaking in the following of transformations and Q-transformations we mean those in the categories **HLAPN** and **QHLAPN**.

Definition 21 (Fusion). *The fusion of two morphisms $f_1, f_2 : F \longrightarrow G$ in* **HLAPN** *between nets F and G is the coequalizer $(g : G \longrightarrow G', G')$ of f_1 and f_2. The fusion is denoted by $G \models O \Rightarrow G'$ via (F, f_1, f_2, g) short $G \overset{F}{\models O} \Rightarrow G'$.*

The intuitive idea is that two occurrences of a subnet $f_1(F)$ and $f_2(F)$ in G are fused together within the given net G leading to a new net G'. Several occurrences of subnets can be fused by iterated fusion, that is by iteration of coequalizers: Given fusion $G \models O \Rightarrow G'$ via (F, f_1, f_2, g) and $G \models O \Rightarrow G''$ via (F', f_1', f_2', g') the iterated fusion $G \models O \Rightarrow G'''$ via $(F + F', f_1, f_1', f_2, f_2', g'')$ can be constructed directly as the pushout $G' \longrightarrow G''' \longleftarrow G''$ of $G' \longleftarrow G \longrightarrow G''$, where (f_1, f_1') and (f_2, f_2') are induced by coproduct constructions.

Examining the relation between transformations and fusion, it becomes clear that a transformation and a fusion applied to one net cannot be compatible if some part of the subnets to be fused is deleted by the transformation. The following parallel independence condition excludes this possibility.

Definition 22 (Independence of Fusion and Transformation). *A fusion $G \models O \Rightarrow G'$ via (F, f_1, f_2, g) is parallel independent from a transformation $G \overset{p}{\Longrightarrow} H$ with $p = (L \longleftarrow K \longrightarrow R)$ and given by pushouts* **(1)** *and* **(2)***, if there are morphisms $k_1, k_2 : F \longrightarrow C$ so that the triangle* **(3)** *commutes component-wise:*

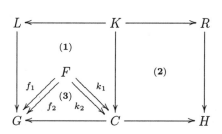

Theorem 7 (Fusion).

1. *Given a fusion $G \overset{F}{\models O} \Rightarrow G'$, which is independent from a transformation $G \overset{p}{\Longrightarrow} H$ there is a net H', obtained by fusion $H \overset{F}{\models O} \Rightarrow H'$ and also by transformation $G' \overset{p}{\Longrightarrow} H'$. That means, we have*

$$G \overset{p}{\Longrightarrow} H \overset{F}{\models O} \Rightarrow H' = G \overset{F}{\models O} \Rightarrow G' \overset{p}{\Longrightarrow} H'$$

in the sense that both sides are defined by the same diagram.

2. *Given a fusion $G \models_{\bigcirc}^{F} \Rightarrow G'$ independent of the Q-transformation $G \overset{(p,q)}{\Longrightarrow} H$,*

   we have in addition $G \overset{(p,q)}{\Longrightarrow} H \models_{\bigcirc}^{F} \Rightarrow H'$ and $G \models_{\bigcirc}^{F} \Rightarrow G' \overset{(p,q')}{\Longrightarrow} H'$ where (p,q') is the induced Q-transformation.

This means that transformations and fusions are compatible, provided we have independence.

Definition 23 (Union). *The union of a pair of nets (G_1, G_2) in **HLAPN** via some interface I with the morphisms $i_1 : I \longrightarrow G_1$ and $i_2 : I \longrightarrow G_2$ is given by the pushout (1) and denoted by $(G_1, G_2) \models_{\square}\Rightarrow G$ via (I, i_1, i_2), short $(G_1, G_2) \overset{I}{\models_{\square}}\Rightarrow G$.*

$$
\begin{array}{ccc}
I & \overset{i_1}{\longrightarrow} & G_1 \\
{\scriptstyle i_2}\downarrow & (1) & \downarrow \\
G_2 & \longrightarrow & G
\end{array}
$$

Definition 24 (Independence of Union and Transformation). *A union $(G_1, G_2) \overset{I}{\models_{\square}}\Rightarrow G$ is called parallel independent from transformations $G_i \overset{p_i}{\Longrightarrow} H_i$ for $i \in \{1,2\}$ given by the pushouts (1) and (2) if there are morphisms $I \longrightarrow C_i$ so that the triangle (3) commutes for $i \in \{1,2\}$.*

$$
\begin{array}{ccc}
L_i & \longleftarrow K_i & \longrightarrow R_i \\
\downarrow \quad (1) & \downarrow \quad (2) & \downarrow \\
G_i & \longleftarrow C_i & \longrightarrow H_i \\
 & \nwarrow \ (3) \ \uparrow & \\
 & I &
\end{array}
$$

Theorem 8 (Union).

1. *Given a union $(G_1, G_2) \overset{I}{\models_{\square}}\Rightarrow G$ independent from the transformations $G_i \overset{p_i}{\Longrightarrow} H_i$ for $i = 1,2$, then there is a net H obtained by the union $(H_1, H_2) \overset{I}{\models_{\square}}\Rightarrow H$ and by the transformation $G \overset{p_1+p_2}{\Longrightarrow} H$ via the parallel rule $p_1 + p_2$, such that we have*

$$
(G_1, G_2) \overset{I}{\models_{\square}}\Rightarrow G \overset{p_1+p_2}{\Longrightarrow} H = (G_1, G_2) \overset{(p_1,p_2)}{\Longrightarrow} (H_1, H_2) \overset{I}{\models_{\square}}\Rightarrow H
$$

   where $(G_1, G_2) \overset{(p_1,p_2)}{\Longrightarrow} (H_1, H_2)$ denotes the tupling of the separate transformations of $G1 \overset{p1}{\Longrightarrow} H_1$ and $G_2 \overset{p_2}{\Longrightarrow} H_2$.

2. *Given a union $G1, G_2 \overset{I}{\models_{\square}}\Rightarrow G$, independent from the Q-transformations $G1 \overset{p1,q1}{\Longrightarrow} H_1$ and $G_2 \overset{p_2,q_2}{\Longrightarrow} H2$ then we have*

$$
(G1, G_2) \overset{I}{\models_{\square}}\Rightarrow G \overset{(p_1+p_2,q)}{\Longrightarrow} H = (G_1, G2) \overset{(p_1,q_1),(p_2,q_2)}{\Longrightarrow} (H1, H2) \overset{I}{\models_{\square}}\Rightarrow H
$$

   such that

$$
I \longrightarrow G_1 \overset{q_1}{\longrightarrow} H_1 \longrightarrow H = I \longrightarrow G_1 \longrightarrow G \overset{q}{\longrightarrow} H
$$

$$
and \ I \longrightarrow G_2 \overset{q_2}{\longrightarrow} H_2 \longrightarrow H = I \longrightarrow G_1 \longrightarrow G \overset{q}{\longrightarrow} H.
$$

9   Conclusion

We have suggested a uniform frame for the classification and systematization of Petri nets. Abstract Petri nets constitute the formal basis of such a uniform frame. They provide an abstract description of Petri nets together with a substantial body of theoretical results, concerning operational behavior, horizontal and vertical structuring. In this paper we have given a uniform approach to different types of Petri nets, including low-level nets, like elementary and place/transition nets, as well as high-level nets, like algebraic high-level or predicate/transition nets. The main idea of this approach is to introduce parameterized net classes, which are based on a net structure parameter for low-level nets and in addition a data type formalism parameter for high-level nets. By instantiation of these parameters we obtain many well-known net classes studied in the literature but also several new interesting net classes in a uniform way. In particular we have achieved in the following results:

- Abstract Petri nets are a uniform approach to capture different kinds of low- and high-level net classes, well-known ones like place/transition nets, elementary nets, S-graphs, algebraic high-level nets, predicate/transition nets, colored nets and several new ones. We have covered the basic notions of Petri nets, like net structure, operational behavior, and marking in a uniform way for low-level abstract Petri nets and for high-level abstract Petri nets as well.
- We make rule-based refinement available for all instantiated net classes. Hence, we have obtained this new, general refinement for place/transition nets, variants of elementary nets, S-graphs, algebraic high-level nets, predicate transition nets and colored nets.
- Horizontal structuring techniques, union and fusion have been introduced at an abstract level in order to make these techniques available to all instantiations.
- We have proven the compatibility of our notion of refinement with the horizontal structuring techniques union and fusion. This compatibility makes use of different notions of independence.

Moreover we have sketched the application of these results in two examples of different instantiations.

References

[AHS90]   J. Adamek, H. Herrlich, and G. Strecker. *Abstract and Concrete Categories*. Series in Pure and Applied Mathematics. John Wiley and Sons, 1990.

[BBD01]   E. Badouel, M. Bednarczyk, and P. Darondeau. Generalized Automata and their Net Representations. In H. Ehrig, G. Juhàs, J. Padberg, and G. Rozenberg, editors, *Advances in Petri Nets: Unifying Petri Nets*, Advances in Petri Nets. Springer, 2001.

[BCM88]   E. Battiston, F. De Cindio, and G. Mauri. OBJSA nets: a class of high-level nets having objects as domains. In G. Rozenberg, editor, *Advances in Petri nets*, volume 340 of *Lecture Notes in Computer Science*, pages 20–43. Springer Verlag Berlin, 1988.

[BD95]   E. Badouel and Ph. Darondeau. Dualities between nets and automata in-
           duced by schizofrenic objects. In *6th International Conference on Category
           Theory and Computer Science*, pages 24–43. Springer, LNCS 953, 1995.

[BD98]   E. Badouel and Ph. Darondeau. Theory of regions. In W. Reisig and
           G. Rozenberg, editors, *Lectures on Petri Nets: Basic Models*, pages 529–
           586. Springer, LNCS 1491, 1998.

[BDC92]   L. Bernardinello and F. De Cindio. A survey of basic net models and
           modular net classes. In *Advances in Petri Nets'92*, pages 304–351. Springer
           LNCS 609, 1992.

[BDH92]   E. Best, R. Devillers, and J. Hall. The Box Calculus: a new causal algebra
           with multi-label communication. In *Advances in Petri Nets*, pages 21–69.
           Lecture Notes in Computer Science 609, 1992.

[BGV90]   W. Brauer, R. Gold, and W. Vogler. A Survey of Behaviour and Equiva-
           lence Preserving Refinements of Petri Nets. *Advances in Petri Nets*, Lec-
           ture Notes in Computer Science 483:1–46, 1990.

[CBEO99]   F. Cornelius, M. Baldamus, H. Ehrig, and F. Orejas. Abstract and be-
           haviour module specifications. *Mathematical Structures in Computer Sci-
           ence*, 9:21–62, 1999.

[CHL95]   F. Cornelius, H. Hußmann, and M. Löwe. The KORSO Case Study for Soft-
           ware Engineering with Formal Methods: A Medical Information System.
           In M. Broy and S. Jähnichen, editors, *KORSO: Methods, Languages, and
           Tools for the Construction of Correct Software*, pages 417–445. Springer
           LNCS 1009, 1995. Also appeared as technical report 94-5, TU Berlin.

[DA94]   R. David and H. Alla. Petri nets for modelling of dynamic systems - a
           survey. *Automatica*, 30(2):175–202, 1994.

[DG91]   W. Deiters and V. Gruhn. Software Process Model Analysis Based on
           FUNSOFT Nets. *Mathematical Modelling and Simulation*, 8, May 1991.

[DG94]   W. Deiters and V. Gruhn. The FUNSOFT Net Approach to Software Pro-
           cess Management. *International Journal on Software Engineering and
           Knowledge Engineering*, 4(2):229–256, June 1994.

[DG98]   W. Deiters and V. Gruhn. Process Management in Practice - Applying
           the FUNSOFT Net Approach to Large-Scale Processes. *Automated Software
           Engineering*, 5:7–25, 1998.

[DJL01]   J. Desel, G. Juhás, and R. Lorenz. Petri Nets over Partial Algebras. In
           H. Ehrig, G. Juhás, J. Padberg, and G. Rozenberg, editors, *Advances in
           Petri Nets: Unifying Petri Nets*, LNCS. Springer, 2001.

[DM93]   M. Droste and Shrott R. M. Petri nets and automata with concurrency
           relation - an adjunction. In M. Droste and Y. Gurevich, editors, *Semantics
           of Programming Languages and Model Theory*, pages 69–97. Gordon and
           Breach Sc. Publ., 1993.

[DS01]   M. Droste and R.M. Shortt. Continuous Petri Nets and Transition Systems.
           In H. Ehrig, G. Juhás, J. Padberg, and G. Rozenberg, editors, *Advances
           in Petri Nets: Unifying Petri Nets*, LNCS. Springer, 2001.

[EBCO91]   H. Ehrig, M. Baldamus, F. Cornelius, and F. Orejas. Theory of algebraic
           module specification including behavioural semantics and constraints and
           aspects of generalized morphisms. In *Proc. AMAST '91, Iowa City*. Uni-
           versity of Iowa, 1991.

[EBO91]   H. Ehrig, M. Baldamus, and F. Orejas. New concepts for amalgamation
           and extension in the framework of specification logics. Technical Report
           91-05, Technical University of Berlin, 1991.

[EG94]   H. Ehrig and M. Große-Rhode. Functorial theory of parameterized specifications in a general specification framework. *Theoretical Computer Science*, (135):221–266, 1994.

[EGL+97]   H. Ehrig, M. Gajewsky, S. Lembke, J. Padberg, and V. Gruhn. Reverse Petri Net Technology Transfer: On the Boundary of Theory and Application. In Hartmut Ehrig, Wolfgang Reisig, and Herbert Weber, editors, *Move-On-Workshop der DFG-Forschergruppe Petrinetz-Technologie*. Forschergruppe PETRINETZ-TECHNOLOGIE, 1997. Technical Report TR 97-21, Technische Universität Berlin.

[EGLP97]   H. Ehrig, M. Gajewsky, S. Lembke, and J. Padberg. Reverse Petri Net Technology Transfer: On the Boundary of Theory and Application. In Lindsay Groves and Steve Reeves, editors, *Formal Methods Pacific '97*, pages 297–298. Springer - Verlag Singapore Pte. Ltd, 1997.

[EGW96]   H. Ehrig, M. Große-Rhode, and U. Wolter. On the role of category theory in the area of algebraic specifications. In *LNCS , Proc. WADT11, Oslo*. Springer Verlag, 1996.

[EHKP91a] H. Ehrig, A. Habel, H.-J. Kreowski, and F. Parisi-Presicce. From graph grammars to High Level Replacement Systems. In H. Ehrig, H.-J. Kreowski, and G. Rozenberg, editors, *4th Int. Workshop on Graph Grammars and their Application to Computer Science, LNCS 532*, pages 269–291. Springer Verlag, 1991. Lecture Notes in Computer Science 532.

[EHKP91b] H. Ehrig, A. Habel, H.-J. Kreowski, and F. Parisi-Presicce. Parallelism and concurrency in high-level replacement systems. *Math. Struct. in Comp. Science*, 1:361–404, 1991.

[EM85]   H. Ehrig and B. Mahr. *Fundamentals of Algebraic Specification 1: Equations and Initial Semantics*, volume 6 of *EATCS Monographs on Theoretical Computer Science*. Springer Verlag, Berlin, 1985.

[EP97]   H. Ehrig and J. Padberg. A Uniform Approach to Petri Nets. In Ch. Freksa, M. Jantzen, and R. Valk, editors, *Foundations of Computer Science: Potential - Theory - Cognition*. Springer, LNCS 1337, 1997.

[EPE96]   C. Ermel, J. Padberg, and H. Ehrig. Requirements Engineering of a Medical Information System Using Rule-Based Refinement of Petri Nets. In D. Cooke, B.J. Krämer, P. C-Y. Sheu, J.P. Tsai, and R. Mittermeir, editors, *Proc. Integrated Design and Process Technology*, pages 186–193. Society for Design and Process Science, 1996. Vol.1.

[EPR94a]   H. Ehrig, J. Padberg, and L. Ribeiro. Algebraic High-Level Nets: Petri Nets Revisited. In *Recent Trends in Data Type Specification*, pages 188–206. Springer Verlag, 1994. Lecture Notes in Computer Science 785.

[EPR94b]   H. Ehrig, J. Padberg, and G. Rozenberg. Behaviour and realization construction for Petri nets based on free monoid and power set graphs. In *Workshop on Concurrency, Specification & Programming*. Humboldt University, 1994. Extended version as Technical Report of University of Leiden.

[ER96]   H. Ehrig and W. Reisig. Integration of Algebraic Specifications and Petri Nets. *Bulletin EATCS, Formal Specification Column*, (61):52–58, 1996.

[Erm96]   C. Ermel. Anforderungsanalyse eines medizinischen Informationssystems mit Algebraischen High-Level-Netzen. Technical Report 96-15, TU Berlin, 1996.

[FHMO91]   E. Fleck, H. Hansen, B. Mahr, and H. Oswald. Systementwicklung für die Integration und Kommunikation von Patientendaten und -dokumenten. Forschungsbericht 02-91, PMI am DHZB, 1991.

[GB84]   J.A. Goguen and R.M. Burstall. Introducing institutions. *Proc. Logics of Programming Workshop, Carnegie-Mellon*, Springer LNCS 164:221 – 256, 1984.

[Gen91]   H.J. Genrich. Predicate/Transition Nets. In *High-Level Petri Nets: Theory and Application*, pages 3–43. Springer Verlag, 1991.

[GL81]   H.J. Genrich and K. Lautenbach. System Modelling with High-Level Petri Nets. *Theoretical Computer Science*, 13:109–136, 1981.

[GL98]   Volker Gruhn and Sabine Lembke. Integration of Petri Net Based Process Description with Different Data Modelling Techniques. In *Third World Conference on Integrated Design and Process Technology*, pages 105–112. Society for Process and Design, 1998.

[GL99]   Volker Gruhn and Sabine Lembke. Flexible Integration of Petri Net Based Process Description with User-Specific Data Descriptions. *Journal of Integrated Design and Process Technology*, 1999. To appear.

[GR83]   U. Goltz and W. Reisig. The Non-Sequential Behaviour of Petri Nets. In *Information and Computation*, pages 125–147. Academic Press, 1983.

[Hum89]   U. Hummert. *Algebraische High-Level Netze*. PhD thesis, Technische Universität Berlin, 1989.

[Jen81]   K. Jensen. Coloured Petri Nets and the Invariant Method. *Theoretical Computer Science*, 14:317–336, 1981.

[Jen92]   K. Jensen. *Coloured Petri Nets. Basic Concepts, Analysis Methods and Practical Use*, volume 1: Basic Concepts. Springer Verlag, EATCS Monographs in Theoretical Computer Science edition, 1992.

[JR91]   K. Jensen and G. Rozenberg, editors. *High-Level Petri-Nets: Theory and Application*. Springer Verlag, 1991.

[Juh98a]   Gabriel Juhás. *Algebraically generalised Petri nets*. PhD thesis, Institute of Control Theory and Robotics, Slovak Academy of Sciences, 1998.

[Juh98b]   Gabriel Juhás. The essence of Petri nets and transition systems through Abelian groups. *Electronic Notes in Theoretical Computer Science*, 18, 1998.

[Juh99]   Gabriel Juhás. Reasoning about algebraic generalisation of Petri nets. In *Proc. of 20th Conference on Theory and Application of Petri nets*, pages 324–343. Springer, LNCS 1639, 1999.

[KP95]   H. Klaudel and E. Pelz. Communication as unification in the Petri Box Calculus. Technical report, LRI, Universite de Paris Sud, 1995.

[KW98]   Ekkart Kindler and Michael Weber. The dimensions of Petri nets: The Petri net cube. Informatik-Bericht, Humboldt-Universität zu Berlin, 1998. To appear.

[Lil95]   J. Lilius. *On the Structure of High-Level Nets*. PhD thesis, Helsinki University of Technology, 1995. Digital Systems Laoratory, Research Report 33.

[Mah89]   B. Mahr. Empty carriers: the categorical burden on logic. In H. Ehrig, H. Herrlich, H.J. Kreowski, and G. Preuß, editors, *Categorical Methods in Computer Science – with Aspects from Topology*, pages 50–65. Springer LNCS 393, 1989.

[Mes92]   J. Meseguer. Conditional rewriting logic as a unified model of concurrency. *Theoretical Computer Science*, 96:73–155, 1992.

[MM90]   J. Meseguer and U. Montanari. Petri Nets are Monoids. *Information and Computation*, 88(2):105–155, 1990.

[MP92]   Zohar Manna and Amir Pnueli. *The Temporal Logic of Reactive and Concurrent Systems, Specification*. Springer Verlag, 1992.

[Pad93]   J. Padberg. Survey of high-level replacement systems. Technical Report 93/8, Technical University of Berlin, 1993.

[Pad96]   J. Padberg. *Abstract Petri Nets: A Uniform Approach and Rule-Based Refinement.* PhD thesis, Technical University Berlin, 1996. Shaker Verlag.

[Pad98]   Julia Padberg. Classification of Petri Nets Using Adjoint Functors. *Bulletin of EACTS 66,* 1998.

[Pad99a]   J. Padberg. The Petri Net Baukasten: An Application-Oriented Petri Net Technology. In Weber et al. [WER99], pages 191–209.

[Pad99b]   Julia Padberg. Categorical Approach to Horizontal Structuring and Refinement of High-Level Replacement Systems. *Applied Categorical Structures,* 7(4):371–403, December 1999.

[PER95]   J. Padberg, H. Ehrig, and L. Ribeiro. Algebraic high-level net transformation systems. *Mathematical Structures in Computer Science,* 5:217–256, 1995.

[PER01]   J. Padberg, H. Ehrig, and G. Rozenberg. Behaviour and Realization Construction for Petri Nets Based on Free Monoid and Power Set Graphs. In H. Ehrig, G. Juhás, J. Padberg, and G. Rozenberg, editors, *Advances in Petri Nets: Unifying Petri Nets,* LNCS. Springer, 2001.

[PGH99]   Julia Padberg, Maike Gajewsky, and Kathrin Hoffmann. Incremental Development of Safety Properties in Petri Net Transformations. In G. Engels and G. Rozenberg, editors, *Theory and Application of Graph Transformations(TAGT'98), Lecture Notes in Computer Science 1764,* pages 410–425. Springer Verlag, 1999.

[PR91]   H. Plünnecke and W. Reisig. Bibliography of Petri Nets 1990. *Springer LNCS 524,* 1991.

[Rei85]   W. Reisig. *Petri Nets,* volume 4 of *EATCS Monographs on Theoretical Computer Science.* Springer Verlag, 1985.

[Rei91]   W. Reisig. Petri Nets and Algebraic Specifications. *Theoretical Computer Science (Fundamental Studies),* (80):1–34, April 1991.

[RR98a]   Reisig, W. and Rozenberg, G., editors. *Lectures on Petri Nets I: Applications ,* volume 1492 of *Lecture Notes in Computer Science.* Springer-Verlag, 1998.

[RR98b]   Reisig, W. and Rozenberg, G., editors. *Lectures on Petri Nets I: Basic Models,* volume 1491 of *Lecture Notes in Computer Science.* Springer-Verlag, 1998.

[RT86]   G. Rozenberg and P.S. Thiagarajan. Petri nets: Basic notions, structure, behaviour. In *Current Trends in Concurrency,* pages 585–668. Lecture Notes in Computer Science 224, Springer, 1986.

[Sas98]   V. Sassone. An axiomatization of the category of Petri net computations. *Mathematical Structures in Computer Science,* 8:117–151, 1998.

[ST84]   D. T. Sannella and A. Tarlecki. Building specifications in an arbitrary institution. In *Proc. Int. Symposium on Semantics of Data Types, LNCS 173,* pages 337–356. Springer, 1984.

[TBG87]   A. Tarlecki, R.M. Burstall, and J.A. Goguen. Some fundamental algebraic tools for the semantics of computation. Part III: Indexed categories. Technical report, University of Edinburgh, 1987.

[Vau86]   J. Vautherin. Parallel Specification with Coloured Petri Nets and Algebraic Data Types. In *Proc. of the 7th European Workshop on Application and Theory of Petri nets,* pages 5–23, Oxford, England, jul. 1986.

[WER99]   H. Weber, H. Ehrig, and W. Reisig, editors. *Int. Colloquium on Petri Net Technologies for Modelling Communication Based Systems, Part II: The »Petri Net Baukasten«*. Fraunhofer Gesellschaft ISST, October 1999.

[Win87]   G. Winskel. Petri nets, algebras, morphisms, and compositionality. *Information and Computation*, 72:197–238, 1987.

[Wol95]   U. Wolter. Institutional frames. In E. Astesiano, G. Reggio, and A. Tarlecki, editors, *Recent Trends in Data Type Specification*, pages 469–482. 10th Workshop on Specification of Abstract Data Types joint with the 5th COMPASS Workshop, S. Margherita Italy, May/June 1994, Selected papers, Springer, LNCS 906, 1995.

A   Review of High-Level Replacement Systems

Here we briefly review the concepts of high-level replacement (HLR) systems in the sense of [EHKP91b], a categorical generalization of graph grammars and of Q-conditions according to [Pad96, Pad99a]. High-level replacement systems are formulated for an arbitrary category **CAT** with a distinguished class $\mathcal{M}$ of morphisms.

A.1   High-Level Replacement System and HLR-Conditions

Definition 25 (High-Level Replacement System). *Given a category* **CAT** *together with a distinguished class of morphisms* $\mathcal{M}$ *then* $(\mathbf{CAT}, \mathcal{M})$ *is called a HLR-category if* $(\mathbf{CAT}, \mathcal{M})$ *satisfies the HLR-Conditions.*

Variants of the following HLR-conditions have been stated in [EHKP91b] in order to prove local Church-Rosser and Parallelism Theorems in the framework of high-level replacement systems. In fact the conditions below imply those in [EHKP91b] and they are referred to as HLR1*-conditions [Pad93].

Definition 26 (HLR-Conditions). *Given a category* **CAT** *(of high-level structures) and a distinguished class* $\mathcal{M}$ *of morphisms in* **CAT** *the following conditions 1 – 8 are called HLR-conditions:*

1. *Existence of* $\mathcal{M}$ *pushouts*

   For objects A,B,C and morphisms $A \longrightarrow B$ and $A \longrightarrow C$, where at least one is in $\mathcal{M}$ there exists a pushout $C \longrightarrow D \longleftarrow B$.

$$
\begin{array}{ccc}
A & \longrightarrow & B \\
\downarrow & (1) & \downarrow \\
C & \longrightarrow & D
\end{array}
$$

2. *Existence of* $\mathcal{M}$-*pullbacks*
   For objects B,C,D and morphisms $B \longrightarrow D$, $C \longrightarrow D$ as in diagram **(1)** *above, where both morphisms are in* $\mathcal{M}$ *there exists a pullback $C \longleftarrow A \longrightarrow B$.*

3. *Inheritance of* $\mathcal{M}$
   − *under pushouts :*
     For each pushout diagram **(1)** *as above the morphism $A \longrightarrow B \in \mathcal{M}$ implies $C \longrightarrow D \in \mathcal{M}$.*

– *under pullbacks :*
   For each pullback diagram **(1)** *as above the morphism* $B \longrightarrow D \in \mathcal{M}$
   and $C \longrightarrow D \in \mathcal{M}$ *implies* $A \longrightarrow B \in \mathcal{M}$ *and* $A \longrightarrow C \in \mathcal{M}$

4. *Existence of binary coproducts and compatibility with* $\mathcal{M}$
   – *For each pair of objects* A, B *there is a coproduct* $A+B$ *with the universal morphisms* $A \longrightarrow A+B$ *and* $B \longrightarrow A+B$.
   – *For each pair of morphisms* $A \xrightarrow{f} A'$ *and* $B \xrightarrow{g} B'$ *in* $\mathcal{M}$ *the coproduct morphism* $A+B \xrightarrow{f+g} A'+B'$ *is also in* $\mathcal{M}$.

5. *Monomorphism condition:* $\mathcal{M}$ *is a class of monomorphisms in* **CAT**.
6. *Existence of initial object:* **CAT** *has an initial object.*
7. $\mathcal{M}$-*pushouts are pullbacks: Pushouts of* $\mathcal{M}$-*morphisms are pullbacks.*
8. $\mathcal{M}$-*pushout-pullback-decomposition*
   For each diagram to the right, we have:
   If **(1+2)** *is a pushout ,* **(2)** *is a pullback and* $A \longrightarrow C$, $B \longrightarrow D$, $E \longrightarrow F$, $B \longrightarrow E$ *and* $D \longrightarrow F$ *are* $\mathcal{M}$-*morphisms, then also* **(1)** *is a pushout.*

$$
\begin{array}{ccccc}
A & \longrightarrow & B & \longrightarrow & E \\
\downarrow & & \downarrow & & \downarrow \\
 & (1) & & (2) & \\
C & \longrightarrow & D & \longrightarrow & F
\end{array}
$$

A.2 $\mathcal{Q}$-Conditions

The main idea in the following definition is to enlarge the given HLR-category in order to include morphisms, that are adequate for refinement. The $\mathcal{Q}$-conditions [Pad99b] state additional requirements, that an HLR-category has to satisfy for the extension to refinement morphisms.

Definition 27 ($\mathcal{Q}$-Conditions [Pad99b]). *Let* **QCAT** *be a category, so that* **CAT** *is a subcategory* **CAT** $\subseteq$ **QCAT** *and* $\mathcal{Q}$ *a class of morphisms in* **QCAT**.

1. *The morphisms in* $\mathcal{Q}$ *are called* $\mathcal{Q}$-*morphisms, or refinement morphisms.*
2. *Then we have the following* $\mathcal{Q}$-*conditions:*
   Closedness: $\mathcal{Q}$ *has to be closed under composition.*
   Preservation of Pushouts: *The inclusion functor* $I : $ **CAT** $\longrightarrow$ **QCAT** *preserves pushouts, that is, given* $C \xrightarrow{f'} D \xleftarrow{g'} B$ *a pushout of* $B \xleftarrow{f} A \xrightarrow{g} C$ *in* **CAT**, *then* $I(C) \xrightarrow{I(f')} I(D) \xleftarrow{I(g')} I(B)$ *is a pushout of* $I(B) \xleftarrow{I(f)} I(A) \xrightarrow{I(g)} I(C)$ *in* **QCAT**.
   Inheritance of $\mathcal{Q}$-**morphisms under Pushouts:** *The class* $\mathcal{Q}$ *in* **QCAT** *is closed under the construction of pushouts in* **QCAT**, *that is, given* $C \xrightarrow{f'} D \xleftarrow{g'} B$ *a pushout of* $B \xleftarrow{f} A \xrightarrow{g} C$ *in* **QCAT**, *then* $f \in \mathcal{Q} \implies f' \in \mathcal{Q}$.
   Inheritance of $\mathcal{Q}$-**morphisms under Coproducts:** *The class* $\mathcal{Q}$ *in* **QCAT** *is closed under the construction of coproducts in* **QCAT**, *that is, for* $A \xrightarrow{f} B$ *and* $A' \xrightarrow{f'} B'$ *we have* $f, f' \in \mathcal{Q} \implies f + f' \in \mathcal{Q}$ *provided the coproduct* $A+A' \xrightarrow{f+f'} B+B'$ *of* f *and* f' *exists in* **QCAT**.

3. *A Q-rule (r, q) is given by a rule $r = L \xleftarrow{u} K \xrightarrow{v} R$ in **CAT** and a Q-morphism $q : L \longrightarrow R$, so that $K \xrightarrow{u} L \xrightarrow{q} R = K \xrightarrow{v} R$ in **QCAT**.*

B   Proofs of the Main Results

We now give the proofs of Theorem 1 to Theorem 8, that have been presented in Section 5 to Section 8 in Part II.

B.1   Proofs of Theorem 1 and Theorem 2

Proof of Theorem 1 is a special case of Theorem 2 that is proven below.
Proof of Theorem 2
First, we have to define the abstract assignment using amalgamation (see Definition 7): $AAS_2 = AAS_1 +_{M_1} M_2$ which is well-defined due to $V_{f_\Sigma}(M_2) = M1 = V_\phi(AAS_1)$.
Because of the compatibility of firing conditions we have for each $\varphi_2 \in cond_2$ $(f_T(t))$ there is $\varphi_1 \in cond_1(t)$ with $Cond(f_\Sigma)(\varphi_1) = \varphi_2$ and due to the satisfaction condition we can conclude: $AAS_1 \models \varphi_1 \Leftrightarrow AAS_2 \models \varphi_2$
Next, note that for each $t_1 \in T_1$ and $f_T(t_1) = t_2 \in T_2$ the following diagram commutes:

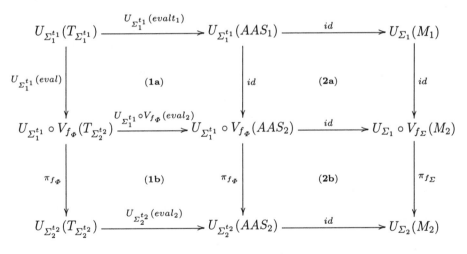

(**1a**) commutes due to initiality of $T_{\Sigma_1^{t_1}}$ in $Cat(\Sigma_1^{t_1})$
(**1b**) commutes due to natural transformation π_{f_Φ} $Cat(\Sigma_1^{t_1})$
(**2a**) and (**2b**) commute due to the amalgamation $AAS_2 = AAS_1 +_{M_1} M_2$

We have $aas_1 = F(U_{\Sigma_1^{t_1}}(eval_1) \times id_{P_1})$, $aas_2 = F(U_{\Sigma_2^{t_2}}(eval_2) \times id_{P_2})$. The cartesian product and functors preserve commutativity of diagram (**1a + 1b + 2a + 2b**), so we conclude that (**3**) commutes in **Struct**.

$$F(U_{\Sigma_1^{t_1}}(T_{\Sigma_1^{t_1}}) \times P_1) \xrightarrow{\quad aas_1 \quad} F(U_{\Sigma_1}(M_1) \times P_1)$$

$$\Big\downarrow F(f_{ins}) \qquad\qquad (3) \qquad\qquad \Big\downarrow F(\pi_{f_\Sigma} \times f_P) = F(f_m)$$

$$F(U_{\Sigma_2^{t_2}}(T_{\Sigma_2^{t_2}}) \times P_2) \xrightarrow{\quad aas_2 \quad} F(U_{\Sigma_2}(M_2) \times P_2)$$

with f_{ins} as defined in Definition 14.2.

This leads to the desired propositions.

1. $F(f_T)(v_t)$ is aas_2-enabled:

$F(f_m)(m)$

$\quad = F(f_m)(\overline{m} + aas_1(\overline{pre_1}(v_t)))$ m is AAS_1-enabled

$\quad = F(f_m)(\overline{m}) + F(f_m)(aas_1(\overline{pre_1}(v_t)))$ due to homomorphisms
                                                  in **Struct**

$\quad = F(f_m)(\overline{m}) + aas_2(F(f_{ins})(\overline{pre_1}(v_t)))$ as **(3)** commutes

$\quad = F(f_m)(\overline{m}) + aas_2(\overline{pre_2}(F(f_T)(v_t)))$ due to **HLAPN**-morphisms
                                                  (Definition 14)

2. the follower marking after firing $F(f_T)(v_t)$ is preserved:

$F(f_m)(m')$

$\quad = F(f_m)(\overline{m} + aas_1(\overline{post_1}(v_t)))$ m' is follower marking

$\quad = F(f_m)(\overline{m}) + F(f_m)(aas_1(\overline{post_1}(v_t)))$ due to homomorphisms
                                                  in **Struct**

$\quad = F(f_m)(\overline{m}) + aas_2(F(f_{ins})(\overline{post_1}(v_t)))$ as **(3)** commutes

$\quad = F(f_m)(\overline{m}) + aas_2(\overline{post_2}(F(f_T)(v_t)))$ due to **HLAPN**-morphisms
                                                  (Definition 14)

B.2   Proof of Theorem 3

(Proof of Theorem 3)
The proof is based on the fact that arbitrary finite colimits can be constructed if the category has initial objects and pushouts (by dualization of Theorem 12.4 in [AHS90]).

Initial Object:
    The initial object $N_\emptyset$ is given by:
    $N_\emptyset = (P_\emptyset, T_\emptyset, \Sigma_\emptyset, pre_\emptyset, post_\emptyset, cond_\emptyset)$ with
    – $P_\emptyset = T_\emptyset = COND_\emptyset = \emptyset$ the initial object in **Sets**
    – $SPEC_\emptyset$ is the initial object in **ASPEC**
    Given a high-level abstract Petri net $N = (P, T, SPEC, Var, pre, post, cond)$
    the unique **HLAPN**-morphism $f = (f_P^\emptyset, f_T^\emptyset, f_\Sigma^\emptyset) : N_\emptyset \longrightarrow N$ is given by
    – $f_P^\emptyset = f_T^\emptyset$ is the empty function
    – $f_\Sigma^\emptyset$ is given by initiality of $SPEC_\emptyset$.

The conditions 1 to 3 for **HLAPN**-morphisms (see Definition 14) hold, due to emptyness of $T_\emptyset$.

Pushout:

Given $N_i = (P_i, T_i, SPEC_i, Var_i, pre_i, post_i, cond_i)$ for $0 \leq i \leq 2$ and
$N_1 \xleftarrow{\ f\ } N_0 \xrightarrow{\ g\ } N_2$ with $f = (f_P, f_T, f_\Sigma)$ and $g = (g_P, g_T, g_\Sigma)$.

Then the pushout $N_3 = (P_3, T_3, SPEC_3, Var_3, pre_3, post_3, cond_3)$ with
$N_1 \xrightarrow{\ g'\ } N_3 \xleftarrow{\ f'\ } N_2$ and $g' = (g'_P, g'_T, g'_\Sigma)$ and $f' = (f'_P, f'_T, f'_\Sigma)$ is
constructed by :

- $P_1 \xrightarrow{\ g'_P\ } P_3 \xleftarrow{\ f'_P\ } P_2$ is pushout of $P_1 \xleftarrow{\ f_P\ } P_0 \xrightarrow{\ g_P\ } P_2$ in **Sets**.
- $T_1 \xrightarrow{\ g'_T\ } T_3 \xleftarrow{\ f'_T\ } T_2$ is pushout of $T_1 \xleftarrow{\ f_T\ } T_0 \xrightarrow{\ g_T\ } T_2$ in **Sets**.
- $SPEC_1 \xrightarrow{\ g'_\Sigma\ } SPEC_3 \xleftarrow{\ f'_\Sigma\ } SPEC_2$ is pushout of
  $$SPEC_1 \xleftarrow{\ f_\Sigma\ } SPEC_0 \xrightarrow{\ g_\Sigma\ } SPEC_2 \text{ in } \mathbf{ASPEC}.$$
- Signatures with variables are given for each transition, so we have to construct them with respect to the pushout T_3 of transitions. Because pushouts in **Sets** are jointly surjective we have:
  1. $t_3 \in T_3$ is obtained from T_2 and is not in the interface T_0
  2. $t_3 \in T_3$ is obtained from T_1 and is not in the interface T_0
  3. $t_3 \in T_3$ is obtained from the interface T_0

  For each $t_3 \in T_3$ we construct $\Sigma_3 \xrightarrow{\ \phi 3_{t_3}\ } \Sigma_3^t$ using pushouts in **ASIG**:
  For $g'_T(t_1) = t_3 \in T_3$ and $t_1 \notin f_T(T_0)$ we have the pushout **PO 1**.
  For $f'_T(t_2) = t_3 \in T_3$ and $t_2 \notin g_T(T_0)$ we have the pushout **PO 2**.
  For $g'_T \circ f_T(t_0) = f'_T \circ g_T(t_0) = t_3 \in T_3$ we have the pushout **PO 3**.

$$
\begin{array}{ccc}
\Sigma_1 \xrightarrow{\ \phi_1^{t_1}\ } \Sigma_1^{t_1} &
\Sigma_2 \xrightarrow{\ \phi_2^{t_2}\ } \Sigma_2^{t_2} &
\Sigma_0 \xrightarrow{\ \phi_0^{t_0}\ } \Sigma_0^{t_0} \\[2pt]
\Big\downarrow g'_\Sigma \quad \mathbf{PO\ 1} \quad \Big\downarrow g'_\Phi &
\Big\downarrow f'_\Sigma \quad \mathbf{PO\ 2} \quad \Big\downarrow f'_\Phi &
\Big\downarrow g'_\Sigma \circ f'_\Sigma \quad \mathbf{PO\ 3} \quad \Big\downarrow g'_\Phi \circ f'_\Phi \\[2pt]
\Sigma_3 \xrightarrow[\ \phi_3^{t_3}\ ]{} \Sigma_3^{t_3} &
\Sigma_3 \xrightarrow[\ \phi_3^{t_3}\ ]{} \Sigma_3^{t_3} &
\Sigma_3 \xrightarrow[\ \phi_3^{t_3}\ ]{} \Sigma_3^{t_3}
\end{array}
$$

$\phi_3^{t_3}$ is well-defined due to the pushout properties of T_3.

- Pushout properties of T_3 yield
  $$pre_3, post_3 : T_3 \longrightarrow Net(U_{\Sigma_3^{t_3}}(T_{\Sigma_3^{t_3}}) \times P_3) \text{ and}$$
  $$cond_3 : T_3 \longrightarrow \mathcal{P}_{fin}(Cond(\Sigma_3^{t_3})).$$
- $f' = (f'_P, f'_T, f'_\Sigma)$ and $g' = (g'_P, g'_T, g'_\Sigma)$ are well-defined, that is they satisfy the conditions 1 to 3 for **HLAPN**-morphisms in Definition 14 :

  1. preservation of variables is due to the construction of $\Sigma_3 \xrightarrow{\ \phi_3^{t_3}\ } \Sigma_3^{t_3}$
  2. compatibility of pre and post domain
     is due to the induced pushout morphisms pre_3 and $post_3$
  3. compatibility of firing conditions
     is due to the induced pushout morphisms $cond_3$

Next we check the pushout properties:

Given a high-level abstract Petri net N_4 and two morphisms $f^* = (f_P^*, f_T^*, f_\Sigma^*)$ and $g^* = (g_P^*, g_T^*, g_\Sigma^*)$, such that $f^* \circ g = g^* \circ f$, then the induced pushout morphism $k = (k_P, k_T, k_\Sigma)$, where k_P, k_T and k_Σ are the induced morphisms of the components P_3, T_3 and $SPEC_3$.

k is well-defined with respect to the conditions 1 to 3 for **HLAPN**-morphisms in Definition 14.

1. Preservation of variables:

   Let $t_2 \in T_2$, $f_T'(t_2) = t_3 \in T_3$, and $f_T^*(t_2) = t_4 \in T_4$, then for each t_2 we have in **ASIG**:

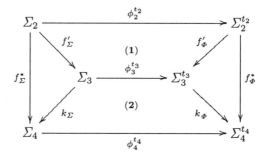

   (1) is pushout due to morphism f' and k_Φ is induced by pushout (1). The outer square is pushout due to morphism f^*. (2) commutes, thus by pushout decomposition property the square (2) is pushout as well. The same holds for each $t_1 \in T_1$.

2. Compatibility of pre- and postdomain:

   $k_{ins} = (\pi_{k_\Phi} \circ U_{\Sigma_3^{t_3}}(eval)) \times k_P$ with $eval : T_{\Sigma_3^{t_3}} \longrightarrow V_{k_\Phi}(T_{\Sigma_4^{t_4}})$ is given and commutes the squares, because T_3 is pushout.

3. Compatibility with firing conditions:

   $k_{cond} = \mathcal{P}(Cond(k_\Phi))$ is given and commutes the squares, because T_3 is pushout.

Uniqueness of k is obtained by uniqueness of k_P, k_T and k_Σ.

B.3   Proofs of Theorems Concerning Rule-Based Refinement

These proofs rely on the one hand on those theorems stating that the HLR-conditions are sufficient to show the results for HLR-systems in general. On the other hand they rely on the satisfaction of the HLR-conditions in Definition 26 by abstract Petri nets, i.e. for the category **HLAPN**.

(Proof of Theorems 4, 5, 6 7, and 8)

Church-Rosser Theorem I

   In Theorem 4.3 in [EHKP91b] and Theorem 4.5 in [Pad99b]

Church-Rosser Theorem II

   In Theorem 4.4 in [EHKP91b] and Theorem 4.7 in [Pad99b]

Parallelism Theorem
    In Theorem 4.6 in [EHKP91b] and Theorem 4.11 in [Pad99b]
Fusion Theorem In Theorem 3.4 in [PER95] and Fact 5.4 in [Pad99b]
Union Theorem In Theorem 3.7 in [PER95] and Fact 5.8 in [Pad99b]

It remains to verify the HLR-conditions (see Definition 26) for the category
HLAPN.

Lemma 5 (HLAPN satifies HLR-conditions).

Proof. The proof uses the existence and decomposition of pushouts (see Theorem 3 and Corollary 1), the assumed existence of pullbacks of $\mathcal{M}$-morphisms (see Lemma 6) and the satisfaction of the HLR-conditions by the high-level replacement system $(\mathbf{ASPEC}, \mathcal{M}_{\mathbf{ASPEC}})$.

1. *Existence of $\mathcal{M}$-pushouts* is due to finite cocompleteness of **HLAPN** (see Theorem 3).
2. *Existence of $\mathcal{M}$-pullbacks* due to assumption.
3. *Inheritance of $\mathcal{M}$* is due to the corresponding HLR-property of the components, that is given in **Sets** and is assumed in **ASPEC**.
4. *Existence of binary coproducts* is due to finite cocompleteness and the corresponding HLR-property of the components, that is given in **Sets** and is assumed in **ASPEC**.
5. *Monomorphism condition* is due to the corresponding HLR-property of the components, that is given in **Sets** and is assumed in **ASPEC**.
6. *$\mathcal{M}$-pushouts are pullbacks* is due to the corresponding HLR-property of the components, that is given in **Sets** and is assumed in **ASPEC**, and the decomposition of pullbacks of high-level abstract Petri nets (see Lemma 6).
7. *$\mathcal{M}$-pushout-pullback-decomposition* is due to the corresponding HLR-property of the components, that is given in **Sets** and is assumed in **ASPEC**, and the decomposition of pullbacks of high-level abstract Petri nets (see Lemma 6).

Lemma 6 (Pullbacks of $\mathcal{M}$-Morphisms in HLAPN). *The category* **HLAPN** *has $\mathcal{M}$-pullbacks, that are pullbacks of $\mathcal{M}$-morphisms, if*

- *the cube-pushout-pullback-lemma (see [EHKP91b]) holds in* **ASIG**,
- *The left-adjoint $F \dashv G$ with $Net = G \circ F$ preserves pullbacks of injective morphisms*
- **ASIG** *has $\mathcal{M}_{\mathbf{ASIG}}$-pullbacks, we have inheritance of $\mathcal{M}_{\mathbf{ASIG}}$ under pushouts and pullbacks, and*
- *the construction of the term algebra T_Σ, the functors $U_\Sigma : \mathbf{Cat}(\Sigma) \longrightarrow$* **Sets** *and $Cond : \mathbf{ASIG} \longrightarrow$* **Sets** *preserves pullbacks of injective morphisms.*

Proof. Given the **HLAPN**-morphims $f : N_1 \longrightarrow N_0$ and $g : N_2 \longrightarrow N_0$ with $f = (f_P, f_T, f_\Sigma) \in \mathcal{M}$ and $g = (g_P, g_T, g_\Sigma) \in \mathcal{M}$:

Then there are the pullbacks

$P_1 \xleftarrow{\ g'_P\ } P_3 \xrightarrow{\ f'_P\ } P_2$ of $P_1 \xrightarrow{\ f_P\ } P_0 \xleftarrow{\ g_P\ } P_2$ in **Sets**

$T_1 \xleftarrow{\ g'_T\ } T_3 \xrightarrow{\ f'_T\ } T_2$ of $T_1 \xrightarrow{\ f_T\ } T_0 \xleftarrow{\ g_T\ } T_2$ in **Sets**

$\Sigma_1 \xleftarrow{\ g'_\Sigma\ } \Sigma_3 \xrightarrow{\ f'_\Sigma\ } \Sigma_2$ of $\Sigma_1 \xrightarrow{\ f_\Sigma\ } \Sigma_0 \xleftarrow{\ g_\Sigma\ } \Sigma_2$ in **ASIG**

Thus we can construct the following diagram in **ASIG** for each $t_3 \in T_3$ with $g'_T(t_3) = t_1 \in T_1$, $f'_T(t_3) = t_2 \in T_2$, and $f_T \circ g'_T(t_3) = g_T \circ f'_T(t_3) = t_0 \in T_0$ Where the front and the right square are pushouts due to the translation of variables, and the back and the left square are constructed below. All morphisms of the top square are in $\mathcal{M}_{\textbf{ASIG}}$, due to the inheritance of $\mathcal{M}_{\textbf{ASIG}}$ under pullbacks. Since the front and the right squares are pushouts, and we have inheritance of $\mathcal{M}_{\textbf{ASIG}}$-morphisms under pushouts,

also $\Sigma_2^{t_2} \longrightarrow \Sigma_0^{t_0}$ and $\Sigma_1^{t_1} \longrightarrow \Sigma_0^{t_0}$ are in $\mathcal{M}_{\textbf{ASIG}}$. Due to existence of $\mathcal{M}_{\textbf{ASIG}}$ pullbacks in **ASIG**, there is $\Sigma_3^{t_3}$ and due to inheritance of $\mathcal{M}_{\textbf{ASIG}}$ the bottom square consists of $\mathcal{M}_{\textbf{ASIG}}$ morphisms. Summarizing we have: The top-square is a pullback and all morphisms are in $\mathcal{M}_{\textbf{ASIG}}$. The front and right diagram are pushouts and the bottom square is also a pullback. Using the cube-pushout-pullback-lemma (see [EHKP91b, Pad93]) we conclude, that the left and the back square are pushouts as well. Thus the construction $\Sigma_3 \longrightarrow \Sigma_3^{t_3}$ is well-defined for each $t_3 \in T_3$.

This leads to the situation depicted in the following diagram. Due to the fact, that F preserves $\mathcal{M}$-pullbacks, also Net preserves $\mathcal{M}$-pullbacks. Furthermore, we assume that the term algebra construction T_Σ and the functors U_Σ : **Cat**$(\Sigma) \longrightarrow$ **Sets** preserves injective pullbacks. Thus the right square is a pullback in **Sets** and pre_3 and $post_3$ are the induced pullback morphisms, so they are well-defined.

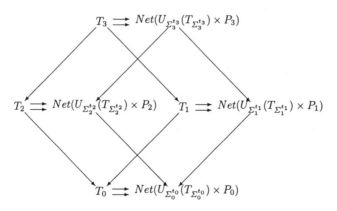

Furthermore we have in the following diagram the construction of the term algebra T_Σ, the functors $U_\Sigma : \mathbf{Cat}(\Sigma) \longrightarrow \mathbf{Sets}$ and $Cond : \mathbf{ASIG} \longrightarrow \mathbf{Sets}$ preserves pullbacks of injective morphisms.

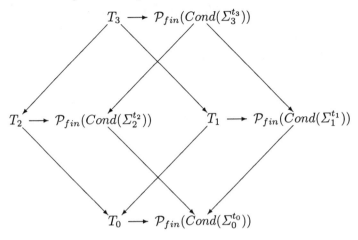

As the $\mathcal{P}_{fin}$ functor preserves injective pullbacks and $Cond : \mathbf{ASIG} \longrightarrow$ $\mathbf{Sets}$ preserves pullbacks of injective morphisms by assumption, $\mathcal{P}_{fin}(Cond(\Sigma_3^{t_3}))$ is pullback in $\mathbf{Sets}$ and $cond3$ is the induced pullback morphism.
Thus we can construct the pullback of $f : N_1 \longrightarrow N_0$ and $g : N_2 \longrightarrow N_0$:

$$N_3 = (P_3, T_3, SPEC_3, Var_3, pre_3, post_3, cond_3)$$

with the components as defined above.
The pullback morphisms are given by the pullback morphisms of the components, $f' = (f'_P, f'_T, f'_\Sigma) : N_3 \longrightarrow N_2$ and $g' = (g'_P, g'_T, g'_\Sigma) : N_3 \longrightarrow N_1$.
The pullback properties are obtained from the pullback properties of the components.

Corollary 2 (Decomposition of Pullbacks). *Given a pullback of $\mathcal{M}$-morphisms in* **HLAPN** *the components are pullbacks of injective morphisms in* **Sets** *and pullbacks of $\mathcal{M}_{\mathbf{ASIG}}$-morphisms in* **ASIG**.

Behavior and Realization Construction for Petri Nets Based on Free Monoid and Power Set Graphs

Julia Padberg[1], Hartmut Ehrig[1], and Grzegorz Rozenberg[2]

[1] Technische Universität Berlin
{padberg,ehrig}@cs.tu-berlin.de
[2] Reijksuniversiteit Leiden
rozenber@cs.leidenuniv.nl

Abstract Starting from the algebraic view of Petri nets as monoids (as advocated by Meseguer and Montanari in [MM90]) we present the marking graphs of place transition nets as free monoid graphs and the marking graphs of specific elementary nets as powerset graphs. These are two important special cases of a general categorical version of Petri nets based on a functor M, called M-nets. These nets have a compositional marking graph semantics in terms of F-graphs, a generalization of free monoid and powerset graphs. Moreover we are able to characterize those F-graphs, called reflexive F-graphs, which are realizable by corresponding M-nets. The main result shows that the behavior and realization constructions are adjoint functors leading to an equivalence of the categories **MNet** of M-nets and **RFGraph** of reflexive F-graphs. This implies that the behavior construction preserves colimits so that the marking graph construction using F-graphs is compositional.

In addition to place transition nets and elementary nets we provide other interesting applications of M-nets and F-graphs. Moreover we discuss the relation to classical elementary net systems. The behavior and realization constructions we have introduced are compatible with corresponding constructions for elementary net systems (with initial state) and elementary transition systems in the sense of [NRT92].

Keywords: place/transition nets, elementary nets, uniform approach, realization problem, category theory

1   Introduction

Petri nets are the most common approach to model concurrent and distributed systems within the *true concurrency* paradigm. Concurrency is considered to be a primitive notion modeled by the concurrent firing of transitions. This behavior is given by marking, case, or occurrence graphs. These graphs describe the relation between markings and the firing of transitions. The corresponding graph consists of nodes that represent markings and arcs that represent the firing of enabled transitions. On the other hand it is useful to characterize those graphs which realize the behavior of Petri nets. This realization problem was solved for

H. Ehrig et al. (Eds.): Unifying Petri Nets, LNCS 2128, pp. 230–249, 2001.
© Springer-Verlag Berlin Heidelberg 2001

elementary transition system, a restricted kind of graphs, yielding elementary net systems in [NRT92]. The behavior and realization construction have been given as adjoint functors between the corresponding categories.

In this paper we take the general idea of [NRT92] but consider Petri nets without initial state. We use category theory to define an abstract notion of Petri nets based on a functor M. For these abstract Petri nets, called M-nets[1], we can give a functorial semantics by F-graphs, a suitable generalization of the marking graph. On the other hand we can characterize those F-graphs that are realizable as M-nets. The main result shows that the behavior construction for M-nets and the realization construction for F-graphs is given by a pair of adjoint functors. Since the behavior functor is left adjoint it preserves pushouts and general colimits. This means that the marking graph construction using F-graphs is compositional. Note, that this fails to be true using ordinary graphs.

In the next section we present place/transition nets as well as F-graphs and state the main results concerning behavior and realization. In fact, these results and their proofs are given in a more abstract way in Section 3 where M-nets and their behavior as F-graphs are introduced. The construction of adjoint functors and the compositionality results are shown. In Section 4 we offer different kinds of place/transition nets and elementary nets being special cases of M-nets and F-graphs. This leads to new behavior and realization results as a corollary of the general theory. Compatibility between this approach concerning elementary nets and the approach using regional constructions in [NRT92] is discussed in Section 5.

Related work follows mainly two lines of research. Starting from the fundamental work in[ER90] and later in [DR96] the realization problem has been studied in several net classes. An overview is given in [BD98]. The basic idea is to use regions of graphs in order to represent extensions of places of the corresponding nets. Recent work [Dar00] emphases its application to distributed software and distributed control. In this volume the paper [BB01] is concerned with the realization problem for reactive systems.

The other line of research concerns approaches to a uniform description of Petri nets, where this volume is expecially devoted to. These approaches stem from the influential paper by Meseguer and Montanari [MM90]. There have been extensions towards partial algebras [Juh99,DJL01], and the combination with institutions [Pad96,EP97,PE01]. In [DKPS91] a similar approach to elementary nets has been introduced.

Acknowledgments. This paper is based on a technical report [EPR94] which was the result of the cooperation between Leiden and Berlin due to the Kloosterman-Professorship of Hartmut Ehrig at the University of Leiden in 1993. This paper was a first step towards the concept of abstract Petri nets presented in [Pad96] and parameterized net classes in [EP97] and in this volume [PE01].

---

[1] Not to be mixed with Mnets in the sense of [BFF+95], that are a high-level net extension of the Petri Box Calculus.

For historical reasons we have preserved the main structure of the report [EPR94]. But the material is presented in a condensed form and a comparison with recent related work has been provided.

2   Place/Transition Nets and Free Monoid Graphs

Subsequently we introduce an algebraic formulation of place/transition nets in the sense of Meseguer and Montanari [MM90], and graphs with free monoids of edges and nodes, called free monoid graphs, short FM-graphs. In both cases we define morphisms leading to categories **PT** and **FMGraph**. We give an example how to compose place/transition nets using pushouts in the category **PT** and how to construct the marking graph semantics using pushouts in **FMGraph**. However, these are not pushouts in the category **Graph** of graphs. Moreover we characterize those FM-graphs, called reflexive FM-graphs, which are realizable as place/transition nets. The concept of place/transition nets and FM-graphs is a special case of M-nets and F-graphs in the theory of categorical Petri nets developed in Section 3. As special cases of the general results we obtain:

1. There is a compositional construction of place/transition nets using pushouts in **PT**.
2. The marking graph construction of place/transition nets is defined in terms of FM-graphs.
3. The marking graph construction is compositional, that is a functor $MG :$ **PT** $\rightarrow$ **FMGraph** preserving pushouts and coproducts.
4. There is a subcategory **RFMGraph** of reflexive FM-graphs and a realization functor $R :$ **RFMGraph** $\rightarrow$ **PT** that is left and right inverse to the marking graph construction.
5. An F-graph G is realizable by a place/transition net N, i.e. $MG(N) \cong G$, if and only if G is reflexive.

Definition 1 (Category PT of Place/Transition Nets).

The category **PT** *consists of place/transition nets* $N = (T \underset{post}{\overset{pre}{\rightrightarrows}} P^{\oplus})$

as objects, where T (transitions) and P (places) are sets, $P^{\oplus}$ is the free commutative monoid over P and $pre, post : T \rightarrow P^{\oplus}$ denote the pre- and post-domain of each transition. Morphisms $f : N_1 \rightarrow N_2$ are tuples of functions $f = (f_T : T_1 \rightarrow T_2, f_P : P_1 \rightarrow P_2)$ such that the subsequent diagram commutes:

$$
\begin{array}{ccc}
T_1 & \overset{pre_1}{\underset{post_1}{\rightrightarrows}} & P_1^{\oplus} \\
f_T \downarrow & & \downarrow f_P^{\oplus} \\
T_2 & \overset{pre_2}{\underset{post_2}{\rightrightarrows}} & P_2^{\oplus}
\end{array}
$$

Remark 1 (Morphisms). In contrast to [MM90] we only admit functions $f_P : P_1 \rightarrow P_2$. Hence we use freely generated homomorphisms $f_P^{\oplus} : P_1^{\oplus} \rightarrow P_2^{\oplus}$ instead of general homomorphisms $f_P : P_1^{\oplus} \rightarrow P_2^{\oplus}$. This restriction allows arbitrary colimits in the category **PT** that are constructed componentwise in the category **Set**.

Remark 2 (Marking Graph). Given a place/transition net $N = (T \underset{post}{\overset{pre}{\rightrightarrows}} P^\oplus)$
then the marking graph $\mathrm{MG}(N)$ of N is a free monoid graph

$$\mathrm{MG}(N) = ((T + P)^\oplus \underset{t}{\overset{s}{\rightrightarrows}} P^\oplus)$$

with edges $(T+P)^\oplus$ and vertices $P^\oplus$. Then for $e \in (T+P)^\oplus$ given by the linear sum $e = \sum_{i=1}^n \lambda_i\, e_i$ with $e_i \in T \uplus P$ and $\lambda_i \in \mathbb{N}$ we have $s(e) = \sum_{i=1}^n \lambda_i\, s(e_i)$ with $s(e_i) = pre(e_i)$ for $e_i \in T$ and $s(e) = e_i$ for $e_i \in P$.

Next we give the category **FMGraph** of free monoid graphs, consisting of graphs whose nodes and edges have a free monoid structure.

Definition 2 (Category FMGraph).

*The category **FMGraph** of free monoid graphs consists of free monoid graphs*
$G = (E^\oplus \underset{t}{\overset{s}{\rightrightarrows}} V^\oplus)$ *where E and V are sets of basic edges and vertices.*

*$E^\oplus$ and $V^\oplus$ are free commutative monoids over
E and V defining general edges and vertices.
And $s, t : E^\oplus \to V^\oplus$ are homomorphisms defining source and target of edges.
The morphisms in FM-graphs $f : G_1 \to G_2$ are
given by a tuple of functions $f = (f_E : E_1 \to
E_2, f_V : V_1 \to V_2)$ such that the following dia-
gram commutes with $f_E^\oplus$ and $f_V^\oplus$ being induced
by the free construction $(\_)^\oplus$:*

$$
\begin{array}{ccc}
E_1^\oplus & \overset{s_1}{\underset{t_1}{\rightrightarrows}} & V_1^\oplus \\
{\scriptstyle f_E^\oplus}\downarrow & & \downarrow{\scriptstyle f_V^\oplus} \\
E_2^\oplus & \overset{s_2}{\underset{t_2}{\rightrightarrows}} & V_2^\oplus
\end{array}
$$

For the formal descriptions and theorems concerning behavior and realization as presented at the beginning of this subsection we refer to Sections 3 and 4. The following example illustrates the compositionality of the behavior constructions using free monoid graphs.

Example 1 (Compositionality of Marking Graph Construction).
In Figure 1 the pushout in the category **PT** describes the union N4 of the nets N2 and N3 with respect to the interface N1 where the arrows are the obvious inclusions for the transitions and places.

The marking graphs M1, ..., M4 of N1, ..., N4 are given in Figure 2. Note that the marking graph, i.e. the graph of all possible markings of a place/transition net is infinite. Thus Figure 2 merely illustrates a finite part of the corresponding marking graphs. Nodes without adjacent arcs are simply markings enabling no transition. Arcs with inscriptions like p3 $\oplus$ t2 denote the firing of t2 where one token remains on p3. Inscriptions like t2 $\oplus$ t3 denote the concurrent firing of t2 and t3. All places are equipped with loops inscribed with their name. For place p1 this loop has been made explicit.

The marking graphs M1, ..., M4 of N1, ..., N4 form a pushout in the category **FMGraph** of free monoid graphs. But this is *not* a pushout in the category **Graph** of graphs $E \underset{t}{\overset{s}{\rightrightarrows}} V$ in the classical sense where morphisms are given

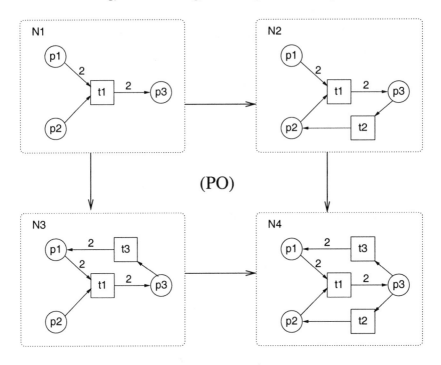

Figure 1. Pushout of place/transition nets

by functions $f_E : E_1 \to E_2$ and $f_V : V_1 \to V_2$ defined on all edges and nodes and not only on the generators.

In fact the edge t2 $\oplus$ t3 in M4 has neither a preimage in M2 nor in M3 as it would be the case for a pushout in the category **Graph**.

3   The Behavior and Realization Problem for Petri Nets in a Categorical Framework

In this section we introduce a categorical version of Petri nets, called M-nets, because these nets are based on a functor M, which is the composition of a free functor F and a forgetful functor U. Place/Transition nets and elementary nets introduced in Section 2 are special cases of M-nets. On the other hand we introduce F-graphs generalizing free monoid- and powerset graphs. The behavior problem for M-nets is solved by giving a compositional marking graph construction for M-nets in terms of F-graphs. Vice versa we characterize those F-graphs which admit a realization in terms of M-nets. These F-graphs are called reflexive F-graphs. The realization problem is solved by showing that we obtain a realization functor being a left and right inverse to the marking graph construction. The marking graph construction for M-nets is closely related to the free functor between the categories **Petri** (Petri nets) and **CMonRPetri** (reflexive commu-

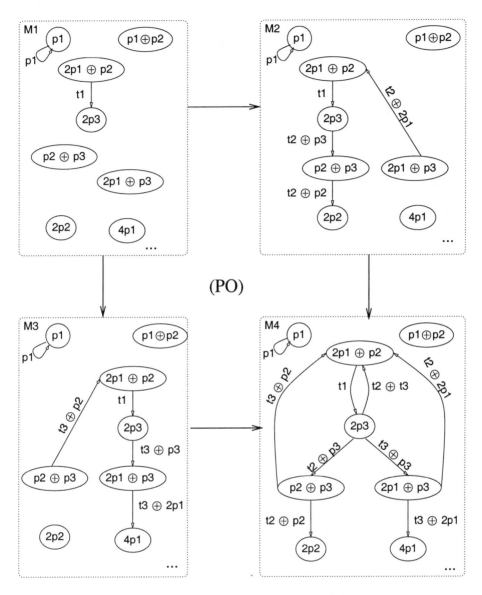

Figure 2. Pushout of marking graphs

tative monoid Petri nets) given in [MM90], but our realization functor is essentially different from the corresponding forgetful functor from **CMonRPetri** to **Petri**.

Applications of M-nets and F-graphs to various kinds of place/transition and elementary nets, especially those considered in Section 2, are given in Section 4 and summarized in Subsection 4.10.

Definition 3 (General Assumption for BASE and CAT).

We assume to have two categories **BASE** *and* **CAT** *, called base and structured category, and functors* $U : $ **CAT** $\to$ **BASE** *and* $F : $ **BASE** $\to$ **CAT** *, called forgetful and free functor, where F is left adjoint to U with universal morphisms* $u_B : B \to U \circ F(B)$ *for all objects B in* **BASE**. *The composition* $M = U \circ F : $ **BASE** $\to$ **BASE** *is called marking functor, because $M(B)$ corresponds to the set of all markings over B in all our examples of Petri nets.*

Moreover, we assume that the base category **BASE** *has colimits, especially pushouts and coproducts.*

Based on thes categories and functors we now can define M-nets.

Definition 4 (M-net).

An M-net $N = (T, P, pre, post)$ is given by objects T and P in **BASE**, *called transitions and places, and* **BASE** *morphisms $T \overset{pre}{\underset{post}{\rightrightarrows}} M(P)$, called pre- and post-domain of each $t \in T$.*

Definition 5 (Category MNet).

Given M-nets $N_i = (T_i, P_i, pre_i, post_i)$ with $(i = 1, 2)$ an M-net morphism $f : N_1 \to N_2$ is given by a pair $f = (f_T, f_P)$ of **BASE** *morphisms $f_T : T_1 \to T_2$, $f_P : P_1 \to P_2$ such that the following diagram commutes separately for pre- and post-domain, i.e. $M(f_P) \circ pre_1 = pre_2 \circ f_T$ and $M(f_P) \circ post_1 = post_2 \circ f_T$.*

$$\begin{array}{ccc} T_1 & \overset{pre_1}{\underset{post_1}{\rightrightarrows}} & M(P_1) \\ f_T \downarrow & & \downarrow M(f_P) \\ T_2 & \overset{pre_2}{\underset{post_2}{\rightrightarrows}} & M(P_2) \end{array}$$

Lemma 1 (Cocompleteness of MNet).

The category **MNet** *has colimits which are constructed componentwise in the base category* **BASE**.

Proof. Follows directly from the fact that **BASE** has colimits which allows componentwise construction of colimits in **MNet**.
Note that it is not necessary to assume that M preserves colimits.

Next we introduce F-graphs and the corresponding category.

Definition 6 (F-graphs).

1. *An F-graph $G = (F(E), F(V), s, t)$ is given by objects $F(E)$ and $F(V)$ in* **CAT**, *called edges and vertices, for objects E and V in* **BASE**, *called base edges and base vertices, and* **CAT**-*morphisms $F(E) \overset{s}{\underset{t}{\rightrightarrows}} F(V)$, called source and target of G.*
2. *Given F-graphs $G_i = (F(E_i), F(V_i), s_i, t_i)$ for $(i = 1, 2)$ then a F-graph morphism $f : G_1 \to G_2$ is given by $f = (f_E, f_V)$ of* **BASE** *morphisms $f_E : E_1 \to E_2$ and $f_V : V_1 \to V_2$, such that the following diagram commutes separately for source and target.*

$$F(E_1) \xrightarrow[t_1]{s_1} F(V_1)$$

$$F(f_E) \downarrow \qquad \qquad \downarrow F(f_V)$$

$$F(E_2) \xrightarrow[t_2]{s_2} F(V_2)$$

3. *The category* **FGraph** *consists of F-graphs as objects and F-graph morphisms as morphisms.*

4. *The category of base grapphs* **BaseGraph** *is the category* **FGraph** *with F as the identity functor* $F = Id : \textbf{BASE} \to \textbf{BASE}$. *The forgetful functor* $U_F : \textbf{FGraph} \to \textbf{BaseGraph}$ *is defined by* $U_F(\ (F(E), F(V), s, t)\ ) = (U(F(E)), U(F(V)), U(s), U(t))$ *and* $U_F(\ (f_E, f_V)\ ) = (U(f_E), U(f_V))$.

For **BASE** = **Set** we obtain the usual graphs as base graphs **BaseGraph** = **Graph**.

Lemma 2 (Cocompleteness of F-graphs).

The category **FGraph** *has colimits that are constructed componentwise in the base category* **BASE**, *but in general they are not preserved by the forgetful functor* $U_F : \textbf{FGraph} \to \textbf{BaseGraph}$.

Proof. Follows directly from the fact that **BASE** has colimits and that the free functor F preserves colimits. $U \circ F$ does not preserve colimits in general, if $U : \textbf{CAT} \to \textbf{BASE}$ does not preserve them. Since U is a right adjoint functor we cannot expect that colimits are preserved.

Definition 7 (Marking Graph of M-nets).

Given an M-net $N = (T, P, pre, post)$ *the marking graph* $MG(N)$ *of N is the following F-graph:*

$$MG(N) = (F(T + P), F(P), s, t)$$

where $T + P$ *is the coproduct of T and P in* **BASE** *and* $s = (\overline{pre}, id_{F(P)})$: $F(T + P) \to F(P)$, *and* $t = (\overline{post}, id_{F(P)}) : F(T + P) \to F(P)$ *are uniquely defined by:*

$$U(\overline{pre}) \circ u_T = pre \text{ and } U(\overline{post}) \circ u_T = post$$

Remark 3. $MG(N)$ is well-defined because the free functor F preserves coproducts so that we have $F(T + P) \cong F(T) + F(P)$. This means that s and t are uniquely defined by the universal morphisms $\overline{pre}, \overline{post} : F(T) \to F(P)$ and $id_{F(P)} : F(P) \to F(P)$, where $\overline{pre}, \overline{post}$ are induced morphisms of $pre, post$: $T \to U(F(P))$.

Now we are able to formulate and proof the first main theorem stating the marking graph functor $MG(N)$ to be cocontious.

Theorem 1 (Compositionality of Marking Graph Construction).

The marking graph construction $MG(N)$ *of an M-net N can be extended to a functor* $MG : \textbf{MNet} \to \textbf{FGraph}$ *that preserves colimits.*

Proof. The construction MG is extended to morphisms in the following way:

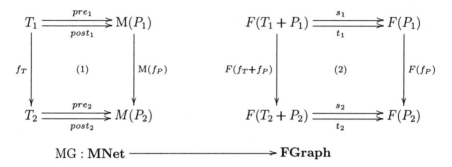

MG : **MNet** $\longrightarrow$ **FGraph**

We have to show that commutativity of (1) implies that of (2).

Let $in_{1T} : F(T_1) \to F(T_1) + F(P_1) = F(T_1 + P_1)$ and $in_{1P} : F(P_1) \to F(T_1) + F(P_1) = F(T_1 + P_1)$ be the coproduct injections then standard reasoning allows showing in diagram (2)

1. $F(f_P) \circ s_1 \circ in_{1T} = s_2 \circ F(f_T + f_P) \circ in_{1T}$, and
2. $F(f_P) \circ s_1 \circ in_{1P} = s2 \circ F(f_T + f_P) \circ in1_P$

Then we can conclude $F(f_P) \circ s_1 = s_2 \circ F(f_T + f_P)$.

A similar reasoning for the T-component implies commutativity of (2). Hence MG is a functor. Colimits in the category **MNet** and in the category **FGraph** are constructed componentwise in the base category **BASE** (see Lemma 1 and Lemma 2). And colimits in **BASE** are preserved by the coproduct $T + P$, hence MG preserves colimits.

Now we introduce reflexive F-graphs to be those F-graphs, where the edges E are given as an coproduct of the basic vertices V and some other basic edges E_0. This corresponds to the transitions in Definition 7 of marking graphs. In fact, all the marking graphs in Figure 2 are reflexive free monoid graphs.

Definition 8 (Reflexive F-graphs).
    *An F-graph $G = (F(E), F(V), s, t)$ is called reflexive, if there is a unique (up to isomorphism) coproduct complement E_0 of E in category **BASE** such that $E_0 + V = E$ and there are unique functions $s', t' : F(E_0) \to F(V)$ with $s = (s', id_{F(V)})$ and $t = (t', id_{F(V)})$.*

Example 2 (Reflexive F-graphs).
    The notion of reflexive F-graphs coincides with reflexive free monoid graphs defined in Section 2 if we specialize F-graphs as shown in Example 4.1. Note that coproduct complements E_0 of E in the base category **Set** are unique up to isomorphism in general, and in the case $V \subseteq E$ we have a unique canonical choice $E_0 = E \setminus V$. In general we only need an injective function $i : V \to E$ in order to have a coproduct complement $E_0 = E \setminus i(V)$.

Theorem 2 (Realization of Reflexive F-graphs).

There is a subcategory **RFGraph** *of F-graphs and a realization functor* $R : $ **RFGraph** $\to$ **MNet** *which is left and right inverse to the marking graph construction.*

Proof. Let **RFGraph** be the subcategory of F-graphs consisting of reflexive F-graphs (see Definition 8) and *reflexive F-graph morphisms* $f = (f_E, f_V) :$ $G_1 \to G_2$ which means that we have a unique (up to isomorphism) coproduct complement f_{E_0} of f_E in **BASE** such that $f_{E_0} + f_V = f_E$. From the marking graph functor construction in Definition 7 it follows that MG can be restricted to a functor $MG_0 :$ **MNet** $\to$ **RFGraph**.

The following construction of $R :$ **RFGraph** $\to$ **MNet** is left and right inverse

to MG_0 with $\qquad$ $R : \quad F(E_0 + V) \underset{t}{\overset{s}{\rightrightarrows}} F(V) \quad \longmapsto \quad E_0 \underset{post}{\overset{pre}{\rightrightarrows}} M(V)$

$$(f_{E_0} + f_V, f_E) \qquad\qquad \longmapsto \quad (f_{E0}, f_E)$$

where $s = (\overline{pre}, id_{F(V)})$ and $t = (\overline{post}, id_{F(v)})$ as well as *pre* and *post* are defined by

$\qquad pre = U(\overline{pre}) \circ u_T$ and
$\qquad post = U(\overline{post}) \circ u_T$

Similar to the proof of Theorem 1 we can show that R is a well-defined functor.

Moreover, we have $R(MG(N)) = N$ for each M-net N and $MG(R(G)) = G$ for each reflexive F-graph G and a similar properties for morphisms in the case of unique coproduct complements. This shows that R is left and right inverse to MG_0. If the coproduct complements are unique up to isomorphism the equalities above have to be replaced by isomorphisms.

Corollary 1 (Unique Realization).

An F-graph G is realizable by an M-net N, i.e. $MG(N) \cong G$, if and only if F is reflexive.

This follows from Theorem 2 and the fact that $MG(N)$ is reflexive (see Definition 7). The realization functor $R :$ **RFGraph** $\to$ **MNet** is left and right inverse to $MG_0 :$ **MNet** $\to$ **RFGraph** according to the proof of Theorem 2. This means especially that R is right adjoint and right inverse to MG_0 and hence a minimal realization in the sense of [Gog73]. Otherwise R is also left adjoint and right inverse to MG_0, i.e. a maximal realization. The fact that R is also left inverse to MG_0, i.e. $R \circ MG_0 \cong ID_{\mathbf{MNet}}$, shows that the realization is unique up to isomorphism, i.e. $MG(N_1) \cong MG(N_2)$ implies $N_1 \cong N_2$.

This fact shows that the marking graph construction does not forget any kind of property of M-nets, similar to the construction of all processes as behavior for an elementary net system (see [Roz87]). In categorical terms this means that the categories **MNet** and **RFGraph** are equivalent or even isomorphic in the case of unique coproducts and coproduct complements in **BASE**.

On the other hand there is also a natural isomorphism between the categories **MNet** and **FGraph** defined by the adjunction of morphisms $T \to M(P)$ in category **BASE** and $F(T) \to F(P)$ in category **CAT**. This leads to the interesting conclusion that also the categories **FGraph** and **RFGraph** are equivalent, although the inclusion $I : \textbf{RFGraph} \rightarrow \textbf{FGraph}$ is not an equivalence in general. This is no contradiction, but similar to the fact that the set $2\mathbb{N}$ of all even natural numbers is properly included in the set $\mathbb{N}$ of all natural numbers but also in bijective correspondence with $\mathbb{N}$.

4   Behavior and Realization of Place/Transition Nets and Elementary Nets

In this section we investigate several different Petri nets that fit into the framework of M-nets and F-graphs. Especially we reconsider place/transition nets introduced in Section 2 and obtain the behavior and realization constructions proposed in Section 2 as special case of the general theory in Section 3.

4.1   Place/Transition Nets as M-nets

Let $\textbf{BASE} = \textbf{Set}$, the category of sets and functions, $\textbf{CAT} = \textbf{CMon}$ where **CMon** is the category of commutative monoids, $U : \textbf{CMon} \rightarrow \textbf{Set}$ the forgetful functor and $F : \textbf{Set} \rightarrow \textbf{CMon}$ the free commutative monoid construction. Then $M(P) = P^{\oplus}$ is the base set of the free commutative monoid over P. So an M-net is a place/transition net in the sense of [MM90] and of Definition 1. Clearly the base category has colimits, especially pushouts (gluing of sets via functions) and coproducts. The marking graph of a place/transition net is given by the marking graph functor $\text{MG} : \textbf{PT} \rightarrow \textbf{FMGraph}$, where the marking graph construction has already been discussed in Section 2.

4.2   Construction of the Marking Graph of Place/Transition Nets

The marking graph functor $\text{MG} : \textbf{PT} \rightarrow \textbf{FMGraph}$ is given for a place/transition net $N = (P, T, pre, post)$ with:

$$\text{MG}(N) = (\ (T + P)^{\oplus} \underset{(\overline{post}, id)}{\overset{(\overline{pre}, id)}{\rightrightarrows}} P^{\oplus}\ )$$

where $T + P$ is binary coproduct in **Set** and $(\overline{pre}, id)$ (respectively $(\overline{post}, id)$ is the universal coproduct morphism, consisting of the extension to the free monoid $\overline{pre}$ (respectively $\overline{post}$) and the identity.

The next important result of M-nets is the characterization of realizable free monoid graphs.

4.3   Realizable Free Monoid Graphs

The subcategory **RFMGraph** of reflexive free monoid graphs is realizable. A free monoid graph $G = (E, V, s, t)$ is reflexive if there exists a set E_0 of edges so that:

$$(E^\oplus \underset{r}{\overset{s}{\Longrightarrow}} V^\oplus) = ((E_0 + V)^\oplus \underset{(t_0, id)}{\overset{(s_0, id)}{\Longrightarrow}} V^\oplus)$$

with $(E^\oplus \underset{t_0}{\overset{s_0}{\Longrightarrow}} V^\oplus)$.

Next we consider another variant of place/transition nets, called place/transition group nets, short PT-G nets. These nets allow negative tokens. This variant is especially interesting when computing invariants in a categorical way [MM90]. Using capacities PT-G nets clearly can be reduced to usual place/transition nets.

4.4 PT-G nets as M-nets

Let **BASE** = **Set** and **CAT** = **CGroup** where **CGroup** is the category of commutative groups, $U : \mathbf{CGroup} \to \mathbf{Set}$ the forgetful functor and $F : \mathbf{Set} \to \mathbf{CGroup}$ the free commutative group construction. Thus we obtain PT-G nets $T \underset{post}{\overset{pre}{\Longrightarrow}} P^\circledast$ where P is the base set of the free commutative group $F(P) = P^\circledast$. These have been studied in [MM90]. We so obtain the category **PTG** of PT-G nets.

The marking graph construction is given by MG : **PTG** $\to$ **FGGraph**, where **FGGraph** is the category of free group graphs with objects $G = (E^\circledast \underset{t}{\overset{s}{\Longrightarrow}} P^\circledast)$ and is constructed like the marking graph of place/transition nets.

Realizable free group graphs are characterized by the subcategory **RFGGraph** of reflexive free group graphs.

4.5 Main Results for Specific Place/Transition Nets

Application of the general results in Section 3 leads to the following main result.

Corollary 2. *Main Results for Place/Transition Nets and PT-G nets*

1. *There is a compositional construction of place/transition nets (and PT-G nets) using pushouts in the category* **PT** *(and* **PTG***).*
2. *The marking graph construction of place/transition nets (and PT-G nets) is defined in terms of free monoid graphs (and of free group graphs).*
3. *The marking graph construction is compositional, i.e. a functor* $MG : \mathbf{PT} \to$ **FMGraph** *(and* $MG : \mathbf{PTG} \to \mathbf{FGGraph}$*) that preserves pushouts and coproducts.*
4. *For the category* **FMGraph** *(and* **FGGraph***) there is a subcategory* **RFMGraph** *(and* **RFGGraph***) of reflexive place/transition nets (and PT-G nets) and a realization functor* $R : \mathbf{RFMGraph} \to \mathbf{PT}$ *(and* $R : \mathbf{RFGGraph} \to \mathbf{PTG}$*) that is left and right inverse to the marking graph construction.*

5. *A free monoid graph (free group graph) G is realizable by a place/transition net (PT-G net) N, i.e. MG(N) ≅ G, if and only if G is reflexive.*

Proof. Assumption 3 is satisfied for place/transition nets (see Definition 1) as well as PT-G nets (see Definition 4.4). Thus Lemma 1 yields Item 1, Definition 7 yields Item 2, Theorem 1 yields Item 3, Definition 8 and Theorem 2 yield Item 4, and Corollary 1 yields Item 5.

4.6 Specific Elementary Nets as M-nets

We now introduce two variants of elementary nets, which are less restricted than the usual elementary nets. For this purpose we first give the notions of additive and distinct additive functions.

Definition 9 (Additive and Distinct Additive Functions).

An additive *function* $f : \mathcal{P}(A) \to \mathcal{P}(B)$, *where* $\mathcal{P}$ *is the powerset construction, is a function so that for all* $A' \subseteq A$ *we have* $f(A') = \bigcup_{a \in A'} f(a)$.

A distinct additive *function* $f : \mathcal{P}(A) \to \mathcal{P}(B)$ *is a function so that for all* $A' \subseteq A$ *we have:*

$$f(A') = \begin{cases} \bigcup_{a \in A'} f(a) & ; \textit{if } \forall a, a' \in A' : a \neq a' \wedge f(a) \cap f(a') = \emptyset \\ \emptyset & ; \textit{else} \end{cases}$$

This gives rise to two categories of powersets with additive and with distinct additive functions respectively.

Definition 10 (Category PSet of Powersets).

The category **PSet** *of powersets consists of the class of all powersets* $\mathcal{P}(A)$ *as objects and of additive functions* $f : \mathcal{P}(A) \to \mathcal{P}(B)$ *as morphisms.*

Definition 11 (Category DPSet of Distinct Powersets).

The category **DPSet** *of distinct powersets consists of the class of powersets* $\mathcal{P}(A)$ *as objects and of distinct additive functions* $f : \mathcal{P}(A) \to \mathcal{P}(B)$ *as morphisms.*

To obtain the results achieved for M-nets and F-graphs in Section 3, we now have to show that $\mathcal{P} : \textbf{Set} \to \textbf{PSet}$ and $\mathcal{P} : \textbf{Set} \to \textbf{DPSet}$ are left adjoint with respect to the corresponding inclusion functors.

Lemma 3 (Left Adjointness of Powerset Functors).

1. *The powerset functor* $\mathcal{P} : \textbf{Set} \to \textbf{PSet}$ *is left adjoint to the inclusion functor* $I : \textbf{PSet} \to \textbf{Set}$.
2. *The distinct powerset functor* $\mathcal{P} : \textbf{Set} \to \textbf{DPSet}$ *is left adjoint to the inclusion functor* $I : \textbf{DPSet} \to \textbf{Set}$.

Proof. 1. For each $f : A \to \mathcal{P}(B)$ in **Set** there is $incl_A : A \to \mathcal{P}(A)$ with $incl_A(a) = \{a\}$ and we obtain $\overline{f} : \mathcal{P}(A) \to \mathcal{P}(B)$ in **PSet** with:
$$\overline{f}(A') = \bigcup_{a \in A'} f(a) \text{ for each } A' \subseteq A$$
so that $\overline{f}(incl_A(a)) = \overline{f}(a) = f(a)$ for all $a \in A$.

2. Analogously with $\overline{f} : \mathcal{P}(A) \to \mathcal{P}(B)$ in **DPSet** is defined for each $A' \subseteq A$:

$$\overline{f}(A') = \begin{cases} \bigcup_{a \in A'} f(a) & ; \forall a, a' \in A' : a \neq a' \implies f(a) \cap f(a') = \emptyset \\ \emptyset & ; \text{else} \end{cases}$$

4.7 Elementary Nets with Loops

Let the categories **BASE** = **Set**, and **CAT** = **DPSet** be given as well as the functors $U = I : $ **DPSet** $\to$ **Set**, and $F = \mathcal{P} : $ **Set** $\to$ **DPSet**. Then $M(P) =$ $\mathcal{P}(P)$ so that an M-net is given by $N = (\ T \overset{pre}{\underset{post}{\rightrightarrows}} \mathcal{P}(P) \ )$. As in elementary net systems (see Definition 12) the amount of tokens on each place is less or equal one. The corresponding F-graph is given by $G = (\ \mathcal{P}(E) \overset{s}{\underset{t}{\rightrightarrows}} \mathcal{P}(V) \ )$ where s and t are additive functions. In this case the category M-nets is called category **ENL** of elementary nets with loops [2] and the category of F-graphs is called category **DPGraph** of distinct powerset graphs.

This kind of elementary nets has less or equal one token on each place and the marking graph models the usual firing. But in contrast to usual elementary net systems replacing a token on a place that is in the pre- and post-domain of a transition is admitted. Hence we have loops. The net of Figure 3 in case p0, p1 does not admit firing in usual elementary net systems. But in elementary nets with loops firing of the transition t is allowed where p0 can be considered as a loop. For this reason these nets are called elementary nets with loops.

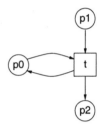

Fig.3. Elementary net with loops

4.8 Unsafe Elementary Nets

Let the categories **BASE** = **Set**, and **CAT** = **PSet** be given as well as the functors $U = I : $ **PSet** $\to$ **Set**, and $F = \mathcal{P} : $ **Set** $\to$ **PSet** the powerset functor.

However, the marking graph construction in this case (see Definition 7) models unsafe firing, i.e. firing of a transition even if there are tokens on some place in the post-domain of the transition. Moreover, transitions can fire in parallel even if they share some places in the pre-domain. For this reason M-nets are called unsafe elementary nets in this case and the corresponding category is called category of unsafe elementary nets, written **UEN**. The F-graphs in this case are called powerset graphs and the category is denoted by **PGraph**.

---

[2] In [EPR94] these net have been called contextual elementary nets, leading to confusion with contextual nets in the sense of [MR94,MR95]. Concurrent use of a token on a contextual place is forbidden in our notion of nets, whereas it is the main feature of contextual nets by Montanari and Rossi.

4.9 Main Results for Specfic Elementary Nets

Since the above defined elementary nets are a special cases of M-nets we obtain the following results:

Corollary 3 (Main Results for Elementary Nets with Loops and Unsafe Elementary Nets).

1. *There is a compositional construction for elementary nets with loops and for unsafe elementary nets using pushouts in the category* **ENL** *(and* **UEN***).*
2. *The marking graph construction of elementary nets with loops and unsafe elementary nets is defined in terms of distinct powerset graphs (and powerset graphs).*
3. *The marking graph construction is compositional, i.e. a functor MG :* **ENL** $\to$ **DPGraph** *(and MG :* **UEN** $\to$ **PGraph***) that preserves pushouts and coproducts.*
4. *For the category* **DPGraph** *(and* **PGraph***) there is a subcategory* **RDPGraph** *(and* **RPGraph***) of reflexive distinct powerset graphs (and reflexive powerset graphs) and a realization functor R :* **RDPGraph** $\to$ **ENL** *(and R :* **RPGraph** $\to$ **UEN***) that is left and right inverse to the marking graph construction.*
5. *A powerset distinct graph (powerset graph) G is realizable by an elementary net with loops (and a unsafe elementary net) N, i.e. $MG(N) \cong G$, if and only if G is reflexive.*

Proof. The assumptions in Definition 3 are satisfied for elementary nets with loops (see Subsection 4.7), and for unsafe elementary nets (see Subsection 4.8). Thus Lemma 1 yields Item 1, Definition 7 yields Item 2, Theorem 1 yields Item 3, Definition 8 and Theorem 2 yield Item 4, and Corollary 1 yields Item 5.

4.10 Summary (Applications of M-nets and F-graphs)

The constructions and results of Section 2 and Section 4 as applications of Section 3 are summarized together with their in Table 1 below:

Table 1. Applications of M-nets and F-graphs to place/transition nets and elementary nets

M-net	F-graph	**BASE**	**CAT**	Free functor	Results
place/transition net	free monoid graph	**Set**	**CMon**	free com. monoid	Corollary 2
PT-G net	free group graph	**Set**	**CGroup**	free com. group	Corollary 2
elementary nets with loops	distinct powerset graph	**Set**	**DPSet**	distinct powerset	Corollary 3
unsafe elementary net	powerset graph	**Set**	**PSet**	powerset	Corollary 3

5   Comparison of Constructions for Elementary Nets and Elementary Transition Systems

In this section we relate our constructions of a marking graph for elementary nets to the construction of a case graph for elementary net systems in the sense of [NRT92], where the case graph is defined to be an elementary transition system. The behavior with respect to an initial marking of a net is given by an elementary transition system. Vice versa the realization of an elementary transition system is given by an elementary net system using regions and the Set Representation Theorem in the sense of [ER90].

A marking graph includes all possible markings and all possible (and even parallel) transitions. Thus the marking is merely dependent on the net structure, but it does not take into account the initial marking. On the other hand the case graph represents all markings reachable from a given initial marking. In fact, the case graph is a subgraph of the marking graph. We show the compatibility of our construction with the construction of elementary net system and elementary transition system. It is interesting, that although the constructions in [NRT92] are quite different we obtain the same kind of results, especially the adjunction between the category of nets and the category of the behavior graphs. The compatibility is given by the fact that the behavior of the nets and the realization of the graphs is the same w.r.t. a given initial marking.

We now briefly review the notions of elementary net systems and transition systems as far as we need them to relate these approaches (for more details see [NRT92]). In [NRT92] two categories are established, the category **ENS** of elementary net systems and **ETS** the category of elementary transition systems. Between these categories there are adjoint functors $H : \mathbf{ENS} \to \mathbf{ETS}$ and $J : \mathbf{ETS} \to \mathbf{ENS}$. **ETS** consists of a set of states, a set of events, a transition relation and an initial state. Furthermore, some conditions have to be satisfied. The realization of an elementary transition system via the functor J is achieved using regions. The regional construction is discussed in [NRT92].

Elementary net systems consist of a set of conditions, a set of events, an initial marking and a flow relation.

Definition 12 (Elementary Net System).

An elementary net system N is a quadruple $N = (B, E, F, c_{in})$ where
  B is a set of conditions,
  E is a set of events and $B \cap E = \emptyset$
  $c_{in} \subseteq B$ is the initial marking, and
  $F \subseteq (B \times E) \cup (E \times B)$ the flow relation
such that the following conditions are satisfied:

1. $\forall x \in B \cup E : \exists y \in B \cup E : (x, y) \in F \vee (y, x) \in F$
2. $\forall x, y \in B \cup E : {}^{\bullet}x = {}^{\bullet}y \wedge x^{\bullet} = y^{\bullet} \Rightarrow x = y$

Remark 4 (Relationship to Elementary Nets with Loops). In contrast to elementary nets with loops (see Subsection 4.7) elementary net systems have an initial marking. Condition 1 ensures that there are nor isolated elements. Condition 2 allows neither parallel transitions nor parallel places.

We have considered in [EPR94] a subcategory $\mathbf{EN} \subset \mathbf{ENL}$, that consists of elementary nets (i.e. without an initial marking) satisfying the above restrictions. This category $\mathbf{EN}$ corresponds to the category of elementary net systems $\mathbf{ENS}$.

Definition 13 (Case Graph of an Elementary Transition System N).

Given an elementary net system $N = (B, E, F, c_{in})$ we define the case graph to be an elementary transition system $ETS = (C_N, E_N, \to_N, c_{in})$ given by the following conditions:

1. *$\to_{NN} \subseteq \mathcal{P}(B) \times E \times \mathcal{P}(B)$ is the transition relation with*
$$\to_{NN} = \{(c, e, c') | c - c' = {}^\bullet e \wedge c' - c = e^\bullet\}$$
2. *C_N is the state space of N, it is the least subset of $\mathcal{P}(B)$ that contains the initial state c_{in} and satisfies:*
$$(c, e, c') \in \to_N \wedge c \in C_N \implies c' \in C_N$$
3. *$\to_N$ is $\to_{NN}$ restricted to C_N.*
4. *$E_N = \{e | e \in E \wedge (c, e, c') \in \to_N\}$ is the set of active events.*

Remark 5 (Relationship to Distinct Powerset Graphs). In contrast to distinct powerset graphs (see Subsection 4.6) elementary transition systems associated with an elementary net system N satisfy a number of restrictions. In [EPR94] it is shown that this leads to a subcategory $\mathbf{EGraph}$ of $\mathbf{DPGraph}$ and $\mathbf{REGraph}$ of $\mathbf{RDPGraph}$, respectively.

However the constructions of the marking graph functor MG and the realization functor R remain the same. So there is the marking graph functor MG : $\mathbf{EN} \to \mathbf{EGraph}$ and the realization functor R : $\mathbf{REGraph} \to \mathbf{EN}$, where similar to Corollary 3 R is left and right inverse to MG.

The case graph construction in Definition 13 has been extended in [NRT92] to a case graph functor H : $\mathbf{ENS} \to \mathbf{ETS}$. And the realization construction based on regions has been extended to a realization functor J : $\mathbf{ETS} \to \mathbf{ENS}$ that is adjoint to H with $H \circ J \cong ID$.

Getting from a marking graph to a case graph means to distinguish one initial marking and forgetting all other potential behaviors of a net. Thus information is lost and the case graph is smaller and easier to cope with. The marking graph represents the elementary net without loosing information. This difference is detectable as the category of elementary nets $\mathbf{EN}$ and reflexive elementary powerset graphs $\mathbf{REGraph}$ are equivalent, whereas the category of elementary transition systems $\mathbf{ETS}$ and elementary net systems $\mathbf{ENS}$ are only related by adjoint functors H and J with $H \circ J \cong ID$ and $J \circ H \not\cong ID$.

Summarizing we have:

1. **Adjoint Functors for Elementary Net Systems and Case Graphs**

$$\mathbf{ENS} \underset{J}{\overset{H}{\rightleftarrows}} \mathbf{ETS}$$

with $H \circ J \cong ID$ and $J \circ H \not\cong ID$ (see [NRT92])

2. Adjoint Functors for Elementary Nets and Marking Graphs

$$\mathbf{EN} \underset{R}{\overset{MG}{\rightleftarrows}} \mathbf{REGraph}$$

with $MG \circ R \cong ID$ and $R \circ MG \cong ID$ (see [EPR94])

In [EPR94] we have defined constructions $init_m$ and $case_m$, that transfer objects from the category **EN** to the category **ENS** and from the category **REGraph** to the category **ETS**. In both cases the construction is depended on a chosen initial marking m of an elementary net N, i.e. we obtain an indexed construction, where the index set is the powerset of places of N.

Furthermore we can simply forget the initial marking m and thus transform an elementary net system into an elementary net given by the construction $forget_m$.

The constructions $init_m$, $forget_m$ and $case_m$ are no functors in general. But they admit to prove explicitly (see [EPR94]) the compatibility of the case graph construction in [NRT92] and the marking graph construction in this paper.

3. Functional Relationship Between the Case Graph and the Marking Graph Construction

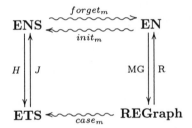

The case graph, the marking graph, and the realization constructions are compatible in the following sense:

(a) $case_m \circ MG = H \circ init_m$ for all initial markings m
(b) $case_m \circ MG \circ forget_m \circ J \cong ID$ for all initial markings m

Interpretation:

(a) Reducing the marking graph of a given net with respect to an initial marking yields the case graph of the net system consisting of the given net and the same initial marking.
(b) Given an elementary transition system the marking graph of the realization of this elementary transition system yields again the given elementary transition system. This is constructed as the marking graph's subgraph that is induced by the initial marking.

6 Conclusion

In this paper we have presented an abstract notion of Petri nets, called M-nets, based on a free functor F between a base category **BASE** and a category **CAT** with higher structured objects. This yields a category of F-graphs, so that we can define an abstract notion of marking graphs. The category of M-nets has colimits and thus a suitable construction for composition. As the marking graph construction is a left adjoint functor MG, composition of M-nets is preserved. So we obtain a compositional semantics. Furthermore we have been able to characterize the subcategory of realizable F-graphs, namely the reflexive F-graphs. These F-graphs describe the behaviour of a M-net, that is constructed on the basis of the given F-graph.

We have introduced several different notions of place/transition and elementary nets that are instantiations of M-nets.

As the realization problem was solved already for elementary net systems (with initial state) in [NRT92] we have shown compatibility of our approach in the special case of elementary nets (without initial state).

References

[BB01]   M. A. Bednarczyk and A. M. Borzyszkowski. On Concurrent Realization of Reactive Systems and Their Morphisms. In H. Ehrig, G. , Juhás, J. Padberg, and G. Rozenberg, editors, *Advances in Petri Nets: Unifying Petri Nets*, LNCS. Springer, 2001.

[BD98]   E. Badouel and Ph. Darondeau. Theory of regions. In W. Reisig and G. Rozenberg, editors, *Lectures on Petri Nets: Basic Models*, pages 529–586. Springer, LNCS 1491, 1998.

[BFF+95]   E. Best, H. Fleischhack, W. Fraczak, R.P. Hopkins, H. Klaudel, and E. Pelz. An M-Net Semantics of $B(PN)^2$. In J. Desel, editor, *Structures in Concurrency Theory*, pages 85–100. Springer-Verlag, 1995.

[Dar00]   P. Darondeau. Region Based Synthesis of P/T-Nets and Its Potential Applications. In M. Nielsen and D. Simpson, editors, *21st International Conference on Application and Theory of Petri Nets (ICATPN 2000)*, volume 1825 of *Lecture Notes in Computer Science*, pages 16–23. Springer-Verlag, 2000.

[DJL01]   J. Desel, G. Juhás, and R. Lorenz. Petri Nets over Partial Algebras. In H. Ehrig, G. Juhás, J. Padberg, and G. Rozenberg, editors, *Advances in Petri Nets: Unifying Petri Nets*, LNCS. Springer, 2001.

[DKPS91]   C. Diamantini, S. Kasangian, L. Pomello, and C. Simone. Elementary Nets and 2-Categories. In E. Best and et al., editors, *GMD-Studien Nr. 191; Hildesheimer Informatik-Berichte 6/91; 3rd Workshop on Concurrency and Compositionality, 1991, Goslar, Germany*, pages 83–85. Gesellschaft für Mathematik und Datenverarbeitung mbH — Universität Hildesheim (Germany), Institut für Informatik, 1991.

[DR96]   J. Desel and W. Reisig. The synthesis problem of Petri nets. *Acta Informatica*, 33:297–315, 1996.

[EP97]   H. Ehrig and J. Padberg. A Uniform Approach to Petri Nets. In Ch. Freksa, M. Jantzen, and R. Valk, editors, *Foundations of Computer Science: Potential - Theory - Cognition*. Springer, LNCS 1337, 1997.

[EPR94]   H. Ehrig, J. Padberg, and G. Rozenberg. Behaviour and Realization Con-
           struction for Petri NetsBased on Free Monoid and Power Set Graphs. Tech-
           nical report, Technical University Berlin TR 94-15, 1994.
[ER90]   A. Ehrenfeucht and G. Rozenberg. Partial (Set) 2-Structures, Part I and
           II. *Acta Informatica*, 27:315–368, 1990.
[Gog73]   J.A. Goguem. Realization is universal. *Mathematical systems Theory*, 6,
           1973.
[Juh99]   Gabriel Juhás. Reasoning about algebraic generalisation of Petri nets. In
           Proc. of 20th Conference on Theory and Application of Petri nets, pages
           324–343. Springer, LNCS 1639, 1999.
[MM90]   J. Meseguer and U. Montanari. Petri Nets are Monoids. *Information and
           Computation*, 88(2):105–155, 1990.
[MR94]   U. Montanari and F. Rossi. Contextual occurrence nets and concurrent
           constraint programming. In H.-J. Schneider and H. Ehrig, editors, *Proceed-
           ings of the Dagstuhl Seminar 9301 on Graph Transformations in Computer
           Science*, volume 776 of *Lecture Notes in Computer Science*. Springer Verlag,
           1994.
[MR95]   U. Montanari and F. Rossi. Contextual nets. *Acta Informatica*, 32, 1995.
           Also as Technical Report TR 4-93, Department of Computer Science, Uni-
           versity of Pisa, February 1993.
[NRT92]   M. Nielsen, G. Rozenberg, and P.S. Thiagarajan. Elementary transition
           systems. *TCS*, 96:3–33, 1992.
[Pad96]   J. Padberg. *Abstract Petri Nets: A Uniform Approach and Rule-Based
           Refinement*. PhD thesis, Technical University Berlin, 1996. Shaker Verlag.
[PE01]   J. Padberg and H. Ehrig. Introduction to Parametrized Net Classes. In
           H. Ehrig, G. Juhás, J. Padberg, and G. Rozenberg, editors, *Advances in
           Petri Nets: Unifying Petri Nets*, LNCS. Springer, 2001.
[Roz87]   G. Rozenberg. Behaviour of elementary net systems. In W. Brauer,
           W. Reisig, and G. Rozenberg, editors, *Advances in Petri nets 1986*, pages
           60–94. Springer Verlag Berlin, 1987.

Rewriting Logic as a Unifying Framework for Petri Nets

Mark-Oliver Stehr, José Meseguer, and Peter Csaba Ölveczky

SRI International, Menlo Park, CA 94025, USA

Abstract. We propose rewriting logic as a unifying framework for a wide range of Petri nets models. We treat in detail place/transition nets and important extensions of the basic model by individual tokens, test arcs, and time. Based on the idea that "Petri nets are monoids" suggested by Meseguer and Montanari we define a rewriting semantics that maps place/transition nets into rewriting logic specifications. We furthermore generalize this result to a general form of algebraic net specifications subsuming also colored Petri nets as a special case. The soundness and completeness results we state relate the commutative process semantics of Petri nets proposed by Best and Devillers to the model-theoretic semantics of rewriting logic in the sense of natural isomorphisms between suitable functors. In addition we show how place/transition nets with test arcs and timed Petri nets can be equipped with a rewriting semantics and discuss how other extensions can be treated along similar lines. Beyond the conceptual unification of quite different kinds of Petri nets within a single framework, the rewriting semantics can provide a guide for future extensions of Petri nets and help to cope with the growing diversity of models in this field. On the practical side, a major application of the rewriting semantics is its use as a logical and operational representation of Petri net models for formal verification and for the efficient execution and analysis using a rewriting engine such as Maude, which also allows us to specify different execution and analysis strategies in the same rewriting logic language by means of reflection.

1 Introduction

This paper attempts to contribute to the general goal of unifying Petri net models by studying in detail the unification of a wide range of such models within rewriting logic [48], which is used as a logical and semantic framework. Specifically, we show how place/transition nets, nets with test arcs, algebraic net specifications, colored Petri nets, and timed Petri nets can all be naturally represented within rewriting logic. Our work extends in substantial ways previous work on the rewriting logic representation of place/transition nets [48], nets with test arcs [50], algebraic net specifications [69], and timed Petri nets [62].

The representations in question associate a rewrite specification to each net in a given class of Petri net models in such a way that concurrent computations in the

H. Ehrig et al. (Eds.): Unifying Petri Nets, LNCS 2128, pp. 250–303, 2001.
© Springer-Verlag Berlin Heidelberg 2001

original net naturally coincide with concurrent computations in the associated rewrite specification. That is, we exhibit appropriate bijections between Petri net computations and rewriting logic computations, viewed as equivalence classes of proofs, that is, as elements of the free model associated to the corresponding rewrite specification [48].

Furthermore, for certain classes of nets, namely place/transition nets and a general form of algebraic net specifications, which subsume the well-known class of colored Petri nets, we show that the representation maps into rewriting logic are *functorial*; that is, that they map in a functorial way net morphisms to rewrite specification morphisms. In addition, such functorial representations can be further extended to the level of *semantic models*, yielding *semantic equivalence theorems* (in the form of natural isomorphisms of functors) between well-known semantic models for the given class of Petri nets and the free models of the corresponding rewrite theories or, more precisely, models obtained from such free models by forgetting some structure.

As we further explain in the body of the paper, this work, including the above-mentioned functorial semantics and the semantic equivalences, generalizes in some ways, and complements in others, a substantial body of work initiated by the second author in joint work with Ugo Montanari under the motto "Petri nets are monoids" [52, 45, 46, 55, 21, 53, 54, 56, 12, 13], in which categorical models are naturally associated as semantic models to Petri nets, and are shown to be equivalent to well-known "true concurrency" models. Our work is also related to linear logic representations of Petri nets [45, 46, 4, 11, 10, 26]. All this is not surprising, since, as explained in [48], both the categorical place/transition net models of [52] and the linear logic representations of place/transition nets inspired rewriting logic as a generalization of both formalisms. But, as shown in this paper, the extra algebraic expressiveness of rewriting logic is very useful to model in a simple and natural way not only place/transition nets, but also *high-level nets*, such as algebraic net specifications, colored Petri nets, and timed Petri nets.

Our proposed unification of Petri net models is not only of conceptual interest. Given that, under reasonable assumptions, rewrite theories can be executed, the representation maps that we propose provide a uniform operational semantics in terms of efficient logical deduction. Furthermore, using a rewriting logic language implementation such as Maude [19, 18], or the Real-Time Maude tool in the timed case [61, 60], it is possible to use the results of this paper to create execution environments for different classes of Petri nets. In addition, because of Maude's reflective capabilities [17], the Petri nets thus represented cannot only be executed, but they can also be formally analyzed and model checked by means of *rewriting strategies* that explore and analyze at the metalevel the different rewriting computations of a given rewrite specification.

The general way of representing Petri nets within rewriting logic that we propose is by no means limited to the net classes explicitly discussed in this paper. We believe that it can be similarly applied to other important classes of nets that we cannot discuss in detail due to space limitations. We briefly address how similar representations could be defined for other Petri net classes, such as colored Petri nets based on (higher-order) programming languages [39], nets with macroplaces [2, 3], nets with FIFO places [30, 40, 29, 27], object-oriented variants of Petri nets [67, 44], and object nets [72, 73, 28, 74] where nets are viewed as token objects.

We conclude this introduction with a brief overview of the paper: After introducing rewriting logic together with the underlying membership equational logic in the following section, we introduce in Section 3 a category of place/transition nets together with a functor that associates the process semantics of Best and Devillers [7] with each place/transition net. We then define the rewriting specification associated with a place/transition net and we establish a semantic connection in terms of a natural isomorphism at the level of symmetric monoidal categories. We conclude the section on place/transition nets by showing how test arcs can be incorporated using a slightly richer state space that satisfies certain symmetries. In Section 4 we generalize the rewriting semantics for place/transition nets to algebraic net specifications, which we view as colored net specifications over membership equational logic. As it is the case for rewriting logic, the concept of colored net specifications is quite general, since it is parameterized over an underlying logic. However, for the sake of concreteness we only deal with rewriting logic and colored net specifications over membership equational logic in this paper. As in the previous section we relate the Best-Devillers process semantics and the model-theoretic semantics obtained via rewriting logic in terms of a natural isomorphism. In Section 5 we deal with timed Petri nets, an extension of place/transition nets by a notion of real time. The model we use is closely related to the model of interval timed colored Petri nets proposed by van der Aalst [1], but for the purpose of a simpler exposition we deal with the corresponding uncolored model and focus on the essential real-time aspects. Finally, in Section 6 we conclude by discussing how our approach can be generalized or extended to the other models of Petri nets like those mentioned before.

2 Preliminaries

A *finite multiset* over a set S is a function m from S to N such that its support $S(m) = \{s \in S \mid m(s) > 0\}$ is finite. We denote by $S^{\oplus}$ the set of finite multisets over S, by $\emptyset_S$ the empty multiset over S (we usually omit S if it is clear from the context), and we use the standard definitions of multiset membership $\in$, multiset inclusion $\sqsubseteq$, multiset union $\oplus$, and multiset difference $-$. Sometimes we write x instead of the singleton multiset containing x.

A *list* of length n over a set S is a function l from the interval $[1, n]$ of N to S. We denote by $\mathcal{L}(S)$ the set of lists of arbitrary length over S. Concatenation of lists u and v is written as uv. Sometimes we write x instead of the singleton list containing x. If x is a variable ranging over elements, we often use the variable $\bar{x}$ to range over lists of such elements.

Often we implicitly lift functions $f : X \to Y$ to sets $f : \mathcal{P}(X) \to \mathcal{P}(Y)$, finite multisets $f : X^{\oplus} \to Y^{\oplus}$, and lists $f : \mathcal{L}(X) \to \mathcal{L}(Y)$ in the natural homomorphic way. Given a finite set S we sometimes assume a *canonical enumeration* of S, i.e. a list $\bar{x}$ of n distinct elements such that $S = \{\bar{x}_1, \ldots, \bar{x}_n\}$ which is fixed thoughout the paper. In order to ensure the existence of a canonical enumeration of certain sets we could assume that all their elements are drawn from a single total order that we do not make explicit in this paper.

2.1   Membership Equational Logic

Membership equational logic (MEL) [9, 51] is a many-sorted logic with subsorts and overloading of function symbols. It can express partiality very directly by defining membership in a sort by means of membership equational conditions. In accordance with the terminology introduced in the references above we refer to the types of the logic as *kinds*, and we view the *sorts* for each kind as unary predicates. The atomic sentences are *equalities* $M = N$ for terms M, N of the same kind, and *memberships* $M : s$ for M a term and s a sort, both of the same kind. Sentences of MEL are universally quantified Horn clauses on the atoms.

Definition 1. A *memberhip equational signature* Ω consists of a set of *kinds* K_{Ω}, a set of sorts S_{Ω}, a function $\pi_{\Omega} : S_{\Omega} \to K_{\Omega}$ that associates to each sort its kind, and a family $(OP^{\bar{k},k}_{\Omega})_{\bar{k} \in K_{\Omega}^{*}, k \in K_{\Omega}}$ of *operator symbols* such that the following *overloading restriction* holds: If $OP^{\bar{k},k}_{\Omega} \cap OP^{\bar{k}',k'}_{\Omega} \neq \emptyset$ then $\bar{k} = \bar{k}'$ implies $k = k'$. Instead of $o \in OP^{\bar{k},k}_{\Omega}$ we simply write $o : \bar{k} \to k$. If $\bar{k}$ is empty we write $o : \to k$, and o is called a *constant symbol*, otherwise o is called a *function symbol*.

Given membership equational signatures Ω and Ω' a *membership equational signature morphism* $H : \Omega \to \Omega'$ consists of functions $H_K : K_{\Omega} \to K_{\Omega'}$, $H_S : S_{\Omega} \to S_{\Omega'}$ and $H_{OP} : OP_{\Omega} \to OP_{\Omega'}$ such that (1) $H_K(\pi_{\Omega}(s)) = \pi_{\Omega'}(H_S(s))$ for each sort $s \in S_{\Omega}$, and (2) $f : \bar{k} \to k$ in Ω implies $H_{OP}(f) : H_K(\bar{k}) \to H_K(k)$ in Ω'. We usually omit the indices of H if there is no danger of confusion. Membership equational signatures together with their morphisms form a category **MESign**.

A *kinded variable set* is a family $(X_k)_{k \in K}$ of pairwise disjoint sets which are also disjoint from the operator symbols in OP_{Ω}. Given a kinded variable set X, the kinded set of Ω-*terms* over X, written $\mathbf{T}_{\Omega}(X) = (\mathbf{T}_{\Omega}(X)_k)_{k \in K}$, is inductively defined as follows: (1) each variable $x \in X_k$ is in $\mathbf{T}_{\Omega}(X)_k$; (2) each constant symbol c with $c : \to k$ is in $\mathbf{T}_{\Omega}(X)_k$ for $k \in K$; (3) each *function application* of the form $f(\bar{M}_1, \ldots, \bar{M}_n)$ is in $\mathbf{T}_{\Omega}(X)_k$ for $f : \bar{k} \to k$ and $\bar{M}_1 \in \mathbf{T}_{\Omega}(X)_{\bar{k}_1}, \ldots, \bar{M}_n \in \mathbf{T}_{\Omega}(X)_{\bar{k}_n}$ where $\bar{k} = \bar{k}_1 \ldots \bar{k}_n$. If X is the empty variable set the terms above are called *ground terms* and we write $\mathbf{T}_{\Omega}$ and $\mathbf{T}_{\Omega,k}$ instead of $\mathbf{T}_{\Omega}(X)$ and $\mathbf{T}_{\Omega}(X)_k$, respectively.

We define *atomic Ω-formulae* over X as either (1) Ω-*memberships* over X of the form $M : s$ for $M \in \mathbf{T}_{\Omega}(X)_{\pi(s)}$, or (2) Ω-*equations* over X of the form $M = N$ for $M, N \in \mathbf{T}_{\Omega}(X)_k$ for some kind k. Furthermore, Ω-*conditions* over

X are of the form $\bar{\phi}_1 \wedge \ldots \wedge \bar{\phi}_n$, where $\bar{\phi}_1, \ldots, \bar{\phi}_n$ are atomic formulae over X. Given an Ω-condition $\bar{\phi}_1 \wedge \ldots \wedge \bar{\phi}_n$ over X an Ω-*axiom* can be either (1) a *membership axiom* of the form $\forall\, X\, .\, M : s$ if $\bar{\phi}_1 \wedge \ldots \wedge \bar{\phi}_n$, where $M : s$ is an Ω-membership over X, or (2) an *equational axiom* of the form $\forall\, X\, .\, M = N$ if $\bar{\phi}_1 \wedge \ldots \wedge \bar{\phi}_n$, where $M = N$ is an Ω-equation over X. We usually omit the quantifier if X is empty.

A *membership equational theory (MET)* $\mathcal{T}$ consists of a signature $\Omega_{\mathcal{T}}$ and a set of $\Omega_{\mathcal{T}}$-axioms $E_{\mathcal{T}}$.

The *algebraic semantics* of membership equational logic is a standard model-theoretic one [9, 51]. *Models* of a membership equational theory are suitable algebras satisfying the axioms.

Definition 2. Let Ω be a signature. An Ω-*algebra* A consists of a *kind interpretation* $[\![k]\!]_A$ for each $k \in K$, a *sort interpretation* $[\![s]\!]_A \subseteq [\![k]\!]_A$ for each $s \in \pi^{-1}(k)$, an *operator interpretation* $[\![o_{\bar{k},k}]\!]_A$ for each $o : \bar{k} \to k$ such that $[\![c]\!]_A \in [\![k]\!]_A$ for $c :\to k$ and $[\![f]\!]_A \in [\![\bar{k}]\!]_A \to [\![k]\!]_A$ for $f : \bar{k} \to k$ where $[\![\bar{k}]\!]_A = [\![\bar{k}_1]\!]_A \times \ldots \times [\![\bar{k}_n]\!]_A$ if $\bar{k} = \bar{k}_1 \ldots \bar{k}_n$. For better readability we often write $[\![c]\!]_A$ and $[\![f]\!]_A$ instead of $[\![c_k]\!]_A$ and $[\![f_{\bar{k},k}]\!]_A$ assuming that the subscripts are clear from the context. To simplify some constructions we assume in this paper without loss of generality that $[\![k]\!]_A \cap [\![k']\!]_A = \emptyset$ for all kinds $k \neq k'$.

Let A, B be Ω-algebras. A Ω-*morphism*, written $h : A \to B$, is a kinded function $h = (h_k)_{k \in K}$ such that $h_k : [\![k]\!]_A \to [\![k]\!]_B$ for all $k \in K$ and the following conditions hold: (1) $h_k([\![s]\!]_A) \subseteq [\![s]\!]_B$ for $s \in \pi^{-1}(k)$; (2) $h_k([\![c_k]\!]_A) = [\![c_k]\!]_B$ for $c :\to k$; and (3) $h_k([\![f_{\bar{k},k}]\!]_A(\bar{a}_1, \ldots, \bar{a}_n)) = [\![f_{\bar{k},k}]\!]_B(h_{\bar{k}_1}(\bar{a}_1), \ldots, h_{\bar{k}_n}(\bar{a}_n))$ for $f : \bar{k} \to k$ with $\bar{k} = \bar{k}_1 \ldots \bar{k}_n$ and $\bar{a}_i \in [\![\bar{k}_i]\!]_A$. Ω-algebras together with Ω-morphisms constitute a category $\mathbf{Mod}(\Omega)$.

Definition 3. Let A be an Ω-algebra. An *assignment* $\beta : X \to A$ is a kinded function $\beta = (\beta_k)_{k \in K}$ associating to each $x \in X_k$ an element $\beta_k(x) \in [\![k]\!]_A$. It is extended to terms over X as follows: (1) $\beta_k(c) = [\![c_k]\!]_A$ for $c :\to k$; and (2) $\beta_k(f(\bar{M}_1, \ldots, \bar{M}_n)) = [\![f_{\bar{k},k}]\!]_A(\beta_{\bar{k}_1}(\bar{M}_1), \ldots, \beta_{\bar{k}_n}(\bar{M}_n))$ for $f : \bar{k} \to k$ and $\bar{M}_i \in \mathbf{T}_\Omega(X)_{\bar{k}_i}$ where $\bar{k} = \bar{k}_1 \ldots \bar{k}_n$. Instead of $\beta_k(M)$ for $M \in \mathbf{T}_\Omega(X)_k$ we also use the notation $\beta(M)$ or $[\![M]\!]_{A,\beta}$.

Let A be an Ω-algebra, let $\beta : X \to A$ be an assignment, and let $M, N \in \mathbf{T}_\Omega(X)_k$. We define validity of formulae starting with atomic formulae: an Ω-membership $M : s$ over X is *valid* under β iff $[\![M]\!]_{A,\beta} \in [\![s]\!]_A$; and an Ω-equation $M = N$ over X is *valid* under β iff $[\![M]\!]_{A,\beta} = [\![M]\!]_{A,\beta}$. We write $A, \beta \models \phi$ iff an atomic formula ϕ is valid under β. Furthermore, an Ω-condition $\bar{\phi}_1 \wedge \ldots \wedge \bar{\phi}_n$ over X is *valid* under β iff $A, \beta \models \bar{\phi}_i$ for each $i \in \{1 \ldots n\}$, in which case we also write $A, \beta \models \bar{\phi}_1 \wedge \ldots \wedge \bar{\phi}_n$. An Ω-axiom $\forall\, X\, .\, \phi$ if $\bar{\phi}_1 \wedge \ldots \wedge \bar{\phi}_n$ is *valid* iff for each assignment $\beta : X \to A$ we have $A, \beta \models \phi$ whenever $A, \beta \models \bar{\phi}_1 \wedge \ldots \wedge \bar{\phi}_n$. We also write $A \models \forall\, X\, .\, \phi$ if $\bar{\phi}_1 \wedge \ldots \wedge \bar{\phi}_n$ in this case. Given a set E of Ω-axioms we write $A \models E$ iff $A \models \psi$ for each $\psi \in E$. Given a MET $\mathcal{T}$ we say that A is a $\mathcal{T}$-*algebra* iff $A \models E_{\mathcal{T}}$. We write $\mathcal{T} \models \psi$ iff $A \models \psi$ for each $\mathcal{T}$-algebra

A and given a set E of Ω-axioms we write $\mathcal{T} \models E$ iff $\mathcal{T} \models \psi$ for each $\psi \in E$. We furthermore say that M and N are E-equivalent iff $[\![M]\!]_{A,\beta} = [\![N]\!]_{A,\beta}$ for all Ω-algebras A satisfying $A \models E$ and assignments $\beta : X \to A$.

Given METs $\mathcal{T}$ and $\mathcal{T}'$, a *MET morphism* $H : \mathcal{T} \to \mathcal{T}'$ is a membership equational signature morphism $H : \Omega_{\mathcal{T}} \to \Omega_{\mathcal{T}'}$ such that $\mathcal{T}' \models H(E_{\mathcal{T}})$, where H is lifted to terms and axioms in the natural homomorphic way. We say that $\mathcal{T}$ is a *subtheory* of $\mathcal{T}'$, written $\mathcal{T} \hookrightarrow \mathcal{T}'$, iff there is a MET morphism $J : \mathcal{T} \hookrightarrow \mathcal{T}'$ that is an inclusion.

METs together with their morphisms form a category **MET**, and given a MET $\mathcal{T}$ the class of $\mathcal{T}$-algebras together with their Ω-morphisms constitutes a full subcategory of $\mathbf{Mod}(\Omega)$ denoted by $\mathbf{Mod}(\mathcal{T})$. Each MET morphism $H : \mathcal{T} \to \mathcal{T}'$ induces an obvious forgetful functor $\mathbf{Mod}(H) : \mathbf{Mod}(\mathcal{T}') \to \mathbf{Mod}(\mathcal{T})$ that we also write as $\mathbf{U}_H$. In fact, we have a contravariant functor $\mathbf{Mod} : \mathbf{MET} \to \mathbf{Cat}^{\mathrm{op}}$. Given an inclusion $I : \mathcal{T} \hookrightarrow \mathcal{T}'$ and a $\mathcal{T}'$-algebra A we also write $A|\mathcal{T}$ instead of $\mathbf{U}_I(A)$.

METs have initial and free models [9, 51]. In fact, given a MET $\mathcal{T}'$ there exists an initial $\mathcal{T}'$-algebra, written $\mathbf{I}(\mathcal{T}')$. More generally, given a MET morphism $H : \mathcal{T} \to \mathcal{T}'$ between METs $\mathcal{T}$ and $\mathcal{T}'$ there exists a free functor $\mathbf{F}_H : \mathbf{Mod}(\mathcal{T}) \to \mathbf{Mod}(\mathcal{T}')$, i.e. a functor that is left adjoint to $\mathbf{U}_H$. In the following we write η_H and ϵ_H for unit and counit, respectively, of this adjunction, i.e., we have natural transformations $\eta_H(A) : A \to \mathbf{U}_H(\mathbf{F}_H(A))$ for $\mathcal{T}$-algebras A and $\epsilon_H(A') : \mathbf{F}_H(\mathbf{U}_H(A')) \to A'$ for $\mathcal{T}'$-algebras A'.

In contrast to an entirely loose or entirely initial semantics of membership equational theories, in practice a mixed specification style is used, where certain subtheories are intended to be equipped with initial interpretations or certain subtheories are interpreted freely over their parameter specifications. To make such restrictions on the models explicit in the specification we enrich a membership equational theory by initiality and freeness constraints [23, 35], and refer to these enriched theories as *membership equational logic specifications (MES)*.

From a model-theoretic point of view, constraints are axioms that are treated in full analogy to membership or equational axioms, i.e., as sentences that have to be valid in all models. Hence, the models of a MES are algebras which satisfy all the given initiality and freeness constraints. Given a MES model, a model of a subspecification is obtained by its associated forgetful functor $\mathbf{U}_K$ for K the corresponding subspecification inclusion. In particular, this means that a model induces a unique interpretation for each subspecification, which is the justification for the condition on ϵ below. The notion of constraint we use here is a special case of the notion proposed in [35], where initiality constraints are seen as a special case of freeness constraints.

Definition 4. Let $J : \mathcal{T}'' \hookrightarrow \mathcal{T}'$ and $I : \mathcal{T}' \hookrightarrow \mathcal{T}$ be MET inclusions. A *constraint* for $\mathcal{T}$ can take one of the following two forms: (1) $\mathcal{T}'$ `initial` or (2) $\mathcal{T}'$ `free` over $\mathcal{T}''$. A *membership equational specification (MES)* $\mathcal{S}$ is a MET $\mathcal{T}_{\mathcal{S}}$ together with a set $C_{\mathcal{S}}$ of constraints for $\mathcal{T}_{\mathcal{S}}$.

Let A be a $\mathcal{T}$-algebra. We define *validity* of a constraint as follows: (1) the constraint $\mathcal{T}'$ `initial` is *valid* iff the unique morphism from $\mathbf{I}(\mathcal{T}')$ to $A|\mathcal{T}'$ is an isomorphism, and (2) the constraint $\mathcal{T}'$ `free over` $\mathcal{T}''$ is *valid* iff $\epsilon_J(A|\mathcal{T}')$: $\mathbf{F}_J(A|\mathcal{T}'') \to A|\mathcal{T}'$ is an isomorphism. Given a MES $\mathcal{S}$ with an underlying MET $\mathcal{T}_S$, an *$\mathcal{S}$-algebra* is a $\mathcal{T}_S$-algebra A such that each constraint in C_S is valid in A.

In complete analogy to METs we define: Given MESs $\mathcal{S}$ and $\mathcal{S}'$, a *MES morphism* $H : \mathcal{S} \to \mathcal{S}'$ is a morphism $H : \mathcal{T} \to \mathcal{T}'$ such that $\mathcal{S}' \models H(C_S)$, i.e. the constraints $H(C_S)$ are valid in all $\mathcal{S}'$-algebras, where H is lifted to constraints in the natural way. $\mathcal{S}$ is a *subspecification* of $\mathcal{S}'$, written $\mathcal{S} \hookrightarrow \mathcal{S}'$, iff there is a MES morphism $J : \mathcal{S} \hookrightarrow \mathcal{S}'$ that is an inclusion. MESs together with their morphisms form a category **MES** and the category of $\mathcal{S}$-algebras $\mathbf{Mod}(\mathcal{S})$ is the full subcategory of $\mathbf{Mod}(\mathcal{T}_S)$ that contains only $\mathcal{S}$-algebras. Each MES morphism $H : \mathcal{S} \to \mathcal{S}'$ induces an obvious forgetful functor $\mathbf{Mod}(H) : \mathbf{Mod}(\mathcal{S}') \to \mathbf{Mod}(\mathcal{S})$ that we also write as $\mathbf{U}_H$. Again, we have a contravariant functor $\mathbf{Mod} : \mathbf{MES} \to \mathbf{Cat}^{\mathrm{op}}$ that generalizes $\mathbf{Mod} : \mathbf{MET} \to \mathbf{Cat}^{\mathrm{op}}$. Given an inclusion $I : \mathcal{S}' \hookrightarrow \mathcal{S}$ and an $\mathcal{S}$-algebra A we also write $A|\mathcal{S}'$ instead of $\mathbf{U}_I(A)$.

Furthermore, we introduce interpreted specifications together with a general notion of morphism that reflects a transformation of the specification as well as a transformation of the algebras possibly associated with different specifications.

Definition 5. An *interpreted MES* $(\mathcal{S}, A)$ consists of a MES $\mathcal{S}$ and a $\mathcal{S}$-algebra A. The category **IMES** of interpreted MES is given by the Grothendiek construction $\Sigma(\mathbf{Mod})$ where $\mathbf{Mod} : \mathbf{MES} \to \mathbf{Cat}^{\mathrm{op}}$. Recall that a morphism (H, h) : $(\mathcal{S}, A) \to (\mathcal{S}', A')$ in $\Sigma(\mathbf{Mod})$ consists of morphisms $H : \mathcal{S} \to \mathcal{S}'$ and $h : A \to \mathbf{U}_H(A')$ satisfying the conditions of the Grothendiek construction [71].

Given a MES $\mathcal{S}$, the *operational semantics* [9], that can be used to efficiently execute a specification under certain assumptions, is explained using a refinement of $\mathcal{S}$, namely by viewing E_S as composed of a set E_S^S of *structural axioms* and a set E_S^C of *computational axioms*, i.e., $E = E_S^S \cup E_S^C$. Assuming that the computational axioms in E_S^C satisfy the variable restriction explained below the equational axioms in E_S^C can be seen as reduction rules that operate modulo the equational theory induced by E_S^S. Identifying E_S^S-equivalent terms, we write $M \Rightarrow M'$ to express that M can be *reduced* to M' by applying an equation in E_S^C to a subterm of M. The *variable restriction* requires that all variables occurring in the righthand side or in the condition of an equational axiom also appear in the lefthand side, and for membership axioms that all variables occurring in the condition also appear in the conclusion. A MET $\mathcal{S}$ is said to be *executable* iff the variable restriction[1] is satisfied for all axioms in E_S^C and the following conditions hold after identifying E_S^S-equivalent terms: the equations in E_S^C are confluent, equational and membership axioms in E_S^C are terminating, equational axioms in E_S^C are sort-decreasing and satisfy the regularity condition. For formal details of

---

[1] In its most recent version Maude imposes an even weaker restriction for executability due to the admissibility of conditions with *matching equations* [20].

these conditions we refer to [9]. In particular, these conditions imply that each term M has a unique normal form w.r.t. $\Rightarrow$ which is denoted by $\mathrm{NF}(M)$.

2.2 Rewriting Logic

In the simplified setting of [48] a rewrite specification $\mathcal{R}$ consists of a single-sorted signature $\Omega_{\mathcal{R}}$, a set $E_{\mathcal{R}}$ of equations over $\Omega_{\mathcal{R}}$, and a set $R_{\mathcal{R}}$ of labelled rewrite rules of the form $\forall\, X \;.\; l : M \to N$ if $\bar{\phi}_1 \wedge \ldots \wedge \bar{\phi}_n$, where l is a label, M and N are $\Omega_{\mathcal{R}}$-terms, and $\bar{\phi}_1 \wedge \ldots \wedge \bar{\phi}_n$ is a $\Omega_{\mathcal{R}}$-condition[2] over the variable set X. The rewrite rules in $R_{\mathcal{R}}$ are applied *modulo* the equations $E_{\mathcal{R}}$. *Rewriting logic (RWL)* has rules of deduction to infer all rewrites, i.e., those sentences of the form $P : M \to N$ that are valid in a given rewrite specification [48]. A *rewrite* $P : M \to N$ means that the term M *rewrites* to the term N modulo $E_{\mathcal{R}}$, and this rewrite is witnessed by the *proof term* P. Apart from general (concurrent) rewrites $P : M \to N$ that are generated from identity and atomic rewrites by parallel and sequential composition, rewriting logic classifies its most basic rewrites as follows: a *one-step (concurrent) rewrite* is generated by parallel composition from identity and atomic rewrites and contains at least one atomic rewrite, and a *one-step sequential rewrite* is a one-step rewrite containing exactly one atomic rewrite.

From a more general point of view, rewriting logic is parameterized by the choice of its underlying equational logic, which can be single-sorted, many-sorted, order-sorted and so on. In the design of the Maude language [19, 18], *membership equational logic* has been chosen as the underlying equational logic. To introduce rewriting logic over membership equational logic, abbreviated as $\mathrm{RWL}_{\mathrm{MEL}}$ or just RWL, we assume an underlying MES $\mathcal{S}_{\mathcal{R}}$ with a distinguished *data subspecification* $\mathcal{S}_{\mathcal{R}}^{D}$. The data subspecification specifies the static data part of the system whereas the remaining part of $\mathcal{S}_{\mathcal{R}}$ specifies the *state space* by introducing the *rewrite kinds*, i.e., kinds whose terms correspond to states and therefore can be rewritten, together with their algebraic structure, which characterizes the possibilities of parallel composition. In the context of this paper the state space is always specified in a purely equational way.

Definition 6. A *rewrite specification (RWS)* $\mathcal{R}$ consists of a MES $\mathcal{S}_{\mathcal{R}}$ with a distinguished *data subspecification* $\mathcal{S}_{\mathcal{R}}^{D}$, a set of *labels* $L_{\mathcal{R}}$, and a set of *rules* $R_{\mathcal{R}}$ of the form $\forall\, X \;.\; l : M \to N$ if $\bar{\phi}_1 \wedge \ldots \wedge \bar{\phi}_n$ where $l \in L_{\mathcal{R}}$, $\bar{\phi}_1 \wedge \ldots \wedge \bar{\phi}_n$ is a $\mathcal{S}_{\mathcal{R}}$-condition over X, and $M, N \in \mathbf{T}_{\mathcal{R}}(X)_k$ in $\mathcal{S}_{\mathcal{R}}$ for a rewrite kind k. To simplify the exposition we identify either $E_{\mathcal{R}}$-equivalent or $E_{\mathcal{R}}^{S}$-equivalent terms in the context of a RWS $\mathcal{R}$ whenever we are concerned with the algebraic semantics or the operational semantics, respectively.

Given two RWSs $\mathcal{R}$ and $\mathcal{R}'$, a RWS morphism $H : \mathcal{R} \to \mathcal{R}'$ consists of a MES morphism $H_{\mathcal{S}} : \mathcal{S}_{\mathcal{R}} \to \mathcal{S}_{\mathcal{R}'}$ and a function $H_L : L_{\mathcal{R}} \to L_{\mathcal{R}'}$ such that $H_{\mathcal{S}}$ has a

---

[2] Rewriting logic as presented in [48] admits rewrites in conditions of rules, but we do not exploit this possibility in the present paper.

restriction $H_D : \mathcal{S}_{\mathcal{R}}^D \to \mathcal{S}_{\mathcal{R}'}^D$ to the data subspecification and for each rule $r \in R_{\mathcal{R}}$ there is a rule in $R_{\mathcal{R}'}$ that is $E_{\mathcal{R}'}$-equivalent to $H(r)$ up to a renaming of the variables, where H is lifted to rules in the obvious homomorphic way. RWSs together with their morphisms form a category that is denoted by **RWS**.

The *algebraic semantics* of rewriting logic is defined as follows. A model of a rewrite specification (RWS) $\mathcal{R}$ is a model A of the underlying MES $\mathcal{S}_{\mathcal{R}}$ together with an enriched categorical structure for each set $[\![k]\!]_A$, where k is a rewrite kind. The interpretation of $\_ : \_ \to \_$, which can be regarded as a ternary predicate, is given by the arrows of the category. Sequential composition of rewrite proofs is interpreted by arrow composition, and parallel composition operators are interpreted by enriching the category with an algebraic structure as it has been specified for the rewrite kinds in $\mathcal{S}_{\mathcal{R}}$. In order to be a model, the category has to satisfy a number of natural requirements, namely, functoriality w.r.t. the algebraic structure that is relevant for the rewrite kinds, the equations in $\mathcal{S}_{\mathcal{R}}$ that are relevant for the rewrite kinds lifted to arrows, and for each rule in $\mathcal{R}$ the so-called exchange and decomposition laws. For a detailed description of these requirements we refer to [48]. The models of a RWS we consider in this paper are freely generated over models of the data subspecification $\mathcal{S}_{\mathcal{R}}^D$. In the important case where $\mathcal{S}_{\mathcal{R}}^D$ is interpreted initially, we obtain precisely the initial model described in [48]. A more precise definition of the algebraic semantics of rewriting logic will be given in Sections 3.2 and 4.4 for the particular forms of underlying specifications that are relevant for Petri nets.

The *operational semantics* of RWSs extends the operational semantics of MESs by applying computational equations $E_{\mathcal{R}}^C$ *and* rewrite rules $R_{\mathcal{R}}$ modulo the structural equations $E_{\mathcal{R}}^S$. In this way we can achieve the effect of rewriting modulo $E_{\mathcal{R}}$ provided that a suitable coherence requirement between equations and rules is satisfied. In particular, we say that a RWS is *weakly executable* iff the underlying MES is executable, and the equations in $E_{\mathcal{R}}^C$ are coherent with the rules in $R_{\mathcal{R}}$ modulo $E_{\mathcal{R}}^S$. Identifying terms that are $E_{\mathcal{R}}^S$-equivalent and identifying proof terms that are equivalent in the sense of [48], *coherence* means that if $P : M \to N$ then there is a term N' such that $NF(P) : NF(M) \to N'$ and $N' \Rightarrow^* NF(N)$ (this is stronger than coherence in [77] since we take proofs into account). A RWS is *strongly executable* iff additionally the *variable restriction for rules* is satisfied, i.e., all variables occurring in the righthand side or in the condition of a rule also appear in the lefthand side. In this case matching is sufficient for finding instantiations for the variables, whereas in the case of weak executability a strategy is needed to take care of this.

3   Place/Transition Nets

Place/transition nets (PTNs) are a model of concurreny in which behaviour is governed by local state changes in a distributed state space. The global distributed state of the system is represented by a *marking*, which assigns a number of indistinguishable *tokens* to each *place*. State changes that may occur in the system are specified by *transitions*. Each transition can only affect the part of the marking that is local to the transition, i.e., present in the places the transition is connected to. More precisely, a local state change corresponds to the atomic occurrence of a transition which removes tokens from its *input places* and produces tokens on its *output places*. The number of tokens that are transported by an arc is specified by its *inscription*.

As an example consider the PTN modeling an instance of the well-known banker's problem depicted in Fig. 1, which models the situation of a bank loaning money to (in this case two) clients. As usual, places and transitions are drawn as circles and rectangles, respectively. The flow relation and the weight function are given by arrows and their inscriptions. An additional initial marking is specified by place inscriptions. The money available for clients is modeled by the number of tokens in the place BANK. Furthermore, each client n has an individual credit limit modeled by a place CLAIM-n. The fact that client n requests and receives money is modeled by a transition GRANT-n and we assume that after exhausting the credit limit client n returns all the money via the transiton RETURN-n.

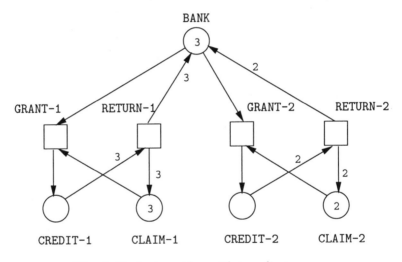

Fig. 1. Banker's problem with two clients

We now give formal definitions of basic nets and define a PTN as a particular form of an inscribed net. Instead of just finite nets we admit infinite nets, but we restrict our attention to nets with transitions that can affect only a finite part of the marking (locality principle) so that each transition can be represented in a finitary way.

Definition 7. A *net* N consists of a set of *places* P_N, a set of *transitions* T_N disjoint from P_N, and a *flow relation* $F_N \subseteq (P_N \times T_N) \cup (T_N \times P_N)$ such that $^\bullet t = \{p \mid p \ F_N \ t\}$ and $t^\bullet = \{p \mid p \ F_N \ t\}$ are finite for each $t \in T_N$ (local finiteness). A net is *finite* iff the sets P_N and T_N are finite. Given nets N and N', a *net morphism* $H : N \to N'$ consists of functions $H_P : P_N \to P_{N'}$ and $H_T : T_N \to T_{N'}$ such that $H_P(^\bullet t) = {}^\bullet H_T(t)$ and $H_P(t^\bullet) = H_T(t)^\bullet$. Nets together with their morphisms form a category **Net**.

A place/transition net is essentially a net with arcs inscribed by natural numbers.

Definition 8. A *place/transition net (PTN)* $\mathcal{N}$ consists of: (1) a net $N_{\mathcal{N}}$ and (2) an *arc inscription* $W_{\mathcal{N}} : F_{\mathcal{N}} \to \mathrm{N}$. $W_{\mathcal{N}}$ is extended to $W_{\mathcal{N}} : (P_{\mathcal{N}} \times T_{\mathcal{N}}) \cup (T_{\mathcal{N}} \times P_{\mathcal{N}}) \to \mathrm{N}$ in such a way that $(x, y) \notin F_{\mathcal{N}}$ implies $W_{\mathcal{N}}(x, y) = 0$. Given PTNs $\mathcal{N}$ and $\mathcal{N}'$, a *PTN morphism* $H : \mathcal{N} \to \mathcal{N}'$ is a net morphism $H : N_{\mathcal{N}} \to N_{\mathcal{N}'}$ such that:

1. $W_{\mathcal{N}'}(p', t') = W_{\mathcal{N}}(p_1, t) + \ldots + W_{\mathcal{N}}(p_n, t)$
   for all $p' \in P_{\mathcal{N}'}$, $t' \in T_{\mathcal{N}'}$, $t \in H^{-1}(t')$,
   and $\{p_1, \ldots, p_n\} = H^{-1}(p') \cap {}^\bullet t$ with distinct p_i, and
2. $W_{\mathcal{N}'}(t', p') = W_{\mathcal{N}}(t, p_1) + \ldots + W_{\mathcal{N}}(t, p_n)$
   for all $p' \in P_{\mathcal{N}'}$, $t' \in T_{\mathcal{N}'}$, $t \in H^{-1}(t')$,
   and $\{p_1, \ldots, p_n\} = H^{-1}(p') \cap t^\bullet$ with distinct p_i.

PTNs together with their morphisms form a category **PTN**. Each net N can be conceived as a PTN $\mathcal{N}$ with $N_{\mathcal{N}} = N$ and $W_{\mathcal{N}}(x, y) = 1$ iff $x\ F_N\ y$.

The notion of net morphism we use here is more restrictive than the (topological) net morphisms used in [63] and close to, but slightly stronger than, the (algebraic) net morphisms used in [52]. The justification for our definition is that net morphisms should be morphisms in the sense of [63] and should preserve the behaviour in the strongest reasonable sense. Given a net morphism $H : \mathcal{N} \to \mathcal{N}'$ the intention is that the behaviour of $\mathcal{N}$ is subsumed by the behaviour of $\mathcal{N}'$, although $\mathcal{N}'$ may exhibit a richer behaviour. In this paper we focus on a description of behaviour by Best-Devillers processes in a way that generalizes the well-known step semantics. Indeed, not only the interleaving semantics but also the step semantics and the process semantics can be regarded as labelled transition systems where the states are markings and the labels are steps or processes, respectively. In the case of Best-Devillers processes, the labelled transition system is equipped with additional algebraic structure which will be made explicit by regarding the transition system as a symmetric monoidal category.

Definition 9. Let $\mathcal{N}$ be a PTN. A *marking* is a multiset of places. A *(concurrent) step* is a nonempty finite multiset of transitions. The *set of markings* and the *set of steps* are denoted by $\mathcal{M}_{\mathcal{N}}$ and $\mathcal{ST}_{\mathcal{N}}$, respectively. We define *preset* and *postset functions* $\partial_0, \partial_1 : T_{\mathcal{N}} \to \mathcal{M}_{\mathcal{N}}$ by $\partial_0(t)(p) = W(p, t)$ and $\partial_1(t)(p) = W(t, p)$, respectively. The *(concurrent) step semantics* of a place/transition net $\mathcal{N}$ is given by the labelled transition system which has $\mathcal{M}_{\mathcal{N}}$ as its set of states, $\mathcal{ST}_{\mathcal{N}}$ as its set of labels and a transition relation $\to\ \subseteq \mathcal{M}_{\mathcal{N}} \times \mathcal{ST}_{\mathcal{N}} \times \mathcal{M}_{\mathcal{N}}$

defined by $m_1 \xrightarrow{e} m_2$ iff there is a marking m such that, for all $p \in P_{\mathcal{N}}$,

$$m_1(p) = m(p) + \partial_0(e)(p) \text{ and}$$
$$m_2(p) = m(p) + \partial_1(e)(p) .$$

Writing the occurrence rule in the way given above makes it evident that the occurrence of an action replaces its preset by its postset, whereas the remainder of the marking, here denoted by m, is not involved in this process. This is an important fact that will be made formally explicit in the process semantics that we review subsequently in a somewhat informal style. For details we refer to [21] and [7].

Definition 10. An *occurrence net* $\mathcal{N}$ is a net such that $F_{\mathcal{N}}$ is acyclic and $|{}^\bullet t|$, $|t^\bullet| \leq 1$ for each $t \in T_{\mathcal{N}}$. Given an occurrence net $\mathcal{N}$, $F_{\mathcal{N}}$ induces a partial order $(<) = F_{\mathcal{N}}{}^+$ on $P_{\mathcal{N}} \cup T_{\mathcal{N}}$ and its minimal and maximal elements are denoted by $Max(\mathcal{N})$ and $Min(\mathcal{N})$, respectively.

Let $\mathcal{N}$ be a PTN. Then a *finite process* $\mathcal{P}$ of $\mathcal{N}$ with *origin* marking m_1 and *destination* marking m_2 consists of a finite occurrence net $N_{\mathcal{P}}$ and a PTN morphism $L_{\mathcal{P}} : N_{\mathcal{P}} \to \mathcal{N}$ (where $N_{\mathcal{P}}$ is viewed as a PTN) such that $L_{\mathcal{P}}(Min(N_{\mathcal{P}})) = m_1$ and $L_{\mathcal{P}}(Max(N_{\mathcal{P}})) = m_2$. Given finite processes $\mathcal{P}$ and $\mathcal{P}'$ the *parallel composition* of $\mathcal{P}$ and $\mathcal{P}'$ is defined as the disjoint union of the underlying nets and label functions. Given processes $\mathcal{P}$ and $\mathcal{P}'$ such that the destination of $\mathcal{P}$ is equal to the origin of $\mathcal{P}'$, a *sequential composition* of $\mathcal{P}$ and $\mathcal{P}'$ is obtained by disjoint union (as above) pairwise identifying maximal places of $\mathcal{P}$ with minimal places of $\mathcal{P}'$, where every two places to be identified must have the same label. Notice that in general the result of sequential composition is not unique [21].

Intuitively, a process of a PTN is generated by "temporal unfolding" starting from a marking that becomes the origin of the process. Observe that for a given finite process $\mathcal{P}$ of $\mathcal{N}$, not only $Min(\mathcal{P})$ and $Max(\mathcal{P})$ but each snapshot (S-cut in the sense of [8]) of $\mathcal{P}$ corresponds to a marking of $\mathcal{N}$ by virtue of $L_{\mathcal{P}}$. The ambiguity of the result of sequential composition is caused by a snapshot corresponding to a marking with several identical tokens in some place, say p. Consider a transition in $\mathcal{N}$ that removes one token from p. A single firing of this transition gives rise to two different processes, since identical tokens are represented by different places in the process net. An obvious solution to avoid this ambiguity is to restrict our attention to *safe processes*, i.e., processes that take place in the safe part of the state space where such situations do not occur. A marking m is said to be a *safe* marking iff all markings m' reachable from m in the step semantics satisfy $m'(p) \leq 1$ for all $p \in P$. A process is said to be *safe* iff its origin is safe. Safe processes coincide with the classical notion of processes if we consider 1-safe PTNs which are equivalent to contact-free elementary net systems [8,63]. Our definition of safe processes is restrictive enough to ensure that the class of safe finite processes is always closed under sequential composition, a property that is not shared by the subclass of finite processes with the weaker property that all markings m corresponding to snapshots (S-cuts) satisfy $m(p) \leq 1$ for each $p \in P$.

Definition 11. A *(strict) monoidal category (MC)* $\mathbf{C}$ is a category equipped with a monoidal operation $\_ \otimes_{\mathbf{C}} \_$ and an identity object $id_{\mathbf{C}}$ such that $\_ \otimes_{\mathbf{C}} \_$ is an associative bifunctor with left and right identity $id_{\mathbf{C}}$. A monidal category morphism $h : \mathbf{C} \to \mathbf{C}'$ is a functor that preserves $\_ \otimes \_$ and id, i.e., $h(u \otimes_{\mathbf{C}} v) = h(u) \otimes_{\mathbf{C}'} h(v)$ and $h(id_{\mathbf{C}}) = id_{\mathbf{C}'}$. If in addition $\_ \otimes_{\mathbf{C}} \_$ is commutative, then we say that $\mathbf{C}$ is a *(strictly) symmetric (strict) monoidal category (SMC)*. The category of SMCs is denoted by **SMC**.

A variation of an SMC is a *partial SMC* $\mathbf{C}$ where $\_ \otimes_{\mathbf{C}} \_$ is a partial functor and each equation in the definition of SMCs is only required to be satisfied iff both sides are defined. The category of partial SMCs is denoted by **PSMC**. Clearly, **SMC** is a subcategory of **PSMC**.

Definition 12. The *safe process semantics* $\mathbf{SP}(\mathcal{N})$ of a PTN $\mathcal{N}$ is given by a partial SMC that has safe markings as objects and safe processes as arrows. Arrow composition is given by sequential composition, the partial monoidal operation is given by parallel composition, and the identity for an object m is given by the finite process without transitions with origin m and destination m. **SP** can be extended to a functor $\mathbf{SP} : \mathbf{SPTN} \to \mathbf{PSMC}$, where **SPTN** is the subcategory of **PTN** obtained by restricting morphisms to safe PTN morphisms. Here, a PTN morphism $H : \mathcal{N} \to \mathcal{N}'$ is *safe* iff it maps each safe marking in $\mathcal{N}$ to a safe marking in $\mathcal{N}'$. Now **SP** lifts each safe PTN morphism $H : \mathcal{N} \to \mathcal{N}'$ to a functor $\mathbf{SP}(H) : \mathbf{SP}(\mathcal{N}) \to \mathbf{SP}(\mathcal{N}')$ defined in the obvious way.

If we restrict our attention to safe markings there is a close correspondence between the step semantics and the process semantics: Each step sequence, i.e., each computation w.r.t. the step semantics, generates a unique process, and a process determines a set of step sequences that contains the original one. As a consequence processes are more abstract than step sequences. A similar correspondence exists for the interleaving semantics, i.e., if we restrict steps to single transitions. Both correspondences are investigated in [7]. On the other hand, the authors of [7] observe that step sequences and processes become incomparable when we admit markings that are not safe, which means that the natural view of processes as an abstraction of step sequences does not hold anymore. In order to recover this correspondence a more abstract notion of process is needed, and in fact Best-Devillers processes [7], which became also known as commutative processes [52, 21], provide such a notion. In contrast to processes which adhere to the *individual token philosophy*,[3] Best-Devillers processes share with step sequences the *collective token philosophy*, meaning that identical tokens on a place in the system are not distinguished in the process. This allows us to define an

---

[3] A functorial semantics following the individual token philosophy has recently been given in [14] by using pre-nets, a refinement of PTNs.

operation of sequential composition that has a unique result whenever sequential composition is possible. The following definition of Best-Devillers processes is equivalent to the definition given in [7], except for the fact that [7] does not make explicit the algebraic and categorical structure.

Definition 13. Let $\mathcal{P}$ and $\mathcal{P}'$ be finite processes and let $p_1, p_2 \in P_{\mathcal{P}}$ with $L_{\mathcal{P}}(p_1) = L_{\mathcal{P}}(p_2)$. We define a predicate swap$(\mathcal{P}, \mathcal{P}', p_1, p_2)$ which holds iff $P_{\mathcal{P}'} = P_{\mathcal{P}}$, $T_{\mathcal{P}'} = T_{\mathcal{P}}$, $L_{\mathcal{P}'} = L_{\mathcal{P}}$ and:

1. $t \; F_{\mathcal{P}'} \; p \Leftrightarrow t \; F_{\mathcal{P}} \; p$,
2. $p \; F_{\mathcal{P}'} \; t \Leftrightarrow p \; F_{\mathcal{P}} \; t$ if $p \neq p_1$ and $p \neq p_2$,
3. $p_1 \; F_{\mathcal{P}'} \; t \Leftrightarrow p_2 \; F_{\mathcal{P}} \; t$,
4. $p_2 \; F_{\mathcal{P}'} \; t \Leftrightarrow p_1 \; F_{\mathcal{P}} \; t$.

We define an *equivalence* on finite processes as the smallest equivalence relation that contains $(\mathcal{P}, \mathcal{P}')$ if there are $p_1, p_2 \in P_{\mathcal{P}}$ such that $L_{\mathcal{P}}(p_1) = L_{\mathcal{P}}(p_2)$ and swap$(\mathcal{P}, \mathcal{P}', p_1, p_2)$ holds. The equivalence classes are called *Best-Devillers processes*.

The notions of *origin*, *destination*, *parallel* and *sequential composition* of processes are lifted to Best-Devillers processes in the obvious way. At this level the result of sequential composition becomes unique, since all potentially different results obtained by composing two processes fall into the same equivalence class.

Definition 14. The *Best-Devillers process semantics* **BDP**$(\mathcal{N})$ of a PTN $\mathcal{N}$ is given by an SMC that has markings as objects and Best-Devillers processes as arrows. Arrow composition is given by sequential composition, the monoidal operation is given by parallel composition, and the identity for an object m is given by the Best-Devillers process without transitions with origin m and destination m. **BDP** can be extended to a functor **BDP** : **PTN** $\rightarrow$ **SMC** that sends each PTN morphism $H : \mathcal{N} \rightarrow \mathcal{N}'$ to a functor **BDP**(H) : **BDP**$(\mathcal{N}) \rightarrow$ **BDP**$(\mathcal{N}')$.

The above definition is also equivalent to the one given in [21], although we define Best-Devillers processes as a quotient of (classical) processes as in [7] rather than as a quotient of *concatenable processes* as in [21]. Concatenable processes are a slight refinement of finite (classical) processes: a concatenable process is a finite process together with a total ordering of $\{p \in Min(\mathcal{N}) \mid L(p) = p'\}$ for each place p' in the origin and a total ordering of $\{p \in Max(\mathcal{N}) \mid L(p) = p'\}$ for each place p' in the destination. Using this refined notion of process the obvious definition of sequential composition, where places are only identified if they have the same position in this order, yields a unique result, which allows us to view the class of concatenable processes as a category.

Since a safe process is only equivalent to itself, it corresponds to a Best-Devillers process given by a singleton equivalence class. Hence each safe process can be regarded as a Best-Devillers process giving rise to an injection $\iota(\mathcal{N}) : \mathbf{SP}(\mathcal{N}) \rightarrow \mathbf{BDP}(\mathcal{N})$. Actually we can state the following stronger

Remark 1. $\mathbf{SP}$: $\mathbf{SPTN} \to \mathbf{PSMC}$ is a subfunctor of $\mathbf{BDP}$: $\mathbf{SPTN} \to$ $\mathbf{PSMC}$ (the obvious restriction of $\mathbf{BDP}$: $\mathbf{PTN} \to \mathbf{SMC}$) as witnessed by ι : $\mathbf{SP} \to \mathbf{BDP}$ which is in fact a natural transformation.

We take this remark as a justification for focusing primarily on the Best-Devillers processes in the following, keeping in mind that classical safe processes form an important subcategory. In the context of nets with individual tokens we shall give some additional arguments for the relevance of this subcategory.

3.1   Rewriting Semantics: An Example

Rewriting logic can provide a direct semantics of PTNs following the motto "Petri nets are monoids" advocated in [52]. In fact, the categorical semantics presented in that work and also the relation between PTNs and linear logic explained in [46] inspired the development of rewriting logic.

The PTN of the banker's problem can be represented by the following RWS given in Maude syntax [19, 18], which consists of a MES specification and a set of rewrite rules. As usual in Maude, the rewrite kind [Marking] is implicitly introduced by introducing a sort Marking of this kind.[4]

```
sort Marking .

op empty : -> Marking .
op __ : Marking Marking -> Marking [assoc comm id: empty] .

ops BANK CREDIT-1 CREDIT-2 CLAIM-1 CLAIM-2 : -> Marking .

rl [GRANT-1]  : BANK CLAIM-1 => CREDIT-1 .

rl [RETURN-1] : CREDIT-1 CREDIT-1 CREDIT-1 =>
                BANK BANK BANK CLAIM-1 CLAIM-1 CLAIM-1 .

rl [GRANT-2]  : BANK CLAIM-2 => CREDIT-2 .

rl [RETURN-2] : CREDIT-2 CREDIT-2 =>
                BANK BANK CLAIM-2 CLAIM-2 .
```

Here we have applied the translation of PTNs into rewriting logic suggested in [48], which is closely related to the translation of PTNs into linear logic [46]. A marking is represented as an element of the finite multiset sort Marking. The constant empty represents the empty marking and __ is the corresponding multiset union operator. Associativity, commutativity and identity laws are specified

---

[4] In fact, here and in the rest of the paper Marking and [Marking] can be identified, since the latter does not contain any additional (error) elements (cf. [9, 51]).

as structural equations by the operator attributes in square brackets. For each place p there is a constant p, called *token constructor*, representing a single token residing in that place. In fact, under the initial semantics Marking is a multiset sort over tokens generated by these token constructors. For each transition t there is a rule, called *transition rule*, labelled by t and stating that its preset marking may be replaced by its postset marking.

As clearly demonstrated by the use of rewrite rules in the above RWS, there is an important difference between the reduction rules induced by computational equations of a MES and the rewrite rules of a RWS: The relation induced by one-step rewrites is in general neither terminating nor confluent, although there may be situations where this is the case. Only terminating systems where for each initial state there is a unique final state can be described by terminating and confluent rewrite rules. Hence this generalization is a practical necessity to represent general system models. For instance, the PTN model of the banker's problem has not only infinite executions but also finite ones due to the possibility of deadlock. Therefore, the transition system is neither terminating nor confluent in this case.

In order to control the execution of a RWS the user can specify a strategy which successively selects rewrite rules and initiates rewriting steps. For instance, in the case of the banker's example it is possible to define an execution strategy that avoids states which are necessarily leading to a deadlock such that the banker stays always in the "safe" part of the state space. In applications such as net execution and analysis the choice of a strategy will be guided by the need to explore the behaviour of the system under certain conditions. Strategies are well-supported by the Maude engine via reflection [19, 18], i.e. the capability to represent rewrite specifications as objects and control their execution at the meta-level, which makes Maude a suitable tool not only for executing place-transition nets but also for analyzing such nets using strategies for (partial) state-space exploration and model checking.

3.2   Rewriting Semantics in the General Case

The rewriting semantics that has been explained in terms of the banker's example in the previous section can be conceived as a functor from the category **PTN** of place/transitions nets to the category **SMRWS** of symmetric monoidal RWSs (SMRWSs) that will be introduced next. The characteristic feature of SMRWSs is that their underlying specification has a single rewrite kind [Marking] that is specified to be a free commutative monoid over a set of constants. The definition of SMRWSs given below is quite restrictive, but is sufficient for the rewriting semantics of PTNs. In Section 4.4 SMRWSs will be generalized to provide a rewriting semantics for nets with individual tokens.

Definition 15. A RWS $\mathcal{R}$ is a *symmetric monoidal RWS (SMRWS)* iff the following conditions are satisfied:

1. $\mathcal{S}_{\mathcal{R}}^{D}$ is empty.
2. $\mathcal{S}_{\mathcal{R}}$ contains precisely the following:
   (a) a kind [Marking] together with operator symbols

   $$\texttt{empty} : \to \texttt{[Marking]}, \quad \_\_ : \texttt{[Marking] [Marking]} \to \texttt{[Marking]};$$

   (b) any number of operator symbols of the general form

   $$p : \to \texttt{[Marking]};$$

   (c) the parallel composition axioms

   $$\forall\ u, v, w : \texttt{[Marking]}\ .\ u\ (v\ w) = (u\ v)\ w,$$
   $$\forall\ u, v : \texttt{[Marking]}\ .\ u\ v = v\ u,$$
   $$\forall\ u : \texttt{[Marking]}\ .\ \texttt{empty}\ u = u.$$

3. Rules in $R_{\mathcal{R}}$ do not have conditions and do not contain any variables.

Given two SMRWSs $\mathcal{R}$ and $\mathcal{R}'$, a SMRWS morphism $H : \mathcal{R} \to \mathcal{R}'$ is a RWS morphism that preserves [Marking], empty and $\_\_$. SMRWSs together with their morphisms form a subcategory of **RWS** denoted **SMRWS**.

In order to obtain a precise definition of the initial model-theoretic semantics $\mathbf{I}(\mathcal{R})$ of a SMRWS $\mathcal{R}$, it is convenient to define the model-theoretic semantics of $\mathcal{R}$ by means of a MES $\mathbf{E}(\mathcal{R})$ which has a standard model-theoretic semantics in terms of $\mathbf{E}(\mathcal{R})$-algebras. Having done that, we then define $\mathbf{I}(\mathcal{R})$ as $\mathbf{I}(\mathbf{E}(\mathcal{R}))$, i.e., as the initial model of $\mathbf{E}(\mathcal{R})$.

Definition 16. The *membership equational presentation* of a SMRWS $\mathcal{R}$ is a MES $\mathbf{E}(\mathcal{R})$ that extends $\mathcal{S}_{\mathcal{R}}$, the underlying MES of $\mathcal{R}$, by the following:

1. a new kind [RawProc] together with new operator symbols called *proof constructors*

   $$\texttt{id} : \texttt{[Marking]} \to \texttt{[RawProc]},$$
   $$\_\_ : \texttt{[RawProc] [RawProc]} \to \texttt{[RawProc]},$$
   $$\_;\_ : \texttt{[RawProc] [RawProc]} \to \texttt{[RawProc]};$$

2. a new operator symbol called *atomic proof constructor*

   $$t :\to \texttt{[RawProc]}$$

   for each rule $t : M \to N$ in $R_{\mathcal{R}}$;
3. a kind [Proc] with a sort Proc and an operator symbol

   $$\_:\_\to\_ : \texttt{[RawProc] [Marking] [Marking]} \to \texttt{[Proc]};$$

4. a membership axiom

$$t : M \to N$$

for each rule $t : M \to N$ in $R_{\mathcal{R}}$, where we introduce the notation

$$P : M \to N \quad \text{as a shorthand for} \quad (P : M \to N) : \texttt{Proc};$$

5. membership axioms corresponding to the standard inference rules of rewriting logic, namely:
   (a) *identity*:

   $$\texttt{id}(u) : u \to u$$

   (b) *composition*:

   $$\alpha;\beta : u_1 \to u_3 \text{ if } \alpha : u_1 \to u_2 \wedge \beta : u_2 \to u_3$$

   (c) *compatibility* of parallel composition:

   $$\alpha_1 \; \alpha_2 : u_1 \; u_2 \to u_1' \; u_2' \text{ if } \alpha_1 : u_1 \to u_1' \wedge \alpha_2 : u_2 \to u_2'$$

6. equational axioms corresponding to the standard rewriting logic axioms, namely:
   (a) *identity*:

   $$\texttt{id}(u);\alpha = \alpha \text{ if } \alpha : u \to u'$$
   $$\alpha;\texttt{id}(u') = \alpha \text{ if } \alpha : u \to u'$$

   (b) *associativity*:

   $$\alpha;(\beta;\gamma) = (\alpha;\beta);\gamma$$
   $$\text{if } \alpha : u_1 \to u_2 \wedge \beta : u_2 \to u_3 \wedge \gamma : u_3 \to u_4$$

   (c) *functoriality* of the parallel composition operator:

   $$\texttt{id}(u_1) \; \texttt{id}(u_2) = \texttt{id}(u_1 \; u_2)$$
   $$(\alpha_1;\beta_1)(\alpha_2;\beta_2) = (\alpha_1 \; \alpha_2);(\beta_1 \; \beta_2)$$
   $$\text{if } \alpha_1 : u_1 \to v_1 \wedge \beta_1 : v_1 \to w_1 \wedge$$
   $$\alpha_2 : u_2 \to v_2 \wedge \beta_2 : v_2 \to w_2$$

   (d) *inherited equations* for the parallel composition operator:

   $$\alpha_1 \; (\alpha_2 \; \alpha_3) = (\alpha_1 \; \alpha_2) \; \alpha_3$$
   $$\text{if } \alpha_1 : u_1 \to u_1' \wedge \alpha_2 : u_2 \to u_2' \wedge \alpha_3 : u_3 \to u_3'$$
   $$\alpha_1 \; \alpha_2 = \alpha_2 \; \alpha_1$$
   $$\text{if } \alpha_1 : u_1 \to u_1' \wedge \alpha_2 : u_2 \to u_2'$$
   $$\texttt{id}(\texttt{empty}) \; \alpha = \alpha \text{ if } \alpha : u \to u'$$

For better readability we leave universal quantifiers implicit: $u, u', v, w, u_i, u'_i, v_i,$ w_i are distinct variables of kind [Marking] and $\alpha, \beta, \gamma, \alpha_i, \beta_i, \gamma_i$ are distinct variables of kind [RawProc].

$\mathbf{E}$ can be extended to a functor $\mathbf{E} : \mathbf{SMRWS} \to \mathbf{MES}$ in the obvious way. Furthermore, composing $\mathbf{E} : \mathbf{SMRWS} \to \mathbf{MES}$ with the functor $\mathbf{Mod} : \mathbf{MES} \to \mathbf{Cat}^{\mathrm{op}}$ we obtain $\mathbf{Mod} \circ \mathbf{E} : \mathbf{SMRWS} \to \mathbf{Cat}^{\mathrm{op}}$ which is also denoted $\mathbf{Mod} :$ $\mathbf{SMRWS} \to \mathbf{Cat}^{\mathrm{op}}$. As usual we write $\mathbf{U}_H$ for $\mathbf{Mod}(H)$ given a SMRWS morphism H.

In this paper we are not interested in the entire algebraic structure of SM-RWS models. Instead, our first goal is to relate two different semantics of PTNs, namely, the Best-Devillers process semantics and the rewriting semantics of Definition 14, in terms of SMCs. In other words, the category $\mathbf{SMC}$ will serve as a common basis and suitable level of abstraction to compare different descriptions. Below, the initial models of SMRWSs, that are defined in terms of a functor $\mathbf{I}$, will be uniformly mapped into the same domain via a forgetful functor $\mathbf{V}$.

Definition 17. Let $\Sigma(\mathbf{Mod})$ be the Grothendiek construction for the functor $\mathbf{Mod} : \mathbf{SMRWS} \to \mathbf{Cat}^{\mathrm{op}}$ and let $\pi_1 : \Sigma(\mathbf{Mod}) \to \mathbf{SMRWS}$ be the obvious projection functor that sends $(\mathcal{R}, A)$ to $\mathcal{R}$. Given a SMRWS $\mathcal{R}$ we define $\mathbf{I}(\mathcal{R})$ as $\mathbf{I}(\mathbf{E}(\mathcal{R}))$ and $\Sigma\mathbf{I}(\mathcal{R})$ as $(\mathcal{R}, \mathbf{I}(\mathcal{R}))$. Given a SMRWS morphism $H : \mathcal{R} \to \mathcal{R}'$ we define $\Sigma\mathbf{I}(H)$ as the morphism $(H, \mathbf{I}(H)) : (\mathcal{R}, \mathbf{I}(\mathcal{R})) \to (\mathcal{R}', \mathbf{I}(\mathcal{R}'))$ with $\mathbf{I}(H)$ the unique morphism $\mathbf{I}(H) : \mathbf{I}(\mathcal{R}) \to \mathbf{U}_H(\mathbf{I}(\mathcal{R}'))$ guaranteed by the fact that $\mathbf{I}(\mathcal{R})$ and $\mathbf{U}_H(\mathbf{I}(\mathcal{R}'))$ are objects in $\mathbf{Mod}(\mathcal{R})$ with the former being initial. In this way we have defined a functor $\Sigma\mathbf{I} : \mathbf{SMRWS} \to \Sigma(\mathbf{Mod})$ that is left adjoint to π_1.

Let $\mathbf{V} : \Sigma(\mathbf{Mod}) \to \mathbf{SMC}$ be the forgetful functor which sends $(\mathcal{R}, \hat{A})$ to the SMC defined as follows: The sets of objects and arrows are $[\![\,[\texttt{Marking}]\,]\!]_{\hat{A}}$ and $[\![\texttt{Proc}]\!]_{\hat{A}}$, respectively. Arrow composition is $[\![\_;\_]\!]_{\hat{A}}$ and identities are $[\![\texttt{id}]\!]_{\hat{A}}(m)$ for $m \in [\![\,[\texttt{Marking}]\,]\!]_{\hat{A}}$. The monoidal operation and its identity are given by $[\![\_\_]\!]_{\hat{A}}$ and $[\![\texttt{empty}]\!]_{\hat{A}}$, respectively. Given a morphism $(H, h) : (\mathcal{R}, \hat{A}) \to (\mathcal{R}', \hat{A}')$ in $\Sigma(\mathbf{Mod})$ we define $\mathbf{V}(H, h)$ as the SMC morphism given by the obvious restriction of h.

The rewriting semantics of PTNs is then defined as follows:

Definition 18. Given a PTN $\mathcal{N}$ the *rewriting semantics* of $\mathcal{N}$ is the smallest SMRWS $\mathbf{R}(\mathcal{N})$ such that:

1. $\mathcal{S}_{\mathbf{R}(\mathcal{N})}$ contains a *token constructor*

    $p : \to$ [Marking]

    for each place $p \in P_{\mathcal{N}}$;

2. $\mathbf{R}(\mathcal{N})$ has a label t and a rule called a *transition rule*, namely,

$$t : \underbrace{p_1 \ldots p_1}_{W(p_1,t)} \cdots \underbrace{p_m \ldots p_m}_{W(p_m,t)} \rightarrow \underbrace{p_1 \ldots p_1}_{W(t,p_1)} \cdots \underbrace{p_m \ldots p_m}_{W(t,p_m)}$$

for each transition $t \in T_{\mathcal{N}}$ assuming $P_{\mathcal{N}} = \{p_1, \ldots, p_m\}$ with distinct p_i.

$\mathbf{R}$ can be extended to a functor $\mathbf{R} : \mathbf{PTN} \rightarrow \mathbf{SMRWS}$ that maps each PTN morphism $H : \mathcal{N} \rightarrow \mathcal{N}'$ to the unique SMRWS morphism $G : \mathbf{R}(\mathcal{N}) \rightarrow \mathbf{R}(\mathcal{N}')$ with $G_L(t) = H(t)$ for each $t \in T_{\mathcal{N}}$ and $G_S(p) = H(p)$ for each $p \in P_{\mathcal{N}}$.

The main result in this section states that for a PTN $\mathcal{N}$ the Best-Devillers process semantics $\mathbf{BDP}(\mathcal{N})$ coincides with the initial semantics of $\mathbf{R}(\mathcal{N})$ in the strongest possible categorical sense of a natural isomorphism.

In fact, this theorem is closely related to and can be proved using a result in [21] (Theorem 27), which states that the monoidal category $\mathcal{CP}(\mathcal{N})$ of concatenable processes and a monoidal category $\mathcal{P}(\mathcal{N})$ defined by an inductive equational definition are isomorphic. Both $\mathcal{CP}(\mathcal{N})$ and $\mathcal{P}(\mathcal{N})$ are not symmetric, but they still enjoy certain symmetries. For an exact definition of $\mathcal{CP}(\mathcal{N})$ and $\mathcal{P}(\mathcal{N})$ we refer to [21].

The difference between Theorem 27 in [21] and Theorem 1 below is that: (1) Theorem 1 is about Best-Devillers processes which are more abstract than concatenable processes, (2) it uses rewriting logic instead of giving a direct inductive equational definition, and (3) it states a natural isomorphism instead of just an isomorphism, that is, we use not only categories in the small, but we also aim at a systematic categorical treatment in the large.

Theorem 1. There is a natural isomorphism $\hat{\tau} : \mathbf{BDP} \rightarrow \mathbf{V} \circ \Sigma \mathbf{I} \circ \mathbf{R}$ between the functors $\mathbf{BDP} : \mathbf{PTN} \rightarrow \mathbf{SMC}$ and $\mathbf{V} \circ \Sigma \mathbf{I} \circ \mathbf{R} : \mathbf{PTN} \rightarrow \mathbf{SMC}$ (with $\mathbf{R} : \mathbf{PTN} \rightarrow \mathbf{SMRWS}$ and $\mathbf{V} \circ \Sigma \mathbf{I} : \mathbf{SMRWS} \rightarrow \mathbf{SMC}$).

In particular, the previous theorem entails that for each individual PTN we have precisely characterized Best-Devillers processes in rewriting logic via $\mathbf{R}$ as stated by the corollary below. As a byproduct we have obtained a corresponding characterization in membership equational logic via $\mathbf{E}$.

Corollary 1. The rewrite specification $\mathbf{R}(\mathcal{N})$ provides a sound and complete axiomatization of the Best-Devillers processes of the PTN $\mathcal{N}$.

Again, this is closely related to Corollary 33 in [21], which states that the presentation of an SMC denoted by $\mathcal{T}(\mathcal{N})$ provides a complete and sound axiomatization of Best-Devillers processes. Similar to the category $\mathcal{P}(\mathcal{N})$ mentioned before, $\mathcal{T}(\mathcal{N})$ is given by a direct inductive equational definition, whereas here we use the SMRWS $\mathbf{R}(\mathcal{N})$ to express the same category. In other words we use rewriting logic to equip the presentation of $\mathcal{T}(\mathcal{N})$ itself with a first-class formal status.

3.3   Petri Nets with Test Arcs

In this section we illustrate how the techniques for giving a rewriting logic semantics to place/transition nets can be extended to deal with the important class of place/transition nets with *test arcs* [16, 58, 78, 15]. Petri nets have been equipped with test arcs (also called *read arcs*, or *positive contexts* in contextual nets [58]) to naturally model cases where a certain resource may be *read without being consumed* by a transition, such as in a database system where multiple users are allowed to simultaneously read the same piece of data. In contrast to ordinary arcs, several test arcs are allowed to access the same token in the same concurrent step, but a token accessed by a test arc may not be accessed by an ordinary arc in the same step.[5] Test arcs cannot change the marking of a place.

Formally, a place/transition net with test arcs $\mathcal{N}$ is a place/transition net together with a set of *test arcs* $TA_{\mathcal{N}} \subseteq P_{\mathcal{N}} \times T_{\mathcal{N}}$. We define the *context function* $\partial_{TA} : T_{\mathcal{N}}^{\oplus} \to \mathcal{M}$ on finite multisets e of transitions by $\partial_{TA}(e)(p) = 1$ if there is a transition $t \in e$ with $(p, t) \in TA_{\mathcal{N}}$, and by $\partial_{TA}(e)(p) = 0$ otherwise. The *step semantics* of a place/transition net with test arcs is defined as for place/transition nets (see Section 3) with the modification that for $m_1 \overset{e}{\to} m_2$ to hold we require additionally that, for each place $p \in P_{\mathcal{N}}$, $\partial_{TA}(e)(p) \leq m(p)$.

We propose a rewriting semantics for a place/transition net with test arcs, defined in terms of a rewrite specification $\mathbf{R}(\mathcal{N})$ similar to the one in Definition 18, but specifying tokens by means of a kind [Place] and two operators $[\_], \langle\_\rangle$: [Place] $\to$ [Marking] so that a token residing at place p is represented by the term $[p]$. An occurrence of $[p]$ may not be shared by more than one rewrite at the same time; to allow simultaneous rewrites with *read-only* access to a token at place p, we consider a token $[p]$ to be equivalent to an arbitrary number of read-only tokens of the form $\langle p \rangle$. This can be accomplished, using a technique described in [50], by adding to our specification $\mathbf{R}(\mathcal{N})$ an operator $\{\_ \mid \_\}$: [Marking] [Nat] $\to$ [Marking] and two "copying" axioms[6]

$$[p] = \{p \mid 0\} \quad \text{and} \quad \{p \mid n\} = \{p \mid n+1\}\, \langle p \rangle,$$

where p and n are variables ranging, respectively, over [Place] and [Nat].

A transition t which consumes the tokens $a_1, \ldots, a_n$, produces the tokens $b_1, \ldots, b_m$, and "reads" the tokens $c_1, \ldots, c_k$, is modeled by a rewrite rule

$$t : [a_1] \ldots [a_n]\, \langle c_1 \rangle \ldots \langle c_k \rangle \longrightarrow [b_1] \ldots [b_m]\, \langle c_1 \rangle \ldots \langle c_k \rangle.$$

The database example in Figure 2, taken from [16], where multiple users may

---

[5] This last restriction is omitted in some definitions of Petri nets with test arcs (see e.g. [78]).

[6] The counting of the read-only copies and their read-only use guarantee that all the copies must have been "folded back together" in order for the original token to be engaged in a transition that consumes the token.

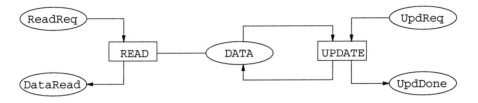

Fig. 2. Small database example using test arcs.

read some data simultaneously, but where only one at a time is allowed to update the data, is, therefore, modeled in rewriting logic by the following rules:

READ :   [ReadReq] ⟨Data⟩ ⟶ [DataRead] ⟨Data⟩
UPDATE : [UpdReq] [Data] ⟶ [UpdDone] [Data].

Let $\mathbf{R}(\mathcal{N})$ be the rewrite specification representing a place/transition net with test arcs $\mathcal{N}$ as explained above, and for any marking m in $\mathcal{N}$, let $m^{\sharp}$ denote the term of kind [Marking] which contains exactly $m(p)$ occurrences of the term [p] for each place p in $\mathcal{N}$. Then, there is a step $m_1 \xrightarrow{e} m_2$ in $\mathcal{N}$ iff there is a one-step concurrent rewrite $\alpha : m_1^{\sharp} \longrightarrow m_2^{\sharp}$ in $\mathbf{R}(\mathcal{N})$, where, in addition, the step e can be extracted from the proof α. Furthermore, as in Definition 17 we can define a functor that associates with $\mathbf{R}(\mathcal{N})$ a symmetric monoidal category determined by the initial semantics. This provides a categorical semantics for all the concurrent computations of the net $\mathcal{N}$ that is closely related to the one recently proposed by Bruni and Sassone in [15].

4   High-Level Petri Nets

We use the term *high-level Petri nets* to refer to a range of extensions of PTNs by individual tokens, a line of research that has been initiated by the introduction of *predicate/transition nets* in [33, 34, 32]. High-level Petri nets make use of an underlying formalism, such as first-order logic in the case of predicate/transition nets, to describe the information that is associated with each token and its transformation. *Colored nets*[7] introduced in [38] are another quite general model of this kind with a more set-theoretic flavour. They generalize PTNs in such a way that tokens can be arbitrary set-theoretic objects. Quite different from, but closely related to, colored nets are high-level Petri nets that use an algebraic specification language as an underlying formalism [75, 6, 76, 66, 64, 65, 22, 5]. In this paper we subsume such approaches under the general notion of *algebraic net specifications*, parameterized over an underlying equational

---

[7] In fact, the nets introduced in [38] are called *colored Petri nets (CPNs)*, but this name has later been used for the more syntactic version introduced in [39], which is also the sense for which we would like to reserve this term (see below).

specification language. The main feature that algebraic net specifications have in common with predicate/transition nets is that an algebraic net specification does not necessarily specify a single colored net, but instead denotes a *class* of colored nets that satisfy the specification. In the following we first define colored nets, and then we introduce algebraic net specifications over MEL, a straightforward generalization of algebraic net specifications over many-sorted equational-logic (MSA). Both, algebraic net specifications and rewriting logic are specification formalisms that admit a variety of models. From an even more general point of view that is only briefly sketched in this paper, one can define colored net specifications parameterized over an underlying logic. In fact, predicate/transition nets can essentially be regarded as colored net specifications over first-order logic. From this more general point of view we restrict our attention in this paper to the particular class of colored net specifications over MEL, that we also call algebraic net specifications (over MEL), to establish a systematic connection to rewriting logic (over MEL). Later, in Section 6, we will discuss how other high-level Petri net extensions can be covered as generalizations or variants of our approach.

4.1   Colored Nets and Colored Net Specifications

Algebraic net specifications will be introduced later as a formal specification language for colored nets. In the following we define the most general set-theoretic version of colored nets [38]. We also give a suitable notion of colored net morphism and we use **CN** to abbreviate the resulting category of colored nets.

Colored nets are nets with places, transitions, and arcs inscribed with additional information given by functions C and W. The color set $C(p)$ of a place p is the set of possible objects p can carry. The color set $C(t)$ of a transition t can be seen as a set of modes in which t may occur. The arc inscription W defines a multiset of objects ("colored" tokens) that are transported by an arc when the associated transition occurs. In fact, this multiset may depend on the mode in which the transition occurs, which is why $W(p, t)$ and $W(t, p)$ take the form of functions in the definition below.

Definition 19. A *colored net (CN)* $\mathcal{N}$ consists of:

1. a finite net $N_{\mathcal{N}}$;
2. a set of *color sets* $CS_{\mathcal{N}}$;
3. a *color function* $C_{\mathcal{N}} : P_{\mathcal{N}} \cup T_{\mathcal{N}} \to CS_{\mathcal{N}}$; and
4. an *arc inscription* $W_{\mathcal{N}}$ on $F_{\mathcal{N}}$ such that

$$W_{\mathcal{N}}(p, t) : C_{\mathcal{N}}(t) \to C_{\mathcal{N}}(p)^{\oplus}, \text{ and } W_{\mathcal{N}}(t, p) : C_{\mathcal{N}}(t) \to C_{\mathcal{N}}(p)^{\oplus}.$$

$W_{\mathcal{N}}$ is extended to a function on $(P_{\mathcal{N}} \times T_{\mathcal{N}}) \cup (T_{\mathcal{N}} \times P_{\mathcal{N}})$ in such a way that $(p, t) \notin F_{\mathcal{N}}$ implies $W_{\mathcal{N}}(p, t)(b) = \emptyset$ and $(t, p) \notin F_{\mathcal{N}}$ implies $W_{\mathcal{N}}(t, p)(b) = \emptyset$ for each $b \in C_{\mathcal{N}}(t)$.

Let $\mathcal{N}$ and $\mathcal{N}'$ be CNs. A *CN morphism* $H : \mathcal{N} \to \mathcal{N}'$ consists of a net morphism $H_N : N_{\mathcal{N}} \to N_{\mathcal{N}'}$, and functions $H_x : C_{\mathcal{N}}(x) \to C_{\mathcal{N}'}(H_N(x))$ for each $x \in P_{\mathcal{N}} \cup T_{\mathcal{N}}$ such that:

1. $W_{\mathcal{N}'}(p', t')(H_t(b)) = H_{p_1}(W_{\mathcal{N}}(p_1, t)(b)) \oplus \cdots \oplus H_{p_n}(W_{\mathcal{N}}(p_n, t)(b))$
   for all $p' \in P_{\mathcal{N}'}$, $t' \in T_{\mathcal{N}'}$, $t \in H_N^{-1}(t')$, $b \in C(t)$,
   and $\{p_1, \ldots, p_n\} = H_N^{-1}(p') \cap {}^{\bullet}t$ with distinct p_i;
2. $W_{\mathcal{N}'}(t', p')(H_t(b)) = H_{p_1}(W_{\mathcal{N}}(t, p_1)(b)) \oplus \cdots \oplus H_{p_n}(W_{\mathcal{N}}(t, p_n)(b))$
   for all $p' \in P_{\mathcal{N}'}$, $t' \in T_{\mathcal{N}'}$, $t \in H_N^{-1}(t')$, $b \in C(t)$,
   and $\{p_1, \ldots, p_n\} = H_N^{-1}(p') \cap t^{\bullet}$ with distinct p_i.

CNs together with their morphisms form a category denoted by **CN**.

CNs generalize PTNs. The two dual objects of generalization are places and transitions. PTNs arise as the special case in which $C(x)$ is a singleton set for each $x \in P \cup T$. This gives rise to an obvious inclusion functor $\iota : \mathbf{PTN} \to \mathbf{CN}$.

Although CNs can be seen as a generalization of PTNs, there is a more fundamental justification for introducing CNs, namely, that a CN is just a convenient abbreviation for a typically rather complex PTN [39, 32]. Indeed, this connection can be exploited to lift low-level concepts such as markings, safe processes, and Best-Devillers processes to the higher level. This is achieved by the following flattening functor $(\_)^{\flat} : \mathbf{CN} \to \mathbf{PTN}$ which associates to each CN the PTN obtained by "spatial unfolding." We call this operation *flattening* to clearly distinguish it from "temporal unfolding" which generates the processes of a PTN as we defined them earlier.

Definition 20. Given a CN $\mathcal{N}$, we define the *flattening* $\mathcal{N}^{\flat}$ of $\mathcal{N}$ as the unique PTN that satisfies:

1. $P_{\mathcal{N}^{\flat}} = \{(p, c) \mid p \in P_{\mathcal{N}}, c \in C_{\mathcal{N}}(p)\}$;
2. $T_{\mathcal{N}^{\flat}} = \{(t, b) \mid t \in T_{\mathcal{N}}, b \in C_{\mathcal{N}}(t)\}$;
3. $W_{\mathcal{N}^{\flat}}((p, c), (t, b)) = W_{\mathcal{N}}(p, t)(b)(c)$; and
4. $W_{\mathcal{N}^{\flat}}((t, b), (p, c)) = W_{\mathcal{N}}(t, p)(b)(c)$

for $p \in P_{\mathcal{N}}, c \in C_{\mathcal{N}}(p), t \in T_{\mathcal{N}}, b \in C_{\mathcal{N}}(t)$.

Flattening is extended to a functor $(\_)^{\flat} : \mathbf{CN} \to \mathbf{PTN}$ as follows: Given a CN morphism $H : \mathcal{N} \to \mathcal{N}'$, the PTN morphism $H^{\flat} : \mathcal{N}^{\flat} \to \mathcal{N}'^{\flat}$ is given by

1. $H^{\flat}((p, c)) = (H_N(p), H_p(c))$,
2. $H^{\flat}((t, b)) = (H_N(t), H_t(b))$

for $p \in P_{\mathcal{N}}, c \in C_{\mathcal{N}}(p), t \in T_{\mathcal{N}}$, and $b \in C_{\mathcal{N}}(t)$.

It is important to point out that although we have defined the notion of a colored net, we have not yet introduced a notion of finite specification of colored nets. This is unsatisfactory if we want to reason about colored net specifications instead of just reasoning about colored nets. It is also unsatisfactory if we want to apply tools for execution, analysis and verification of colored nets, since such tools rely on a finitary, formal specification. Although a formal inscription language can be obtained by a formalization of set theory, such an enterprise is cumbersome and is of little help when we are interested in effective net execution and analysis. Also, the direct use of formalized set theory for specification and verification purposes is not very convenient and could be compared with the use of a low-level programming language.

Colored Petri nets, a more syntactic, finitary version of colored nets based on an underlying programming language, are proposed in [39]. A remarkable point is that this definition leaves open the particular choice of the underlying programming language. We use $\mathbf{CPN}_\mathcal{L}$ to abbreviate the class of colored Petri nets over a programming language $\mathcal{L}$. A quite well known instance of this definition is $\mathbf{CPN}_{\mathrm{ML}}$, the class supported by the execution and analysis tool Design/CPN [39] that employs the functional programming language ML. Appart from their operational flavor, the essential characteristic of colored Petri nets is that each colored Petri net denotes a single well-defined colored net in the above sense. A more logic-oriented view of colored nets (which emphasizes classes of models) is given by colored net specifications that are introduced subsequently.

As a useful concept, we informally introduce *colored net specifications (CNS)* which capture the essential idea shared by predicate/transition nets and algebraic net specifications, namely, that they denote an entire class of colored nets instead of just a single one. In fact, there is a general concept of CNSs that is parameterized by an underlying logic. A logic has a deductive system and a model-theoretic semantics, a concept that can be formalized by general logics [47] which contain institutions [35] as the model-theoretic component. We denote by $\mathbf{CNS}_\mathcal{L}$ the class of colored net specifications over the underlying logic $\mathcal{L}$. Possible candidates for $\mathcal{L}$ include equational logics such as many-sorted equational logic (MSA), order-sorted equational logic (OSA), or membership equational logic (MEL). We refer to CNSs over such equational logics also as *algebraic net specifications (ANS)*, and we denote by $\mathbf{ANS}_\mathcal{L}$ the class of algebraic net specifications over $\mathcal{L}$. Obviously, there are other possible choices for the underlying logic, such as full first-order logic (as in predicate/transition nets), a version of higher-order logic, or a higher-order algebraic specification language (as in [37]).

4.2 Algebraic Net Specifications

In the following we use the term *algebraic net specification (ANS)* to specifically refer to ANSs over MEL, since MEL is sufficiently expressive to cover other commonly used algebraic specification languages such as MSA and OSA [51]. The use of MEL is particularly attractive, because it is weak enough to admit initial models. Indeed, under the initial semantics (which can be internally specified using constraints in the data subspecification) an ANS denotes a unique

CN. Another benefit of the use of membership equational logic is that, under the restrictions mentioned in Section 2.1, it comes with a natural operational semantics (which is actually implemented in the Maude engine) so that it can be used directly as a programming language or, more generally, as a metalanguage to specify the logical and operational semantics of other specification or programming languages. As a consequence, colored Petri nets in $\mathbf{CPN}_\mathcal{L}$ which use $\mathcal{L}$ as a programming language can be seen as a special case of algebraic net specifications in $\mathbf{ANS}_{\mathrm{MEL}}$ if the semantics of $\mathcal{L}$ can be specified in MEL.

Due to the fact that MEL generalizes MSA in an obvious way, ANSs over MEL are a straightforward generalization of ANSs over MSA, i.e., many-sorted algebraic net specifications. Disregarding the issue of the underlying specification language, the definition we give below is equivalent to the one in [41, 43], generalizing [64] by so-called *flexible arcs*, which transport variable multisets of tokens in the sense that the number of tokens transported by an arc is not fixed but can depend on the mode in which the associated transition occurs. Later, in Section 4.3 we will illustrate by means of an example how an executable subset of the specification language can be used to obtain executable specifications of net models.

An ANS presupposes an underlying specification that has a multiset kind for each place domain. Hence we introduce a generic notion of multiset specification first.

Definition 21. A *MES of finite multisets* over a kind k consists of:

1. a MET having kinds k and $[\mathrm{FMS}_k]$ with operator symbols

   $\mathrm{empty}_k : [\mathrm{FMS}_k]$,
   $\mathrm{single} : k \to [\mathrm{FMS}_k]$,
   $\_\_ : [\mathrm{FMS}_k]\ [\mathrm{FMS}_k] \to [\mathrm{FMS}_k]$;

   equational axioms

   $\forall\, a, b, c : [\mathrm{FMS}_k]\ .\ a\ (b\ c) = (a\ b)\ c$,
   $\forall\, a, b : [\mathrm{FMS}_k]\ .\ a\ b = b\ a$,
   $\forall\, a : [\mathrm{FMS}_k]\ .\ \mathrm{empty}_k\ a = a$;

2. and a constraint stating that this theory is free over k.

To simplify notation we write M instead of $\mathrm{single}(M)$. To further simplify the exposition we assume without loss of generality that $[\![[\mathrm{FMS}_k]]\!] = [\![k]\!]^{\oplus}$, i.e., $[\mathrm{FMS}_k]$ is interpreted in the standard way, and the operator symbols are interpreted accordingly.

The subsequent definition of algebraic net specifications should be regarded as an instance of CNSs over a logic $\mathcal{L}$ choosing MEL for $\mathcal{L}$. In fact, the only requirements that $\mathcal{L}$ has to meet is that it has a notion of type and that it is expressive enough to axiomatize multisets.

Definition 22. An *algebraic net specification (ANS)* $\mathcal{N}$ consists of:

1. a MES $\mathcal{S}_\mathcal{N}$;
2. a finite net $N_\mathcal{N}$;
3. a *place declaration*, i.e., a function $D_\mathcal{N} : P_\mathcal{N} \to K_{\mathcal{S}_\mathcal{N}}$ assigning a kind $D_\mathcal{N}(p)$ to each place $p \in P_\mathcal{N}$ such that $\mathcal{S}_\mathcal{N}$ includes a MES of finite multisets over $D_\mathcal{N}(p)$;
4. a *variable declaration*, i.e., a function $V_\mathcal{N}$ on $T_\mathcal{N}$ associating to each transition $t \in T_\mathcal{N}$ a kinded variable set $V_\mathcal{N}(t)$;
5. an *arc inscription*, i.e., a function $W_\mathcal{N}$ on $F_\mathcal{N}$ such that for $p \in P_\mathcal{N}$, $t \in T_\mathcal{N}$,
   (a) $(p, t) \in F_\mathcal{N}$ implies $W_\mathcal{N}(p, t) \in \mathbf{T}_{\mathcal{S}_\mathcal{N}}(V_\mathcal{N}(t))_{[\mathrm{FMS}_{D_\mathcal{N}(p)}]}$ and
   (b) $(t, p) \in F_\mathcal{N}$ implies $W_\mathcal{N}(t, p) \in \mathbf{T}_{\mathcal{S}_\mathcal{N}}(V_\mathcal{N}(t))_{[\mathrm{FMS}_{D_\mathcal{N}(p)}]}$;
6. a *guard definition*, i.e., a function $G_\mathcal{N}$ on $T_\mathcal{N}$ with $G_\mathcal{N}(t)$ being an $\mathcal{S}_\mathcal{N}$-condition over $V_\mathcal{N}(t)$.

$W_\mathcal{N}$ is extended to a function on $(P_\mathcal{N} \times T_\mathcal{N}) \cup (T_\mathcal{N} \times P_\mathcal{N})$ such that $(p, t) \notin F_\mathcal{N}$ implies $W_\mathcal{N}(p, t) = \mathtt{empty}_{D_\mathcal{N}(p)}$ and $(t, p) \notin F_\mathcal{N}$ implies $W_\mathcal{N}(t, p) = \mathtt{empty}_{D_\mathcal{N}(p)}$ for $p \in P_\mathcal{N}$ and $t \in T_\mathcal{N}$.

Let $\mathcal{N}$ and $\mathcal{N}'$ be ANSs. An *ANS morphism* $H : \mathcal{N} \to \mathcal{N}'$ consists of a MES morphism $H_\mathcal{S} : \mathcal{S}_\mathcal{N} \to \mathcal{S}_{\mathcal{N}'}$ of the underlying MESs, a net morphism $H_N : N_\mathcal{N} \to N_{\mathcal{N}'}$, and a function $H_V^t : V_\mathcal{N}(t) \to V_{\mathcal{N}'}(t)$ for each $t \in T_\mathcal{N}$ such that $x \in V_\mathcal{N}(t)_k$ implies $H_V^t(x) \in V_{\mathcal{N}'}(t)_{H_\mathcal{S}(k)}$ for $k \in K_{\mathcal{S}_\mathcal{N}}$ and the following conditions are satisfied:

1. $H_\mathcal{S}(D_\mathcal{N}(p)) = D_{\mathcal{N}'}(H_N(p))$ for each $p \in P_\mathcal{N}$;
2. $\mathcal{S}_{\mathcal{N}'} \models \forall V_{\mathcal{N}'}(t) . H_\mathcal{S}^t(G_\mathcal{N}(t)) \Rightarrow G_{\mathcal{N}'}(H_N(t))$ for each $t \in T_\mathcal{N}$;
3. $\mathcal{S}_{\mathcal{N}'} \models \forall V_{\mathcal{N}'}(t) . H_\mathcal{S}^t(G_\mathcal{N}(t)) \Rightarrow$
   $$W_{\mathcal{N}'}(p', t') = H_\mathcal{S}^t(W_\mathcal{N}(p_1, t)) \dots H_\mathcal{S}^t(W_\mathcal{N}(p_n, t))$$
   for all $p' \in P_{\mathcal{N}'}$, $t' \in T_{\mathcal{N}'}$, $t \in H_N^{-1}(t')$,
   and $\{p_1, \dots, p_n\} = H_N^{-1}(p') \cap {}^\bullet t$ with distinct p_i;
4. $\mathcal{S}_{\mathcal{N}'} \models \forall V_{\mathcal{N}'}(t) . H_\mathcal{S}^t(G_\mathcal{N}(t)) \Rightarrow$
   $$W_{\mathcal{N}'}(t', p') = H_\mathcal{S}^t(W_\mathcal{N}(t, p_1)) \dots H_\mathcal{S}^t(W_\mathcal{N}(t, p_n))$$
   for all $p' \in P_{\mathcal{N}'}$, $t' \in T_{\mathcal{N}'}$, $t \in H_N^{-1}(t')$,
   and $\{p_1, \dots, p_n\} = H_N^{-1}(p') \cap t^\bullet$ with distinct p_i;

where $H_\mathcal{S}^t : \mathbf{T}_{\mathcal{S}_\mathcal{N}}(V_\mathcal{N}(t)) \to \mathbf{T}_{\mathcal{S}_{\mathcal{N}'}}(V_{\mathcal{N}'}(t))$ is the common extension of $H_\mathcal{S}$ and H_V^t to terms. We assume for the above definition that validity $\models$ has been extended to first-order formulae in the standard way.

ANSs together with their morphisms form a category **ANS**.

A typical ANS admits several colored nets as models. Since we want to state our results for an arbitrary but fixed model we also consider interpreted ANSs, i.e., ANSs together with distinguished data models. We furthermore equip interpreted ANS with a notion of morphism that allows us to express simultaneous transformations at the level of the ANSs and at the level of the data models.

Definition 23. An *interpreted ANS* $\mathcal{IN} = (\mathcal{N}, A)$ consists of an ANS $\mathcal{N}$ and a $\mathcal{S}_{\mathcal{N}}$-algebra A. An *interpreted ANS morphism* $(H, h) : (\mathcal{N}, A) \to (\mathcal{N}', A')$ consists of an ANS morphism $H : \mathcal{N} \to \mathcal{N}'$ and an interpreted MES morphism $(H_S, h) : (\mathcal{S}_{\mathcal{N}}, A) \to (\mathcal{S}_{\mathcal{N}'}, A')$. Interpreted ANSs together with their morphisms form a category **IANS**.

Interpreted ANSs are considerably richer than CNs, since they contain their specification together with a model equipped with a corresponding algebraic structure. In this sense they are similar to concrete predicate/transition nets [33, 34, 32] and algebraic high-level nets [25]. In fact, interpreted ANS, concrete predicate/transition nets [31] and algebraic-high-level nets [25] can be regarded as instances of a general notion of *interpreted CNSs*.[8] The transition from interpreted ANSs to CNs can be described by a forgetful functor as follows.

Definition 24. Given an interpreted ANS $(\mathcal{N}, A)$, the *CN semantics* of $(\mathcal{N}, A)$ is given by the CN $\mathbf{CN}(\mathcal{N}, A)$ defined as follows:

1. the underlying net $N_{\mathbf{CN}(\mathcal{N},A)}$ is precisely $N_{\mathcal{N}}$;
2. the color function $C_{\mathbf{CN}(\mathcal{N},A)}$ is defined by
   $C_{\mathbf{CN}(\mathcal{N},A)}(p) = [\![D_{\mathcal{N}}(p)]\!]_A$ for $p \in P_{\mathcal{N}}$ and
   $C_{\mathbf{CN}(\mathcal{N},A)}(t) = B_{\mathcal{N},A}(t)$ for $t \in T_{\mathcal{N}}$,
   where $B_{\mathcal{N},A}(t)$ is the set of *valid bindings* of $t \in T_{\mathcal{N}}$, i.e.,
   the set of assignments $\beta : V_{\mathcal{N}}(t) \to A$ satisfying $G_{\mathcal{N}}(t)$;
3. the set of color sets $CS_{\mathbf{CN}(\mathcal{N},A)}$ is the smallest set that
   contains all $C_{\mathbf{CN}(\mathcal{N},A)}(x)$ for $x \in P_{\mathcal{N}} \cup T_{\mathcal{N}}$; and
4. the arc inscription $W_{\mathbf{CN}(\mathcal{N},A)}$ is defined by
   $W_{\mathbf{CN}(\mathcal{N},A)}(p, t)(\beta) = [\![W_{\mathcal{N}}(p, t)]\!]_{A,\beta}$ and
   $W_{\mathbf{CN}(\mathcal{N},A)}(t, p)(\beta) = [\![W_{\mathcal{N}}(t, p)]\!]_{A,\beta}$
   for $p \in P_{\mathcal{N}}$, $t \in T_{\mathcal{N}}$ and assignments $\beta : V_{\mathcal{N}}(t) \to A$.

CN is extended to a functor $\mathbf{CN} : \mathbf{IANS} \to \mathbf{CN}$ that maps each morphism $(H, h) : (\mathcal{N}, A) \to (\mathcal{N}', A')$ to the morphism $G : \mathbf{CN}(\mathcal{N}, A) \to \mathbf{CN}(\mathcal{N}', A')$ satisfying $G_N = H_N$ and $G_x = h_{D_{\mathcal{N}}(x)}$ for $x \in P_{\mathcal{N}} \cup T_{\mathcal{N}}$.

We lift the flattening functor $(\_)^\flat : \mathbf{CN} \to \mathbf{PTN}$ to interpreted ANS, denoting also by $(\_)^\flat : \mathbf{IANS} \to \mathbf{PTN}$ the composition $(\_)^\flat \circ \mathbf{CN}$. Using flattening we furthermore lift $\mathbf{BDP} : \mathbf{PTN} \to \mathbf{SMC}$ by defining $\mathbf{BDP} : \mathbf{IANS} \to \mathbf{SMC}$ as $\mathbf{BDP} \circ (\_)^\flat$.

4.3   A Case Study

In the following we generalize the rewriting semantics from PTNs to ANSs. Before dealing with the general case we try to convey the main ideas using

---

[8] To be precise, arc inscriptions have to be restricted, since flexible arcs are not available in predicate/transition nets and algebraic high-level nets.

a distributed network algorithm as a running example, and we show how the rewriting semantics is obtained in this particular but typical case.

An algorithm which admits a very natural presentation as an algebraic net specification is the well-known echo algorithm, also called PIF algorithm (where PIF stands for propagation of information with feedback). The algebraic net model we use here has been developed and verified in [42].

Given a network of agents with bidirectional channels the echo problem can be informally described as follows. A distinguished agent initiates the transmission of a piece of information which should be propagated (possibly using other agents) to all agents participating in the network. After that the initiator should receive feedback about the succesful completion of this task, i.e., that each agent has received the information transmitted.

A possible solution to this problem is modeled by the algebraic net specification described below. To focus on the algorithm itself, the model abstracts from the concrete information that is transmitted. This information can be easily added by refining the messages without major changes to the algorithm.

We assume that the agents can be distinguished in terms of their identifiers, which are modeled by a sort Id. The network of agents is represented as a directed multigraph, i.e., as a finite multiset of (directed) channels, where each channel is a pair of agent identifiers. In the specification fragment below, Pair is the sort of pairs of identifiers and FMS-Pair is the sort of finite multisets over such pairs. Finite multisets are equationally axiomatized as discussed before. The obvious initiality and freeness constraints for Id, FMS-Id, Pair, and FMS-Pair can be specified using (parameterized) functional modules in Maude [19, 18], but for the sake of brevity we omit the details here.

```
sort Id FMS-Id
     Pair FMS-Pair .

op (_,_) : Id Id -> Pair .

op empty-Id : -> FMS-Id .
op single : Id -> FMS-Id .
op __ : FMS-Id FMS-Id -> FMS-Id
   [assoc comm id: empty-Id] .

op empty-Pair : -> FMS-Pair .
op single : Pair -> FMS-Pair .
op __ : FMS-Pair FMS-Pair -> FMS-Pair
   [assoc comm id: empty-Pair] .

var x y x' y' : Id .
var fmsp fmsp' : FMS-Pair .
var p p' : Pair .
```

To work with a concrete example we assume agent identifiers and a network as specified below. Actually, the algorithm is parametric in the choice of agent identifiers and in the network topology, the only assumptions being that there is a distinguished initiator and that the network is a strongly connected network with bidirectional channels. Again this parameterization could be made explicit in Maude by viewing the entire specification as a parameterized module which can be instantiated, for instance, by the following choices for Id and network.[9]

```
ops i a b c d e : -> Id .

op  sym : Id Id -> FMS-Pair .
eq  sym(x,y) = ((x,y) (y,x)) .

op  network : -> FMS-Pair .
eq  network = (sym(i,a) sym(i,b) sym(e,b) sym(e,d)
               sym(c,d) sym(c,i) sym(c,a) sym(a,b)) .
```

Now we equationally specify three auxiliary functions operating on finite multisets of pairs. The first one _-_ removes one occurrence of a given pair from a multiset of pairs. The other functions out and in will be used with network as a first argument: out(network,x) denotes the multiset of messages to be sent to neighbours of x and, correspondingly, in(network,x) denotes the multiset of messages to be received from neighbours of x.

```
op  _-_ : FMS-Pair Pair -> FMS-Pair .
eq  empty-Pair - p = empty-Pair .
eq  (p fmsp) - p = fmsp .
ceq (p' fmsp) - p = (p' (fmsp - p)) if p =/= p' .

op  in : FMS-Pair Id -> FMS-Pair .
eq  in(empty-Pair,y') = empty-Pair .
eq  in(((x,y) fmsp),y) = ((x,y) in(fmsp,y)) .
ceq in(((x,y) fmsp),y') = in(fmsp,y') if y =/= y' .

op  out : FMS-Pair Id -> FMS-Pair .
eq  out(empty-Pair,x') = empty-Pair .
eq  out(((x,y) fmsp),x) = ((x,y) out(fmsp,x)) .
ceq out(((x,y) fmsp),x') = out(fmsp,x') if x =/= x' .
```

This concludes the MES. We are now ready to define the ANS on top of it. Its inscribed net is depicted in Fig. 3. In the center we have a message pool MESSAGES modeling messages in transit. The net elements at the top model the

---

[9] If we were interested in (abstract) formal verification rather than (concrete) execution we would leave open the interpretation of Id and network and in this way obtain an ANS admitting a rich variety of quite different models.

activity of the initiating agent i, which is initially in a state QUIET, whereas
the net elements at the bottom model the activities of all the remaining agents
which are initially UNINFORMED. More precisely, the activities of initiators and
non-initiators are the following:

- After the initiator i sends out a message to all its neighbours (transition
  ISEND) it will remain in the WAITING state until it receives an acknowledgement message from all its neighbours. If this happens, it will go into
  the TERMINATED state (transition IRECEIVE), i.e., the initiator has locally
  detected that all agents have received a message.
- After a non-initiator x receives a message from an agent y, it sends messages
  to all its neighbours except for y (transition SEND), and goes into a PENDING
  state, where it remembers that it is pending after receiving a message from
  y. As soon it receives acknowledgement messages from all neigbours except
  for y it goes into the ACCEPTED state (transition RECEIVE).

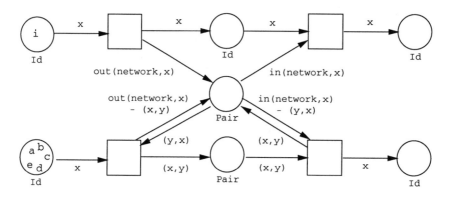

Fig. 3. Echo Algorithm

The initial marking specificaton m_0 for our concrete choice of the network is
given by the terms inside places. It is

$$m_0(\text{QUIET}) = \text{i} \qquad\qquad m_0(\text{UNINFORMED}) = \text{a b c d e}$$
$$m_0(\text{WAITING}) = \text{empty-Id} \qquad\qquad m_0(\text{PENDING}) = \text{empty-Pair}$$
$$m_0(\text{TERMINATED}) = \text{empty-Id} \qquad\qquad m_0(\text{ACCEPTED}) = \text{empty-Id}$$

In Section 3 we have already discussed a rewriting semantics for the PTN of
the banker's problem. Using the echo algorithm we will demonstrate how the
rewriting semantics generalizes to ANSs. It is worth mentioning that our semantics is designed to cope with flexible arcs as the ones connected with the place
MESSAGES in the echo algorithm.

The rewriting semantics associated to a RWS extends but does not modify the underlying MES of the net, the advantage being that properties established for the equational logic specification are preserved and their proofs remain valid.

As in the PTN case we represent a marking as an element of the kind [Marking], which is equipped with a monoidal structure via the operations empty and __. For each place p we have a *token constructor*, also written as p, representing the fact that a single token resides in place p. A difference with respect to the PTN rewriting semantics is that tokens carry data, which is reflected in the fact that token constructors are functions instead of being constants. For instance, a token MESSAGES(msg) represents a token carrying the data msg residing in the place MESSAGES. So the token constructor can be seen as a function *tagging* a data object with information about the place in which it is currently located.

```
sort Marking .

op empty : -> Marking .
op __ : Marking Marking -> Marking
   [assoc comm id: empty] .

ops MESSAGES PENDING : Pair -> Marking .
ops QUIET WAITING TERMINATED UNINFORMED ACCEPTED : Id -> Marking .
```

When formulating the transition rule for ISEND we are faced with the problem of how to translate the flexible arc between ISEND and MESSAGES appropriately. Of course we would like to express that the multiset out(network, x) is added to the place MESSAGES, but this presupposes an interpretation of places as containers of objects which is different from our current one, where tokens are tagged objects "mixed up in a soup together with other tokens."

An elegant solution is the linear extension of MESSAGES to multisets. For this purpose we *generalize* the token constructor MESSAGES which has been declared above to

```
op MESSAGES : FMS-Pair -> Marking .
```

and we add two equations expressing linearity of MESSAGES, which will also be called *place linearity equations*:

```
seq MESSAGES(empty-Pair) = empty .
seq MESSAGES(fmsp fmsp') = MESSAGES(fmsp) MESSAGES(fmsp') .
```

The place linearity equations express the equivalence of different ways of looking at the same marking of an ANS. So, as indicated by the keyword seq, from a high-level specification point of view it is reasonable to assign them to the class of structural equations expressing symmetries of the state representation.

For reasons of uniformity we generalize the remaining token constructors correspondingly and we impose corresponding place linearity equations that we omit here.

Now the translation of transitions into rewrite rules can be done in full analogy with the rewriting semantics for PTN. Each transition is represented as a rewrite rule, also called a *transition rule*, replacing its preset marking by its postset marking. If the transition has a guard, then that guard becomes a condition of the rewrite rule. In this way we obtain the following rules:

```
rl  [ISEND]:    QUIET(x) =>
                WAITING(x) MESSAGES(out(network,x))) .

rl  [IRECEIVE]: WAITING(x) MESSAGES(in(network,x)) =>
                TERMINATED(x) .

rl  [SEND]:     UNINFORMED(x) MESSAGES((y,x)) =>
                PENDING((x,y)) MESSAGES(out(network,x)-(x,y)) .

rl  [RECEIVE]:  PENDING((x,y)) MESSAGES(in(network,x)-(y,x)) =>
                ACCEPTED(x) MESSAGES((x,y)) .
```

According to our initial explanation a place can be seen as the tag of an object which indicates the place the token resides in. This is what we call the *tagged-object view*. The place linearity equations suggest a complementary view which is encountered more often in the context of Petri nets: a place is simply a container of objects. We call this the *place-as-container view*. The place linearity equations express our intention to consider both views as equivalent.

4.4   Rewriting Semantics in the General Case

Generalizing the above example, we now define for an arbitrary ANS its associated rewriting semantics. We also show in which sense the rewriting semantics is equivalent to the Best-Devillers process semantics of ANSs, which we have defined by lifting the Best-Devillers process semantics of PTNs to ANSs via the flattening construction. First we generalize symmetric monoidal RWSs (SMR-WSs) to extended symmetric monoidal RWSs (ESMRWSs), which will serve as a suitable domain for the rewriting semantics. Notice that in ESMRWSs the data subspecification is not required to be empty. A second difference w.r.t. SMRWSs is that token constructors are extended to multisets and place linearity equations are added.

Definition 25. A RWS $\mathcal{R}$ is an *extended symmetric monoidal RWS (ESMRWS)* iff the the following conditions are satisfied:

1. $\mathcal{S}_{\mathcal{R}}$ extends $\mathcal{S}_{\mathcal{R}}^{D}$ precisely by:

(a) a new kind [Marking] and new operator symbols

$$\text{empty} : \text{[Marking]}, \qquad \_\_ : \text{[Marking]} \text{ [Marking]} \to \text{[Marking]};$$

(b) any number of new operator symbols of the general form

$$p : \text{[FMS}_k] \to \text{[Marking]},$$

where k is a kind in $\mathcal{S}_{\mathcal{R}}^{D}$ such that
$\mathcal{S}_{\mathcal{R}}^{D}$ includes a MES of finite multisets over k;

(c) the axioms for parallel composition

$$\forall\ u, v, w : \text{[Marking]} . u\ (v\ w) = (u\ v)\ w,$$
$$\forall\ u, v : \text{[Marking]} . u\ v = v\ u,$$
$$\forall\ u : \text{[Marking]} . \text{empty}\ u = u; \text{ and}$$

(d) the *place linearity equations*

$$p(\text{empty}_k) = \text{empty},$$
$$\forall\ a, b : \text{[FMS}_k] . p(a\ b) = p(a)\ p(b)$$

for each operator $p : \text{[FMS}_k] \to \text{[Marking]}$ introduced above.

2. Rules in $R_{\mathcal{R}}$ contain only variables with kinds in $\mathcal{S}_{\mathcal{R}}^{D}$ and have $\mathcal{S}_{\mathcal{R}}^{D}$-conditions.

Given two ESMRWSs $\mathcal{R}$ and $\mathcal{R}'$, an ESMRWS morphism $H : \mathcal{R} \to \mathcal{R}'$ is a RWS morphism that preserves [Marking], empty and $\_\_$. ESMRWSs together with their morphisms form a subcategory of **RWS** that is denoted by **ESMRWS**.

Definition 26. The *membership equational presentation* of an ESMRWS $\mathcal{R}$ is a MES $\mathbf{E}(\mathcal{R})$ that extends $\mathcal{S}_{\mathcal{R}}$, the underlying MES of $\mathcal{R}$, as explained in Definition 16, but modifying items 2 and 4 we have:

2. a new operator symbol called *atomic proof constructor*

$$t : \bar{k} \to \text{[RawProc]},$$

4. a membership axiom

$$\forall\ X . t(\bar{x}) : M \to N \text{ if } \bar{\phi}_1 \wedge \dots \wedge \bar{\phi}_n$$

for each rule $\forall\ X . t : M \to N$ if $\bar{\phi}_1 \wedge \dots \wedge \bar{\phi}_n$ in $R_{\mathcal{R}}$, assuming that $\bar{x} : \bar{k}$ is a canonical enumeration of the variables X.

As in Definition 16, $\mathbf{E}$ can be extended to a functor $\mathbf{E} : \textbf{ESMRWS} \to \textbf{MES}$ in the obvious way.

Definition 27. An *interpreted ESMRWS* $(\mathcal{R}, A)$ consists of an ESMRWS $\mathcal{R}$ and a $\mathcal{S}_{\mathcal{R}}^{D}$-model A. An interpreted ESMRWS morphism $(H, h) : (\mathcal{R}, A) \to (\mathcal{R}', A')$ consists of an ESMRWS morphism $H : \mathcal{R} \to \mathcal{R}'$ and an interpreted MES morphism $(H_D, h) : (\mathcal{S}_{\mathcal{R}}^{D}, A) \to (\mathcal{S}_{\mathcal{R}'}^{D}, A')$. Interpreted ESMRWSs together with their morphisms form a category **IESMRWS**.

Definition 28. For an interpreted ESMRWS $(\mathcal{R}, A)$ we define $\mathbf{Mod}(\mathcal{R}, A)$ as the subcategory of $\mathbf{Mod}(\mathcal{R})$ (i.e. $\mathbf{Mod}(\mathbf{E}(\mathcal{R}))$), with objects being $\mathcal{R}$-algebras (i.e. $\mathbf{E}(\mathcal{R})$-algebras) $\hat{A}$ satisfying $\hat{A}|_{\mathcal{S}_{\mathcal{R}}^{D}} = A$. In fact, this gives rise to a functor $\mathbf{Mod} : \mathbf{IESMRWS} \to \mathbf{Cat}^{\mathrm{op}}$, and again we write $\mathbf{U}_H$ for $\mathbf{Mod}(H)$ given a ESMRWS morphism H.

Lemma 1 (Protection Lemma).

Let $(\mathcal{R}, A)$ be an interpreted ESMRWS and consider the obvious inclusion $K : \mathcal{S}_{\mathcal{R}}^{D} \hookrightarrow \mathbf{E}(\mathcal{R})$. Then $\eta_K(A) : A \to \mathbf{U}_K(\mathbf{F}_K(A))$ is an isomorphism.

To simplify the exposition assume that the free functor $\mathbf{F}_K$ has been defined in such a way that $\eta_K(A)$ becomes the identity and therefore $\mathbf{U}_K(\mathbf{F}_K(A)) = A$ for all $(\mathcal{R}, A)$ and K as above. The protection lemma ensures that this is possible without loss of generality.

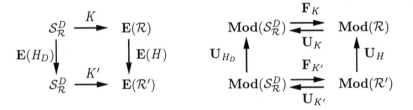

Fig. 4. Morphisms in Definition 29

Definition 29. Let $\Sigma(\mathbf{Mod})$ be the Grothendieck construction for the functor $\mathbf{Mod} : \mathbf{IESMRWS} \to \mathbf{Cat}^{\mathrm{op}}$ and let $\pi_1 : \Sigma(\mathbf{Mod}) \to \mathbf{IESMRWS}$ be the obvious projection functor that sends $((\mathcal{R}, A), \hat{A})$ to $(\mathcal{R}, A)$. Furthermore, let $(\mathcal{R}, A)$ and $(\mathcal{R}', A')$ be interpreted ESMRWSs and let $K : \mathcal{S}_{\mathcal{R}}^{D} \hookrightarrow \mathbf{E}(\mathcal{R})$ and $K' : \mathcal{S}_{\mathcal{R}'}^{D} \hookrightarrow \mathbf{E}(\mathcal{R}')$ be the obvious inclusions (cf. Fig. 4). We then define $\mathbf{F}(\mathcal{R}, A)$ as $\mathbf{F}_K(A)$ and $\Sigma\mathbf{F}(\mathcal{R}, A)$ as $((\mathcal{R}, A), \mathbf{F}(\mathcal{R}, A))$. Given an interpreted ESMRWS morphism $(H, h) : (\mathcal{R}, A) \to (\mathcal{R}', A')$ with $H : \mathcal{R} \to \mathcal{R}'$ and $h : A \to \mathbf{U}_{H_D}(A')$ we define $\Sigma\mathbf{F}(H, h)$ as the morphism $((H, h), \mathbf{F}(H) \circ \mathbf{F}_K(h)) : ((\mathcal{R}, A), \mathbf{F}_K(A)) \to ((\mathcal{R}', A'), \mathbf{F}_{K'}(A'))$ where $\mathbf{F}_K(h) : \mathbf{F}_K(A) \to \mathbf{F}_K(\mathbf{U}_{H_D}(A'))$ and $\mathbf{F}(H)$ is the unique morphism $\mathbf{F}(H) : \mathbf{F}_K(\mathbf{U}_{H_D}(A')) \to \mathbf{U}_H(\mathbf{F}_{K'}(A'))$ guranteed by the fact that $\mathbf{F}_K(\mathbf{U}_{H_D}(A'))$ and $\mathbf{U}_H(\mathbf{F}_{K'}(A'))$ are objects in $\mathbf{Mod}(\mathcal{R}, \mathbf{U}_{H_D}(A'))$, since using Lemma 1 we find $\mathbf{U}_K(\mathbf{F}_K(\mathbf{U}_{H_D}(A'))) = \mathbf{U}_{H_D}(A')$ and $\mathbf{U}_K(\mathbf{U}_H(\mathbf{F}_{K'}(A'))) = \mathbf{U}_{H_D}(\mathbf{U}_{K'}(\mathbf{F}_{K'}(A'))) = \mathbf{U}_{H_D}(A')$, and by the fact that $\mathbf{F}_K(\mathbf{U}_{H_D}(A'))$ is initial. In this way we have defined a functor $\Sigma\mathbf{F} : \mathbf{IESMRWS} \to \Sigma(\mathbf{Mod})$ that is left adjoint to π_1.

Furthermore, let $\mathbf{V} : \Sigma(\mathbf{Mod}) \to \mathbf{SMC}$ be the forgetful functor, which sends $((\mathcal{R}, A), \hat{A})$ to a SMC, defined as in Def. 17.

Definition 30. Given an ANS $\mathcal{N}$, the *rewriting semantics* of $\mathcal{N}$ is the smallest ESMRWS $\mathbf{R}(\mathcal{N})$ with an underlying data specification $\mathcal{S}_{\mathbf{R}(\mathcal{N})}^D = \mathcal{S}_{\mathcal{N}}$ such that:

1. $\mathcal{S}_{\mathbf{R}(\mathcal{N})}$ contains a *token constructor*

   $$p : [\mathtt{FMS}_{D_{\mathcal{N}}(p)}] \to [\mathtt{Marking}]$$

   for each place $p \in P_{\mathcal{N}}$; and
2. $\mathbf{R}(\mathcal{N})$ has a label t and a rule called *transition rule*, namely,

   $$\forall\ V_{\mathcal{N}}(t)\ .\ t : (p_1(W_{\mathcal{N}}(p_1, t)) \ \cdots \ p_m(W_{\mathcal{N}}(p_m, t))) \to$$
   $$(p_1(W_{\mathcal{N}}(t, p_1)) \ \cdots \ p_m(W_{\mathcal{N}}(t, p_m))) \ \mathtt{if}\ G_{\mathcal{N}}(t)$$

   for each transition $t \in T_{\mathcal{N}}$, assuming $P_{\mathcal{N}} = \{p_1, \ldots, p_m\}$ with distinct p_i.

$\mathbf{R}$ can be extended to a functor $\mathbf{R} : \mathbf{ANS} \to \mathbf{ESMRWS}$ that maps each ANS morphism $H : \mathcal{N} \to \mathcal{N}'$ to the unique ESMRWS morphism $G : \mathbf{R}(\mathcal{N}) \to \mathbf{R}(\mathcal{N}')$ with $G_{\mathcal{S}}(p) = H_N(p)$ for each $p \in P_{\mathcal{N}}$ and $G_L(t) = H_N(t)$ for each $t \in T_{\mathcal{N}}$.

The functor $\mathbf{R} : \mathbf{ANS} \to \mathbf{ESMRWS}$ is naturally extended to a functor $\mathbf{R} : \mathbf{IANS} \to \mathbf{IESMRWS}$ sending each interpreted ANS $(\mathcal{N}, A)$ to the interpreted ESMRWS $(\mathbf{R}(\mathcal{N}), A)$. Furthermore, $\mathbf{R}$ sends each interpreted ANS morphism $(H, h) : (\mathcal{N}, A) \to (\mathcal{N}', A')$ to the interpreted ESMRWS morphism $(\mathbf{R}(H), h) : \mathbf{R}(\mathcal{N}, A) \to \mathbf{R}(\mathcal{N}', A')$.

Definition 31. Given an interpreted ESMRWS $(\mathcal{R}, A)$ we define the *flattening* of $(\mathcal{R}, A)$ as the smallest SMRWS $(\mathcal{R}, A)^{\flat}$ satisfying the following conditions:

1. For each operator $p : [\mathtt{FMS}_k] \to [\mathtt{Marking}]$ in $\mathcal{S}_{\mathcal{R}}$ and for each $a \in [\![k]\!]_A$ there is a constant $p^a : \to [\mathtt{Marking}]$ in $\mathcal{S}_{(\mathcal{R}, A)^{\flat}}$.
2. For each rule $\forall\ X\ .\ t : M \to N$ if $\bar{\phi}_1 \wedge \ldots \wedge \bar{\phi}_n$ in $R_{\mathcal{R}}$ and for each assignment $\beta : X \to A$ with $A, \beta \models \bar{\phi}_1 \wedge \ldots \wedge \bar{\phi}_n$ we define functions σ and σ_p for each operator p as above by

   $$\sigma(\mathtt{empty}) = \mathtt{empty},\ \sigma(p(M)) = \sigma_p([\![M]\!]_{A,\beta}),\ \sigma(M\ N) = \sigma(M)\ \sigma(N),$$
   $$\sigma_p([\![\_]\!]([\![\mathtt{single}]\!](a_1), \ldots, [\![\mathtt{single}]\!](a_m))) = p^{a_1} \ldots p^{a_m}$$

   ($[\![\_]\!]$ is naturally extended to an arbitrary number of arguments) and we have a rule

   $$t^{\beta(\bar{x})} : \sigma(M) \to \sigma(N)$$

   to $R_{(\mathcal{R}, A)^{\flat}}$, assuming that $\bar{x} : \bar{k}$ is a canonical enumeration of X.

$(\_)^{\flat}$ is extended to a functor $(\_)^{\flat} : \mathbf{IESMRWS} \to \mathbf{SMRWS}$ as follows: $(\_)^{\flat}$ sends each interpreted ESMRWS morphism $(H, h) : (\mathcal{R}, A) \to (\mathcal{R}', A')$ with $H : \mathcal{R} \to \mathcal{R}'$ and $h : A \to \mathbf{U}_{H_D}(A')$ to a SMRWS morphism $(H, h)^{\flat} : (\mathcal{R}, A)^{\flat} \to (\mathcal{R}', A')^{\flat}$ defined such that $(H, h)^{\flat}(p^a) = H_{\mathcal{S}}(p)^{h(a)}$ and $(H, h)^{\flat}(t^{\bar{a}}) = H_L(t)^{h(\bar{a})}$ for $\bar{a} = \beta(\bar{x})$ and all $p, t, a, \beta, \bar{x}$ as above.

The theorem and the corollary below are stated in complete analogy to the corresponding results for PTNs. Indeed the former results can be seen as special cases of the latter via an inclusion $\iota : \mathbf{PTN} \to \mathbf{ANS}$ which is the counterpart of $\iota : \mathbf{PTN} \to \mathbf{CN}$ on the specification level. However, for the proof we exploit the opposite direction, namely that Theorem 2 can be reduced to Theorem 1 via the flattening constructions introduced earlier. This can be done by a combination of commutative diagrams using the following two lemmas.

The first lemma essentially states that the rewriting semantics is compatible with flattening. Notice the overloading of $\mathbf{R}$ and $(\_)^\flat$.

Lemma 2. There is a natural isomorphism $\sigma : (\_)^\flat \circ \mathbf{R} \to \mathbf{R} \circ (\_)^\flat$ between the functors $(\_)^\flat \circ \mathbf{R} : \mathbf{IANS} \to \mathbf{SMRWS}$ (with $\mathbf{R} : \mathbf{IANS} \to \mathbf{IESMRWS}$ and $(\_)^\flat : \mathbf{IESMRWS} \to \mathbf{SMRWS}$) and $\mathbf{R} \circ (\_)^\flat : \mathbf{IANS} \to \mathbf{SMRWS}$ (with $(\_)^\flat : \mathbf{IANS} \to \mathbf{PTN}$ and $\mathbf{R} : \mathbf{PTN} \to \mathbf{SMRWS}$).

The second lemma expresses that flattening preserves models at the level of abstraction given by SMCs.

Lemma 3. There is a natural isomorphism $\rho : \mathbf{V} \circ \Sigma\mathbf{F} \to \mathbf{V} \circ \Sigma\mathbf{I} \circ (\_)^\flat$ between the functors $\mathbf{V} \circ \Sigma\mathbf{F} : \mathbf{IESMRWS} \to \mathbf{SMC}$ and $\mathbf{V} \circ \Sigma\mathbf{I} \circ (\_)^\flat : \mathbf{IESMRWS} \to \mathbf{SMC}$ (with $(\_)^\flat : \mathbf{IESMRWS} \to \mathbf{SMRWS}$ and $\mathbf{V} \circ \Sigma\mathbf{I} : \mathbf{SMRWS} \to \mathbf{SMC}$).

Now the main result follows from Lemma 2, Lemma 3 and Theorem 1:

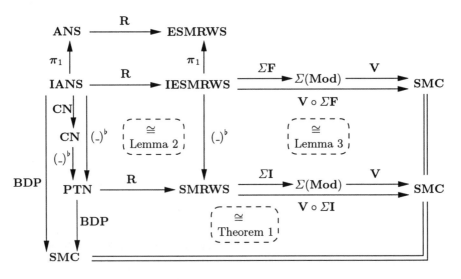

Fig. 5. Proof of Theorem 2

Theorem 2. There is a natural isomorphism $\tilde{\tau} : \mathbf{BDP} \to \mathbf{V} \circ \Sigma\mathbf{F} \circ \mathbf{R}$ between the functors $\mathbf{BDP} : \mathbf{IANS} \to \mathbf{SMC}$ and $\mathbf{V} \circ \Sigma\mathbf{F} \circ \mathbf{R} : \mathbf{IANS} \to \mathbf{SMC}$ (with $\mathbf{R} : \mathbf{IANS} \to \mathbf{IESMRWS}$ and $\mathbf{V} \circ \Sigma\mathbf{F} : \mathbf{IESMRWS} \to \mathbf{SMC}$).

Proof   By composition of natural isomorphisms (see Fig. 5).   □

In analogy to Corollary 1 we obtain:

Corollary 2. The interpreted RWS $\mathbf{R}(\mathcal{N}, A)$ provides a sound and complete axiomatization of the Best-Devillers processes of the interpreted ANS $(\mathcal{N}, A)$.

Remember that the models we consider here do not only contain Best-Devillers processes. They also contain safe processes as an important special case. Safe processes are not only a special case of the classical notion of process in Petri net theory, but they seem to be sufficient in practice as witnessed by [65], which presents a methodology for modeling and verification of distributed algorithms based on a version of ANSs that only admits safe processes.

Another related issue, namely the gap between the individual token philosophy and the collective token philosophy which clearly exists at the level of PTNs seems to become less relevant at the level of CNs, because of the increase of expressivity. We argue that interpreting CNs under the collective token philosophy is not only simpler and less dependent on the structure of the state space but also sufficient in principle, since by a suitable transformation of the CN we can equip tokens with unique identities in such a way that each original process corresponds to a safe process of the resulting CN.[10] As we discussed earlier, individual and collective token philosophies coincide for safe processes. Non-safe processes of the resulting CN are not considered any more. In this sense, the individual token philosophy can be seen as a special case of the collective token philosophy. Beyond that it may well be adequate for certain applications to mix the individual and the collective token views in the same system model, and indeed this is possible with the approach that we propose, namely by adopting the collective token semantics as a framework semantics and equipping tokens with additional identity attributes whenever needed for modeling purposes. Indeed this view reveals that individual and collective tokens semantics are just two extreme levels of abstraction and there are many intermediate levels that can be covered in this way. A good example of a very similar experience giving support to this point of view is the work [57] on a partial order semantics for object-oriented systems that, although typical of the individual token philosophy, is shown to be isomorphic to the rewriting semantics typical of the collective token philosophy, thanks to the unique identities of objects and messages.

---

[10] One policy to maintain unique identities is to encode the local history, i.e. the information about all events in the past cone of a token, in the identity of the token itself, and to ensure locally that the identities of the tokens produced by a transition are distinct. Of course, it is easy to imagine interesting classes of nets, e.g. object-oriented versions of high-level nets, where tokens are already equipped with unique identities so that this transformation is not needed at all.

4.5   Execution of Algebraic Net Specifications

First of all we lift the notion of executability from rewriting logic to net specifications. We say that a net specification is *weakly/strongly executable* iff its rewriting semantics is weakly/strongly executable. To actually execute a specification it is necessary to have an implementation of a matching algorithm for all combinations of structural equations used in the specification. A typical rewrite engine such as Maude supports matching modulo all combinations of the laws of associativity, commutativity and identity (ACU) [19, 18]. Since the place linearity equations belong to a class of equations that are typically not supported by standard rewrite engines we distinguish in the following between direct execution using ACUL-matching (L stands for linearity) and an alternative approach, namely execution via ACU-matching, which makes use of a simple semantics-preserving translation that can achieve executability without structural linearity equations.

Direct Execution via ACUL-Matching. It is easy to verify that the underlying MES in our example is already in executable form when the place linearity equations are seen as structural equations. Still the rewriting specification is not coherent and, as a consequence, the net specification is not executable as given.

A subterm of the form `in(network,x)` which occurs in the lefthand side of the rewrite axiom `IRECEIVE` can be reduced using the equations for `in`, so that the rewrite axiom is not applicable anymore. An obvious solution is to replace the arc inscription `in(network,x)` of the transition `IRECEIVE` by a variable `fmsmsg` and to add the guard `fmsmsg == in(network,x)` to this transition. A corresponding modification of the net specification has to be carried out for the transition `RECEIVE`. In the rewriting semantics these changes are reflected by the modified rewrite rules given below.

```
var fmsmsg : FMS-Message .

crl [IRECEIVE]: WAITING(x) MESSAGES(fmsmsg) =>
                TERMINATED(x)
    if fmsmsg == in(network,x) .

crl [RECEIVE]:  PENDING((x,y)) MESSAGES(fmsmsg) =>
                ACCEPTED(x) MESSAGES((x,y))
    if fmsmsg == in(network,x)-(y,x) .
```

After this simple semantics-preserving transformation the rewrite specification is indeed coherent and therefore strongly executable. To execute the RWS it is sufficient to use rewriting modulo associativity, commutativity, identity and linearity for the representation of markings.

Execution Using ACU-Matching. We show in the following that, given an executable ANS such as the one we have just obtained, there is an alternative approach to net execution by regarding the place linearity equations as computational equations instead of as structural equations. Of course, from the viewpoint of the abstract algebraic semantics nothing will change. An immediate consequence is, however, that the net specification can be executed using a standard rewriting engine such as Maude, without the need for a new matching algorithm.

The first step is to regard the place linearity equations as reduction rules, i.e.,

```
eq  MESSAGES(empty-Pair) = empty .
ceq MESSAGES(fmsp fmsp') = (MESSAGES(fmsp) MESSAGES(fmsp'))
    if fmsp =/= empty-Pair and fmsp' =/= empty-Pair .
```

After applying this modification to all place linearity equations the reduction rules are terminating (the condition avoids potential non-terminating computations) and confluent, yielding an executable equational part of the specification.

However, as a consequence of the use of place-linearity equations as reduction rules instead of as structural equations, the rewrite specification is not coherent anymore, because of the rules for IRECEIVE and RECEIVE and the new equations above. Again, we can carry out a simple semantics-preserving translation by introducing a variable mmsg ranging over markings containing only tokens on MESSAGES and satisfying the equality condition mmsg == MESSAGES(fmsmsg). By introducing the inverse inv-MESSAGES of MESSAGES this condition becomes inv-MESSAGES(mmsg) == fmsmsg. Therefore, inv-MESSAGES(mmsg) gives us access to the flexible arc inscription fmsmsg. As a result we replace these two rules by the following, which make the specification coherent and, hence, strongly executable:

```
sorts empty MESSAGES-Marking Marking .
subsorts empty < MESSAGES-Marking < Marking .

vars mmsg,mmsg' : MESSAGES-Marking .

op  empty : -> empty .
op  __ : Marking Marking -> Marking
    [assoc comm id: empty] .
op  __ : MESSAGES-Marking MESSAGES-Marking -> MESSAGES-Marking
    [assoc comm id: empty] .
op  __ : empty empty -> empty
    [assoc comm id: empty] .

op MESSAGES : FMS-Pair -> MESSAGES-Marking .

op  inv-MESSAGES : MESSAGES-Marking -> FMS-Pair .
```

```
eq  inv-MESSAGES(empty) = empty-Pair .
eq  inv-MESSAGES(MESSAGES(fmsp)) = fmsp .
ceq inv-MESSAGES(mmsg mmsg') =
    (inv-MESSAGES(mmsg) inv-MESSAGES(mmsg'))
    if mmsg =/= empty and mmsg' =/= empty .

crl [IRECEIVE]: WAITING(x) mmsg =>
                TERMINATED(x)
    if inv-MESSAGES(mmsg) == in(network,x) .

crl [RECEIVE]:  PENDING((x,y)) mmsg =>
                ACCEPTED(x) MESSAGES((x,y))
    if inv-MESSAGES(mmsg) == in(network,x)-(y,x) .
```

It should be clear from this example how the general translation works. It takes the form of a conservative theory transformation from the original RWS of an ANS executable by ACUL matching to a logically equivalent RWS executable by ACU matching. The transformation can be applied to any executable ANS satisfying the mild condition that flexible arcs are inscribed by variables, as it is the case in the executable version of the echo algorithm.[11]

Even though the resulting RWS is strongly executable, a strategy to execute the specification or to partially explore the state space can be useful, because of the highly nondeterminstic nature of the algorithm. A strategy of this kind can be seen as restricting the possible rewrites leading to a subcategory of the original category of all rewrites. If the RWS is only weakly executable, as in the example discussed in [69], the strategy can play an additional role, namely to find suitable instantiations for the variables that cannot be determined by matching.

5   Timed Petri Nets

This section illustrates how an important class of *timed* Petri nets can be given a rewriting logic semantics. Petri nets have been extended to model real-time systems in different ways (see e. g. [1, 59, 36]). Three of the most frequently used time extensions are the following [59], from which other timed versions of Petri nets can be obtained either as special cases or by combining the extensions:

1. Each *transition t* has an associated time interval $[l_t, u_t]$. A transition fires as soon as it can, but the resulting tokens are delayed, that is, when a transition t fires, the resulting tokens are produced after some time delay $r \in [l_t, u_t]$.
2. Each *place p* has a duration d_p. A token at place p cannot participate in a transition until it has been at p for at least time d_p.

---

[11] A more general transformation is possible if we use conditions with matching equations, a feature supported by the most recent version of Maude [20].

3. Each transition t is associated with a time interval $[l_t, u_t]$, and the transition t cannot fire before it has been continuously enabled for at least time l_t. Also, the transition t cannot have been enabled continuously for more than time u_t without being taken.

We will not treat the third case in this paper. We will instead cover the first two cases as special cases of the *interval timed colored Petri net* (ITCPN) model proposed by van der Aalst [1]. ITCPNs appear in the context of colored nets, but to simplify the exposition and focus on real-time features, we abstract from the colors of the tokens and instead have atomic tokens (with timestamps).

5.1 Interval Timed Petri Nets

We define a new model called *interval timed Petri nets* (ITPNs). Our model is similar to the interval timed colored Petri net model proposed in [1], but with two differences: (1) ITPNs ignore the coloring of the tokens, and (2) ITPNs have a notion of concurrent firing of multisets of transitions.

An ITPN is a PTN where the outgoing arcs are inscribed by time intervals denoting the range of possible firing delays of the produced tokens. The set *TI* of all time intervals, in a time domain *Time*, is the set $TI = \{ [r_1, r_2] \mid r_1, r_2 \in Time \wedge r_1 \leq r_2 \}$.

Definition 1. *An interval timed Petri net (ITPN) $\mathcal{N}$ is a PTN together with a delay inscription $D_\mathcal{N} : F_\mathcal{N} \cap (T_\mathcal{N} \times P_\mathcal{N}) \to TI^\oplus$ verifying $|D_\mathcal{N}(t,p)| = W_\mathcal{N}(t,p)$. The preset function $\partial_0 : T_\mathcal{N} \to P_\mathcal{N}^\oplus$ is defined, as for PTNs, by $\partial_0(t)(p) = W_\mathcal{N}(p,t)$, and the postset function $\partial_1^* : T_\mathcal{N} \to (P_\mathcal{N} \times TI)^\oplus$, where each resulting token is equipped with its delay interval, is defined by $\partial_1^*(t)(p,\Delta) = D_\mathcal{N}(t,p)(\Delta)$.*

The ITPN in Fig. 6 models a setting where each process performs transition a, followed by transition b within time 5 to 10, again followed by transition a time 4 to 8 thereafter, and so on. Furthermore, each process *forks* when performing transition a (this is modeled by having two arcs from *a* to *q*).

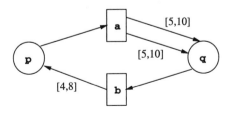

Fig. 6. An interval timed Petri net.

In the ITPN model, as in the ITCPN model, we attach a *timestamp* to each token. This timestamp indicates the time when a token becomes available. The *enabling time* of a transition is the maximum timestamp of the tokens to be consumed. Transitions are *eager* to fire (i.e., they fire as soon as possible), therefore the transition with the smallest enabling time will fire first. Firing is an atomic action, producing tokens with a timestamp equal to the firing time plus some *firing delay* specified by the delay inscription.

In the following, let $\mathcal{N}$ be an ITPN. The set of *markings* $\mathcal{M_N} = (P_\mathcal{N} \times Time)^\oplus$ is the set of all finite multisets of pairs (p, r) representing the presence of a token at place p with timestamp r. The function $places : \mathcal{M_N} \to P_\mathcal{N}^\oplus$ which removes the timestamps from a marking is defined by $places(m)(p) = \Sigma_{(p,r) \in \mathcal{S}(m)}\, m(p, r)$. The function $max : (\mathcal{M_N} - \{\emptyset\}) \to Time$ which finds the maximal timestamp in a non-empty marking is given by $max(m) = \max\{r \in Time \mid \exists p\,.\,(p, r) \in m\}$. The earliest enabling time of any transition in a marking is given by a function $EET : \mathcal{M_N} \to Time \cup \{\infty\}$ defined by $EET(m) = \min\{max(m') \mid \exists t \in T_\mathcal{N}, m' \in \mathcal{M_N}\,.\,m' \sqsubseteq m \wedge places(m') = \partial_0(t)\}$ with $\min(\emptyset) = \infty$. The function $\hat{+} : (P_\mathcal{N} \times TI)^\oplus \times Time \to (P_\mathcal{N} \times TI)^\oplus$ adds a delay to all the intervals in a multiset and is defined by $(m\hat{+}r)(p, [r_1 + r, r_2 + r]) = \Sigma_{(p,[r_1,r_2]) \in \mathcal{S}(m)}\, m(p, [r_1, r_2])$, and $(m\hat{+}r)(p, [r', r'']) = 0$ if $r' < r$. Finally, to relate multisets of tokens with timestamps with multisets of tokens with time intervals, we define the specialization relation $\vartriangleleft \subseteq (P_\mathcal{N} \times Time)^\oplus \times (P_\mathcal{N} \times TI)^\oplus$, where $m \vartriangleleft m'$ holds if and only if each token in m corresponds to one token in m', such that they are in the same place and the timestamp of the token in m is in the interval of the corresponding token in m'. That is, $m \vartriangleleft m'$ if and only if either $(m = \emptyset \wedge m' = \emptyset)$ or $\exists (p, r) \in m,\, (p, [r_1, r_2]) \in m'\,.\,r_1 \le r \le r_2 \wedge (m - (p, r)) \vartriangleleft (m' - (p, [r_1, r_2]))$.

An ITPN makes computational progress by applying transitions, thereby consuming and producing multisets of timestamped tokens. A nonempty finite multiset of transitions firing at the same time constitutes a *(concurrent) step*. The *(concurrent) step semantics* of an ITPN $\mathcal{N}$ is given by the labelled transition system which has $\mathcal{M_N}$ as states, $\mathcal{ST_N} = T_\mathcal{N}^\oplus - \{\emptyset\}$ as steps, and where the transition relation $\to \, \subseteq \mathcal{M_N} \times \mathcal{ST_N} \times \mathcal{M_N}$ is defined inductively by the following rules:

$$\frac{places(m) = \partial_0(t) \quad max(m) = EET(m) \quad m' \vartriangleleft \partial_1^*(t)\hat{+}EET(m)}{m \xrightarrow{t} m'}$$

$$\frac{m \xrightarrow{e} m' \quad m'' \in \mathcal{M_N} \quad EET(m \oplus m'') = EET(m)}{(m \oplus m'') \xrightarrow{e} (m' \oplus m'')}$$

$$\frac{m_1 \xrightarrow{e_1} m_1' \quad m_2 \xrightarrow{e_2} m_2' \quad EET(m_1 \oplus m_2) = EET(m_1) = EET(m_2)}{(m_1 \oplus m_2) \xrightarrow{e_1 \oplus e_2} (m_1' \oplus m_2')}.$$

A *step sequence* in $\mathcal{N}$ is a (finite or infinite) sequence

$$\varsigma: \quad m_0 \xrightarrow{e_1} m_1 \xrightarrow{e_2} m_2 \xrightarrow{e_3} \cdots ,$$

where each step e_i represents the simultaneous firing of its transitions at time $EET(m_{i-1})$. The set of all step sequences of an ITPN $\mathcal{N}$ is denoted $\mathcal{C}_{\mathcal{N}}^{\infty}$.

Timed Petri nets of type (1) described above, where transitions have durations, can be seen as a special case of ITPNs as follows. A transition t with time interval $[l_t, u_t]$ which consumes the tokens m and produces the tokens m' can be simulated in an ITPN by adding a new place p_t, and having a transition t_1 which consumes the tokens m and produces one token at place p_t in some time in the interval $[l_t, u_t]$, and another transition t_2 which consumes one token from p_t and produces the tokens m' in zero time. Timed Petri nets of type (2), where each place has a duration d_p, corresponds to the special case of ITPNs where each token produced at place p has firing delay d_p.

5.2   Representing ITPNs in Rewriting Logic

Representing Real-Time Theories in Rewriting Logic. We have proposed in [62] a framework for modeling real-time and hybrid systems in rewriting logic by means of *real-time rewrite theories*, and have shown that a number of well-known models of real-time and hybrid systems can naturally be specified as such theories. Essentially, a real-time rewrite theory should include a sort *Time* and an operator $\{\_\}$ which encloses the global state of the system and is used to ensure that time advances uniformly in all parts of a system. In addition to ordinary rewrite rules modeling instantaneous change in a system, a real-time rewrite theory may contain *tick rules* of the form $l : \{t\} \xrightarrow{\tau_l} \{t'\}$ **if** C, which model time elapse in a system, and where the term τ_l of sort *Time* denotes the duration of the rule. The total time elapse $\tau(\alpha)$ of a rewrite proof $\alpha : \{u\} \longrightarrow \{u'\}$ is defined as the sum of the time elapsed in each tick rule application in α. Even though it is useful to highlight the real-time aspects of a system using real-time rewrite theories, we have shown that, by adding an explicit clock, such theories are reducible to ordinary rewrite theories in a way that preserves all their expected properties.

In real-time systems, some actions are *eager*, that is, their application should take precedence over the application of time-advancing tick rules. We divide the rules of a real-time rewrite theory into *eager* and *lazy* rules, and define the *admissible rewrites* [62] to be the subset of all rewrites satisfying the additional requirement that a lazy rule may only be applied when no eager rule is applicable. The Real-Time Maude language and tool [60, 61] supports the specification and analysis of real-time rewrite theories, including the possibility to define eager and lazy rules.

Specifying ITPNs as Real-Time Rewrite Theories. The rewriting logic semantics of interval timed Petri nets is based on the rewriting logic semantics of untimed place/transition nets given in Section 3.2. For the sake of simplicity of the rewriting logic representation of ITPNs, we choose not to carry the times-tamps in the tokens at all times. Instead, a term $[p]$ of sort VisibleMarking represents a occurrence of a token at place p that is "visible", i.e., available for consumption. A token that *will be* visible at place p in time r is represented by the term $\mathtt{dly}(p, r)$, which has the sort DelayedMarking whenever $r \neq 0$.[12] A token with delay 0 is visible, i.e., $\mathtt{dly}(p, 0) = [p]$. The sort Marking is a super-sort of the sorts VisibleMarking and DelayedMarking, and denotes multisets of these two forms of tokens, where multiset union is represented by juxtapo-sition. The function mte takes a term of sort DelayedMarking and returns the time that can elapse until the next delayed token becomes visible. The function delta models the effect of the passage of time on delayed tokens by decreasing their delays according to the time elapsed.

The rewriting logic semantics of an ITPN $\mathcal{N}$ with $P_{\mathcal{N}} = \{p_1, \ldots, p_n\}$ is given by a real-time rewrite specification $\mathbf{R}(\mathcal{N})$, where the underlying MES contains an axiomatization of the sort Time of the time domain [62], a sort TimeInf for the time domain extended with ∞, together with the functions $+$, $\leq$, min, and $\overset{.}{-}$ ("monus"), and the following declarations and axioms:

```
sorts Place EmptyMarking VisibleMarking DelayedMarking Marking System .
subsorts EmptyMarking < VisibleMarking DelayedMarking < Marking .

ops p₁ ... pₙ : -> Place .
op [_] : Place -> VisibleMarking .
op dly : Place Time -> Marking .
op empty : -> EmptyMarking .
op __ : Marking Marking -> Marking [assoc comm id: empty] .
op __ : DelayedMarking DelayedMarking -> DelayedMarking
                                   [assoc comm id: empty] .
op __ : VisibleMarking VisibleMarking -> VisibleMarking
                                   [assoc comm id: empty] .
op __ : EmptyMarking EmptyMarking -> EmptyMarking
                                   [assoc comm id: empty] .
op {_} : Marking -> System .
op delta : DelayedMarking Time -> Marking .
op mte : DelayedMarking -> TimeInf .

vars DM DM' : DelayedMarking .  var VM : VisibleMarking .
var P : Place .  vars X Y : Time .
cmb dly(P, X) : DelayedMarking if X =/= 0 .
eq dly(P, 0) = [P] .
eq delta(empty, X) = empty .
eq delta(dly(P, X), Y) = dly(P, X - Y) .
ceq delta(DM DM', X) = delta(DM, X) delta(DM', X)
```

---

[12] We will see later that no interesting information about time is lost by this simplifi-cation, since the time when a firing of a transition occurs can always be extracted from the proof term.

```
          if DM =/= empty and DT' =/= empty .
eq mte(empty) = ∞ .
ceq mte(dly(P, X)) = X if X =/= 0 .
ceq mte(DM DM') = min(mte(DM), mte(DM'))
          if DM =/= empty and DM' =/= empty .
```

The rewrite semantics of an ITPN $\mathcal{N}$ is a real-time rewrite theory $\mathbf{R}(\mathcal{N})$ whose signature Ω and axioms E define the sort Marking and the functions delta and mte. The set of rules of $\mathbf{R}(\mathcal{N})$ consists of a *lazy* tick rule modeling time elapse and, for each transition t in $T_{\mathcal{N}}$, an *eager* rule

$$t : \overbrace{[p_1] \ldots [p_1]}^{W(p_1,t)} \ldots \overbrace{[p_n] \ldots [p_n]}^{W(p_n,t)} \longrightarrow$$
$$\underbrace{\mathrm{dly}(p_1, x_{1,1}) \ldots \mathrm{dly}(p_1, x_{1, W(t,p_1)})}_{W(t,p_1)} \ldots \underbrace{\mathrm{dly}(p_n, x_{n,1}) \ldots \mathrm{dly}(p_n, x_{n, W(t,p_n)})}_{W(t,p_n)}$$
$$\textbf{if } (l_{1,1} \leq x_{1,1} \leq u_{1,1}) \wedge \ldots \wedge (l_{1,W(t,p_1)} \leq x_{1,W(t,p_1)} \leq u_{1,W(t,p_1)}) \wedge \ldots$$
$$\wedge (l_{n,1} \leq x_{n,1} \leq u_{n,1}) \wedge \ldots \wedge (l_{n,W(t,p_n)} \leq x_{n,W(t,p_n)} \leq u_{n,W(t,p_n)})$$

where $P_{\mathcal{N}} = \{p_1, \ldots, p_n\}$, with p_i distinct, $D(t, p_i)$ is the multiset $\{[l_{i,1}, u_{i,1}], \ldots, [l_{i,W(t,p_i)}, u_{i,W(t,p_i)}]\}$ for each $p_i \in P_{\mathcal{N}}$, and the $x_{i,j}$'s are distinct variables of sort *Time*. The following lazy tick rule advances time until the first delayed token becomes visible:

$$\text{tick} : \{VM\ DM\} \xrightarrow{\mathrm{mte}(DM)} \{VM\ \mathrm{delta}(DM, \mathrm{mte}(DM)\} \text{ if } \mathrm{mte}(DM) \neq \infty.$$

For example, the translation of the ITPN in Fig. 6 contains the above tick rule and the following two instantaneous eager rules:

$$\text{a} : [p] \longrightarrow \mathrm{dly}(q,X)\ \mathrm{dly}(q,Y) \text{ if } (5 \leq X \leq 10) \wedge (5 \leq Y \leq 10)$$
$$\text{b} : [q] \longrightarrow \mathrm{dly}(p,X) \text{ if } 4 \leq X \leq 8.$$

The tick rule only needs to compute the time until the next delayed token becomes visible and advances time by that amount. After such a tick, the tick rule is again enabled but, due to its being lazy, it will not be applied if the new visible token(s) enable some transition(s) (whose firing in turn could immediately trigger further instantaneous transitions).

Since a step $m \xrightarrow{e} m'$ of an ITPN does not depend on the firing delays of the individual transitions taken in the step, two *one-step* rewrites $\{\alpha\} : u \longrightarrow v$ and $\{\beta\} : u \longrightarrow v$ should be considered equal — in the sense that we add the equivalence $t(r_1, \ldots, r_n) = t(r'_1, \ldots, r'_n)$, for each $t \in T_{\mathcal{N}}$, as a further equality identifying rewrite proofs — if the multisets of rule labels in α and β are the same. A *timed computation* in $\mathbf{R}(\mathcal{N})$ is a finite or infinite sequence

$$\tilde{\varsigma} : \quad u_0 \xrightarrow{\gamma_1; \delta_1} u_1 \xrightarrow{\gamma_2; \delta_2} u_2 \xrightarrow{\gamma_3; \delta_3} \cdots$$

with admissible rewrite proofs $\gamma_i; \delta_i : u_{i-1} \longrightarrow u_i$ in $\mathbf{R}(\mathcal{N})$, such that each γ_i corresponds either to the identity proof or to a sequence of tick applications, each δ_i corresponds to a one-step concurrent rewrite using instantaneous rules, and u_0 is a term $\{w_0\}$ with w_0 a term of sort Marking. The set of timed computations in $\mathbf{R}(\mathcal{N})$ is denoted $\mathcal{C}(\mathbf{R}(\mathcal{N}))$. It follows from the factorization property of proofs in rewriting logic [48] that each non-identity admissible ground rewrite $\alpha : \{w\} \longrightarrow \{w'\}$ in $\mathbf{R}(\mathcal{N})$, is equivalent to a rewrite $\gamma; \delta$, such that γ can be rearranged as a finite timed computation, and δ corresponds to the identity proof or to a sequence of tick applications. Furthermore, each (infinite) computation of $\mathbf{R}(\mathcal{N})$, consisting of admissible rewrites involving ground terms of sort System, which contains an infinite number of applications of instantaneous rules, can be rearranged as a timed computation.

The fact that ITPNs are faithfully represented in their rewriting logic semantics is made precise in the theorem below, which can be used as the basis of a method to execute and analyze ITPNs in a tool such as Real-Time Maude [60, 61].

Theorem 3. Let $\mathcal{N}$ be an ITPN. Then, there is a bijective function $\widetilde{(\_)} : \mathcal{C}_{\mathcal{N}}^{\infty} \to \mathcal{C}(\mathbf{R}(\mathcal{N}))$ taking a step sequence of the form ς to an timed computation of the form $\widetilde{\varsigma}$ (see above for ς and $\widetilde{\varsigma}$) such that:

- Each u_i is a term equivalent to a term of the form $\{w_i\}$, which consists of $m_i(p, r)$ occurrences of the term $\mathtt{dly}(p, r \dot{-} \tau(\gamma_1; \delta_1; \ldots; \gamma_i; \delta_i))$, for all p and r (recall that $\mathtt{dly}(p, 0)$ is equivalent to $[p]$).
- The transitions fire at the same time in ς and $\widetilde{\varsigma}$, that is, $\tau(\gamma_1; \delta_1; \ldots; \gamma_{i+1}) = EET(m_i)$.
- The transitions taken (concurrently) in each step are the same. That is, each δ_i is equivalent to a proof term of the form $\{\varepsilon_i\}$, where ε_i is a term containing, for each $t \in T_{\mathcal{N}}$, exactly $e_i(t)$ occurrences of proof terms of the form $t(r_1, \ldots, r_n)$.

6 Conclusions

In this paper we have explained in detail how rewriting logic can be used as a semantic framework in which a wide range of Petri net models can be naturally unified. Specifically, we have explored how place/transition nets, nets with test arcs, algebraic net specifications, colored Petri nets, and timed Petri nets can all be naturally expressed in rewriting logic, and how well-known semantic models often coincide with (in the sense of being naturally isomorphic to) the natural semantic models associated to the rewriting logic representations of the given nets. Space limitations do not allow us to explain in detail how other classes of Petri nets could similarly be treated. However, we sketch below a number of extensions of the ideas presented here that could deal with some of these.

A question that deserves some discussion is how colored Petri nets based on higher-order programming languages such as ML can be formally represented and, furthermore, how can they be related to the approach to ANSs presented in this paper. One possible answer is to translate each colored Petri net over a possibly higher-order language $\mathcal{L}$ into an ANS with an initial semantics. This reduces the problem to finding a translation of $\mathcal{L}$ into membership equational logic. The main problem with embedding a higher-order language into a first-order framework is the treatment of bound variables and there are different solutions. Recently, we have developped CINNI [68, 70], a new calculus of names and explicit substitutions, to solve this problem in a systematic way, and we have applied it to obtain executable embeddings of languages such as the lambda calculus and Abadi and Cardelli's object calculus into membership equational logic.

The step from higher-order programming languages to higher-order specification languages can be regarded in some instances as a move from typed lambda calculi to higher-order logics. The use of a specification language with higher-order capabilities seems to be not only attractive for enhancing the modeling and abstraction capabilities, but it can also provide a framework for extensions of algebraic specifications by initiality and freeness constraints [35] or first-order axioms such as those used in [65]. Recent experience with representing an entire family of pure type systems in rewriting logic [70] indicates that, using rewriting logic as a metalanguage, typed lambda calculi and higher-order logics can be naturally expressed. By viewing membership equational logic as a sublogic of a higher-order logic the approach presented in this paper, including the important aspect of executability, naturally extends to the higher-order case.

Apart from generalizations of the underlying specification or programming language, there is another potential source of Petri net generalizations, namely, the structure of the state space. Instead of considering a flat state space as in ordinary Petri nets we could choose a hierarchical one, or we could consider extensions such as *macroplaces* [2, 3] that can be seen as combining several places into a single one from the viewpoint of certain transitions. Also, we could consider different kinds of places. For instance, we could distinguish ordinary high-level places that carry a multiset of tokens from places that are organized as a queue or as a stack. The former idea has been studied in the literature in terms of *FIFO-nets* [30, 40, 29, 27]. Rewriting logic seems to be a suitable formalism to represent and unify such variations of Petri nets, since the state space can be specified by an equational theory that is entirely user-definable. A related approach that allows some freedom in the choice of the state space algebra and specializes to different low-level and high-level Petri net classes is presented in [24]. The approach of rewriting logic has the advantage of being more general, in the sense that it goes beyond Petri-net-like models and hence provides a bridge to formalisms that are quite different from ordinary Petri nets.

Yet another interesting generalization of Petri nets are different variants of *object Petri nets* [72, 73, 28, 74], where tokens can themselves be nets with their own dynamic behaviour. A quite different line of research is the integration of object-oriented techniques with Petri nets. As a result there are a number of variants of object-oriented Petri nets [67, 44], where the tokens are objects according to standard object-oriented terminology. As a unifying generalization of both approaches we propose a notion of *active token nets*. In contrast to object Petri nets, tokens can not only be nets but arbitrary objects with an internal dynamic behaviour. In constrast to *object-oriented approaches to Petri nets*, active tokens are not static but dynamic entities. In particular, they can evolve concurrently with the overall system behaviour, and they can also interact or communicate with each other. It might appear that the complexity of such models is beyond the scope of a rigorous formal treatment. However, a closer look reveals that the approach to Petri nets via rewriting logic is closer to the ideas described above than it might appear at the first sight. In fact, our approach can be easily generalized to active token nets by essentially replacing the underlying MES of a net by a RWS. So far we have employed rewrite rules only to represent transitions of the net. In order to describe tokens with internal activity we could use rewrite rules that transform individual tokens. To capture group activity such as interaction (which corresponds to synchronous communication) and asynchronous communication we have to add rewrite rules that operate on a group of tokens. One possible realization is to view tokens as objects in the sense of the rewriting logic approach to concurrent object-oriented programming [49], where rewrite rules operate on a multiset of objects that are interrelated by object references. As we have already pointed out, recent work on partial-order semantics for object-oriented systems specified in rewriting logic in this manner [57] is in fact very close to the safe process semantics of high-level Petri nets.

In our view, the unification of Petri net models within the rewriting logic logical framework is useful not only for conceptual reasons, but also for purposes of execution, formal analysis, and formal reasoning about Petri net specifications. Using the reflective and metalanguage capabilities of Maude, it is possible to build execution environments for Petri net specifications where the language description provided by the user and the user interaction could all take place at the Petri net level with which the user is familiar. Similarly, the Real-Time Maude tool [61] could offer corresponding capabilities for executing and analyzing timed Petri net models.

Acknowledgements.   Support by DARPA through Rome Laboratories Contract F30602-C-0312 and NASA through Contract NAS2-98073, by Office of Naval Research Contract N00014-99-C-0198, and by National Science Foundation Grant CCR-9900334 is gratefully acknowledged. Part of this work is based on earlier work conducted by the first author at University of Hamburg, Germany and in the scope of the European Community project MATCH (CHRX-CT94-0452). Furthermore, we would like to thank Roberto Bruni, Narciso Martí-Oliet, Rüdiger Valk and the referees for their constructive criticism and many helpful suggestions.

References

1. W. M. P. van der Aalst. Interval timed coloured Petri nets and their analysis. In M. A. Marsan, editor, *Application and Theory of Petri Nets 1993*, volume 691 of *Lecture Notes in Computer Science*, pages 453–472. Springer, 1993.

2. N. A. Anisimov. An algebra of regular macronets for formal specification of communication protocols. *Computers and Artificial Intelligence*, 10(6):541–560, 1991.

3. N. A. Anisimov, K. Kishinski, A. Miloslavski, and P. A. Postupalski. Macroplaces in high level Petri nets: Application for design inbound call center. In *Proceedings of the Int. Conference on Information System Analysis and Synthesis (ISAS'96), Orlando, FL, USA*, pages 153–160, July 1996.

4. A. Asperti. A logic for concurrency. unpublished manuscript, November 1987.

5. E. Battiston, F. De Cindio, and G. Mauri. OBJSA nets: a class of high-level nets having objects as domains. In G. Rozenberg, editor, *Advances in Petri Nets*, volume 340 of *Lecture Notes in Computer Science*. Springer-Verlag, 1988.

6. B. Berthomieu, N. Choquet, C. Colin, B. Loyer, J. M. Martin, and A. Mauboussin. Abstract data nets: Combining Petri nets and abstract data types for high level specifications of distributed systems. In *Proc. of the Seventh Workshop on Applications and Theory of Petri Nets, Oxford, UK*, pages 25–48, 1986.

7. E. Best and R. Devillers. Sequential and concurrent behaviour in Petri net theory. *Theoretical Computer Science*, 55:87–136, 1987.

8. E. Best and C. Fernandez. *Nonsequential Processes—A Petri Net View*, volume 13 of *EATCS Monographs on Theoretical Computer Science*. Springer-Verlag, 1988.

9. A. Bouhoula, J.-P. Jouannaud, and J. Meseguer. Specification and proof in membership equational logic. *Theoretical Computer Science*, 236:35–132, 2000.

10. C. Brown. Relating Petri nets to formulae of linear logic. Technical Report ECS-LFCS-89-87, Laboratory of Foundations of Computer Science, University of Edinburgh, June 1989.

11. C. Brown and D. Gurr. A categorical linear framework for Petri nets. In *Proc. Fifth Annual IEEE Symposium on Logic in Computer Science*, pages 208–218, June 1990.

12. R. Bruni, J. Meseguer, U. Montanari, and V. Sassone. A comparison of Petri net semantics under the collective token philosophy. In J. Hsiang and A. Ohori, editors, *Proceedings of ASIAN'98, 4th Asian Computing Science Conference*, volume 1538 of *Lecture Notes in Computer Science*, pages 225–244. Springer-Verlag, 1998.

13. R. Bruni, J. Meseguer, U. Montanari, and V. Sassone. Functorial semantics for Petri nets under the individual token philosophy. In *Proc. Category Theory and Computer Science, Edinburgh, Scottland, September 1999*, volume 29 of *Electronic Notes in Theoretical Computer Science*. Elsevier, 1999. http://www.elsevier.nl/locate/entcs/volume29.html.

14. R. Bruni, J. Meseguer, U. Montanari, and V. Sassone. Functorial semantics for petri nets under the individual token philosophy. In *Electronic Notes in Theoretical Computer Science: Proceedings of CTCS'99, 8th Conference on Category Theory and Computer Science*, volume 29, pages 1–19. Elsevier Science, 1999.

15. R. Bruni and V. Sassone. Algebraic models for contextual nets. In U. Montanari, J. D. P. Rolim, and E. Welzl, editors, *Automata, Languages and Programming. Proceedings 2000.*, volume 1853. Springer-Verlag, 2000.

16. S. Christensen and N. D. Hansen. Coloured Petri nets extended with place capacities, test arcs and inhibitor arcs. In M. A. Marsan, editor, *Application and Theory of Petri Nets 1993*, volume 691 of *Lecture Notes in Computer Science*, 1993.

17. M. Clavel. Reflection in general logics and in rewriting logic, with applications to the Maude language. Ph.D. Thesis, University of Navarre, 1998.
18. M. Clavel, F. Durán, S. Eker, P. Lincoln, N. Martí-Oliet, J. Meseguer, and J. Quesada. A tutorial on Maude. SRI International, March 2000, http://maude.csl.sri.com.
19. M. Clavel, F. Durán, S. Eker, P. Lincoln, N. Martí-Oliet, J. Meseguer, and J. Quesada. *Maude: Specification and Programming in Rewriting Logic*. Computer Science Laboratory, SRI International, Menlo Park, 1999. http://maude.csl.sri.com.
20. M. Clavel, F. Durán, S. Eker, P. Lincoln, N. Martí-Oliet, J. Meseguer, and J. Quesada. Towards Maude 2.0. In K. Futatsugi, editor, *Third International Workshop on Rewriting Logic and its Applications (WRLA'2000), Kanazawa, Japan, September 18 - 20, 2000*, volume 36 of *Electronic Notes in Theoretical Computer Science*, pages 297 – 318. Elsevier, 2000. http://www.elsevier.nl/locate/entcs/volume36.html.
21. P. Degano, J. Meseguer, and U. Montanari. Axiomizing the algebra of net computations and processes. *Acta Informatica*, 33:641–667, 1996.
22. C. Dimitrovici, U. Hummert, and L. Petrucci. Semantics, composition and net properties of algebraic high-level nets. In G. Rozenberg, editor, *Advances in Petri Nets 1991*, volume 524 of *Lecture Notes in Computer Science*. Springer-Verlag, 1991.
23. F. Durán and J. Meseguer. Structured theories and institutions. In M. Hofmann, G. Rosolini, and D. Pavlović, editors, *Proceedings of CTCS'99, 8th Conference on Category Theory and Computer Science, Edinburgh, Scotland, U.K., September 10-12, 1999*, volume 29, pages 71–90. Elsevier, 1999. http://www.elsevier.nl/locate/entcs/volume29.html.
24. H. Ehrig and J. Padberg. Uniform approach to Petri nets. In C. Freksa, M. Jantzen, and R. Valk, editors, *Foundations of Computer Science: Potential – Theory – Cognition*, volume 1337 of *Lecture Notes in Computer Science*, pages 219–231. Springer-Verlag, August 1997.
25. H. Ehrig, J. Padberg, and L. Ribeiro. Algebraic high-level nets: Petri nets revisited. In *Recent Trends in Data Type Specification*, volume 785 of *Springer-Verlag*, pages 188–206, 1994.
26. U. Engberg and G. Winskel. Petri nets as models of linear logic. In A. Arnold, editor, *CAAP'90*, volume 431 of *Lecture Notes in Computer Science*, pages 147–161. Springer-Verlag, 1990.
27. J. Fanchon. FIFO-net models for processes with asynchronous communication. In G. Rozenberg, editor, *Advances in Petri Nets 1992*, volume 609 of *Lecture Notes in Computer Science*, pages 152–178. Springer-Verlag, 1992.
28. B. Farwer. A linear logic view of object Petri nets. *Fundamenta Informaticae*, 37(3):225–246, 1999.
29. A. Finkel and A. Choquet. FIFO nets without order deadlock. *Acta Informatica*, 25(1):15–36, 1988.
30. A. Finkel and G. Memmi. FIFO nets: New model of parallel computation. In *6th GI-Conference on Theoretical Computer Science, Dortmund*, volume 145 of *Lecture Notes in Computer Science*, pages 111–121. Springer-Verlag, 1982.
31. H. J. Genrich. Equivalence transformation of PrT-nets. In G. Rozenberg, editor, *Advances in Petri Nets 1989*, volume 424, pages 179–208. Springer-Verlag, 1990.
32. H. J. Genrich. Predicate/transition nets. In *High-Level Petri Nets: Theory and Practice*, pages 3–43. Springer-Verlag, 1991.

33. H. J. Genrich and K. Lautenbach. The analysis of distributed systems by means of predicate/transition-nets. In G. Kahn, editor, *Semantics of Concurrent Computation*, volume 70 of *Lecture Notes in Computer Science*, pages 123–146, Berlin, 1979. Springer-Verlag.

34. H. J. Genrich and K. Lautenbach. System modelling with high-level Petri nets. *Theoretical Computer Science*, 13:109–136, 1981.

35. J. Goguen and R. Burstall. Institutions: Abstract model theory for specification and programming. *Journal of the ACM*, 39(1):95–146, 1992.

36. H. M. Hanisch. Analysis of place/transition nets with timed arcs and its application to batch process control. In M. A. Marsan, editor, *Application and Theory of Petri Nets 1993*, volume 691 of *Lecture Notes in Computer Science*, pages 282–299. Springer, 1993.

37. K. Hoffmann. Run time modification of algebraic high level nets and algebraic higher order nets using folding and unfolding construction. In G. Hommel, editor, *Communication-Based Systems, Proceedings of the 3rd International Workshop held at the TU Berlin, Germany, 31 March – 1 April 2000*, pages 55–72. Kluwer Academic Publishers, 2000.

38. K. Jensen. Coloured Petri nets and the invariant-method. *Theoretical Computer Science*, pages 317–336, 1981.

39. K. Jensen. *Coloured Petri Nets, Basic Concepts, Analysis Methods and Practical Use.*, volume 1 of *EATCS monographs on theoretical computer science*. Springer-Verlag, 1992.

40. E. Kettunen, E. Montonen, and T. Tuuliniemi. Comparison of Pr-net based channel models. In *Proc. of the 12th IMACS World Conf.*, volume 3, pages 479–482, 1988.

41. E. Kindler and W. Reisig. Algebraic system nets for modelling distributed algorithms. *Petri Net Newsletter*, (51):16–31, December 1996.

42. E. Kindler, W. Reisig, H. Völzer, and R. Walter. Petri net based verification of distributed algorithms: An example. *Formal Aspects of Computing*, 9:409–424, 1997.

43. E. Kindler and H. Völzer. Flexibility in algebraic nets. In J. Desel and M. Silva, editors, *Application and Theory of Petri Nets 1998, 19th International Conference, ICATPN'98, Lisbon, Portugal, June 1998, Proceedings*, volume 1420 of *Lecture Notes in Computer Science*, pages 345–384. Springer-Verlag, 1998.

44. C. A. Lakos. From coloured Petri nets to object Petri nets. In M. Diaz G. De Michelis, editor, *Application and Theory of Petri Nets*, volume 935 of *Lecture Notes in Computer Science*, pages 278–297, Berlin, 1995. Springer-Verlag.

45. N. Martí-Oliet and J. Meseguer. From Petri nets to linear logic. *Mathematical Structures in Computer Science*, 1:69–101, 1991.

46. N. Martí-Oliet and J. Meseguer. From Petri nets to linear logic through categories: A survey. *International Journal of Foundations of Computer Science*, 2(4):297–399, 1991.

47. J. Meseguer. General logics. In H.-D. Ebbinghaus et al., editors, *Proceedings, Logic Colloquium, 1987*, pages 275–329. North-Holland, 1989.

48. J. Meseguer. Conditional rewriting logic as a unified model of concurrency. *Theoretical Computer Science*, 96:73–155, 1992.

49. J. Meseguer. A logical theory of concurrent objects and its realization in the Maude language. In G. Agha, P. Wegner, and A. Yonezawa, editors, *Research Directions in Concurrent Object-Oriented Programming*. MIT Press, 1993.

50. J. Meseguer. Rewriting logic as a semantic framework for concurrency: a progress report. In U. Montanari and V. Sassone, editors, *Proc. Concur'96*, volume 1119 of *Lecture Notes in Computer Science*, pages 331–372. Springer, 1996.

51. J. Meseguer. Membership algebra as a logical framework for equational speci-fication. In F. Parisi-Presicce, editor, *Recent Trends in Algebraic Development Techniques, 12th International Workshop, WADT '97, Tarquinia, Italy, June 3-7, 1997, Selected Papers*, volume 1376 of *Lecture Notes in Computer Science*, pages 18 – 61. Springer-Verlag, 1998.

52. J. Meseguer and U. Montanari. Petri nets are monoids. *Information and Computation*, 88(2):105–155, October 1990.

53. J. Meseguer, U. Montanari, and V. Sassone. On the semantics of Petri nets. In W.R. Cleaveland, editor, *Proceedings of the Concur'92 Conference, Stony Brook, New York, August 1992*, volume 630 of *Lecture Notes in Computer Science*, pages 286–301. Springer-Verlag, 1992.

54. J. Meseguer, U. Montanari, and V. Sassone. On the model of computation of place/transition Petri nets. In *Proceedings 15th International Conference on Application and Theory of Petri Nets*, volume 815 of *Lecture Notes in Computer Science*, pages 16–38. Springer-Verlag, 1994.

55. J. Meseguer, U. Montanari, and V. Sassone. Process versus unfolding semantics for place/transition Petri nets. *Theoretical Computer Science*, 153(1–2):171–210, 1996.

56. J. Meseguer, U. Montanari, and V. Sassone. Representation theorems for Petri nets. In C. Freska, M. Jantzen, and R. Valk, editors, *Foundations of Computer Science: Potential, Theory, Cognition*, volume 1337 of *Lecture Notes in Computer Science*, pages 239–249. Springer-Verlag, 1997.

57. J. Meseguer and C. Talcott. A partial order event model for concurrent objects. In *Proc. CONCUR'99, Eindhoven, The Netherlands, August 1999*, volume 1664 of *Lecture Notes in Computer Science*, pages 415–430. Springer-Verlag, 1999.

58. U. Montanari and F. Rossi. Contextual nets. *Acta Informatica*, 32:545–596, 1995.

59. S. Morasca, M. Pezzè, and M. Trubian. Timed high-level nets. *The Journal of Real-Time Systems*, 3:165–189, 1991.

60. P. C. Ölveczky. *Specification and Analysis of Real-Time and Hybrid Systems in Rewriting Logic*. PhD thesis, University of Bergen, 2000. Available at http://maude.csl.sri.com/papers.

61. P. C. Ölveczky and J. Meseguer. Real-Time Maude: A tool for simulating and analyzing real-time and hybrid systems. In *Third International Workshop on Rewriting Logic and its Applications*, 2000. To appear in *Electronic Notes in Theoretical Computer Science*.

62. P. C. Ölveczky and J. Meseguer. Specification of real-time and hybrid systems in rewriting logic. To appear in *Theoretical Computer Science*. Available at http://maude.csl.sri.com/papers, September 2000.

63. C. A. Petri. Nets, time and space. *Theoretical Computer Science*, 153(1–2):3–48, 1996.

64. W. Reisig. Petri nets and algebraic specifications. *Theoretical Computer Science*, 80:1–34, 1991.

65. W. Reisig. *Elements of Distributed Algorithms: Modeling and Analysis with Petri Nets*. Springer-Verlag, 1998.

66. W. Reisig and J. Vautherin. An algebraic approach to high level Petri nets. In *Proceedings of the Eighth European Workshop on Application and Theory of Petri Nets*, pages 51–72. Universidad de Zaragoza (Spain), 1987.

67. C. Sibertin-Blanc. Cooperative nets. In R. Valette, editor, *Application and Theory of Petri Nets*, volume 815 of *Lecture Notes in Computer Science*, pages 471–490, Berlin, 1994. Springer-Verlag.

68. M.-O. Stehr. CINNI - A Generic Calculus of Explicit Substitutions and its Application to lambda-, sigma- and pi-calculi. In K. Futatsugi, editor, *Third International Workshop on Rewriting Logic and its Applications (WRLA'2000), Kanazawa, Japan, September 18 - 20, 2000*, volume 36 of *Electronic Notes in Theoretical Computer Science*, pages 71 – 92. Elsevier, 2000. http://www.elsevier.nl/locate/entcs/volume36.html.

69. M.-O. Stehr. A rewriting semantics for algebraic nets. In C. Girault and R. Valk, editors, *Petri Nets for Systems Engineering – A Guide to Modelling, Verification, and Applications*. Springer-Verlag, 2001. To appear.

70. M.-O. Stehr and J. Meseguer. Pure type systems in rewriting logic. In *Proc. of LFM'99: Workshop on Logical Frameworks and Meta-languages, Paris, France, September 28, 1999*.

71. A. Tarlecki, R. M. Burstall, and J. A. Goguen. Some fundamental algebraic tools for the semantics of computation, III: indexed categories. *Theoretical Computer Science*, 79:239–264, 1991.

72. R. Valk. Petri nets as dynamical objects. In *Workshop Proc. 16th International Conf. on Application and Theory of Petri Nets, Torino, Italy*, June 1995.

73. R. Valk. Petri nets as token objects: An introduction to elementary object nets. In J. Desel and M. Silva, editors, *Proceedings of the 19th International Conference on Application and Theory of Petri Nets, Lissabon, June 22-26, 1998*, volume 1420 of *Lecture Notes in Computer Science*, pages 1 – 25. Springer-Verlag, 1998.

74. R. Valk. Relating Different Semantics for Object Petri Nets. Technical report, FBI-HH-B-266/00, Fachbereich Informatik, Universität Hamburg, 2000.

75. J. Vautherin. *Un Modele Algebrique, Base sur les Reseaux de Petri, pour l'Etude des Systemes Paralleles*. These de Docteur Ingenieur, Univ. de Paris-Sud, Centre d'Orsay, June 1985.

76. J. Vautherin. Parallel systems specifications with coloured Petri nets and algebraic specifications. *Lecture Notes in Computer Science: Advances in Petri Nets 1987*, 266:293–308, 1987.

77. P. Viry. Rewriting: An effective model of concurrency. In C. Halatsis, D. Maritsas, G. Philokyprou, and S. Theodoridis, editors, *PARLE'94 – Parallel Architectures and Languages Europe, 6th Int. PARLE Conf. Athes, Greece, July 1994, Proceedings.*, volume 817 of *Lecture Notes in Computer Science*, pages 648–660. Springer-Verlag, 1994.

78. W. Vogler. Partial order semantics and test arcs. In *Proc. MFCS'97*, volume 1295 of *Lecture Notes in Computer Science*. Springer, 1997.

Generalized Automata and Their Net Representations[*]

Eric Badouel[1], Marek A. Bednarczyk[2], and Philippe Darondeau[3]

[1] ENSP, Yaoundé
ebadouel@polytech.uninet.cm
[2] IPI PAN, Gdańsk
M.Bednarczyk@ipipan.gda.pl
[3] IRISA, Rennes
Philippe.Darondeau@irisa.fr

Abstract. We consider two generalizations of the duality between transition systems and Petri nets. In the first, transitions are replaced by *paths*, that is partial functions from a fixed set Δ to states. This allows to model continuous and/or hybrid systems when Δ represents durations. In the second generalization actions are considered to have a structure given by an algebra. This allows to model, for instance, sequential and parallel composition of ordinary actions. In each case the question of the existence of a Galois connection is considered in the framework of ordered sets and in the categorical setting.

1 Preliminaries

Uniform presentations of (extended) Petri nets in which events and markings are interpreted jointly in an algebraic structure, e.g. a monoidal category [14], an ordered group [7], or a partial groupoid [13], have been put forward by several authors. A different, more combinatorial than algebraic, approach to a uniform theory for Petri nets was proposed in [2]. The present paper is a follower of this work, and an informal review of ideas and results given there may be appropriate before telling what we are now looking for. Readers already familiar with [2] or [3] may skip the rest of this section.

It was proposed in [2] to represent classes of Petri nets stable under subnets by the state graph of a particular net, built by gluing on a common place all non-isomorphic nets in the class with one place and one event. For instance, if one allows isolated elements in C/E-nets, then there are three non-isomorphic C/E-nets with one place and one event. By gluing them on a common place, one obtains the net shown in the left part of Fig. 1. The case graph of this net is isomorphic to the transition system $\tau_{C/E}$ shown in the right part of this figure.

This transition system $\tau_{C/E}$ gives full information about the way in which C/E-nets behave, and it allows to reconstruct their firing rule. The two states

---

[*] Supported by programme CATALYSIS within CNRS/PAN cooperation framework.
[2] Partially supported by State Committee for Scientific Research grant 8 T11C 037 16

H. Ehrig et al. (Eds.): Unifying Petri Nets, LNCS 2128, pp. 304–345, 2001.
© Springer-Verlag Berlin Heidelberg 2001

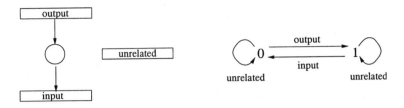

Fig. 1. The representative C/E-net and its case graph

indicate that a condition may take two alternative values in a case: 1 if the condition holds, 0 if it does not hold. The transitions labeled *input* and *output* indicate that an event has concession in a case if and only if all its input conditions and none of its output conditions hold. They indicate moreover that the values of these conditions change under firing. The transitions labeled *unrelated* indicate that the unrelated conditions do not play any direct role for the considered event. If one represents a C/E-net as a flow matrix $F : C \times E \to \{input, output, unrelated\}$, the firing rule of the C/E-nets may now be restated thus:

$$M[e\rangle M' \quad \text{iff} \quad (\forall c \in C) \quad M(c) \xrightarrow{F(c,e)} M'(c) \quad \text{in} \quad \tau_{C/E}$$

All familiar classes of Petri nets, e.g. pure P/T-nets, general P/T-nets, nets with inhibitor arcs, etc., may be accommodated in a similar way. Below, we only sketch how it works in the case of the pure P/T-nets.

Pure P/T-nets with one place and one transition are in an obvious bijective correspondence with $\mathbb{Z}$. By gluing them on a common place π, one obtains a net with flow matrix $F : \{\pi\} \times \mathbb{Z} \to \mathbb{Z}$ such that $F(\pi, z) = z$ for all $z \in \mathbb{Z}$ (this represents an arc with weight $|z|$ from place π to event z, or from event z to place π, according to the sign of z). The state graph $\tau_{PURE-P/T}$ of this net is isomorphic to the induced restriction of the Cayley graph of group $\mathbb{Z}$ on subset of nodes $\mathbb{N}$. The firing rule for a pure P/T-net with flow matrix $F : P \times T \to \mathbb{Z}$ may be restated thus:

$$M[t\rangle M' \quad \text{iff} \quad (\forall p \in P) \quad M(p) \xrightarrow{F(p,t)} M'(p) \quad \text{in} \quad \tau_{PURE-P/T}$$

More generally, one can uniformly define new classes of nets in this way, where each class is induced by a choice of a deterministic transition systems τ. Once such a $\tau = (S, I, \tau)$, called a *type* of nets, has been fixed, places take values in the set S and each net $N = (P, A, F)$ is fit with a flow matrix $F : P \times A \to I$. The firing rule, as one could expect, is the following:

$$M[a\rangle M' \quad \text{iff} \quad (\forall p \in P) \quad M(p) \xrightarrow{F(p,t)} M'(p) \quad \text{in} \quad \tau$$

Moreover, the type τ of the nets may be reconstructed by applying this firing rule to the net $(\{\pi\}, I, F)$ with flow matrix $F(\pi, \iota) = \iota$. Let us stress the fact that we do not require τ to be a group action graph, or whatever kind of algebraic transition system: *every* deterministic transition system τ defines a class of nets and the behaviors of these nets.

What adds interest to this uniform presentation is that it fits nicely with the region based synthesis of nets, first proposed by Ehrenfeucht and Rozenberg in the context of C/E-nets [9], and subsequently extended to P/T-nets by Mukund [15] and by Droste and Shortt [7]. A crucial remark made in [2] is that regions of a transition system $T = (Q, A, T)$ w.r.t. a class of nets defined by a deterministic transition system τ, do coincide with morphisms of transition systems from T to τ. Consider for instance C/E-nets. A morphism from T to $\tau_{C/E}$ is by definition a pair (σ, η) made of two maps $\sigma : Q \to \{0, 1\}$ and $\eta : A \to \{input, output, unrelated\}$ such that $\sigma(q) \xrightarrow{\eta a} \sigma(q')$ in $\tau_{C/E}$ whenever $q \xrightarrow{a} q'$ in T. Therefore, if we set $R = \sigma^{-1}(1)$, then all transitions $q \xrightarrow{a} q'$ with a common label a do simultaneously leave R (case $\eta(a) = input$) or enter R (case $\eta(a) = output$), or they do not cross the bor:der of R (case $\eta(a) = output$). Thus, in terminology of [9], R is a *region*. Conversely, every region coincides with $\sigma^{-1}(1)$ for a unique morphism $(\sigma, \eta) : T \to \tau_{C/E}$. The case of the pure Petri net regions is similar: morphisms $(\sigma, \eta) : T \to \tau_{P/T}$ are in bijective correspondence with maps $\sigma : Q \to \mathbb{N}$ such that $\sigma(q') - \sigma(q)$ has a constant value for all transitions $q \xrightarrow{a} q'$ with the common label a, that is with $\eta(a)$.

Building on the above remark, one may extend Ehrenfeucht and Rozenberg's principle of net synthesis to nets over arbitrary type. Given a transition system $T = (Q, A, T)$, its counterpart in the class of nets over type $\tau = (S, I, \tau)$ is a net $N = (P, A, F)$ with set of places $P = Hom(T, \tau)$ (places are morphisms from T to τ), with flow matrix $F : P \times A \to I$ defined by $F((\sigma, \eta), a) = \eta(a)$ for any place $(\sigma, \eta) \in P$ and event $a \in A$. The construction may be adapted to initialized transition systems (i.e. automata) and nets (i.e. net systems). The counterpart of the automaton (Q, A, T, q_0) is a net system (P, A, F, M_0) with initial marking M_0 defined by $M_0((\sigma, \eta)) = \sigma(q_0)$ for every place $(\sigma, \eta) \in Hom(T, \tau)$.

On this basis, an order theoretic Galois connection between automata and net systems, parametric on the type of nets τ, was established in [1]. Automata and net systems have a fixed alphabet of events. Net systems are ordered by the sub-structure relation, while (deterministic and reachable) automata are ordered by (label preserving) morphisms. It was shown in [1] that

$$\mathcal{A} \leq \mathcal{N}^* \quad iff \quad \mathcal{N} \leq \mathcal{A}^*$$

where $\mathcal{N}^*$ is the reachable state graph of the net system $\mathcal{N}$ while $\mathcal{A}^*$ is the net system, of type τ, synthesized from the automaton $\mathcal{A}$.

A completely symmetric duality between automata and net systems, also parametric on the type of nets τ, was established in [2]. Moreover, it was observed there that state graphs may also be constructed from hom-sets. For this purpose, the underlying set of $\tau = (S, I, \tau)$, i.e. its set of transitions, should also be seen as the underlying set of a net $(\tau, \{enabled\}, F)$, with a single event and with flow matrix F determined thus: $F(\langle s \xrightarrow{\iota} s' \rangle, enabled) = \iota$ for each place $s \xrightarrow{\iota} s'$ in τ. Given any other net (P, A, F') with type τ, the transitions

$M[a\rangle M'$ that may be inferred for this net from the firing rule induced by τ correspond to net morphisms from (P, A, F') to $(\tau, \{enabled\}, F)$. That is to say, every such a morphism is a pair (β, η) which consists of two maps $\beta : P \to \tau$ and $\eta : \{enabled\} \to A$, and such that $F(\beta(p), enabled) = F'(p, \eta(enabled))$ for each place $p \in P$. This correspondence entails that $M(p) = s$ and $M'(p) = s'$ whenever $\beta(p) = (s \xrightarrow{\iota} s')$.

Let us consider C/E-nets for an illustration. Fig. 2 shows a C/E-net (on the left), the net version of $\tau_{C/E}$ (on the right), and a C/E-net morphism (β, η) from the former to the latter. The morphism depicted is defined by $\eta(enabled) = b$, $\beta(p_1) = \beta(p_2) = \langle 1 \xrightarrow{input} 0 \rangle$ and $\beta(p_3) = \langle 1 \xrightarrow{unrelated} 1 \rangle$. This morphism represents the transition $(1, 1, 1) [b\rangle (0, 0, 1)$, or to write it differently, $p_1+p_2+p_3[b\rangle p_3$.

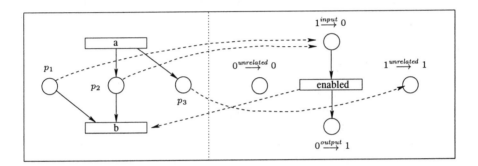

Fig. 2. Transitions of C/E-nets as net morphisms

The transition system $\tau_{C/E}$ and the C/E-net derived from $\tau_{C/E}$ (right part of Fig. 2) are built on the same underlying set. These two objects which live in two different categories form therefore what is called a schizophrenic object. Drawing inspiration from Porst and Tholen's view of concrete dualities induced by schizophrenic objects [16], a dual adjunction between transition systems and nets, parametric on the type of nets τ, was established in [2], namely:

$$Hom(T, N^*) \quad \cong \quad Hom(N, T^*)$$

for any transition system T and net N. A Galois connection between subcategories of initialized transition systems and nets was moreover derived from the dual adjunction. In this more complex setting, generality is gained in that the alphabet of events of nets needs not be fixed once and for all.

2   Objectives

As recalled in the preliminary section, correspondences between automata and net systems, uniform in the type of nets, have been established in [1] and [2], in the order-theoretic setting and in the categorical setting, respectively. The present paper is an attempt to extend these correspondences to generalized transition systems. In doing so, our purpose is twofold. On the one hand, we will show that systems more complex than ordinary automata may also be represented with nets. This fact was already clear from David and Alla's proposal for continuous nets [4], whose correspondence with continuous automata with concurrency is studied elsewhere in this volume [8]. We will examine this and other extensions of automata and nets where continuity does not play any role, and show that all correspondences may be presented uniformly in the type of nets. On the other hand, we will compare the robustness of order-theoretic and categorical correspondences between automata and net systems under the perturbation introduced by considering generalized automata.

The point of view taken in [2] is that in an ordinary transition system it is the set of transitions which is central. Namely, one may see an ordinary transition system (Q, A, T) as a set of transitions T equipped with a map $\lambda : T \to A$ called *labelling*, plus two other maps $\partial^0, \partial^1 : T \to Q$, called *source* and *target*, resp., such that a transition $t = q \xrightarrow{a} q'$ is uniquely characterized by: $\lambda(t) = a$, $\partial^0(t) = q$ and $\partial^1(t) = q'$. In this paper we examine two different ways of enriching this basic model.

One way is to replace both, the source and the target maps ∂^0 and ∂^1, with a single map $\partial : T \to (\Delta \rightharpoonup Q)$ that assigns to each transition $t \in T$ a partial function $\partial(t)$ with co-domain Q. Here, Δ is a *fixed* domain with a distinguished element $\bullet \in \Delta$ such that $\partial(t)(\bullet)$ is always defined. Intuitively, $\partial(t)(\bullet)$ takes place of $\partial^0(t)$. Ordinary transitions are thus generalized to *functional* transitions, labelled on A and parametric on Δ: an action a with actual parameter δ leads from state q to state q' if there exists some transition $t \in T$ such that $\partial(t)(\bullet) = q$, $\lambda(t) = a$, and $\partial(t)(\delta) = q'$. We will not impose any specific interpretation on the set of parameters Δ. The set may represent e.g. *durations*, or *weights*, or *conditions on the environment*, etc. Therefore, we do not impose any algebraic structure on Δ nor any specific constraint on partial functions $\partial(t)$. This is a major difference with the work of Droste and Shortt also reported in this volume. They study the correspondence between continuous automata and nets for a fixed type of Petri nets, while we study a uniform correspondence between functional automata and nets for a many types of nets, including the type of continuous nets. In the specific case of the continuous Petri nets, our results are not as sharp as theirs. In return, our functional definition of nets allows to cover different forms of Petri nets, including notably coloured nets and deterministic vector addition systems with states.

Second way to enrich the basic model of transition systems is to leave the source and target maps ∂^0 and ∂^1 unchanged, but to replace the co-domain of the labelling map $\lambda : T \to A$ by a Σ-algebra of *complex actions* generated from A. If the signature Σ supplies e.g. parallel and sequential composition operators, this gives means to represent concurrent processes of nets as transitions in their state graphs. Conversely, this gives means to fit automata with fine specifications of concurrency, to match when realizing automata by net systems. An illustration may already be found in Mukund's step transition systems and their realization by Petri nets [15]. As we shall see, one may go further in this way, by considering for instance complex transitions labelled with series-parallel pomsets, and still obtain a Galois connection between automata and nets, uniform in the type of nets. The point here is to show that net synthesis is compatible with finer concurrent semantics of nets than the one considered up to now.

While it is easy to define generalized transition systems, it could be a problem to define the associated classes of nets. A major advantage of our uniform view of Petri nets is that it makes this task straightforward in many cases. Once generalized transition systems have been defined in one way or the other, it suffices to equip the type of nets $\tau = (S, I, \tau)$ with an adequate structure of generalized transition system, without modifying the definition of nets in any way. Thus, a net (P, A, F) still comprises a set of places P, a set of atomic actions A, and a flow matrix $F : P \times A \to I$. But the dynamics of nets changes, since now it is induced by generalized transitions in a new transition system τ'. To illustrate this, consider the type $\tau_{PURE-P/T} = (\mathbb{N}, \mathbb{Z}, \tau)$ of the pure Petri nets, with transitions $t \in \tau$ satisfying $\partial^1(t) = \partial^0(t) + \lambda(t)$. One can transform τ into a $\mathbb{R}^+$-transition system $\tau' = (\mathbb{R}^+, \mathbb{Z}, \tau')$ as follows. Given $t' \in \tau'$ put $\partial(t')(\delta) = \partial(t')(\bullet) + \delta \times \lambda(t')$, whenever the right-hand side is non-negative. Then, the induced firing rule for Petri nets is the continuous firing rule of [4] — except that no minimal bound is set on durations of firing. For another illustration, consider the type $\tau_{P/T} = (\mathbb{N}, \mathbb{N} \times \mathbb{N}, \tau)$ of the P/T-nets, with transitions $t \in \tau$ satisfying $\partial^0(t) \geq \pi_1 \circ \lambda(t)$ and $\partial^1(t) = \partial^0(t) - \pi_1 \circ \lambda(t) + \pi_2 \circ \lambda(t)$. If one equips its set of labels $\mathbb{N} \times \mathbb{N}$ with component-wise addition, thus getting an enriched type of nets, the induced firing rule becomes the usual step firing rule, and state graphs become step transition systems as defined by Mukund, with transitions labelled in the free commutative monoid over A.

The remaining sections are organized as follows. Section 3 recalls concrete dualities induced by schizophrenic objects and is largely an adaptation of [16], with notable simplifications. Transition systems parametric on Δ and their correspondence with nets are studied in section 4. Transition systems labelled in Σ-algebras and their correspondence with nets are studied in section 5. Conclusions are briefly indicated in section 6. Readers uninterested in categories may skip sections 3, 4.6 and 4.7.

3   Schizophrenic Objects and Dual Adjunctions

In this section we give a primer on dual adjunctions induced by schizophrenic objects. We shall see in forthcoming sections that many region based dualities between nets and automata fit in this framework. For this reason they appear as close analogies to classical representation theorems, like Birkhoff and Stone representation theorems, most of which do indeed arise from concrete dualities induced by schizophrenic objects.

3.1   Dual Adjunctions Induced by Schizophrenic Objects

A typical instance of duality induced by a schizophrenic object is Birkhoff's duality between finite distributive lattices and finite partial orders. Let us recall it as an illustrative support before coming to general definitions and constructions.

Consider a *schedule*, given as a finite partially ordered set of *tasks* where $a \leq b$ means that a task a should be performed before a task b. A *configuration* is a downward closed set of tasks: it consists of all tasks that have been performed in a particular state of the system. The set of configurations ordered by inclusion is a finite distributive lattice where meet and join are given by set-theoretic intersection and union, respectively. We call it the *lattice of configurations* of the ordered set. The problem is to decide whether a given finite distributive lattice is isomorphic to the lattice of configurations of some finite ordered set. For that purpose, consider the *extension* $\lfloor a \rfloor$ of a task a in the lattice of configurations, i.e. the set of configurations in which it is reported. This extension is a prime filter of the lattice of configurations. Indeed, it is a filter, i.e. a non empty upper-set closed w.r.t. meet, because the whole set belongs to $\lfloor a \rfloor$, $a \in x$ and $x \subseteq y$ imply $a \in y$, and finally $a \in x$ and $a \in y$ imply $a \in x \cap y$. By symmetry, its complement is an ideal (a non-empty down-set closed w.r.t. join). Thus, it is a prime filter, cf. [12]. Moreover, $a \leq b \Rightarrow \lfloor a \rfloor \supseteq \lfloor b \rfloor$, thus a candidate for representing a finite distributive lattice is the set of its prime filters ordered by reverse inclusion. Let us call this ordered set the *schedule of the lattice*. Birkhoff's theorem asserts that

> *any finite ordered set is isomorphic to the schedule of its lattice of configurations and any finite distributive lattice is isomorphic to the lattice of configurations of its schedule.*

Birkhoff's duality between finite distributive lattices and finite partial orders relies on the schizophrenic object $\mathbf{2} = \{0, 1\}$, viewed as a lattice and as an ordered set where $0 \leq 1$. The dual L^* of a distributive lattice L, i.e. its schedule, is the ordered set of the prime filters F, whose characteristic functions are the lattice morphisms $\chi_F : L \to \mathbf{2}$. The dual E^* of an ordered set E, i.e. its lattice of configurations, is the lattice of the downwards closed subsets x, whose characteristic functions are the morphisms of ordered sets $\chi_x : E \to \mathbf{2}$.

More precisely, the dual adjunction asserts that for any ordered set E and finite distributive lattice L, the set of monotone maps from E to L^* is in bijective correspondence with the set of lattice morphisms from L to E^*. In fact, both sets are in bijective correspondence with the set of satisfaction relations $\models \subseteq E \times L$ such that:

$$a \not\models 0 \qquad a \models 1$$
$$a \models x \wedge y \;\; \Leftrightarrow \;\; (a \models x \text{ and } a \models y)$$
$$a \models x \vee y \;\; \Leftrightarrow \;\; (a \models x \text{ or } a \models y)$$
$$(a \leq b \text{ and } b \models x) \;\; \Rightarrow \;\; a \models x$$

The above conditions are indeed equivalent to the requirement that the assignment $a \mapsto \{x \mid a \models x\}$ is a monotone map from E to L^*, and also to the requirement that $x \mapsto \{a \mid a \models x\}$ is a lattice morphism from L to E^*.

If $L = E^*$ is the set of configurations of E, the ordered set E is isomorphic to its double dual: $E \cong E^{**}$ where $a \in E$ is identified with $ev_a \in E^{**}$ such that $\chi_{ev_a}(x) = \chi_x(a)$ for every down-set $x \in E^*$. Symmetrically, if $E = L^*$ the lattice L is isomorphic to its double dual: $L \cong L^{**}$ where $x \in L$ is identified with $ev_x \in L^{**}$ such that $\chi_{ev_x}(F) = \chi_F(x)$ for every prime filter $F \in L^*$. Thus both units of the dual adjunction are morphisms with *evaluation maps* as their underlying maps.

Birkhoff's duality between finite ordered sets and finite distributive lattices is an instance of concrete dualities induced by schizophrenic objects. Let us now explain the general picture in detail in preparation for section 4. There, types of nets play the role of schizophrenic objects).

Definition 1. *A* **Set**-*category (or category over* **Set***) is a pair* (C, U) *where* C *is a category and* $U : C \to$ **Set** *is a functor, called the* underlying functor. *It is a* concrete *category if* U *is faithful.*

In most cases the underlying functor will be left implicit. Given an object C, notation $C \in |C|$, and an arrow $f : C \to C'$, notation $f \in C(C, C')$, if C is a **Set**-category we slightly abuse the notation and write uniformly: $|C|$ and $|f|$ to denote the underlying set of C and the underlying mapping of f.

Let C be a **Set**-category.

We recall that the *initial lift* of a *structured source* $\langle C_i; f_i : X \to |C_i|\rangle_{i \in I}$, where C_i's are objects of C, and the f_i's are mappings from a set X to the underlying sets of C_i's, is a corresponding family of arrows $\tilde{f}_i : C \to C_i$ in C such that $|\tilde{f}_i| = f_i$ (and therefore $|C| = X$) and which is initial in the following sense. Whenever one has an object C' and arrows $g_i : C' \to C_i$ in C such that for some mapping $f : |C'| \to X$ and all indices $|g_i| = f_i \circ f$ holds, then there exists a unique arrow $\tilde{f} : C' \to C$ such that $|\tilde{f}| = f$ and $g_i = \tilde{f}_i \circ \tilde{f}$. The following definition is an adaptation from [16].

Definition 2 (Schizophrenic Object). *A* schizophrenic object *between two* **Set**-*categories* A *and* B *is a pair of objects* $\langle K_A, K_B \rangle \in |A| \times |B|$ *having the same underlying set* $K = |K_A| = |K_B|$ *and such that*

1. *for each object* A *in* A, *the family* $\langle K_B; ev_A(a) : A(A, K_A) \to K \rangle_{a \in |A|}$ *of evaluation mappings defined by:* $ev_A(a)(f) = |f|(a)$ *has an initial lift* $\langle \epsilon_A(a) : A^* \to K_B \rangle_{a \in |A|}$, *and symmetrically*
2. *for each object* B *in* B, *the family* $\langle K_A; ev_B(b) : B(B, K_B) \to K \rangle_{b \in |B|}$ *has an initial lift* $\langle \epsilon_B(b) : B^* \to K_A \rangle_{b \in |B|}$.

Object A^*, called the *dual* of A, is therefore an object of the category B whose underlying set is the set of A-morphisms from A to the *classifying object* K_A. If $K = \{0, 1\}$ and if A is concrete, then the elements of the underlying set of the dual of A can be identified with subsets of the underlying set of A: $|A^*| \subseteq 2^{|A|}$. Of course, as an initial lift the dual of an object is only defined up to (a unique) isomorphism. However, once those lifts are (arbitrarily) chosen, we obtain a functorial correspondence, more precisely:

Lemma 1. *Let* $\langle \mathcal{K}_A, \mathcal{K}_B \rangle$ *be a schizophrenic object between* **Set**-*categories* $\mathcal{A}$ *and* $\mathcal{B}$. *For every morphism* $f : A_1 \to A_2$ *in* $\mathcal{A}$, *the mapping "composing with f" given by* $f^{\bullet} : \mathcal{A}(A_2, \mathcal{K}_A) \to \mathcal{A}(A_1, \mathcal{K}_A)$ *where* $f^{\bullet}(g) = g \circ f$, *is the underlying mapping of an arrow* $f^* : A_2^* \to A_1^*$ *in* $\mathcal{B}$ *such that the functoriality laws:* $(1_A)^* = 1_{A^*}$ *and* $(f \circ g)^* = g^* \circ f^*$ *are satisfied.*

Proof. Consider $g : A_2 \to \mathcal{K}_A$ in $\mathcal{A}$, and $a \in |A_1|$. Then, $|\epsilon_{A_2}(|f|(a))|(g) = ev_{A_2}(|f|(a))(g) = |g|(|f|(a)) = |g \circ f|(a) = ev_{A_1}(a)(g \circ f) = (ev_{A_1}(a) \circ f^{\bullet})(g)$ i.e., $|\epsilon_{A_2}(|f|(a))| = ev_{A_1}(a) \circ f^{\bullet}$. By initiality of $\langle \epsilon_{A_1}(a) \rangle_{a \in |A_1|}$ we obtain a unique $f^* : A_2^* \to A_1^*$ such that *(i)* $\epsilon_{A_2}(|f|(a)) = \epsilon_{A_1}(a) \circ f^*$ and *(ii)* $|f^*| = f^{\bullet}$. Thanks to this characterization of f^*, the functoriality laws immediately follow.   □

Lemma 2. *Let* $\langle \mathcal{K}_A, \mathcal{K}_B \rangle$ *be a schizophrenic object between two* **Set**-*categories* $\mathcal{A}$ *and* $\mathcal{B}$. *The initial lift* $\langle \epsilon_A(a) : A^* \to \mathcal{K}_B \rangle_{a \in A}$ *of the evaluation mappings, viewed as a mapping* $\epsilon_A : |A| \to \mathcal{B}(A^*, \mathcal{K}_B)$ *is the underlying mapping of an arrow* $Ev_A : A \to A^{**}$.

Proof. For $f \in |A^*|$ i.e., $f : A \to \mathcal{K}_A$ in $\mathcal{A}$, and $a \in |A|$ one has

$$ev_{A^*}(f)(\epsilon_A(a)) = |\epsilon_A(a)|(f) = ev_A(a)(f) = |f|(a)$$

i.e., $|f| = ev_{A^*}(f) \circ \epsilon_A$, by initiality of $\langle ev_{A^*}(f) \rangle_{f \in |A^*|}$. This gives a unique morphism $Ev_A : A \to A^{**}$ such that *(i)* $\epsilon_{A^*}(f) \circ Ev_A = f$ and *(ii)* $|Ev_A| = \epsilon_A$.   □

Definition 3 (Span). *Let* $\mathcal{K} = \langle \mathcal{K}_A, \mathcal{K}_B \rangle$ *be a schizophrenic object between two* **Set**-*categories* $\mathcal{A}$ *and* $\mathcal{B}$. *A* $\mathcal{K}$-*span* $\varphi \in \mathbf{Span}_{\mathcal{K}}(A, B)$ *from* $A \in |\mathcal{A}|$ *to* $B \in |\mathcal{B}|$ *consists of families of morphisms* $\langle \varphi_a : B \to \mathcal{K}_B \rangle_{a \in |A|}$ *and* $\langle \varphi^b : A \to \mathcal{K}_A \rangle_{b \in |B|}$ *in* $\mathcal{B}$ *and* $\mathcal{A}$, *respectively, such that for all* $a \in |A|$ *and* $b \in |B|$ *the following holds.*

$$|\varphi_a|(b) = |\varphi^b|(a)$$

If $\mathcal{A}$ and $\mathcal{B}$ are concrete categories, then spans coincide with bimorphisms, i.e., those mappings $\varphi : |A| \times |B| \to K$ such that

1. For all $a \in |A|$ function $\varphi(a, -) : |B| \to K$ is the underlying mapping of a morphism φ_a from B to $\mathcal{K}_B$, and
2. For all $b \in |B|$ function $\varphi(-, b) : |A| \to K$ is the underlying mapping of a morphism φ^b from A to $\mathcal{K}_A$.

For concrete categories, a span is a K-valued relation between the underlying sets of A and B, and it can be represented by a matrix with values in K whose rows and columns are indexed by the sets $|A|$ and $|B|$, respectively. Of course, $\mathbf{Span}_{\mathcal{K}}(A, B) = \mathbf{Span}_{\mathcal{K}}(B, A)$, modulo the transposition of matrices.

Lemma 3. *Let* $\langle \mathcal{K}_A, \mathcal{K}_B \rangle$ *be a schizophrenic object between two* **Set**-*categories* $\mathcal{A}$ *and* $\mathcal{B}$. *There is a bijective correspondence between the hom-set* $\mathcal{A}(A, B^*)$ *and the set of* $\mathcal{K}$-*spans* $\mathbf{Span}_{\mathcal{K}}(A, B)$ *given by the following identities, where* $f \in \mathcal{A}(A, B^*)$ *and* $\varphi \in \mathbf{Span}_{\mathcal{K}}(A, B)$.

$$\forall a \in |A| \ \forall b \in |B| \quad \varphi_a = |f|(a) \ and \ \varphi^b = \epsilon_B(b) \circ f \tag{1}$$

Proof. For the one hand the identities (1) clearly determine φ in terms of the morphism f, and it is indeed a span because

$$|\varphi_a|(b) = ||f|(a)|\,(b) = ev_B(b)(|f|(a)) = |\epsilon_B(b) \circ f|(a) = |\varphi^b|(a)$$

For the converse direction, assume φ is a span. Then $ev_B(b)(\varphi_a) = |\varphi_a|(b) = |\varphi^b|(a)$ i.e., $|\varphi^b| = ev_B(b) \circ \varphi_{(-)}$ for every $b \in |B|$, and the initiality of $\langle \epsilon_B(b)\rangle_{b\in|B|}$ precisely ensures the existence and unicity of a morphism $f : A \to B^*$ verifying the identities (1).   □

Proposition 1 (Dual Adjunction Induced by a Schizophrenic Object).
*Let $\langle \mathcal{K}_A, \mathcal{K}_B \rangle$ be a schizophrenic object between two **Set**-categories $\mathcal{A}$ and $\mathcal{B}$. There is a bijective correspondence $\mathcal{A}(A, B^*) \cong \mathcal{B}(B, A^*)$ given by the following identities, where $f : A \to B^*$ and $g : B \to A^*$,*

$$g = f^* \circ Ev_B \quad and \quad f = g^* \circ Ev_A \tag{2}$$

i.e., the functors $(-)^*$ are adjoint to the right with the evaluations as units.

Proof. By Lemma 3 we have $\mathcal{A}(A, B^*) \cong \mathbf{Span}_{\mathcal{K}}(A, B) = \mathbf{Span}_{\mathcal{K}}(B, A) \cong \mathcal{B}(B, A^*)$ given by the following identities where $f : A \to B^*$ and $g : B \to A^*$.

$$(\forall a \in |A|)(\forall b \in |B|) \quad \epsilon_B(b) \circ f = |g|(b) \quad and \quad |f|(a) = \epsilon_A(a) \circ g \tag{3}$$

In order to establish the proposition it suffices to prove that given a morphism $g : B \to A^*$, the morphism $f = g^* \circ Ev_A$ satisfies the identities (3). For that purpose we recall that (by Lemma 1) g^* is the unique morphism $g^* : A^{**} \to B^*$ such that *(i)* $\epsilon_{A^*}(|g|(b)) = \epsilon_B(b) \circ g^*$ for every $b \in |B|$, and *(ii)* $|g^*| = g^\bullet$. We recall also that (by Lemma 2) Ev_A is the unique morphism $Ev_A : A \to A^{**}$ such that *(i)* $\epsilon_{A^*}(f) \circ Ev_A = f$ for every $f \in |A^*|$, and *(ii)* $|Ev_A| = \epsilon_A$. Now we can proceed to the verification that $f = g^* \circ Ev_A$ satisfies (3). Indeed,

1. $\epsilon_B(b) \circ g^* \circ Ev_A = \epsilon_{A^*}(|g|(b)) \circ Ev_A = |g|(b)$, and
2. $|g^* \circ Ev_A|(a) = (g^\bullet \circ \epsilon_A)(a) = \epsilon_A(a) \circ g$.

as required.   □

We can turn the set of $\mathcal{K}$-spans into a category whose objects are triples (A, φ, B) where $\varphi \in \mathbf{Span}_{\mathcal{K}}(A, B)$, equivalently, (A, f, B) with $f \in \mathcal{A}(A, B^*)$ or $(A, f^\sharp, B)$ where $f^\sharp \in \mathcal{B}(B, A^*)$, and whose morphisms are pairs of *reindexing* morphisms $\alpha \in \mathcal{A}(A_1, A_2)$, and $\beta \in \mathcal{B}(B_2, B_1)$ such that

- $\forall a \in A_1 \;\; \forall b \in B_2 \quad \varphi_1^{|\beta|b} = \varphi_2^b \circ \alpha$ and $(\varphi_2)_{|\alpha|a} = (\varphi_1)_a \circ \beta$,
- or equivalently $\beta^* \circ f_1 = f_2 \circ \alpha$,
- or equivalently $\alpha^* \circ f_2^\sharp = f_1^\sharp \circ \beta$.

For concrete categories this condition on morphisms reduces to:

$$(\forall a \in A_1)(\forall b \in B_2) \quad \varphi_1(a, \beta b) = \varphi_2(\alpha a, b) \tag{4}$$

$\mathcal{A}$ (resp., $\mathcal{B}^{op}$) are co-reflective (resp., reflective) full subcategories of $\mathbf{Span}_K$. The *kernel* of $\mathbf{Span}_K$ is the full subcategory consisting of those spans $\varphi \in \mathbf{Span}_K(A, B)$ such that $B \cong A^*$ and $A \cong B^*$, and it is categorically equivalent to the respective full subcategories of $\mathcal{A}$ and $\mathcal{B}^{op}$ consisting of those objects for which units $Ev_A : A \to A^{**}$ and $Ev_B : B \to B^{**}$ are isomorphisms. This equivalence yields a duality between the considered subcategories of $\mathcal{A}$ and $\mathcal{B}$.

3.2   Galois Connections

Birkhoff's dual adjunction between finite ordered sets and finite distributive lattices is a duality, i.e. the kernel of the dual adjunction is isomorphic to the whole of each category. This is not a common case and it is not always easy to identify the kernel of a dual adjunction. An interesting case of dual adjunctions (not necessarily induced by schizophrenic objects) is when the kernel coincides with the respective images of both categories under the adjoint functors. The following definition comes from [11].

Definition 4 (Galois Connections). *Let $\mathcal{A}(A, B^*) \cong \mathcal{B}(B, A^*)$ be a dual adjunction with units $\langle E_A : A \to A^{**}\rangle_{A\in|\mathcal{A}|}$ and $\langle E_B : B \to B^{**}\rangle_{B\in|\mathcal{B}|}$. Further let $\mathcal{B}^*$ (the image of $\mathcal{B}$) denote the full subcategory of $\mathcal{A}$ with objects B^* for $B \in |\mathcal{B}|$. Let the image of $\mathcal{A}$ be defined similarly. Then the dual adjunction is a Galois Connection whenever one of the following equivalent conditions is satisfied:*

1. *It restricts to a duality between the images: $\mathcal{B}^* \overset{op}{\cong} \mathcal{A}^*$,*
2. *the arrows $(E_A)^*$ are isomorphisms for $A \in |\mathcal{A}|$,*
3. *their left-inverses E_{A^*} are isomorphisms,*
4. *the arrows $(E_B)^*$ are isomorphisms for $B \in |\mathcal{B}|$,*
5. *their left-inverses E_{B^*} are isomorphisms,*
6. *the maps $\langle E_A : A \to A^{**}\rangle_{A\in|\mathcal{A}|}$ constitute a reflection of $\mathcal{A}$ into $\mathcal{B}^*$,*
7. *the maps $\langle E_B : B \to B^{**}\rangle_{B\in|\mathcal{B}|}$ constitute a reflection of $\mathcal{B}$ into $\mathcal{A}^*$.*

For the sake of an illustration, consider the dual adjunction between topological spaces and frames induced by the schizophrenic object $\mathbf{2}$, viewed as a boolean algebra and as a topological space [12, 6]. Recall that a frame is a complete lattice with the generalized distributivity law (finite meets distribute over arbitrary joins: $f \wedge \bigvee_i f_i = \bigvee_i (f \wedge f_i)$). For any frame F, let $pt(F)$ be the set of *points* x of F defined as frame morphisms $x : F \to \mathbf{2}$. The dual F^* of F is the topological space $(pt(F), \Omega)$ with the family Ω of open sets defined by $\Omega = \{O_f \mid f \in F\}$ where $O_f = \{x : F \to \mathbf{2} \mid x(f) = 1\}$. Conversely, the dual X^* of a topological space (X, Ω) is the frame of its open sets $O \in \Omega$, whose characteristic functions χ_O are the continuous maps from (X, Ω) to the Sierpiński space $\mathbf{2}$ (with the open sets $\{0,1\}$, $\{1\}$, and $\emptyset$). Frames and topological spaces are connected by a dual adjunction $\mathbf{Frame}(F, X^*) \cong \mathbf{Top}(X, F^*)$.

By restricting this adjunction at both sides on its kernel, one obtains a duality $\mathbf{Top}^* \overset{op}{\cong} \mathbf{Frame}^*$ between the subcategory $\mathbf{Top}^*$ of *spatial* frames and the subcategory $\mathbf{Frame}^*$ of *sober* spaces. So, a frame F is isomorphic to its double dual F^{**} if and only if F is a spatial frame. Now, spatial frames are characterized by two conditions closely similar to the separation conditions for automata that we shall encounter in section 4, when replacing regions defined there by morphisms $x : F \to \mathbf{2}$. Namely, a frame F is spatial if and only if the following conditions are satisfied for all $f, f' \in F$, where $f \leq f' \Leftrightarrow f = f \wedge f'$:

$$
\begin{array}{ll}
(i) & f \neq f' \Rightarrow \exists x : F \to \mathbf{2} \; : \; x(f) \neq x(f') \\
(ii) & f \not\leq f' \Rightarrow \exists x : F \to \mathbf{2} \; : \; x(f) = 1 \wedge x(f') = 0
\end{array}
$$

Condition *(i)* is the analogue of the state separation property SSP. Condition *(ii)* is the counterpart of the event state separation property ESSP, when replacing the structure of labelled transition system by the structure of partial order.

4   Δ-Parameterized Transition Systems and Nets

In this section an extension of the dynamic behavior of nets is put forward in which the firing of events is parameterized with a fixed set Δ. At each marking M, the partial action of an event e is formalized as a partial function f from Δ to the set of all markings. The intuition is that Δ provides controls or measures on the firing process. To capture the idea that the process of firing e starts at M we stipulate that Δ contains a distinguished element $\bullet$ such that $f(\bullet) = M^1$.

This functional extension covers indeed many forms of nets or high-level nets. In order to give evidence of this and to help intuition before introducing formally Δ-parameterized transition systems and nets, we accommodate below David and Alla's Hybrid Petri Nets [5] (Δ is the set of durations), coloured Petri nets (Δ is the set of colours), and deterministic vector addition systems with states (Δ is the set of control states).

4.1   Hybrid Petri Nets

Consider a Petri net $N = (P, A, F)$ with set of places P, set of actions A, and flow matrix $F : P \times A \to (\mathbb{N} \times \mathbb{N})$, notation $F(p, a) = \langle p^\bullet a, a^\bullet p \rangle$. Consider next a net system $\mathcal{N} = (N, M_0)$ endowed with a marking M_0 that assigns to each place $p \in P$ some initial value $M_0(p)$ in the disjoint union of sets $\mathbb{R}_+ \oplus \mathbb{N}$. A place p such that $M_0(p) \in \mathbb{R}_+$ is a continuous place, a place p such that $M_0(p) \in \mathbb{N}$ is a discrete place. In case of a continuous place p, let $p^\bullet a$, resp., $a^\bullet p$, be interpreted as *speeds* of consumption, resp., production, of the resource stored in p when a is fired. In case of a discrete place p, let $p^\bullet a$ and $a^\bullet p$ be interpreted as fixed numbers of tokens, taken from or put into place p each time a is fired continuously. Assume that firing an action for duration 0 is not significant. Finally assume some positive constant d such that no action a can be fired at a marking M for duration $\delta \neq 0$ unless $M(p) \geq d \times p^\bullet a$ for each continuous place p. The firing rule of our hybrid nets may now be set as follows.

---

[1] The strong interpretation of predicates involving partially defined terms is used throughout. Thus, $f(\bullet) = M$ holds if and only if $f(\bullet)$ is defined and equals M.

Action a may be fired at marking M for duration $\delta \neq 0$ and thereby lead to marking M', notation $M[a, \delta\rangle M'$, iff for every place p the following hold.

- p discrete: $M(p) \geq p^\bullet a$ and $M'(p) = M(p) - p^\bullet a + a^\bullet p$;
- p continuous: $M(p) \geq d \times p^\bullet a$ and $M'(p) = M(p) - \delta \times p^\bullet a + \delta \times a^\bullet p \geq 0$.

By convention, let $M[a, 0\rangle M'$ iff $M = M'$. Thus, an action a induces at each marking M a partial function f from $\Delta = \mathbb{R}_+$ to markings, such that $f(\bullet) = M$, where $\bullet = 0$, and $f(\delta) = M'$ if and only if $M[a, \delta\rangle M'$. The state graph of $\mathcal{N}$ may be identified with the set of parametric transitions $\langle f, a \rangle$ defined in this way.

Let us examine what would be the result of firing an action a for a place p, with $n = p^\bullet a$ and $m = a^\bullet p$, depending on the kind of the place.

Let p be a discrete place with contents $i \in \mathbb{N}$. Then p disables firing of a for duration $\delta \neq 0$ if either $i < n$ or $i - n + m < 0$. Otherwise, p can change its contents to $i - n + m$ as a result of firing a. This may be represented by a δ-parametric transition $\langle g, (n, m) \rangle$ labelled (n, m) over set of states $\mathbb{N}$, where $g : \mathbb{R}_+ \rightharpoonup \mathbb{N}$ is the partial map such that $g(0) = i$ and for $\delta \neq 0$ either $g(\delta)$ is undefined, if $i - n + m < 0$, or $g(\delta) = i - n + m$, otherwise.

Assume now p is a continuous place with contents $x \in \mathbb{R}_+$. Then p disables firing a for duration $\delta \neq 0$ if either $x < d \times n$ or $x - \delta \times n + \delta \times m < 0$. Otherwise, p may change its contents to $x - \delta \times n + \delta \times m$ as a result of firing of a. This may be represented by a δ-parametric transition $\langle h, (n, m) \rangle$ labelled (n, m) over set of states $\mathbb{R}_+$, letting $h : \mathbb{R}_+ \rightharpoonup \mathbb{R}_+$ be the partial map such that $h(0) = x$ and for $\delta \neq 0$ either $h(\delta)$ is undefined when $x - \delta \times n + \delta \times m < 0$ or $x < d \times n$, otherwise $h(\delta) = x - \delta \times n + \delta \times m$.

Let τ be the parametric transition system on $\mathbb{R}_+ \oplus \mathbb{N}$ assembled from all the Δ-parametric transitions of the form $\langle g, (n, m) \rangle$ or $\langle h, (n, m) \rangle$. One can see that whenever $\langle f, a \rangle$ is a parametric transition in the state graph of $\mathcal{N}$, there exists for each place p some parametric transition $\langle f_p, F(p, a) \rangle$ in τ such that $f(\delta)$ defined implies $f_p(\delta)$ defined and $f_p(\delta) = f(\delta)(p)$. Conversely, given a p-indexed family of parametric transitions $\langle f_p, F(p, a) \rangle$ in τ and a duration δ such that $f_p(\delta)$ is defined for all p, there exists a parametric transition $\langle f, a \rangle$ in the state graph of $\mathcal{N}$ such that $f_p(0) = f(0)(p)$ and $f_p(\delta) = f(\delta)(p)$ for all p. This shows that the state graph of $\mathcal{N}$ is just a synchronized product of $|P|$ copies of the parametric transition system τ.

4.2   Coloured Petri Nets

For simplicity, let us focus our attention on C/E-nets. A coloured C/E-net with colour set K is like a C/E-net $N = (P, A, F)$, with set of places P and flow matrix $F : P \times A \to \{input, output, unrelated\}$, except that the set of places contains not only ordinary conditions taking values in $\{0, 1\}$, but also high-level places taking values in $\mathcal{P}(K)$. A high-level place p is just a compact representation for a K-tuple of conditions (p, k) —holding when k belongs to the value of p. Similarly, each event $a \in A$ represents a K-tuple of ordinary events (a, k). The idea is that N should behave like the C/E-net that would be derived by expanding high-level places and events such that $F(p, (a, k)) = F(p, a)$ for an ordinary condition

$p \in P$, and $F((p,k),(a,k)) = F(p,a)$ for a high-level place $p \in P$. The firing rule of coloured C/E-nets must therefore be the following. An event a may be fired at a marking M for a colour k and thereby lead to a marking M', notation $M[a,k\rangle M'$, iff the following hold:

- p is a condition:
  - $F(p,a) = input$: $M(p) = 1$ and $M'(p) = 0$;
  - $F(p,a) = output$: $M(p) = 0$ and $M'(p) = 1$;
  - $F(p,a) = unrelated$: $M(p) = M'(p)$;
- p is a high-level place:
  - $F(p,a) = input$: $k \in M(p)$ and $M'(p) = M(p) \setminus \{k\}$;
  - $F(p,a) = output$: $k \notin M(p)$ and $M'(p) = M(p) \cup \{k\}$;
  - $F(p,a) = unrelated$: $M(p) = M'(p)$.

In addition, let by convention $M[a,\bullet\rangle M'$ iff $M = M'$. Thus, an event a induces at each marking M a partial function f from $\Delta = \{\bullet\} \cup K$ to the set of all markings, such that $f(\bullet) = M$ and $f(k) = M'$ iff $M[a,k\rangle M'$ for every colour $k \in K$. Given an initial marking $M_0 : P \to S$, where $S = \{0,1\} \oplus \mathcal{P}(K)$, the state graph of $\mathcal{N} = (N, M_0)$ may be identified with the reachable restriction of the resulting set of parametric transitions $\langle f, a \rangle$.

Consider now the Δ-parametric transition system $\tau = (S, I, \tau)$ where $S = \{0,1\} \oplus \mathcal{P}(K)$ as above, $I = \{input, output, unrelated\}$, and τ is the set of all Δ-parametric transitions $\langle \phi, \iota \rangle$, with $\phi : \Delta \rightharpoonup S$ and $\iota \in I$, such that $\phi(\bullet) \in S$ and the following conditions are satisfied for all $k \in K$:

- $\iota = unrelated$: $\phi(k) = \phi(\bullet)$;
- $\iota = output$:
  - $\phi(\bullet) \in \{0,1\}$: $\phi(\bullet) = 0$ and $\phi(k) = 1$;
  - $\phi(\bullet) \in \mathcal{P}(K)$: $k \in \mathcal{D}om(\phi)$ iff $k \notin \phi(\bullet)$, and then $\phi(k) = \phi(\bullet) \cup \{k\}$;
- $\iota = input$:
  - $\phi(\bullet) \in \{0,1\}$: $\phi(\bullet) = 1$ and $\phi(k) = 0$;
  - $\phi(\bullet) \in \mathcal{P}(K)$: $k \in \mathcal{D}om(\phi)$ iff $k \in \phi(\bullet)$, and then $\phi(k) = \phi(\bullet) \setminus \{k\}$;

The state graph of $\mathcal{N}$ may be reconstructed as the synchronized product of $|P|$ copies of τ, started in the respective states $M_0(p)$: $M[a,k\rangle M'$ if and only if, for each place p, there exists in τ a corresponding transition $\langle \phi_p, \iota_p \rangle$ such that $\phi_p(\bullet) = M(p)$, $\iota_p = F(p,a)$, and $\phi_p(k) = M'(p)$. More precisely, a parametric transition $\langle f, a \rangle$ of $\mathcal{N}$ such that $f(\bullet) = M$ appears as a p-indexed product of parametric transitions $\langle \phi_p, F(p,a) \rangle$, where $\phi_p(\bullet) = f(\bullet)(p)$ and the domain of f is the intersection of the domains of the ϕ_p.

4.3 Vector Addition Systems with States

Given a deterministic automaton $\mathcal{A} = (S, A, T, s_0)$, an initialized vector addition system controlled by $\mathcal{A}$ may be presented as a net system $\mathcal{N} = (P, A, F, M_0)$ in which P contains a distinguished control place p_c with initial value $M_0(p_c) = s_0$, plus ordinary places p with initial values $M_0(p) \in \mathbb{N}$. Let $I = A \oplus (S \to \mathbb{Z})$.

The entries of the flow matrix $F : P \times A \to I$ are defined thus for all $a \in A$: $F(p_c, a) = a$ and $F(p, a) \in (S \to \mathbb{Z})$ for ordinary places p. Now, let $\Delta = \{\bullet\} \cup S$. Consider the Δ-parametric transition system $\tau = (S \oplus \mathbb{N}, I, \tau)$ where τ is the set of all parametric transitions $\langle \phi, \iota \rangle$, with $\phi : \Delta \to S \oplus \mathbb{N}$ and $\iota \in I$, such that the following conditions are satisfied for all $s \in S$:

- $\iota = a \in A : \phi(\bullet) \in S$, $s \in \mathcal{D}om(\phi)$ iff $s = \phi(\bullet)$, and $s \xrightarrow{a} \phi(s)$ in $\mathcal{A}$;
- $\iota \in (S \to \mathbb{Z}) : \phi(\bullet) \in \mathbb{N}$, and $\phi(s) = \phi(\bullet) + \iota(s)$ iff $\phi(\bullet) + \iota(s) \geq 0$.

Using τ as the type of $\mathcal{N}$, one may equip this net with a set of Δ-parametric transitions $\langle f, a \rangle$, defined as p-indexed products of transitions $\langle \phi_p, \iota_p \rangle$ in τ, such that for every place p: $f(\bullet)(p) = \phi_p(\bullet)$, $\iota_p = F(p, a)$, and $f(s)(p) = \phi_p(s)$ whenever $f(s)$ is defined at $s \in S$. This entails that $f(s)$ is defined at at most one s, viz. at $s = f(\bullet)(p_c)$. A Δ-parametric transition $\langle f, a \rangle$ defined at s may be seen as a more customary transition $(s, M)[a\rangle(s', M')$ where $s \xrightarrow{a} s'$ in $\mathcal{A}$ and for every ordinary place p, $M(p) = f(\bullet)(p)$ and $M'(p) = f(s)(p)$. One retrieves in this way the usual firing rule of vector addition systems with states.

4.4 Δ-Parameterized Transition Systems

We proceed now with general definitions. Let Δ be a set of *controls* with a distinguished element $\bullet \in \Delta$. For any set Q let $\Delta \to Q$ be the set of all partial functions f from Δ to Q such that $\bullet \in \mathcal{D}om(f)$. Note that the above assignment becomes a covariant endofunctor on **Set** when for a total function $\sigma : Q \to Q'$ we put $(\Delta \to \sigma) f = \sigma \circ f$, for $f : \Delta \to Q$. Any $f : \Delta \to Q$ is called a *reaction* in Q. Clearly, each reaction $f : \Delta \to Q$ is defined on $\bullet$, and $f_\bullet$ is called the *source* of the reaction. Often in the sequel, notation f_δ is used in this way to denote $f(\delta)$, and $\Im m(f)$ and $\mathcal{D}om(f)$ are used to denote the respective *range* and *domain* of f. Given two reactions $f, g : \Delta \to Q$ we say that g *extends* f, notation $f \prec g$, when the graph of f is included in the graph of g. That is, $\mathcal{D}om(f) \subseteq \mathcal{D}om(g)$ and $g|_{\mathcal{D}om(f)} = f$. Thus, $f \prec g$ implies $f = g$ whenever $\mathcal{D}om(f) = \mathcal{D}om(g)$.

When a set of *actions* A has been fixed, each pair $\langle f, a \rangle \in (\Delta \to Q) \times A$ consisting of an action and a reaction is called a *process* in Q. The above partial order on reactions extends to processes: $\langle f, a \rangle \prec \langle g, b \rangle$ whenever $a = b$ and $f \prec g$.

Definition 5 (Δ-transition systems). *A system of Δ-parametric transitions, or Δ-transition system, is a structure* $\mathbb{S} = (Q, A, T, \partial, \lambda)$ *where Q is a set of states, A is a set of* actions, *T is a set of* transitions, *$\lambda : T \to A$ assigns to each transition an action called its* label, *and $\partial : T \to (\Delta \to Q)$ assigns to each transition a reaction in Q called its* effect. *The components should satisfy the following condition.*

determinism: $\lambda t_1 = \lambda t_2$ *and* $(\partial t_1)_\bullet = (\partial t_2)_\bullet$ *implies* $t_1 = t_2$.   (5)

$\mathbb{S}$ *is complete if for all $q \in Q$, $a \in A$ there exists $t \in T$ s.t. $(\partial t)_\bullet = q \wedge \lambda t = a$.*
$\mathbb{S}$ *is reduced when $A = \Im m(\lambda)$ and $Q = \bigcup \{\Im m(\partial t) \mid t \in T\}$.*

The idea of one process extending another can be applied to transitions, letting $\langle f, a \rangle \prec t$ whenever $\langle f, a \rangle \prec \langle \partial t, \lambda t \rangle$. Then, (5) is equivalent to: $t_1 = t_2$ whenever there exists a process $\langle f, a \rangle$ with $\langle f, a \rangle \prec t_1$ and $\langle f, a \rangle \prec t_2$. Condition (5) implies that ∂ and λ are *jointly monic*:

$$\lambda t_1 = \lambda t_2 \quad \text{and} \quad \partial t_1 = \partial t_2 \quad \text{implies} \quad t_1 = t_2.$$

This amounts to saying that the set T of transitions of S is, essentially, a subset of processes, while λ and ∂ are the respective projections on actions and reactions. In the sequel Δ-transition systems are often presented in the more convenient form $S = (Q, A, T)$ where it is understood that $T \subseteq (\Delta \rightharpoonup Q) \times A$ and ∂ and λ are the two projections. Clearly, each complete Δ-transition system is reduced.

Given a Δ-transition system S we write Q, λ, etc., to denote its set of states, its labelling, and so on, respectively. Various sub-scripts and super-scripts are carried over to the components, e.g., T_i' denotes the set of transitions of S_i', etc. We let a, b range over A, q range over Q, t range over T, and f, g range over $\Delta \rightharpoonup Q$.

Definition 6 (Δ-transition systems morphisms). *A morphism $\phi : S \rightarrow S'$ of Δ-transition systems S and S' is a triple $\phi = (\sigma, \eta, \theta)$ where $\sigma : Q \rightarrow Q'$, $\eta : A \rightarrow A'$ and $\theta : T \rightarrow T'$ satisfy the following condition.*

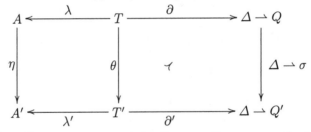

Formally, the (weak) commutativity of the diagram above means the following.

$$\langle \sigma \circ (\partial t), \eta(\lambda t) \rangle \prec \theta t. \tag{6}$$

Thus, we insist that the transition θt be an extension of the image via (σ, η) of the process induced by t.

From (5) it follows that there is at most one extension per process. Thus, given a morphism (σ, η, θ) the component θ is uniquely determined by σ and η. The definition of morphisms can be therefore restated as follows: a morphism is a pair (σ, η) such that there exists θ such that (6) holds.

Conversely, the components σ and η of a morphism are uniquely determined by θ whenever the source of the morphism is reduced. Indeed, to see that in this case σ is determined by θ consider q in Q. Then for some transition t there exists δ such that $q = (\partial t)_\delta$. Then, $\sigma \circ (\partial t) \prec \partial'(\theta t)$ implies $\sigma q = (\sigma \circ (\partial t))_\delta = (\partial'(\theta t))_\delta$, i.e., σq is determined by the value of θ. Thus, the following holds.

Δ-transition systems with their morphisms constitute a **Set**-category Δ-**TS**. The full subcategory of reduced Δ-transition systems, denoted Δ-**TS**°, is concrete. The underlying functor $|\cdot| : \Delta$-**TS** $\rightarrow$ **Set** is given by $|(Q, A, T)| = T$ and $|(\sigma, \eta, \theta)| = \theta$.

4.5 Parameterized Nets and Their State Graphs

Nets over an arbitrary set I of *incidence values* may be defined as follows.

Definition 7. *A net* over the set of incidence values I *is a triple* $N = (P, A, F)$ *where* P *is a set of* places, A *is a set of* actions *or events, and* $F : P \times A \to I$ *is a flow matrix. N is event simple whenever* $(\forall x \in P)$ $F(x, a) = F(x, b)$ *implies* $a = b$. *A* morphism of nets $\phi : N \to N'$ *is a pair* $\phi = (\beta, \eta)$ *where* $\beta : P \to P'$ *and* $\eta : A' \to A$ *are linked by the constraint:* $F(p, \eta a') = F'(\beta p, a')$.

Nets over I and their morphisms form a category **Nets**(I) with the composition of morphisms defined componentwise, and pairs of identity functions as identity morphisms. It is a **Set**-category when $|N| = P$ and $|(\beta, \eta)| = \beta$. It has been noticed in [2] that the full subcategory of event simple nets, denoted **Nets°**(I), is a concrete category. Now, in order to add some dynamic to the static view of nets given so far, let us introduce types of behaviour for these nets, following [2].

Definition 8 (Types of nets). *A complete (and hence reduced)* Δ-transition system $\tau = (S, I, \tau)$ *is called a* type *(of behaviour) for (all) nets in* **Nets**(I) *- notation :* $\tau \in$ **Types**(I).

Let $\tau = (S, I, \tau)$ as above, and let P be a set. Consider a P-indexed family $\mathcal{F} = \{t^p \in \tau \mid p \in P\}$ of transitions of τ. Then, the *synchronized product* of the transitions in $\mathcal{F}$, is a function $\bigwedge \mathcal{F} : \Delta \to S^P$ defined as follows.

$$\mathcal{D}om(\bigwedge \mathcal{F}) = \bigcap \{\mathcal{D}om(\partial t^p) \mid p \in P\} \quad \text{and} \quad (\bigwedge \mathcal{F})_\delta(p) = (\partial t^p)_\delta.$$

The following lemma shows that under suitable conditions the whole family can be reconstructed from its synchronized product.

Lemma 4. *Consider* $\{t_1^p \mid p \in P\}, \{t_2^p \mid p \in P\} \subseteq \tau$ *such that* $\lambda t_1^p = \lambda t_2^p$, *for all* $p \in P$. *Then,* $\bigwedge \{t_1^p \mid p \in P\} = \bigwedge \{t_2^p \mid p \in P\}$ *implies* $t_1^p = t_2^p$, *for all* $p \in P$.

Proof. Given $p \in P$, consider t_1^p and t_2^p. Then,

$$(\partial t_1^p)_\bullet = \bigwedge \{t_1^p \mid p \in P\}_\bullet(p) = \bigwedge \{t_2^p \mid p \in P\}_\bullet(p) = (\partial t_2^p)_\bullet.$$

Thus, since $\lambda t_1^p = \lambda t_2^p$ by assumption, $t_1^p = t_2^p$ follows by (5). □

Now we are ready to formulate the notion of the state graph of a net.

Definition 9 (State graphs relative to type). *A* state graph *of a net* $N = (P, A, F)$, *relative to type* τ *is the* Δ-transition system $SG(N, \tau) = (S^P, A, T)$ *where* $T \subseteq (\Delta \to S^P) \times A$ *is defined as follows.*

$$\langle \xi, a \rangle \in T \quad \text{iff} \quad (\forall p \in P)(\exists t^p \in \tau)\lambda t^p = F(p, a) \wedge \xi = \bigwedge \{t^p \mid p \in P\} \quad (7)$$

It should be now clear that the above definition captures the intuitions put forward in Sect. 4.1, 4.2, and 4.3.

4.6   State Graphs as Duals of Nets

Given a type of nets $\tau = (S, I, \tau)$, we start now to show that τ can be turned into a schizophrenic object $\mathcal{K}^\tau = \langle \mathcal{K}^\tau_{\Delta\text{-TS}^\circ}, \mathcal{K}^\tau_{\text{Nets}^\circ(I)} \rangle$, thus providing a dual adjunction between reduced Δ-transition systems and event simple nets over I (by Prop. 1). We actually want more: we intend to show that the dual functors obtained in this way capture, respectively, the construction of state graphs of nets, and the realization of Δ-transition systems via net synthesis. We go half the way in this section, where we establish one of the two conditions imposed by Def. 2 on schizophrenic objects , namely the existence in Δ-**TS**$^\circ$ of initial lifts of families of maps ev_p evaluating net morphisms from an event simple net N to $\mathcal{K}^\tau_{\text{Nets}^\circ(I)}$ at the corresponding places p of N.

By definition, a type of nets $\tau = (S, I, \tau)$ is a complete, and therefore reduced, Δ-transition system. Thus we take $\mathcal{K}^\tau_{\Delta\text{-TS}^\circ} = \tau$. The underlying set τ of transitions of $\mathcal{K}^\tau_{\Delta\text{-TS}^\circ}$ may at the same time be considered as the underlying set of a net $\mathcal{K}^\tau_{\text{Nets}^\circ(I)} = (\tau, \{enabled\}, \mathbb{F})$ over I, where $\mathbb{F}(t, (enabled)) = \lambda t$. This net with a single event is obviously event simple. Our first observation is that for any N, the homset $\mathbf{Nets}(N, \mathcal{K}^\tau_{\text{Nets}^\circ(I)})$ represents the set of transitions of the state graph of N.

Lemma 5. *There is a bijection between* $\mathbf{Nets}(N, \mathcal{K}^\tau_{\text{Nets}^\circ(I)})$ *and* $|SG(N, \tau)|$.

Proof. Consider a net morphism $(\beta, \eta) : N \to \mathcal{K}^\tau_{\text{Nets}^\circ(I)}$. By construction of $\mathcal{K}^\tau_{\text{Nets}^\circ(I)}$ it follows that $\{\beta p \mid p \in P\}$ is a family of transitions of τ. Put $a = \eta(enabled)$. Then, $F(p, a) = F(p, \eta(enabled)) = \mathbb{F}(\beta p, enabled) = \lambda \beta p$ follows, since (β, η) is a net morphism. Thus, by definition, $\langle \bigwedge \{\beta p \mid p \in P\}, a \rangle$ is a transition of $SG(N, \tau)$. Conversely, any transition $\langle \xi, a \rangle$ of $SG(N, \tau)$ is determined by a P-indexed family of transitions of τ satisfying (7). So, given such a family $\{t^p \in \tau \mid p \in P\}$ we can define (β, η) by $\eta(enabled) = a$ and $\beta p = t^p$. Then, clearly, $(\beta, \eta) : N \to \mathcal{K}^\tau_{\text{Nets}^\circ(I)}$ is a morphism in **Nets**. Moreover, by lemma 4 it follows that there can be only one such family of transitions. Thus, the definition of (β, η) is unambiguous. Finally, by lemma 4, the correspondence is bijective.   □

The above observation paves the way to the following result, showing one of the two conditions in Def. 2.

Proposition 2. *Let N be an event simple net in* **Nets**$^\circ(I)$, *and let $\tau = (S, I, \tau)$ be a type for N. The family of evaluation maps* $\langle ev_p : \mathbf{Nets}(N, \mathcal{K}^\tau_{r\text{Nets}(I)}) \to |\mathcal{K}^\tau_{\text{Nets}^\circ(I)}|\rangle_{p \in P}$ *defined by $ev_p(\beta, \eta) = \beta p$ has an initial lift in Δ-**TS**$^\circ$, and this initial lift is isomorphic to $SG(N, \tau)$.*

Proof. Let $\mathcal{S} = \left(S^P, I^P, \mathbf{Nets}(N, \mathcal{K}^\tau_{\text{Nets}^\circ(I)}), \partial, \lambda \right)$ where ∂ and λ are defined as follows :

$$\partial(\beta, \eta) = \bigwedge \{\beta p \mid p \in P\} \tag{8}$$

$$\lambda(\beta, \eta) = \langle F(p, \eta(enabled)) \rangle_{p \in P} \tag{9}$$

Then, S is a Δ-transition system. Moreover, for each $p \in P$ the pair of projections: $\pi_p : S^P \to S$ and $\varpi_p : I^P \to I$ together with ev_p constitute a morphism $\phi_p = (\pi_p, \varpi_p, ev_p) : S \to \tau$. This follows from the observation that $\pi_p \circ (\bigwedge\{\beta p \mid p \in P\}) \prec \partial(\beta p)$ for all p in P. First, let us verify that S and the ϕ_p's ($p \in P$) constitute an initial lift of the ev_p's ($p \in P$) in the category of Δ-transition systems. Indeed, let $S' = (Q', A', T', \partial', \lambda')$ be another Δ-transition system, and assume there is given a P-indexed family of morphisms $\phi'_p = (\sigma'_p, \eta'_p, \theta'_p) : S' \to \tau$ and a function $\theta' : T' \to \mathbf{Nets}(N, \mathcal{K}^\tau_{\mathbf{Nets}^\circ(I)})$ such that $\theta'_p = ev_p \circ \theta'$ for all $p \in P$. It follows from the universal properties of products that there exist unique $\sigma' : Q' \to S^P$ and $\eta' : A' \to I^P$ such that $\sigma'_p = \pi_p \circ \sigma'$ and $\eta'_p = \varpi_p \circ \eta'$. Using the characteristic property of net morphisms and the definition of $\mathbb{F}$, one can moreover establish the relation $\varpi_p \circ \lambda \circ \theta'(t') = \varpi_p \circ \eta' \circ \lambda'(t')$. Again from the universal properties of products it follows that $\eta' \circ \lambda' = \lambda \circ \theta'$. The remaining part of (6) follows from the following observation, where $\langle \sigma'_p \rangle_{p \in P} : Q' \to S^P$ is the mediating morphism.

$$\langle \sigma'_p \rangle_{p \in P} \circ (\partial' t') \prec \bigwedge\{ev_p(\theta' t') \mid p \in P\}$$

Thus, $\phi' = (\sigma', \eta', \theta') : S' \to S$ is a morphism of Δ-transition systems and S is a lift of the family of maps ev_p. Since there are no alternative choices for σ' and η', the lift obtained is actually an initial lift N^*_τ of the considered maps. Finally, by lemma 5, the transitions of $SG(N, \tau)$ are in bijective correspondence with morphisms in $\mathbf{Nets}(N, \mathcal{K}^\tau_{\mathbf{Nets}^\circ(I)})$. Therefore, N^*_τ is isomorphic to the initial lift of $\langle ev_p : |SG((N, \tau)| \to \tau \rangle_{p \in P}$ where $ev_p(\langle f, a \rangle) = \langle f^p, F(p, a) \rangle$, with $f^p : \Delta \rightharpoonup S$ given by evaluation of f at p, i.e., $(f^p)_\delta = f_\delta p$. It follows from (9) that $\Im m(\lambda)$ is in bijection with A since N is event simple. On the other hand, it follows from the assumption that τ is complete that both S and $SG(N, \tau)$ are complete, so every state in S^P is the source of some transition in both systems. Therefore, $N^*_\tau \simeq SG(N, \tau)$, as required. Finally, N^*_τ is reduced since it is complete, hence it is the initial lift of the ev_p's in Δ-$\mathbf{TS}^\circ$, as required. $\quad\square$

Remark 1. For any net N in the larger category $\mathbf{Nets}(I)$, the above proof shows that N^*_τ is the initial lift of the evaluation maps ev_p in the category Δ-$\mathbf{TS}$.

Remark 2. We could in fact establish a more general result. Namely, we could prove that the category of Δ-transition systems admits lifts of arbitrary jointly monic structured sources, cf. [2, Prop. 12].

4.7   Synthesized Nets as Duals of Systems of Δ-Transitions

We will now show that arbitrary structured sources have initial lifts in $\mathbf{Nets}^\circ(I)$. Hence, in particular, the family of maps ev_t, for $t \in \tau$, each map ev_t evaluating Δ-transition system morphisms from S to $\mathcal{K}^\tau_{\Delta\text{-}\mathbf{TS}^\circ}$ at t, has an initial lift, and $\mathcal{K}^\tau = \langle \mathcal{K}^\tau_{\Delta\text{-}\mathbf{TS}^\circ}, \mathcal{K}^\tau_{\mathbf{Nets}^\circ(I)} \rangle$ is a schizophrenic object (Def. 2). The dual S^*_τ of a reduced Δ-transition system S in the resulting adjunction is the net version of S with places defined as regions of S and with events defined as equivalence classes of actions of S that cannot be distinguished by regions of S.

Proposition 3. *The categories* **Nets**(I), *and* **Nets**$°(I)$ *have initial lifts.*

Proof. (i) The category **Nets**(I) *has initial lifts.*
Let (P_i, A_i, F_i) be a family of nets, together with a corresponding family of mappings $\beta_i : P \to P_i$. The initial structure of net on the set P is defined as follows. The set of events is the coproduct $A = \coprod_i A_i$ with the associated injections $in_i : A_i \to A$. The incidence matrix $F : P \times A \to I$ is the "coproduct" of the matrices F_i, in the sense that

$$\forall x \in P \ \forall a_i \in A_i \quad F(x, in_i(a_i)) = F_i(\beta_i x, a_i)$$

Therefore, each (β_i, in_i) is a morphism of nets from (P, A, F) to (P_i, A_i, F_i). We now verify that this family of morphisms satisfies the required universal property. Let $\{(\beta_i', \eta_i') : (P', A', F') \to (P_i, A_i, F_i)\}$ be another family of morphisms for which there exists a mapping $\beta : P' \to P$ such that $\forall i \ \beta_i' = \beta_i \circ \beta$. We have to prove that there exists a unique morphism $(\beta, \eta) : (P', A', F') \to (P, A, F)$ such that $(\beta_i', \eta_i') = (\beta_i, in_i) \circ (\beta, \eta)$. If a solution exists, it is necessarily unique since, due to the coproduct structure of the family $\{in_i : A_i \to A\}_i$, the mapping η is fully characterized by the equations $\eta_i' = \eta \circ in_i$. We just have to check that the pair of mappings (β, η) is indeed a morphism of nets. Given $a \in A$ and $x' \in P'$, there exists a (unique) pair $\langle i, a_i \rangle$ with $a_i \in A_i$ and $a = in_i(a_i)$. Now, $F(x', \eta(a))$ $= F(x', \eta_i'(a_i)) = F_i(\beta_i'(x'), a_i) = F_i(\beta_i \circ \beta(x'), a_i) = F(\beta(x'), in_i(a_i)) = F(\beta(x'), a)$, as required.

(ii) The category **Nets**$°(I)$ *has initial lifts.*
Let (P_i, A_i, F_i) be a family of nets in **Nets**$°(I)$, together with a corresponding family of mappings $\beta_i : P \to P_i$. The initial structure of a net on the set P is defined in two stages. First, as above, we define the net (P, A, F) where $A = \coprod_i A_i$ with injections $in_i : A_i \to A$ and $\forall x \in P \ \forall a_i \in A_i \ F_i(\beta_i x, a_i) = F(x, in_i(a_i))$. It is generally not an object of **Nets**$°(I)$, we thus define the following equivalence relation on A

$$a \equiv b \quad \textbf{iff} \quad \forall x \in P \ F(x, a) = F(x, b)$$

We let $A_\equiv$ be the set of equivalence classes, with $\pi : A \to A_\equiv$ as the quotient map, and we let $F_\equiv : P \times A_\equiv \to I$ be characterized by

$$\forall x \in P \ \forall a \in A \ F_\equiv(x, \pi(a)) = F(x, a)$$

Hence $(P, A_\equiv, F_\equiv)$ is an object of **Nets**$°(I)$ and the pairs $(\beta_i, \pi \circ in_i)$ are morphisms of nets as composite of two morphisms $(P, A_\equiv, F_\equiv) \xrightarrow{(1, \pi)} (P, A, F) \xrightarrow{(\beta_i, in_i)}$ (P_i, A_i, F_i). We now verify that this family of morphisms satisfies the required universal property in the category **Nets**$°(I)$. Let $\{(\beta_i', \eta_i') : (P', A', F') \to (P_i, A_i, F_i)\}$ be another family of morphisms in **Nets**$°(I)$ for which there exists a mapping $\beta : P' \to P$ such that $\forall i \ \beta_i' = \beta_i \circ \beta$. By initiality of the lift $\{(P, A, F) \xrightarrow{(\beta_i, in_i)} (P_i, A_i, F_i)\}$, we know there exists a unique mapping $\eta : A \to A'$ such that (β, η) is a morphism of nets and $\forall i \ (\beta_i', \eta_i') = (\beta_i, in_i) \circ (\beta, \eta)$. Let $s : A_\equiv \to A$ be an arbitrary section of π, i.e. $\pi \circ s = 1_{A_\equiv}$. As $(1, s) : (P, A, F) \to (P, A_\equiv, F_\equiv)$ is a morphism of nets, so is $(\beta, \eta \circ s)$ from (P', A', F') to $(P, A_\equiv, F_\equiv)$. In order to check $\eta_i' = (\eta \circ s) \circ (\pi \circ in_i)$ we use the fact that (P', A', F') is an object of **Nets**$°(I)$ and notice that $\forall a_i \in A_i$ and $\forall x \in P'$ one has

$$
\begin{aligned}
F'(x, \eta_i'(a_i)) &= F_i(\beta_i'(x), a_i) & \textbf{by } (\beta_i', \eta_i') : (P', A', F') \to (P_i, A_i, F_i) \\
&= F_i(\beta_i \circ \beta(x), a_i) & \text{since } \beta_i' = \beta_i \circ \beta \\
&= F(\beta(x), in_i(a_i)) & \textbf{by } (\beta_i, in_i) : (P, A, F) \to (P_i, A_i, F_i) \\
&= F_{\equiv}(\beta(x), \pi \circ in_i(a_i)) & \textbf{by } (1, \pi) : (P, A_{\equiv}, F_{\equiv}) \to (P, A, F) \\
&= F(\beta(x), s \circ \pi \circ in_i(a_i)) & \textbf{by } (1, s) : (P, A, F) \to (P, A_{\equiv}, F_{\equiv}) \\
&= F'(x, \eta \circ s \circ \pi \circ in_i(a_i)) & \textbf{by } (\beta, \eta) : (P', A', F') \to (P, A, F)
\end{aligned}
$$

We now assume there exists another mapping $\overline{\eta} : A_{\equiv} \to A'$ sharing with $\eta \circ s$ the two following conditions : (i) $(\beta, \overline{\eta})$ is a morphism of nets from (P', A', F') to $(P, A_{\equiv}, F_{\equiv})$ and (ii) $\eta_i' = \overline{\eta} \circ (\pi \circ in_i)$. Then $\overline{\eta} = \eta \circ s$ because for every element $a \in A_{\equiv}$ there exists a pair $\langle i, a_i \rangle$ where $a_i \in A_i$ and $a = \pi \circ in_i(a_i)$ and thus $\overline{\eta}(a) = \overline{\eta} \circ \pi \circ in_i(a_i) = \eta \circ s \circ \pi \circ in_i(a_i) = \eta \circ s(a)$.   □

Remark 3. The family of morphisms $\{(P, A_{\equiv}, F_{\equiv}) \xrightarrow{(1, \pi)} (P, A, F)\}$ where (P, A, F) ranges over all objects of **Nets**(I), constitutes a concrete coreflection of **Nets**(I) into **Nets**°(I). We shall denote $N°$ the coreflection of a net N, i.e. the event-simple net associated with N. This general construction will serve soon to synthesize nets from regions.

At this stage, we have proven that $\mathcal{K}^\tau = \langle \mathcal{K}^\tau_{\Delta\text{-}\mathbf{TS}°}, \mathcal{K}^\tau_{\mathbf{Nets}°(I)} \rangle$ is a schizophrenic object between the respective categories $\Delta\text{-}\mathbf{TS}°$ and **Nets**°(I) of reduced Δ-transition systems and event simple nets, and also between the larger categories $\Delta\text{-}\mathbf{TS}$ and **Nets**(I) (by remark 1). Hence, by Prop. 1, one has dual adjunctions:

$$\Delta\text{-}\mathbf{TS}°(\mathcal{S}, N^*) \cong \mathbf{Nets}°(I)(N, \mathcal{S}^*)$$

$$\Delta\text{-}\mathbf{TS}(\mathcal{S}, N^*) \cong \mathbf{Nets}(I)(N, \mathcal{S}^*)$$

We examine further the former adjunction in the rest of this section, and refine the latter into a Galois connection in the next section.

Proposition 4. *Let $\tau = (S, I, \tau)$ be a type of nets, and let $\mathcal{S} = (Q, A, T)$ be a reduced Δ-transition system. The dual $\mathcal{S}_\tau^*$ of $\mathcal{S}$ w.r.t. the schizophrenic object $\mathcal{K}^\tau$ is isomorphic to the net $(P, A_{\equiv}, F_{\equiv})$ with the set of places $P = \Delta\text{-}\mathbf{TS}°(\mathcal{S}, \tau)$ –the morphisms in P are called the* regions of $\mathcal{S}$– *and with the other components derived as follows from the application $F : P \times A \to I$ given by $F((\sigma, \eta), a) = \eta(a)$: $A_{\equiv}$ is the quotient of A by the equivalence relation*

$$a \equiv b \quad \textbf{iff} \quad \forall x \in P \; F(x, a) = F(x, b)$$

and $F_{\equiv} : P \times A_{\equiv} \to I$ is the map induced from F by this quotient.

Proof. By Prop. (3), $\mathcal{S}_\tau^* = (P, E, F)°$ is the event-simple net associated with the net whose places are the regions of $\mathcal{S}$ (i.e. $P = \Delta\text{-}\mathbf{TS}°(\mathcal{S}, \tau)$), with set of events $E = \coprod_{t \in T} \{*\} \cong T$, and with flow matrix $F : \Delta\text{-}\mathbf{TS}°(\mathcal{S}, \tau) \times T \to I$ given "by evaluation on events" (i.e. $F((\sigma, \eta), q \xrightarrow{a} q') = \eta(a)$). Hence $\mathcal{T}_\tau^* = (P, T_{\equiv}, F_{\equiv})$ where $\equiv$ is the equivalence relation given by

$$(q_1 \xrightarrow{a_1} q_1') \equiv (q_2 \xrightarrow{a_2} q_2') \quad \textbf{iff} \quad (\forall (\sigma, \eta) : \mathcal{S} \to \tau) \; \eta(a_1) = \eta(a_2)$$

Since there exists at least one transition labelled by each event, ($\mathcal{S}$ is reduced), the alphabet $E_{\equiv}$ is a quotient of the alphabet of $\mathcal{S}$ where two events are identified if and only if they are indistinguishable by all regions of $\mathcal{S}$.   □

4.8   Galois Connection Derived from the Dual Adjunction Using Separation Axioms

We plan to show here that the dual adjunction between Δ-transition systems and nets, induced by the type of nets seen as a schizophrenic object, may be turned into a Galois connection by imposing on Δ-transition systems adequate *separation* axioms. More precisely, the above will be done within the framework of *initialized* Δ-transition systems and nets. By associating an initial state with a Δ-transition system we arrive at a Δ-*automaton*. Similarly, associating an initial marking with a net results in a *net system*. The separation axioms should tell us why a Δ-automaton satisfying these axioms is isomorphic to the state graph of some net system. Justifications are given in terms of regions of Δ-automata, seen as potential places of the unknown net system. The separation axioms require that regions explain why events (resp. states) differ from one another, and why Δ-parameterized actions are partially defined at each state.

Before Δ-automata and net systems are formally introduced, let us fix notation. In a deterministic Δ-transition system $\mathcal{S} = (Q, A, T)$, let $q \xrightarrow{a,\delta} q'$ for $q, q' \in Q$ and $\delta \in \Delta$, if there exists a transition $t = \langle f, a \rangle$ in T such that $q = f_\bullet$ and $q' = f_\delta$. Such a transition t, if it exists, is unique due to determinism of $\mathcal{S}$. Hence, the state q' is uniquely determined by: the state q, action a, and parameter δ). State q' is said to be *accessible* from q if $q \xrightarrow{*} q'$ where $\xrightarrow{*}$ is the reflexive and transitive closure of the immediate accessibility relation $\rightarrow = \bigcup_{a,\delta} \xrightarrow{a,\delta}$.

From now on let $\tau = (S, I, \tau)$ be a fixed type of nets in $_\Delta$-**TS**, inducing a dual adjunction $_\Delta$-**TS**$(\mathcal{S}, N^*) \cong$ **Nets**$(I)(N, \mathcal{S}^*)$ between Δ-transition systems and nets over I.

Definition 10 (Δ-automaton). *A Δ-automaton $\mathcal{A}$ consists of a deterministic and complete Δ-transition system $\mathcal{S} = (Q, A, T)$ and an initial state $q_0 \in Q$. It is assumed that every state in Q is accessible from q_0. A morphism of Δ-automata $(\sigma, \eta) : (\mathcal{S}, q_0) \to (\mathcal{S}', q_0')$ is a morphism $(\sigma, \eta) : \mathcal{S} \to \mathcal{S}'$ of the underlying Δ-transition systems, such that $\sigma q_0 = q_0'$ (preservation of the initial state).*

Δ-automata and their morphisms form a category, denoted **Aut** in the sequel.

Definition 11 (Net system). *A net system $\mathcal{N}$ over I consists of a net $N = (P, A, F)$ over I and an initial marking $M_0 : P \to S$. A morphism of net systems $(\beta, \eta) : (N, M_0) \to (N', M_0')$ is a morphism (of nets) $(\beta, \eta) : N \to N'$ such that $M_0 = M_0' \circ \beta$ (preservation of the initial marking).*

Net systems over I and their morphisms form a category denoted **NetSys**.

Definition 12 (Dual of a Net System). *The dual of a net system $\mathcal{N}$ with initial marking M_0 is the Δ-automaton $\mathcal{N}^* = (N^* \downarrow M_0, M_0)$ where $N^* \downarrow M_0$ is the induced restriction of the (complete) Δ-transition system N^* on the subset of all states that coincide with markings accessible from M_0 in N.*

Definition 13 (Dual of a Δ-automaton). *The dual of a Δ-automaton $\mathcal{A} = (\mathcal{S}, q_0)$ is the net system $\mathcal{A}^* = (\mathcal{S}^*, M_0)$ with the initial marking given by $M_0(p) = \sigma(q_0)$ for every place $p = (\sigma, \eta)$ of $\mathcal{S}^*$, i.e., for every region $(\sigma, \eta) \in {}_\Delta$-**TS**$(\mathcal{S}, \tau)$.*

Let $\mathcal{R} = \Delta\text{-}\mathbf{TS}(\mathcal{S}, \tau)$ denote in the sequel the set of regions of $\mathcal{S}$.

Proposition 5. *The dual adjunction between Δ-transition systems and nets induces a dual adjunction between Δ-automata and net systems. A Δ-automaton $\mathcal{A} = (\mathcal{S}, q_0)$ is isomorphic to its double dual $\mathcal{A}^{**}$ if and only if the following axioms of separation are satisfied in $\mathcal{S} = (Q, A, T)$.*

ESP Event Separation Property:

$$a \neq a' \quad implies \quad \exists R = (\sigma, \eta) \in \mathcal{R} \quad \eta(a) \neq \eta(a')$$

i.e., region R separates events a and a'.

SSP State Separation Property:

$$q \neq q' \quad implies \quad \exists R = (\sigma, \eta) \in \mathcal{R} \quad \sigma(q) \neq \sigma(q')$$

i.e., region R separates states q and q'.

ESSP Event/State Separation Property:

$$q \xrightarrow{a, \delta} \quad implies \quad \exists R = (\sigma, \eta) \in \mathcal{R} \quad \sigma(q) \xrightarrow{\eta(a), \delta}$$

i.e., region R inhibits action a at state q for parameter δ.

Proof. Let $\mathcal{N} = (N, M_0) = (P, A, F, M_0)$ be a net system and let $\mathcal{A} = (\mathcal{S}, q_0)$ be a Δ-automaton. Since $\mathcal{S}$ is accessible, and since, by definition, every state of $\mathcal{N}^*$ is accessible from M_0, a morphism $f : \mathcal{A} \to \mathcal{N}^*$ is just the morphism $f : \mathcal{S} \to N^*$ such that $f(q_0) = M_0$ — where we write $f(q) = \sigma(q)$ for $f = (\sigma, \eta)$. From Prop. 1, these morphisms of Δ-transition systems are in bijective correspondence with the morphisms of nets $f^\sharp : N \to \mathcal{S}^*$ such that $f^\sharp(p)(q_0) = M_0(p)$ for all $p \in P$ - where we write $f^\sharp(p) = \beta(p)$ for $f^\sharp = (\beta, \eta)$ and $f(q) = \sigma(q)$ for $f = \beta(p) = (\sigma, \eta)$ in $\Delta\text{-}\mathbf{TS}(\mathcal{S}, \tau)$. Hence the morphisms $f : \mathcal{A} \to \mathcal{N}^*$ are in bijective correspondence with the morphisms $f^\sharp : \mathcal{N} \to \mathcal{A}^*$ and we have obtained a derived adjunction $\mathbf{Aut}(\mathcal{A}, \mathcal{N}^*) \cong \mathbf{NetSys}(\mathcal{N}, \mathcal{A}^*)$ between Δ-automata and net systems. The evaluation map $Ev_\mathcal{A} : \mathcal{A} \to \mathcal{A}^{**}$ represents each state $q \in Q$ by a vector of local states $\langle \sigma(q); R = (\sigma, \eta) \in \mathcal{R} \rangle$, each event $a \in A$ by a vector of local events $\langle \eta(a); R = (\sigma, \eta) \in \mathcal{R} \rangle$, and each transition $t = \langle f, a \rangle$ by a vector of local transitions $\langle \langle \sigma \circ f, \eta(a) \rangle; R = (\sigma, \eta) \in \mathcal{R} \rangle$. Owing to the property of completeness of Δ-automata, this map is an isomorphism of Δ-automata if and only if all three separation axioms are satisfied.   □

Notice that the derived dual adjunction between Δ-automata and net systems is *no longer induced by a schizophrenic object*. Following [2], one could in fact construct a dual adjunction between Δ-automata and net systems based on a schizophrenic object. For this, it would be necessary to replace initial markings with forward closed sets of markings as was done in [2] and worse, to include some partial markings in the present context of Δ-automata. We prefer to stick here to constructions with more intuitive contents. As shown in the proof of the next proposition, places of a net system induce regions of its state graph that suffice to guarantee the validity of the separation axioms **SSP** and **ESSP**. They do not suffice in general to guarantee the validity of the axiom **ESP**. Thus, the evaluation map $Ev_{\mathcal{N}^*}$ is not always an isomorphism of net systems, and the dual adjunction we have obtained is not a Galois connection. Next proposition, put together with Prop. 5, shows that $Ev_{\mathcal{N}^*} : \mathcal{N}^* \to \mathcal{N}^{***}$ is an isomorphism of Δ-automata iff the reachable state graph of $\mathcal{N}$ enjoys separation of events.

Proposition 6. *A Δ-automaton that satisfies axiom* **ESP** *is isomorphic to the reachable state graph of a net system iff it satisfies axioms* **SSP** *and* **ESSP**.

Proof. The condition is sufficient by the preceding proposition. To show that it is necessary, suppose that $\mathcal{A}$ is a Δ-automaton with separated events and that it is isomorphic to the reachable state graph $\mathcal{N}^*$ of some net system $\mathcal{N}$. We show that regions of $\mathcal{N}^*$ derived from places of $\mathcal{N}$ suffice to guarantee that axioms **SSP** and **ESSP** are valid in $\mathcal{A} \cong \mathcal{N}^*$. Let q_1 and q_2 be distinct states of $\mathcal{A}$. The associated markings M_1 and M_2 of $\mathcal{N}$ must differ on some place p of $\mathcal{N}$. By composing the isomorphism on the right of the region (σ, η) induced by p (such that $\sigma(M) = M(p)$ and $\eta(a) = F(p, a)$ for $N = (P, A, F)$), one obtains a region of $\mathcal{A}$ that separates q_1 and q_2. Suppose now that $q \xrightarrow{a,\delta}$ in $\mathcal{A}$, then $M \not\xrightarrow{a,\delta}$ in $\mathcal{N}^*$ where M is the marking of $\mathcal{N}$ associated with q. Hence there must exist some place p of $\mathcal{N}$ such that $M(p) \not\xrightarrow{F(p,a),\delta}$ in τ. By composing the isomorphism on the right of the region (σ, η) induced by p, one obtains a region of $\mathcal{A}$ that inhibits action a at state q for parameter δ. $\qquad\square$

Proposition 7. *Consider a Δ-automaton $\mathcal{A} = (Q, A, T, q_0)$ that satisfies axioms* **SSP** *and* **ESSP**. *Then $\mathcal{A}$ satisfies axiom* **ESP** *if it has no redundant events, meaning: $a \neq a'$ implies $\{f \mid \langle f, a \rangle \in T\} \neq \{f \mid \langle f, a' \rangle \in T\}$.*

Proof. Let $a_1 \neq a_2$ be distinct events, then by the assumption of non redundancy of events, in view of the property of completeness of Δ-automata, there exist transitions $\langle f_1, a_1 \rangle \in T$ and $\langle f_2, a_2 \rangle \in T$ such that $(f_1)_\bullet = (f_2)_\bullet = q$ and $(f_1)_\delta \neq (f_2)_\delta$ for some $\delta \in \Delta$. We proceed with case analysis. If $(f_1)_\delta$ and $(f_2)_\delta$ are both defined, set $q_1 = (f_1)_\delta$ and $q_2 = (f_2)_\delta$; by separation of states, there exists a region (σ, η) of $\mathcal{A}$ such that $\sigma(q_1) \neq \sigma(q_2)$; now $\sigma(q) \xrightarrow{\eta(a_1),\delta} \sigma(q_1)$ and $\sigma(q) \xrightarrow{\eta(a_2),\delta} \sigma(q_2)$; it follows by determinism that $\eta(a_1) \neq \eta(a_2)$, hence region (σ, η) separates a_1 and a_2. If one of the $(f_i)_\delta$ is undefined, e.g. $(f_1)_\delta$ is undefined while $(f_2)_\delta$ is defined, then by event/state separation, there exists some region (σ, η) of $\mathcal{A}$ such that $\sigma(q) \not\xrightarrow{\eta(a_1),\delta}$ while $\sigma(q) \xrightarrow{\eta(a_2),\delta} q_2$ for some q_2; hence $\eta(a_1) \neq \eta(a_2)$ and (σ, η) separates a_1 and a_2. $\qquad\square$

Thus, if we require that Δ-automata have no redundant events, and if we force the identification of redundant events when computing the dual of a net system, then the dual adjunction between Δ-automata and net systems becomes a Galois connection. This is obtained by composition with the coreflection that embeds automata without redundant events into arbitrary automata. Therefore, given a *non-redundant* Δ-automaton $\mathcal{A}$, the following three conditions are equivalent.

1. $\mathcal{A}$ satisfies the separation axioms **SSP** and **ESSP**.
2. $\mathcal{A}$ is isomorphic to the dual of some net system.
3. $\mathcal{A}$ is isomorphic to its double dual.

4.9   Order Theoretic Galois Connection between Δ-Automata and Net Systems

We present in this section an alternative, order-theoretic Galois connection between Δ-automata and net systems over a *fixed* set of events. The new Galois connection coincides up to the identification of isomorphic Δ-automata with the earlier Galois connection between categories of Δ-automata and net systems, when the latter Galois connection is restricted to event separated Δ-automata. The complete symmetry of the dual constructions of state graphs and synthesized nets, which is a strong point of the categorical Galois connection, is lost in this alternative setting: state graphs of nets appear no longer as lifted versions of hom-sets. This loss of symmetry is balanced by mathematical simplicity, which is a strong point of the order theoretic Galois connection.

Given a *fixed* set of actions A and a *fixed* type of nets τ, we are thus looking for an order-theoretic Galois connection

$$\mathcal{A} \leq \mathcal{N}^* \quad \text{iff} \quad \mathcal{N} \leq \mathcal{A}^*$$

between Δ-automata $\mathcal{A} = (Q, A, T, q_0)$ and net systems $\mathcal{N} = (P, A, F, M_0)$, where $\mathcal{N}^*$ is the reachable state graph of $\mathcal{N}$ induced by the type τ and $\mathcal{A}^*$ is the net version of $\mathcal{A} = (\mathcal{S}, q_0)$ with places defined as morphisms $(\sigma, \eta) \in \Delta\text{-}\mathbf{TS}(\mathcal{S}, \tau)$ (regions of $\mathcal{A}$ w.r.t. the type τ). The main point now is to equip Δ-automata and net systems with order relations induced by adequate morphisms.

To start with, we turn the whole family[2] of Δ-automata over A into a skeletal category **Aut** where all isomorphic Δ-automata are identified, and so the whole class can be represented by an arbitrary single representative $\mathcal{A} = (Q, A, T, q_0)$. A similar construction of a skeletal category is more thoroughly described later on in section 5.4. Let **Aut** be equipped with relation $\leq$ defined as follows: $\mathcal{A} \leq \mathcal{A}'$ if there exists some morphism of Δ-automata $(\sigma, \mathrm{id}_A) : \mathcal{A} \to \mathcal{A}'$ with the identity of A as its second component. By Def. 10, every such morphism satisfies the following.

$$q \xrightarrow{a, \delta} q' \quad \text{implies} \quad \sigma(q) \xrightarrow{a, \delta} \sigma(q')$$

The proposition below states that the resulting category is in fact a poset.

Proposition 8. $(\mathbf{Aut}, \leq)$ *is an ordered set.*

Proof. When $q \xrightarrow{a, \delta} q'$ in $\mathcal{A}$, state q' is totally determined by q, a, and δ. As every state of $\mathcal{A}$ is accessible from the initial state, and since morphisms (σ, id_A) preserve the initial state and the labels, the unique morphism of this form from $\mathcal{A}$ to $\mathcal{A}$ is the identity in **Aut**, hence $\leq$ is an order relation on **Aut**.   □

---

[2] To avoid foundational problems we can assume that the states of all automata considered are members of a fixed set of *potential states*. Then, **Aut** is a small category.

It is important to remark here that $(\mathbf{Aut}, \leq)$ is not only a partial order but indeed a *complete lattice*, with greatest lower bounds computed as *synchronized products*, where the synchronized product $\bigwedge_{j\in J} \mathcal{A}_j$ of a family of Δ-automata $\mathcal{A}_j = (Q_j, A, T_j, q_{0,j})$ indexed by $j \in J$ is the Δ-automaton (Q, A, T, q_0) defined as follows. If $J = \emptyset$, take $Q = \{q_0\}$ -any singleton set, and let T be the set of all processes $\langle f, a \rangle$ such that $f_\delta = q_0$ for all $\delta \in \Delta$. If $J \neq \emptyset$, let $q_0 = \langle q_{0,j} \rangle_{j\in J}$ be a vector of the initial states, and define $Q \subseteq \prod_{j\in J} Q_j$ as the forward closure of $\{q_0\}$ w.r.t. the *synchronized transitions* defined as follows.

$$\langle q_j \rangle_{j\in J} \xrightarrow{a,\delta} \langle q'_j \rangle_{j\in J} \quad \text{iff} \quad q_j \xrightarrow{a,\delta} q'_j \text{ in } \mathcal{A}_j, \text{ for all } j \in J$$

Finally, let T be the set of all processes $\langle f, a \rangle$ in Q such that $f(\bullet) = q \in Q$, $f(\delta) = q'$ if $q \xrightarrow{a,\delta} q'$ is a synchronized transition, and $f(\delta)$ is undefined otherwise. This construction preserves determinism, completeness, and the accessibility of all states, hence it produces a Δ-automaton as required.

Nets are treated similarly.

Definition 14. *Given fixed sets A and I of actions, resp. of incidence values, let $\preceq$ denote the preorder on net systems over I with set of events A such that $\mathcal{N} \leq \mathcal{N}'$ if there exists some morphism of net systems $(\beta, id_A) : \mathcal{N} \to \mathcal{N}'$ with the identity of A as its second component. Let $(\mathbf{NetSys}, \leq)$ be the set of equivalence classes of net systems (over I and with set of events A) w.r.t. the equivalence generated by $\preceq$, equipped with relation $\leq$ defined as the quotient of $\preceq$.*

Again, the ordered set $(\mathbf{NetSys}, \leq)$ is a complete lattice, with least upper bounds computed by *amalgamation on events*: let us identify a place p in net system $\mathcal{N} = (P, A, F, M_0)$ with the relevant informations $(M_0(p), F(p, \cdot))$. Then the least upper bound $\bigvee_{j\in J}(P_j, A, F_j, M_{0,j})$ is the net system $\mathcal{N} = (P, A, F, M_0)$ with set of places $P = \cup_{j\in J} P_j$, and with initial marking and flow relations such that, for each $j \in J$ and for each $p \in P_j$: $M_0(p) = M_{0,j}(p)$ and $F(p, a) = F_j(p, a)$ for all $a \in A$.

The order relations have been defined, so it remains to define precisely the state graph $\mathcal{N}^*$ of a net system $\mathcal{N}$, and the net version $\mathcal{A}^*$ of a Δ-automaton $\mathcal{A}$. The reachable state graph of an *atomic* net system $\mathcal{N}_p = (\{p\}, A, F, M_0)$ w.r.t. the fixed type of nets $\tau = (S, I, \tau)$ may be defined as follows. This is the unique Δ-automaton $\mathcal{N}_p^* = (Q, A, T, M_0)$ such that: $Q \subseteq S^{\{p\}}$, the maps $\sigma_p : Q \to S : \sigma_p(M) = M(p)$ and $\eta_p : A \to I : \eta_p(a) = F(p, a)$ define a morphism of Δ-transition systems $(\sigma_p, \eta_p) : (Q, A, T) \to \tau$, and $\sigma_p(Q)$ is the largest subset of states that may be accessed from $M_0(p)$ by executing processes (f, ι) in τ with labels $\iota \in \eta_p(A)$.

The reachable state graph of a net system $\mathcal{N} = (P, A, F, M_0)$ is then the synchronized product of the reachable state graphs of its atomic subnet systems $\mathcal{N}_p = (\{p\}, A, F_p, M_{0,p})$, such that $M_{0,p}(p) = M_0(p)$ and $F_p(p, a) = F(p, a)$ for all $a \in A$. Thus:

$$\mathcal{N} = \bigvee \{\mathcal{N}_p \,|\, p \in P\} \quad \text{and} \quad \mathcal{N}^* = \bigwedge \{\mathcal{N}_p^* \,|\, p \in P\}$$

Symmetrically, one may define net systems $\mathcal{A}^*$ synthesized from Δ-automata $\mathcal{A} = (\mathcal{S}, q_0)$ where $\mathcal{S} = (Q, A, T)$ as follows: $\mathcal{A}^*$ is the net system one obtains by gluing on events $a \in A$ all net systems $\mathcal{N}_p = \mathcal{A}_p^* = (\{p\}, A, F_p, M_{0,p})$ that derive from regions $p = (\sigma, \eta) \in \Delta\text{-}\mathbf{TS}(\mathcal{S}, \tau)$ such that:

$$F_p(a) = \eta(a) \ (\in I) \quad \text{and} \quad M_{(0,p)}((\sigma, \eta)) = \sigma(q_0) \ (\in S)$$

Thus, $\mathcal{A}^* = \bigvee \{\mathcal{A}_p^* \mid p \in \Delta\text{-}\mathbf{TS}(\mathcal{S}, \tau)\}$.

Proposition 9. *The correspondence that sends a Δ-automaton $\mathcal{A}$ to the synthesized net system $\mathcal{A}^*$ and a net system $\mathcal{N}$ to the reachable state graph $\mathcal{N}^*$, is a Galois connection between the ordered sets $(\mathbf{Aut}, \leq)$ and $(\mathbf{NetSys}, \leq)$, i.e. :*

1. $\mathcal{A}_1 \leq \mathcal{A}_2$ *implies* $\mathcal{A}_2^* \leq \mathcal{A}_1^*$,   *for $\mathcal{A}_1, \mathcal{A}_2 \in \mathbf{Aut}$,*
2. $\mathcal{N}_1 \leq \mathcal{N}_2$ *implies* $\mathcal{N}_2^* \leq \mathcal{N}_1^*$,   *for $\mathcal{N}_1, \mathcal{N}_2 \in \mathbf{NetSys}$,*
3. $\mathcal{A} \leq \mathcal{A}^{**}$,   *for $\mathcal{A} \in \mathbf{Aut}$.*
4. $\mathcal{N} \leq \mathcal{N}^{**}$,   *for $\mathcal{N} \in \mathbf{NetSys}$.*

Proof. 1. Let $\mathcal{A}_1 \leq \mathcal{A}_2$ and let $(\sigma, \mathrm{id}_A) : \mathcal{A}_1 \to \mathcal{A}_2$ be the associated morphism, then by composition with (σ, id_A), each region (σ_2, η_2) of $\mathcal{A}_2$ (i.e. each morphism from the underlying Δ-transition system to the type of nets τ) induces a region $(\sigma_2 \circ \sigma, \eta_2)$ of $\mathcal{A}_1$ with the same flow relations, showing that $\mathcal{A}_2^* \leq \mathcal{A}_1^*$.

2. Let $\mathcal{N}_1 \leq \mathcal{N}_2$, and assume w.l.o.g. that $P_1 \subseteq P_2$ (where P_i is the set of places of $\mathcal{N}_i$), then $\mathcal{N}_1^* = \bigwedge \{\mathcal{N}_{1p}^* \mid p \in P_1\} \geq \bigwedge \{\mathcal{N}_{2p}^* \mid p \in P_2\} = \mathcal{N}_2^*$.

3. Suppose $q' \xrightarrow{\delta, a} q''$ in $\mathcal{A}$. Let $p = (\sigma, \eta)$ be a region of $\mathcal{A}$, i.e. a morphism from the underlying Δ-transition system to the type of nets τ, and moreover a place of $\mathcal{A}^*$. As (σ, η) is a morphism of Δ-transition systems, $\sigma(q') \xrightarrow{\delta, \eta(a)} \sigma(q'')$ in τ. By definition of the reachable state graph of the atomic net system $\mathcal{A}_p^*$, there must exist markings M' and M'' of $\mathcal{A}_p^*$ such that $\sigma_p(M') = M'(p) = \sigma(q')$, $\sigma_p(M'') = M''(p) = \sigma(q'')$, and $M' \xrightarrow{\delta, a} M''$ in $\mathcal{A}^{**}$. The map φ_p sending each state q of $\mathcal{A}$ to the unique marking M of $\mathcal{A}_p^*$ such that $\sigma_p(M) = M(p) = \sigma(q)$ defines therefore a morphism $(\varphi_p, \mathrm{id}_A) : \mathcal{A} \to \mathcal{A}_p^{**}$. Thus $\mathcal{A} \leq \mathcal{A}_p^{**}$, and $\mathcal{A} \leq \mathcal{A}^{**}$ as $\mathcal{A}^{**} = \bigwedge \{\mathcal{A}_p^{**} \mid p \in \mathcal{R}\}$ where $\mathcal{R}$ is the set of regions of $\mathcal{A}$ (see Def. 13).

4. Finally, given a net system $\mathcal{N}$, the map β sending each place p of $\mathcal{N}$ to the region (σ_p, η_p) of $\mathcal{N}^*$, such that $\sigma_p(M) = M(p)$ and $\eta_p(a) = F(p, a)$, defines a morphism of net systems $(\beta, \mathrm{id}_A) : \mathcal{N} \to \mathcal{N}^{**}$, hence $\mathcal{N} \leq \mathcal{N}^{**}$. $\square$

Let us briefly recall for further use some basic facts about order-theoretic Galois connections. A Galois connection between two ordered sets X and Y consists of a pair of decreasing maps $(\cdot)^* : X \to Y$ and $(\cdot)^* : Y \to X$ such that

$$x \leq x^{**} \quad \text{and} \quad y \leq y^{**}, \quad \text{for all } x \in X, y \in Y.$$

Equivalently, it may be defined as a pair of maps $(\cdot)^* : X \to Y$ and $(\cdot)^* : Y \to X$ such that

$$\forall x \in X \ \forall y \in Y \quad x \leq y^* \Leftrightarrow y \leq x^*$$

Some important properties of Galois connections are listed below.

GC1 $x^* = x^{***}$ and $y^* = y^{***}$, for all $x \in X$, $y \in Y$.

GC2 Both $(\cdot)^{**}$ maps are closure operators;

GC3 Images via $(\cdot)^*$ coincide with sets of closed elements:

$$\{x^* \mid x \in X\} = \{y \in Y \mid y = y^{**}\}$$
$$\{y^* \mid y \in Y\} = \{x \in X \mid x = x^{**}\}$$

GC4 The two maps $(\cdot)^*$, restricted on the respective subsets of closed elements, are inverse bijections inducing a dual isomorphism between the ordered sets X^* and Y^*;

GC5 For all $x_i \in X$ and $y_i \in Y$ the following are satisfied:

$$(\textstyle\bigvee_i x_i)^* = \textstyle\bigwedge_i x_i^* \text{ and } (\textstyle\bigvee_i y_i)^* = \textstyle\bigwedge_i y_i^*$$
$$(\textstyle\bigwedge_i x_i)^* \geq \textstyle\bigvee_i x_i^* \text{ and } (\textstyle\bigwedge_i y_i)^* \geq \textstyle\bigvee_i y_i^*$$

In the end of the section, we examine some consequences of the above properties for the Galois connection between **Aut** and **NetSys** established in Prop. 9.

Proposition 10. *A Δ-automaton $\mathcal{A} = (\mathcal{S}, q_0)$ is the reachable state graph of some net system, relatively to type τ, if and only if there exists a subset of regions $\mathcal{P} \subseteq \Delta\text{-}\mathbf{TS}(\mathcal{S}, \tau)$ such that the following axioms of separation are satisfied in $\mathcal{S}$.*

SSP State Separation Property:

$$q \neq q' \quad implies \quad \exists p = (\sigma, \eta) \in \mathcal{P} \quad \sigma(q) \neq \sigma(q').$$

ESSP Event/State Separation Property:

$$q \xnrightarrow{a,\delta} \quad implies \quad \exists p = (\sigma, \eta) \in \mathcal{P} \quad \sigma(q) \xnrightarrow{\eta(a),\delta}.$$

When these two axioms are satisfied, $\mathcal{P}$ is said to be an admissible *subset of regions of $\mathcal{A}$, and $\mathcal{A} = (\bigvee_{p \in \mathcal{P}} A_p^*)^*$.*

Proof. Recall that $\mathcal{A} \leq \mathcal{A}^{**}$ holds due to morphism $(\varphi, \mathrm{id}_A) : \mathcal{A} \to \mathcal{A}^{**}$ such that $\varphi(q)(p) = \sigma(q)$ for $p = (\sigma, \eta) \in \Delta\text{-}\mathbf{TS}(\mathcal{S}, \tau)$. This morphism, which is the unique morphism of Δ-automata from $\mathcal{A}$ to $\mathcal{A}^{**}$, acts as a bijection on states if and only if **SSP** is satisfied in $\mathcal{S}$ for some subset of regions $\mathcal{P}'$. Also recall that $\varphi(q) \xrightarrow{a,\delta}$ in $\mathcal{A}^{**}$ if and only if $\varphi_p(q) \xrightarrow{a,\delta}$ in A_p^{**} for each region $p = (\sigma, \eta)$, with $\varphi_p(q)(p) = \sigma(q) = \varphi(q)(p)$. Therefore, if φ is actually a bijection on states, $\mathcal{A}$ and $\mathcal{A}^{**}$ are isomorphic if and only if **ESSP** is satisfied in $\mathcal{S}$ for some subset of regions $\mathcal{P}''$. Suppose $\mathcal{P} \subseteq \Delta\text{-}\mathbf{TS}(\mathcal{S}, \tau)$ is an admissible subset of regions. Then there exists a unique morphism of Δ-automata from $\mathcal{A}$ to $\bigwedge_{p \in \mathcal{P}} A_p^{**}$, derived from (φ, id_A) by restricting $\varphi(q) : \Delta\text{-}\mathbf{TS}(\mathcal{S}, \tau) \to \mathcal{S}$ to $\varphi_{\mathcal{P}}(q) : \mathcal{P} \to \mathcal{S}$ for all $q \in Q$. It is obvious from the first part of the proof that $(\varphi_{\mathcal{P}}, \mathrm{id}_A)$ is an isomorphism, hence $\mathcal{A} = \bigwedge_{p \in \mathcal{P}} A_p^{**}$ in **Aut** and by **GC5**, $\mathcal{A} = (\bigvee_{p \in \mathcal{P}} A_p^*)^*$. By **GC3**, this is equivalent to say that $\mathcal{A}$ is the reachable state graph of some net system. $\square$

Let us elaborate further on the interpretation of **GC4** and **GC5** in terms of Δ-automata and net systems. An Δ-automaton $\mathcal{A}$ equal to its double dual $\mathcal{A}^{**}$ is said to be *separated*. A net system $\mathcal{N}$ isomorphic to its double dual $\mathcal{N}^{**}$ is said to be *saturated*. By **GC4**, taking the induced restrictions on closed elements of the two maps $(\cdot)^*$ defined respectively on Δ-automata and net systems provides a dual isomorphism between separated Δ-automata and saturated net systems. As $(\bigvee_i \mathcal{N}_i)^* = \bigwedge_i \mathcal{N}_i^*$ by **GC5**, the reachable state graph of an amalgamated sum of net systems is the synchronized product of their reachable state graphs. As $(\bigwedge_i \mathcal{A}_i)^* \geq \bigvee_i \mathcal{A}_i^*$ by **GC5**, the net system synthesized from a synchronized product of Δ-automata is generally bigger than the amalgamated sum of the net systems synthesized from the components.

Observe now that if $\mathcal{A} = \bigwedge \mathcal{A}_i$ then by **GC5** and the definition of Galois connections:

$$\mathcal{A} = \overset{(1)}{\bigwedge \mathcal{A}_i \leq} \overset{(2)}{(\bigwedge \mathcal{A}_i)^{**} \leq} \overset{(3)}{(\bigvee_i \mathcal{A}_i^*)^* = \bigwedge_i \mathcal{A}_i^{**} \geq} \bigwedge_i \mathcal{A}_i = \mathcal{A}$$

Relation (1) is equality if and only if $\mathcal{A}$ is separated. Relation (3) is equality if each $\mathcal{A}_i$ is separated, and in this case, relations (1) and (2) are also equalities. It may be observed that relation (2) is equality whenever the Δ-automata $\mathcal{A}_i$ form an *orthogonal* family in the sense that $\bigwedge_i \mathcal{A}_i^{**} = \bigwedge_i \mathcal{A}_i$. Therefore, given $\mathcal{A} = \bigwedge_i \mathcal{A}_i$ such that $\mathcal{A}_i$'s form an orthogonal family, then the Δ-automaton $\mathcal{A}$ must be the reachable state graph of the amalgamated sum of net systems $\mathcal{A}_i^*$ even though the components $\mathcal{A}_i$ need *not* be separated Δ-automata. This remark may be of interest for synthesizing net systems from automata specified as synchronized products of "smaller" automata *without computing explicitly synchronized products* (this was considered in [10] in the context of ordinary automata and pure Petri nets), provided the following admits a positive answer.

Problem 1. Construct a process that transforms a family $\mathcal{A}_i$ of automata into an orthogonal family $\mathcal{A}_j'$ such that $\bigwedge_j \mathcal{A}_j' = \bigwedge_i \mathcal{A}_i$ whenever $\bigwedge_i \mathcal{A}_i$ is a separated automaton.

5   Nets with Complex Transitions

In this section we put forward a new extension of transition graphs of nets of an arbitrary type. Actions/transitions/events of nets are often seen as being *atomic*. Here, we are also interested in *complex* actions formed from the atomic by algebraic means. This, implicitly, is done when considering Petri nets with the *step* semantics: *sums* of actions, interpreted as multisets, are used as labels of step transitions. The idea is to generalize on this example so that, for instance, one may also take the sequential composition of actions into account. This leads us to define algebras of actions as initial (Σ, E)-algebras over a finite set of generators (Σ is a finite signature and E is a finite set of equations between Σ-terms, with implicit universal quantification of variables). Before we state a formal definition of (Σ, E)-transition systems and the corresponding generalization of nets, let us spend some time to introduce motivating examples.

5.1   Complex Actions of Petri Nets

General Petri nets may be seen as nets with type $\tau = (\mathbb{N}, \mathbb{N} \times \mathbb{N}, \tau)$, where $n \xrightarrow{x} n'$ is a transition in τ for $x = (k, l)$ if and only if $n \geq k$ and $n' = n - k + l$. In order to render the step semantics of Petri nets, it suffices to turn $\mathbb{N} \times \mathbb{N}$ into a commutative monoid with neutral element $\mathbb{1} = (0, 0)$ and with composition $+$ given by $(k, l) + (k', l') = (k + l, k' + l')$. Actually, if we are given a Petri net $N = (P, A, F)$, a marking M of N and a finite multiset μ over A (formally, $\mu : A \to \mathbb{N}$, $M : P \to \mathbb{N}$, and $F(p, a) \in \mathbb{N} \times \mathbb{N}$ for $p \in P$ and $a \in A$), then $M[\mu\rangle M'$ is a step of N if and only if there exists for each place $p \in P$ a corresponding transition in τ from source $M(p)$ to target $M'(p)$ labelled with the sum of $\mu(a) \times F(p, a)$ for all $a \in A$. It is easily seen that $\mathbb{N} \times \mathbb{N}$ equipped with the above operations in $\Sigma_1 = \{\mathbb{1}, +\}$ is in fact the free commutative monoid over two generators, hence it is isomorphic to the initial (Σ_1, E_1)-algebra $\mathcal{F}_{\Sigma_1, E_1}(\{\bullet, \diamond\})$ where E_1 is the set of equations:

$$\mathbb{1} + x = x \qquad\qquad x + (y + z) = (x + y) + z \qquad\qquad x + y = y + x$$

A minor extension to the above consists of adding a third unary operator $(\cdot)^\sim$ together with the following equations.

$$(x^\sim)^\sim = x \qquad\qquad\qquad (x + y)^\sim = x^\sim + y^\sim$$

Let Σ_2 and E_2 be the resulting sets of operations and equations. The initial algebra $\mathcal{F}_{\Sigma_2, E_2}(\{\bullet\})$ over one generator is still isomorphic to $\mathbb{N} \times \mathbb{N}$ with $(k, l)^\sim = (l, k)$. There are actually two isomorphisms, sending $\bullet$ to $(0, 1)$ and to $(1, 0)$, respectively. Equipped with a structure of this algebra on labels, the type τ of Petri nets induces now a different "step semantics" of Petri nets in which each action a has an implicit dual $a^\sim$. For atomic a its effect on a place p is defined by $F(p, a^\sim) = F(p, a)^\sim$. In other words the input and output arcs of a are reversed in $a^\sim$. Thus, for instance, the complex actions $(a + b^\sim)^\sim$ and $a^\sim + b$ are equal.

A subsequent, more ambitious, extension consists of introducing sequential composition in the algebra of actions. This may be done by defining the following operation on $\mathbb{N} \times \mathbb{N}$:

$$(k, l); (k', l') = \text{ if } l \geq k' \text{ then } (k, l + l' - k') \text{ else } (k + k' - l, l')$$

The type τ of Petri nets, with set of labels $\mathbb{N} \times \mathbb{N}$ equipped with the operations in $\Sigma_3 = \{\mathbb{1}, \sim, +, ;\}$, induces a new semantics of Petri nets, yet unexplored. Consider two atomic actions a and b of net $N = (P, A, F)$, hence $a, b \in A$, and consider the complex action $a ; b$. The idea is that $a ; b$ can be fired at marking M and lead to marking M' if and only if there exists for each place $p \in P$ some transition in τ from source $M(p)$ to target $M'(p)$ labelled with $F(p, a) ; F(p, b)$. This is actually the case if and only if $M[a\rangle M''$ and $M''[a\rangle M'$ for some intermediate marking M'', hence the extension seems to bring nothing new. However, by applying a similar definition of firing to all complex actions, we obtain complex transitions, such as $M[((a ; b) + c)\rangle M''$, that cannot be inferred from ordinary step transitions. A question that must be answered at this stage is the following: formally speaking, *what is a complex action* ? The answer we propose is to define the complex actions of net $N = (P, A, F)$ as the elements of a free (Σ, E)-algebra $\mathcal{F}_{\Sigma,E}(A)$ generated by A. We also insist that the actions of a type τ of such nets also form a free (Σ, E)-algebra.

We have already argued that $\mathbb{N} \times \mathbb{N}$ can be seen as a free algebra in the first two examples. We claim that $\mathbb{N} \times \mathbb{N}$ is actually isomorphic to $\mathcal{F}_{\Sigma_3,E_3}(\{\bullet\})$ when E_3 is formed of all axioms in E_2 plus the following:

$$(x \; ; \; x^\sim)^\sim = x \; ; \; x^\sim \tag{10}$$

$$(x \; ; \; x^\sim) + (x^\sim \; ; \; x) = x + x^\sim \tag{11}$$

$$(x + y) \; ; \; (x^\sim + y^\sim) = (x \; ; \; x^\sim) + (y \; ; \; y^\sim) \tag{12}$$

$$x \; ; \; ((x^\sim \; ; \; x) + y) = x + y \tag{13}$$

$$x \; ; \; (x^\sim + y) = (x \; ; \; x^\sim) + y \tag{14}$$

$$(x + y) \; ; \; y^\sim = x + (y \; ; \; y^\sim) \tag{15}$$

$$(x + (y \; ; \; y^\sim)) \; ; \; y = x + y \tag{16}$$

$$(x + (y \; ; \; y^\sim)) \; ; \; (z + (y \; ; \; y^\sim)) = (x \; ; \; z) + (y \; ; \; y^\sim) \tag{17}$$

$$(x + y) \; ; \; (y^\sim + z) = (x \; ; \; z) + (y \; ; \; y^\sim) \tag{18}$$

We already saw that $\mathcal{F}_{\Sigma_2,E_2}(\{\bullet\})$ is isomorphic to $\mathbb{N} \times \mathbb{N}$ with the operations $\mathbb{1} = (0,0)$, $(k,l)^\sim = (l,k)$, and $(k,l) + (k',l') = (k + k', l + l')$. We claim that equations 10, 11, and 12 determine $(x \; ; \; x^\sim)$ for every x in $\mathbb{N} \times \mathbb{N}$. To see this, we observe first that $(0,0) \; ; \; (0,0) = (0,0)$ by equation 11, and that $(x \; ; \; x^\sim)$ is determined inductively from $(0,1) \; ; \; (1,0)$ and $(1,0) \; ; \; (0,1)$ for $x \neq (0,0)$ by equation 12. Now $((0,1) \; ; \; (1,0)) + ((1,0) \; ; \; (0,1)) = (1,1)$ by equation 11, and $(0,1) \; ; \; (1,0) = (i,i)$ and $(1,0) \; ; \; (0,1) = (j,j)$ for some unknown $i,j \in \mathbb{N}$ by equation 10. There are two solutions: $i = 0, j = 1$ or $i = 1, j = 0$. Both solutions are equivalent up to the isomorphism $^\sim$ on $\mathbb{N} \times \mathbb{N}$. We choose the first solution, hence $(0,1) \; ; \; (1,0) = (0,0)$ and $(1,0) \; ; \; (0,1) = (1,1)$. Thus $(k,l) \; ; \; (l,k) = (k,k)$ for all k,l by equation 12.

Given this, the equations 13, 14, 15, 16 determine $x \; ; \; y$ in the respective cases: $(x = (k,l), y = (l + k', l + l'))$, and $(x = (k,l), y = (l + k', k + l'))$, and $(x = (k + k', l + l'), y = (l', k'))$, and finally $(x = (k + k', l + k'), y = (k', l'))$, consistently with the definition of the operation $;$ on $\mathbb{N} \times \mathbb{N}$.

The two cases in which $x \; ; \; y$ is left undetermined by the equations 10 to 16, with $x = (k,l)$ and $y = (k',l')$, are when $l' < k \, \& \, l' < l \leq k'$ or $k < l' \, \& \, k < k' \leq l$. The remaining equations 17 and 18 should therefore impose adequate values on the expressions $(j + k, j + l) \; ; \; (j + l + m, j)$ and $(j, j + l + m) \; ; \; (j + l, j + k)$. Seeing that $(j,j) = (j,j) \; ; \; (i,j)$, equation 17 reduces the problem to impose adequate values on $(k,l) \; ; \; (l + m, 0)$ and $(0, l + m) \; ; \; (l, k)$. Using $y = (0,l)$ in equation 18, one reduces the problem to impose adequate values on $(k,0) \; ; \; (m,0)$ and $(0,m) \; ; \; (0,k)$. Now $(k,0) \; ; \; (m,0) = (k + m, 0)$ is forced by equation 13 (since $(0,k) \; ; \; (k,0) = (0,0)$), and $(0,m) \; ; \; (0,k) = (0, m + k)$ is forced by equation 16 (since $(k,0) \; ; \; (0,k) = (0,0)$). Therefore, $\mathcal{F}_{\Sigma_3,E_3}(\{\bullet\})$ is isomorphic to $\mathbb{N} \times \mathbb{N}$.

Note that equations 15 and 16 are dual to equations 13 and 14 and may hence, together with equation 11, be replaced with a single equation $(x \ ; y)\tilde{} = y\tilde{} \ ; x\tilde{}$.

Remark 4. It is worth noting that $(x \ ; y) \ ; z = x \ ; (y \ ; z)$ is valid in $\mathcal{F}_{\Sigma_3, E_3}(\{\bullet\})$ although it is not provable from the equations in E_3. This has some importance for action algebras: the sequential composition of actions in $\mathcal{F}_{\Sigma_3, E_3}(A)$ will *not* be associative, as one may wish, unless equation $(x \ ; y) \ ; z = x \ ; (y \ ; z)$ is explicitly added to E_3.

The proposed extension of Petri net semantics diverges significantly from Meseguer and Montanari's view of Petri nets as monoids [14]. In our approach, there is no need to impose the functoriality law $(a + b); (c + d) = (a; c) + (b; d)$. More importantly, the equational axioms we suggest bear on labels of computations, and not directly on computations as do the axioms of monoidal categories. This allows to embed (extended) state graphs of Petri nets into arbitrary graphs labelled with complex actions and hence to consider the Petri net realization problem for such graphs without assuming any internal structure of their nodes.

5.2   Concurrent Inhibitor C/E-Nets

As a second example, let $\Sigma = \{\mathbb{1}, +\}$ where $\mathbb{1}$ and $+$ have arities 0 and 2, respectively. Let E be the set of axioms of commutative and idempotent monoids, i.e. the set of equations:

$$\mathbb{1} + x = x \qquad x + (y + z) = (x + y) + z \qquad x + y = y + x \qquad x + x = x$$

For any finite set A, the initial algebra $\mathcal{F}_{\Sigma, E}(A)$ is obviously isomorphic to the powerset algebra of A. Now consider the type of Concurrent Inhibitor C/E-nets shown in Fig. 3, where the transitions are labelled in the free (Σ, E)-algebra $\mathcal{F}_{\Sigma, E}(I)$ generated from the incidence values in set $I = \{inhibit, input, output\}$.

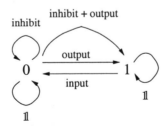

Fig. 3. The type of Inhibitor C/E-nets

An Inhibitor C/E-net may be defined accordingly as a net (P, A, F) equipped with a flow matrix $F : P \times A \to \{inhibit, input, output, \mathbb{1}\}$, where $F(p, a) = inhibit$ means that condition p inhibits event a, and $F(p, a) = \mathbb{1}$ means that condition p and event a are unrelated (see section 1). The firing rule for Inhibitor C/E-nets induced by the above type τ is as follows. For any subset of events $E \subseteq A$, $M[E\rangle M'$ is a step of the net (P, A, F) if and only if, for every condition $p \in P$, there exists in the (Σ, E)-transition system τ a corresponding transition

$$M(p) \xrightarrow{F(p,E)} M'(p)$$

where $F(p, \emptyset) = \mathbb{1}$, $F(p, \{a\}) = F(p, a)$, and $F(p, E_1 \cup E_2) = F(p, E_1) + F(p, E_2)$. According to this rule, two events may be fired in one step even though they share some input condition, or some output condition. Given a condition p, a step may also include simultaneously an event a_1 with p as an inhibiting condition and an event a_2 with p as an output condition (but definitely not as an input condition). Many other variations may of course be envisaged.

5.3 (Σ, E)-Transition Systems and (Σ, E)-Nets

We proceed now with general definitions. In the sequel, Σ is a finite signature, E is a finite set of equations between Σ-terms (over some set of universally quantified variables), and $\mathcal{F}_{\Sigma,E}(A)$ is the initial (Σ, E)-algebra over a finite set A of generators. Recall that any map $\eta : A \to \mathcal{F}_{\Sigma,E}(A')$ extends to a unique morphism of (Σ, E)-algebras $\eta^\star : \mathcal{F}_{\Sigma,E}(A) \to \mathcal{F}_{\Sigma,E}(A')$ such that $\eta^\star(a) = \eta(a)$ for all $a \in A$ (where $a \in A$ is identified with $a \in \mathcal{F}_{\Sigma,E}(A)$ by abuse of notation).

Definition 15 (Transition System). *A (Σ, E)-transition system over A is a subset T of $Q \times \mathcal{F}_{\Sigma,E}(A) \times Q$, where Q is a (possibly infinite) set of states. Transitions in T are represented in the form $t = q \xrightarrow{e} q'$. Let $\lambda : T \to \mathcal{F}_{(\Sigma,E)}(A)$, and $\partial^0, \partial^1 : T \to Q$ be projection maps. i.e., $\lambda(t) = e$, $\partial^0(t) = q$ and $\partial^1(t) = q'$ for $t = q \xrightarrow{e} q'$. The (Σ, E)-transition system is* deterministic *if λ and ∂^0 are jointly monic. It is* reduced *if it is deterministic and if moreover, $\forall q \in Q \; \exists t \in T$ $q \in \{\partial^0(t), \partial^1(t)\}$ and $\forall a \in A \; \exists t \in T \; \lambda(t) = a$.*

A morphism of (Σ, E)-transition systems from $T_1 \subseteq Q_1 \times \mathcal{F}_{\Sigma,E}(A_1) \times Q_1$ to $T_2 \subseteq Q_2 \times \mathcal{F}_{\Sigma,E}(A_2) \times Q_2$ is a pair (σ, η) of maps $\sigma : Q_1 \to Q_2$, $\eta : A_1 \to \mathcal{F}_{\Sigma,E}(A_2)$ such that there exists a map $\theta : T_1 \to T_2$ making the following diagrams commute.

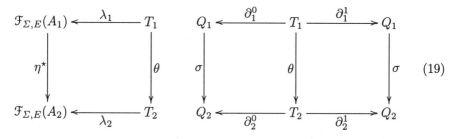

(Σ, E)-transition systems and their morphisms form a category (Σ,E)-**TS** where composition of morphisms is defined as $(\sigma_2, \eta_2) \circ (\sigma_1, \eta_1) = (\sigma_2 \circ \sigma_1, \eta_2^\star \circ \eta_1)$. The full subcategory (Σ,E)-**TS**$^\circ$ of reduced (Σ, E)-transition systems is concrete.

Let I be a finite set of incidence values.

Definition 16 (Nets). *A (Σ, E)-net over I is a triple $N = (P, A, F)$ where A is a finite set of atomic actions, P is a set of places, and $F : P \times A \to \mathcal{F}_{\Sigma, E}(I)$. A (Σ, E)-net $N = (P, A, F)$ is reduced if the following holds for all complex actions $e, e' \in \mathcal{F}_{\Sigma, E}(A)$: $e \neq e'$ implies $F^\star(p, e) \neq F^\star(p, e')$, for some $p \in P$. Above, $F^\star(p, \cdot) : \mathcal{F}_{\Sigma, E}(A) \to \mathcal{F}_{\Sigma, E}(I)$ is the Σ-homomorphism induced by $F(p, \cdot)$.*

Given (Σ, E)-nets $N_1 = (P_1, A_1, F_1)$ and $N_2 = (P_2, A_2, F_2)$, both over I, a morphism $(\beta, \eta) : N_1 \to N_2$ is a pair of maps $\beta : P_1 \to P_2$, $\eta : A_2 \to \mathcal{F}_{\Sigma, E}(A_1)$ such that, for all $p_1 \in P_1$ and $a_2 \in A_2$, $F_2(\beta p_1, a_2) = F^\star(p_1, \eta a_2)$.

(Σ, E)-nets over I and their morphisms form a category (Σ, E)-**Nets**(I), with composition of morphisms given by $(\beta', \eta') \circ (\beta, \eta) = (\beta' \circ \beta, \eta^\star \circ \eta')$. The full subcategory (Σ, E)-**Nets**$^\circ(I)$ of the reduced (Σ, E)-nets over I is concrete, and it is actually the largest concrete subcategory of (Σ, E)-**Nets**(I). Let us shift from structure to semantics and define now the dynamics of (Σ, E)-nets.

Definition 17 (Type of Nets). *A (Σ, E)-type of nets over I is a reduced (Σ, E)-transition system τ over I. Hence, $\tau \subseteq S \times \mathcal{F}_{\Sigma, E}(I) \times S$ for some set S defining a possible range of values for places of (Σ, E)-nets over I.*

Let (Σ, E)-**Types**(I) denote the class of all (Σ, E)-types of nets over I.

Definition 18 (State Graph). *Given a (Σ, E)-net $N = (P, A, F)$ over I and a type of nets τ in (Σ, E)-**Types**(I), the state graph $SG(N, \tau)$ of N relative to $\tau = (S, A, \tau)$ is the (Σ, E)-transition system over A with set of states S^P and with the set T of transitions T defined as follows.*

$$M \xrightarrow{e} M' \in T \quad \text{iff} \quad M(p) \xrightarrow{F^\star(p, e)} M'(p) \in \tau, \quad \text{for all } p \in P$$

for all $M, M' \in S^P$ and for all $e \in \mathcal{F}_{\Sigma, E}(A)$.

The reader may wish to verify that the above definition extends indeed the usual construction of state graphs of nets. In order to complete the machinery, we finally propose a converse construction of (Σ, E)-nets from (Σ, E)-transition systems, based on the regions-as-places analogy.

Definition 19 (Regions and Synthesis). *Given a (Σ, E)-transition system T over A, and given a type of nets $\tau \in (\Sigma, E)$-**Types**(I), a region of T relative to τ is a morphism $(\sigma, \eta) : T \to \tau$ in (Σ, E)-**TS**. The net $SN(T, \tau)$ synthesized from T with respect to τ is defined as $SN(T, \tau) = (P, A, F)$ with all regions as places: $P = (\Sigma, E)$-**TS**(T, τ), and with flow relation given by $F((\sigma, \eta), a) = \eta(a)$, so that $\eta(a) \in \mathcal{F}_{\Sigma, E}(I)$.*

So far so good, and one may expect that for each type $\tau \in (\Sigma, E)$-**Types**(I), a dual adjunction between (Σ, E)-**TS**$^\circ$ and (Σ, E)-**Nets**$^\circ(I)$ would emerge by considering τ as a schizophrenic object, whose net counterpart is the (Σ, E)-net over I defined as $\mathcal{K}^\tau_{(\Sigma, E)\text{-}\mathbf{Nets}} = (\tau, \{enabled\}, F)$, with transitions $s \xrightarrow{e} s' \in \tau$ as places, and with flow relation defined as follows:

$$F(s \xrightarrow{e} s', enabled) = e$$

The above is a correct definition of the net since $e \in \mathcal{F}_{\Sigma, E}(I)$ as required. One may prove that the state graph $SG(N, \tau)$ of a *reduced* net $N = (P, A, F)$, $N \in {}_{(\Sigma, E)}\text{-}\textbf{Nets}^\circ(I)$, is in fact isomorphic to the initial lift in $(\Sigma, E)\text{-}\textbf{TS}^\circ$ of the family $\langle \text{ev}_p \rangle_{p \in P}$ of evaluation maps $\text{ev}_p : (\Sigma, E)\text{-}\textbf{Nets}(N, \mathcal{K}^\tau_{(\Sigma, E)\text{-}\textbf{Nets}}) \to \mathcal{K}^\tau$ defined by:

$$\text{ev}_p(\beta, \eta) = \beta p$$

For the reduced nets, these evaluation maps ev_p do form a jointly monic source.

The proof of this result is not immediate and depends strongly on the assumption that N is reduced. This is a severe restriction, as demonstrated by the following example. Let τ be the type of general Petri nets, with transitions labelled in the (Σ_1, E_1)-algebra $\mathbb{N} \times \mathbb{N}$. A Petri net (P, A, F) is a reduced (Σ_1, E_1)-net if and only if, for all multisets $\mu, \mu' : A \to \mathbb{N}$, $\mu \neq \mu' \Rightarrow \sum_a \mu(a) \times F(p, a) \neq \sum_a \mu'(a) \times F(p, a)$ for some place $p \in P$, if and only if, for all $f : A \to \mathbb{Z}$, $\sum_a f(a) \times F(\cdot, a)$ differs from the null map $\mathbf{0} : P \to \mathbb{Z}$, if and only if the a-indexed family of vectors $F(\cdot, a) : P \to \mathbb{N} \times \mathbb{N}$ ($\cong P + P \to \mathbb{N}$) is linearly independent (in $P + P \to \mathbb{N}$). If we consider alternatively the type τ of general Petri nets as a (Σ_3, E_3)-transition system, the situation becomes dramatic. For instance, the Petri net $N = (\{p\}, \{a\}, F)$ with the flow matrix $F(p, a) = (1, 2)$ is not a reduced (Σ_3, E_3)-net, as $F^*(p, e) = F^*(p, e')$ for $e = (a ; a) + (a^\sim ; a^\sim)$ and $e' = (a ; a^\sim) + a + a^\sim$, whereas $e \neq e'$ in $\mathcal{F}_{\Sigma_3, E_3}(\{a\})$. Actually, $F^*(p, e) = ((1, 2) ; (1, 2)) + ((2, 1) ; (2, 1)) = (1, 3) + (3, 1) = (4, 4)$ and $(4, 4) = (1, 1) + (3, 3) = ((1, 2) ; (2, 1)) + ((1, 2) + (2, 1)) = F^*(p, e')$.

Similar problems strike again when trying to construct synthesized nets $SN(T, \tau)$ as the initial lifts in $(\Sigma, E)\text{-}\textbf{Nets}^\circ(I)$ of the t-indexed families of maps $\text{ev}_t : (\Sigma, E)\text{-}\textbf{TS}(T, \tau) \to \mathcal{K}^\tau$ defined as follows.

$$\text{ev}_t(\sigma, \eta) = \sigma(q) \xrightarrow{\eta(e)} \sigma(q')$$

for $t = q \xrightarrow{e} q'$. The only way to obtain initial lifts in $(\Sigma, E)\text{-}\textbf{Nets}^\circ(I)$ is actually to construct initial lifts in $(\Sigma, E)\text{-}\textbf{Nets}(I)$, which is easy, and to derive from the resulting nets (P, A, F) *canonical reductions* (P, A', F') in $(\Sigma, E)\text{-}\textbf{Nets}^\circ(I)$, such that the two sets of vectors $\{F(\cdot, e) \,|\, e \in \mathcal{F}_{\Sigma_3, E_3}(A)\}$ and $\{F'(\cdot, e') \,|\, e' \in \mathcal{F}_{\Sigma_3, E_3}(A')\}$ are equal. There is no general solution, as the following counter-examples show.

The (Σ_3, E_3)-net $N = (\{p\}, \{a\}, F)$ with flow matrix $F(p, a) = (1, 2)$ cannot be reduced. Assume for contradiction that (P, A', F') is a reduction of N. Then $F'(p, e') = (1, 2)$ for some $e' \in \mathcal{F}_{\Sigma_3, E_3}(A')$. If e' is an atom, let $e' = \alpha \in A'$, then (P, A', F') is not reduced since $F'(p, (\alpha ; \alpha) + (\alpha^\sim ; \alpha^\sim)) = F'(p, (\alpha ; \alpha^\sim) + \alpha + \alpha^\sim$. In the converse case, consider any atom $\alpha \in A'$. Then $F'(p, \alpha) = F(p, e)$ for some $e \in \mathcal{F}_{\Sigma_3, E_3}(\{a\})$, hence $F'(p, \alpha) = F'(p, e[e'/a])$, and we get again a contradiction, as e' is not an atom.

The (Σ_1, E_1)-net $N = (\{p\}, \{a, b, c\}, F)$ with flow matrix given by $F(p, a) = (1, 3)$, $F(p, b) = (2, 2)$, and $F(p, c) = (3, 1)$ cannot be reduced. Observe that $\{(1, 3), (2, 2), (3, 1)\}$ is the set of all minimal vectors different from the null vector in $\{F(p, e) \,|\, e \in \mathcal{F}_{\Sigma_1, E_1}(\{a, b, c\})\}$, and assume for contradiction that (P, A', F') is a reduction of N. The set of all minimal vectors different from the null vector in $\{F'(p, e') \,|\, e' \in \mathcal{F}_{\Sigma_1, E_1}(A')\}$ must then be equal to $\{(1, 3), (2, 2), (3, 1)\}$. As the composition operator $+$ in Σ_1 is interpreted as summation in $\mathbb{N} \times \mathbb{N}$, the only possibility is that $F'(p, a') = (1, 3)$, $F'(p, b') = (2, 2)$, and $F'(p, c') = (3, 1)$ for corresponding atoms $a', b', c' \in A'$. But the considered vectors are not linearly independent, and therefore (P, A', F') was not a reduced (Σ_1, E_1)-net.

The above counter-examples do not show that one cannot define canonical reductions for the particular (Σ, E)-nets T^* derived from homsets (Σ, E)-$\mathbf{TS}(T, \tau)$, but we must nevertheless admit that the use of schizophrenic objects for constructing adjunctions between (Σ, E)-transition systems and (Σ, E)-nets is problematic. There is a way out, if one accepts to loose generality by fixing the alphabet A of atomic actions of nets and transition systems. Then, following the route described in section 4.9, we are able to establish for each type of nets $\tau \in (\Sigma, E)$-$\mathbf{Types}(I)$ an *order theoretic* Galois connection between (Σ, E)-transition systems over A and nets with set of atomic actions A in (Σ, E)-$\mathbf{Nets}(I)$. This construction is presented below.

5.4 Order Theoretic Galois Connection

Henceforth, we assume a fixed type of nets $\tau \subseteq S \times \mathcal{F}_{\Sigma, E}(I) \times S$ and a fixed alphabet of *atomic* actions A. Like we did in section 4.9, we focus on transition systems and nets with set of atomic actions A, and we assign them initial states and markings respectively. In order to alleviate the notation, let us introduce some definitions.

Definition 20 (Automaton). *An automaton $\mathcal{A} = (Q, A, T, q_0)$ is a reduced (Σ, E)-transition system $T \subseteq Q \times \mathcal{F}_{\Sigma, E}(A) \times Q$, with an initial state $q_0 \in Q$ such that Q is the forward closure of $\{q_0\}$ with respect to transitions in T. We let T denote the underlying transition system of $\mathcal{A}$. A morphism of automata $\sigma : \mathcal{A} \to \mathcal{A}'$ is a map $\sigma : Q \to Q'$ such that σ and $\iota_A : A \to \mathcal{F}_{\Sigma, E}(A) : \iota_A(a) = a$ define a morphism (of transition systems) $(\sigma, \iota_A) : T \to T'$, and $\sigma(q_0) = q'_0$ (preservation of the initial state).*

Clearly, automata and their morphisms form a category, denoted **Aut**.

Definition 21 (Net System). *A net system $\mathcal{N} = (P, A, F, M_0)$ is a (Σ, E)-net over I (thus $F : P \times A \to \mathcal{F}_{\Sigma, E}(I)$), with an initial marking $M_0 : P \to S$. We let $N = (P, A, F)$ denote the underlying net of $\mathcal{N}$. A morphism of net systems $\beta : \mathcal{N} \to \mathcal{N}'$ is a map $\beta : P \to P'$ such that (β, ι_A) with $\iota_A : A \to \mathcal{F}_{\Sigma, E}(A)$ given by $\iota_A(a) = a$, defines a morphism (of nets) from N to N', and $M_0(p) = M'_0(\beta p)$ for all $p \in P$ (preservation of the initial marking).*

Again, net systems and their morphisms form a category, denoted **NetSys**.

Definition 22 (Reachable State Graph). *The reachable state graph of a net system $\mathcal{N} = (N, M_0)$ is the automaton $\mathcal{N}^*$ whose underlying transition system T^* is the restriction of $SG(N, \tau)$ (see Def. 18) to the subset of markings reachable from M_0, and whose initial state is M_0.*

Definition 23 (Synthesis of Net Systems). *Given automaton $\mathcal{A} = (T, q_0)$, the net system $\mathcal{A}^*$ synthesized from $\mathcal{A}$ has the underlying net $N^* = SN(T, \tau)$ (see Def. 19) and the initial marking M_0 defined by $M_0(\sigma, \eta) = \sigma(q_0)$ for all regions $(\sigma, \eta) : T \to \tau$. For any region $p \in (\Sigma, E)\text{-}\mathbf{TS}(T, \tau)$, let $\mathcal{A}_p^*$ denote the (atomic) subnet system of $\mathcal{A}^*$ with p as a unique place.*

We will show that the two $(\cdot)^*$ operators, that send an automaton $\mathcal{A}$ to the net system $\mathcal{A}^*$ synthesized from $\mathcal{A}$, and a net system $\mathcal{N}$ to the reachable state graph $\mathcal{N}^*$, respectively, form an order theoretic Galois connection:

$$\mathcal{A} \leq \mathcal{N}^* \quad \text{iff} \quad \mathcal{N} \leq \mathcal{A}^*$$

The development present in the sequel takes the same stages as in section 4.9, with minor adaptations. Firstly, let us equip the automata and net systems with order relations.

Definition 24 (Ordered Set of Automata). *Let $\preceq$ denote the preorder on objects of $\mathbf{Aut}$ defined as $\mathcal{A} \preceq \mathcal{A}'$ if $\exists \sigma : \mathcal{A} \to \mathcal{A}'$. Let $\mathbf{Aut}_\sim$ denote the set of equivalence classes of automata generated by $\preceq$. The Ordered Set of Automata is $(\mathbf{Aut}_\sim, \leq)$ where $\leq$ denotes the order relation induced by $\preceq$.*

Definition 25 (Ordered Set of Net Systems). *Let $\preceq$ denote the preorder on objects of $\mathbf{NetSys}$ defined as $\mathcal{N} \preceq \mathcal{N}'$ if $\exists \beta : \mathcal{N} \to \mathcal{N}'$. Let $\mathbf{NetSys}_\sim$ denote the set of equivalence classes of net systems generated by $\preceq$. The Ordered Set of net systems is $(\mathbf{NetSys}_\sim, \leq)$ where $\leq$ denotes the order relation induced by $\preceq$.*

It follows from the reducedness, determinism, and reachability of automata that $\mathcal{A} \sim \mathcal{A}'$ in $\mathbf{Aut}$ if and only if automata $\mathcal{A}$ and $\mathcal{A}'$ are isomorphic. Similarly, $\mathcal{N} \sim \mathcal{N}'$ in $\mathbf{NetSys}$ if and only if net systems $\mathcal{N}$ and $\mathcal{N}'$ are identical up to a renaming of places which preserves multiplicity assigned to the places by the initial marking. Henceforth, equivalence classes of automata (respectively, of net systems) are denoted with arbitrary representatives $\mathcal{A} = (Q, A, T, q_0) \in \mathbf{Aut}$ (respectively, $\mathcal{N} = (P, A, F, M_0) \in \mathbf{NetSys}$).

$(\mathbf{Aut}_\sim, \leq)$ is a *complete lattice* with greatest lower bounds computed as *synchronized products*, where the synchronized product $\bigwedge_{j \in J} \mathcal{A}_j$ of a family of automata $\mathcal{A}_j = (Q_j, A, T_j, q_{0,j})$ indexed by $j \in J$ is the automaton (Q, A, T, q_0) defined as follows: if $J = \emptyset$, $Q = \{q_0\}$ and T is the set of all transitions $q_0 \xrightarrow{e} q_0$ where e ranges over $\mathcal{F}_{\Sigma,E}(A)$. Otherwise, the initial state $q_0 = \langle q_{0,j} \rangle_{j \in J}$ is the vector of initial states, and the set of states Q is the forward closure of $\{q_0\}$ in the product $\prod_{j \in J} Q_j$ with respect to the *synchronized transitions*:

$$\langle q_j \rangle_{j \in J} \xrightarrow{e} \langle q_j' \rangle_{j \in J} \quad \text{iff} \quad \forall j \in J \quad q_j \xrightarrow{e} q_j' \in T_j$$

where e ranges over $\mathcal{F}_{\Sigma,E}(A)$, and the set of transitions T is the subset of the synchronized transitions with sources in Q. Symmetrically, the ordered set $(\mathbf{NetSys}_\sim, \leq)$ is a complete lattice, with least upper bounds computed by *amalgamation on atomic actions* as it has been described in section 4.9.

By the definition of reachable state graphs, the automaton $\mathcal{N}^*$ generated from a net system $\mathcal{N} = (P, A, F, M_0)$ is the synchronized product of the reachable state graphs of its atomic subnet systems $\mathcal{N}_p = (\{p\}, A, F_p, M_{0,p})$, such that $M_{0,p}(p) = M_0(p)$ and with flow given by: $F_p(p, a) = F(p, a)$, for all $a \in A$. Thus, $\mathcal{N}^* = \bigwedge \{\mathcal{N}_p^* \mid p \in P\}$. By the definition of synthesized net systems, the net system $\mathcal{A}^*$ synthesized from automaton $\mathcal{A} = (Q, A, T, q_0)$ is obtained by gluing on atomic actions all the net systems $\mathcal{N}_p = (\{p\}, A, F_p, M_{0,p})$ that derive from regions $p = (\sigma, \eta) \in {}_{(\Sigma,E)}\text{-}\mathbf{TS}(T, \tau)$ such that:

$$F_p(a) = \eta(a) \text{ (in } \mathcal{F}_{\Sigma,E}(I)) \quad \text{and} \quad M_{(0,p)}((\sigma, \eta)) = \sigma(q_0) \text{ (in } S)$$

Thus, $\mathcal{A}^* = \bigvee \{\mathcal{N}_p \mid p \in {}_{(\Sigma,E)}\text{-}\mathbf{TS}(T, \tau)\}$. The key to establish the intended Galois connection is given by the following result.

Lemma 6. *Regions of $\mathcal{A}$ (i.e. morphisms $(\sigma, \eta) : T \to \tau$) are in bijective correspondence with the atomic net systems $\mathcal{N}$ such that $\mathcal{A} \le \mathcal{N}^*$.*

Proof. Let $\mathcal{A} \le \mathcal{N}^*$, where $\mathcal{A} = (Q, A, T, q_0)$ and $\mathcal{N} = (\{p\}, A, F, M_0)$. Thus $M_0(p) \in S$ and $F(p, a) \in \mathcal{F}_{\Sigma,E}(I)$ for all $a \in A$. By the definition of reachable state graphs, $\mathcal{N}^* = (Q', A, T', M_0)$ where $Q' \subseteq S^{\{p\}}$ and $T' \subseteq Q' \times \mathcal{F}_{\Sigma,E}(A) \times Q'$ are the least sets of states and transitions respectively such that $M_0 \in Q'$ and for each $e \in \mathcal{F}_{\Sigma,E}(A)$: $M' \in Q'$ and $M \xrightarrow{e} M'$ in T' whenever $M \in Q'$ and $M(p) \xrightarrow{F^*(p,e)} M'(p)$ in τ. By definition of the order relation $\le$ on automata, there exists a morphism $\sigma : \mathcal{A} \to \mathcal{N}^*$. By definition of morphisms of automata, (σ, ι_A) is a morphism (of transition systems) from T to T', mapping q_0 to M_0. Now the maps $\iota_p : Q' \to S : \iota_p(M) = M(p)$ and $\eta : A \to \mathcal{F}_{\Sigma,E}(I) : \eta(a) = F(p, a)$ define a morphism (of transition systems) $(\iota_p, \eta) : T' \to \tau$. By composition of morphisms, one obtains a region (σ_p, η_p) of $\mathcal{A}$ such that $\sigma_p = \iota_p \circ \sigma$ and $\eta_p = \eta^* \circ \iota_A$, hence $\sigma_p(q_0) = M_0(p)$ and $\eta_p(a) = F(p, a)$ for all $a \in A$.

Now, let $p = (\sigma_p, \eta_p)$ be an arbitrary region of $\mathcal{A}$, and let $\mathcal{N}_p = \mathcal{A}_p^*$. As an induced restriction of $\mathcal{A}^*$ on a subset of regions (or places) with a single element, $\mathcal{N}_p$ is an atomic net system. Consider the reachable state graph $\mathcal{N}_p^*$ whose initial state is the marking $M_0(p) = \sigma_p(q_0)$. Let q and M be any state and marking of $\mathcal{A}$ and $\mathcal{N}$, respectively, and such that $M(p) = \sigma_p(q)$. If $q \xrightarrow{e} q'$ is a transition in T, there must exist a marking M' of $\mathcal{N}_p$ such that $M'(p) = \sigma_p(q')$ and $M \xrightarrow{e} M'$, because $F^*(p, e) = \eta_p^*(e)$ in $\mathcal{N}_p$ by construction of this net, and $\sigma_p(q) \xrightarrow{\eta_p^*(e)} \sigma_p(q')$ in τ by definition of regions. As all states q of $\mathcal{A}$ are reachable from q_0, it follows by induction that they are all represented by corresponding states M in $\mathcal{N}_p^*$ such that $M(p) = \sigma_p(q)$. Therefore, $\sigma_p : \mathcal{A} \to \mathcal{N}_p^*$ is a morphism of automata.

We are left with proving that $\mathcal{N}^*$ and $\mathcal{N}_p^*$ are isomorphic when $p = (\sigma_p, \eta_p)$ is the region of $\mathcal{A}$ that has been derived from $\mathcal{N}$ in the first part of the proof. As the initial states agree and for all $a \in A$, and $F(p, a) = \eta_p(a)$ in both nets, it follows that the two net systems are isomorphic. Whence, their reachable state graphs are isomorphic, too. $\square$

We can now establish the expected Galois connection between automata and net systems.

Proposition 11. *For all $\mathcal{A} \in \mathbf{Aut}_\sim$ and $\mathcal{N} \in \mathbf{NetSys}_\sim$ the following holds.*

$$\mathcal{A} \leq \mathcal{N}^* \quad iff \quad \mathcal{N} \leq \mathcal{A}^*$$

Thus, the two $(\cdot)^*$ operators, one mapping an automaton $\mathcal{A}$ to the synthesized net system $\mathcal{A}^*$ and the other taking a net system $\mathcal{N}$ to the reachable state graph $\mathcal{N}^*$, establish a Galois connection between ordered sets $(\mathbf{NetSys}_\sim, \leq)$ and $(\mathbf{Aut}_\sim, \leq)$.

Proof. Consider first the case of an atomic net system $\mathcal{N}$. Then by definition of the synthesized net $\mathcal{A}^*$, $\mathcal{N} \leq \mathcal{A}^*$ if and only if the net system $\mathcal{N} = \mathcal{N}_p$ derives from a region $p = (\sigma_p, \eta_p)$ of the automaton $\mathcal{A}$. By lemma 6, this assertion is equivalent to the assertion $\mathcal{A} \leq \mathcal{N}^*$. Consider now an arbitrary net system $\mathcal{N}$. Observe that $\mathcal{N} = \bigvee \{\mathcal{N}_p \mid p \in P\}$ where $\mathcal{N}_p$ is the atomic subnet system of $\mathcal{N}$ with the unique place p, and thus $\mathcal{N}^* = \bigwedge \{\mathcal{N}_p^* \mid p \in P\}$ by definition of reachable state graphs. Therefore, $\mathcal{A} \leq \mathcal{N}^*$ iff $\mathcal{A} \leq \mathcal{N}_p^*$ for all $p \in P$ iff $\mathcal{N}_p \leq \mathcal{A}^*$ for all $p \in P$ (seeing that $\mathcal{N}_p$ is an atomic net system) iff $\mathcal{N} \leq \mathcal{A}^*$.   □

Proposition 12. *A (Σ, E)-automaton $\mathcal{A} = (Q, A, T, q_0)$ is isomorphic to the reachable state graph of some (Σ, E)-net system, relatively to type $\tau = (S, I, \tau)$, if and only if there exists some subset of regions $\mathcal{P} \subseteq (\Sigma, E)\text{-}\mathbf{TS}(T, \tau)$ such that the following axioms of separation are satisfied for all $q, q' \in Q$, and $e \in \mathcal{F}_{\Sigma,E}(A)$.*

SSP *State Separation Property:* $\quad q \neq q' \Rightarrow \exists p = (\sigma, \eta) \in \mathcal{P} \quad \sigma(q) \neq \sigma(q')$.

ESSP *Event/State Separation Property:* $\quad q \stackrel{e}{\nrightarrow} \Rightarrow \exists p = (\sigma, \eta) \in \mathcal{P} \quad \sigma(q) \stackrel{\eta^*(e)}{\nrightarrow}$.

When these two axioms are satisfied, $\mathcal{P}$ is said to be an admissible *subset of regions, and $\mathcal{A} \cong (\bigvee_{p \in \mathcal{P}} \mathcal{A}_p^*)^*$ holds.*

Proof. Let $\mathcal{N}_p = \mathcal{A}_p^*$ for $p \in \mathcal{P}$, and let $\mathcal{N}_\mathcal{P} = \bigvee_{p \in \mathcal{P}} \mathcal{N}_p$. Seeing that $\mathcal{A} \leq \mathcal{N}_p^*$ for every region p, $\mathcal{A} \leq \bigwedge_{p \in \mathcal{P}} \mathcal{N}_p^* = \mathcal{N}_\mathcal{P}^*$. Hence, there exists a morphism of automata $\sigma : \mathcal{A} \to \mathcal{N}_\mathcal{P}^*$. Moreover this morphism is unique. On the other hand, every region $p = (\sigma_p, \eta_p)$ factors into $(\mathrm{id}_S, \eta_p) \circ (\sigma_p, \iota_A)$ where $\iota_A : A \to \mathcal{F}_{\Sigma,E}(A) : \iota_A(a) = a$ and σ_p is the unique morphism from $\mathcal{A}$ to $\mathcal{N}_p^*$. As $\mathcal{N}_\mathcal{P}^*$ is the synchronized product of the net systems $(\mathcal{N}_p^*)_{p \in \mathcal{P}}$, σ must be the map that sends each state q of A to the associated vector $\sigma(q) = \langle \sigma_p(q) \, ; \, p = (\sigma_p, \eta_p) \in \mathcal{P} \rangle$ (the p-component is computed by evaluating region p at state q). Since σ is the unique morphism from $\mathcal{A}$ to $\mathcal{N}_\mathcal{P}^*$, and seeing that automata are accessible and deterministic, the assertion $\mathcal{A} \cong \mathcal{N}_\mathcal{P}^*$ is now equivalent to the conjunction of the following two assertions : (i) σ is an injective map, and (ii) $q \stackrel{e}{\to}$ in $\mathcal{A}$ whenever $\sigma(q) \stackrel{e}{\to}$ in $\mathcal{N}_\mathcal{P}^*$. Thus **SSP** is just another form of assertion (i), and **ESSP** is equivalent to assertion (ii) seeing on the one hand that, for $p = (\sigma_p, \eta_p)$, $\sigma_p(q) \stackrel{\eta_p^*(e)}{\longrightarrow}$ in τ iff $\sigma_p(q) \stackrel{e}{\longrightarrow}$ in $\mathcal{N}_p^*$ (by the definition of state graphs), and on the other hand that $\sigma_p(q) \stackrel{e}{\longrightarrow}$ in $\mathcal{N}_p^*$ for all $p \in \mathcal{P}$ iff $\sigma(q) \stackrel{e}{\longrightarrow}$ in $\mathcal{N}_\mathcal{P}^*$ (by the definition of the synchronized product).   □

In view of the above propositions, all properties of Δ-automata that have been derived from properties **GC1** to **GC5** may be reproduced for (Σ, E)-automata.

We conclude with a remark. One could easily extend (Σ, E)-automata by including arbitrary (Σ, E)-algebras as algebras of complex actions. There is just one point in the technical development where we use the assumption that algebras of actions are initial (Σ, E)-algebras, namely when we infer that $\eta_p(a) = \eta(a)$ for all a from $\eta_p = \eta^* \circ \iota_A$ in the proof of lemma 6, and this inference is also correct in (Σ, E)-algebras. The reason why we stick to initial (Σ, E)-algebras is to keep a good intuition of algebras of actions.

6   Conclusion

We have proposed a theoretical framework for solving the net realization problem, for two kinds of generalized automata. In each case the framework is uniform with respect to the type of nets (given as a generalized automaton). Regions of an automaton, defined as morphisms from this automaton to a fixed type of nets, play a fundamental in net synthesis. Ehrenfeucht-Rozenberg separation axioms, relativized to regions induced by types of nets, supply in each case a plain characterization of automata isomorphic to reachable state graphs of nets.

In an ordinary automaton (Q, A, T, q_0), a transition $t = q \xrightarrow{a} q'$ comes with the source and target maps $\partial^0(t) = q$ and $\partial^1(t) = q'$, and with the labelling map $\lambda(t) = a$. We have considered two kinds of generalized automata, namely Δ-automata and (Σ, E)-automata, which go beyond the above picture.

In a Δ-automaton, a transition t is defined by a labelling action $\lambda(t) = a$ and a reaction $\partial(t) : \Delta \rightharpoonup Q$ parametric on a set of controls Δ. The set of controls includes a distinguished element $\bullet$, such that $\partial(t)(\bullet)$ is always defined and represents the source of the transition. Each type of nets $\tau = (S, I, \tau)$ with the set of transitions τ parametric on Δ induces a corresponding firing rule for all net systems (P, A, F, M_0) with flow matrix $F : (P \times A) \rightarrow I$ and initial marking $M_0 : P \rightarrow S$. According to this firing rule, each net system is fit with a generalized state graph which is a Δ-automaton. Hybrid Petri Nets, Coloured Petri Nets, and Vector Addition Systems with States may be presented uniformly in this style, thus opening the way for their synthesis.

In a (Σ, E)-automaton, a transition $t = q \xrightarrow{e} q'$, with source $\partial^0(t) = q$ and target $\partial^1(t) = q'$, is labelled with a complex action $\lambda(t) = e$ in the initial algebra $\mathcal{F}_{\Sigma,E}(A)$ generated from atomic actions in A (Σ is a signature and E a set of equations between Σ-terms). An admissible (Σ, E)-type of nets $\tau = (S, I, \tau)$ has transitions labelled in $\mathcal{F}_{\Sigma,E}(I)$, that is in the free (Σ, E)-algebra over I as the set of generators. Any such a type induces a corresponding firing rule for nets, so that again each net system is fit with a (Σ, E)-automaton as its generalized state graph. Truly concurrent Petri Nets, synchronous Inhibitor C/E-nets, and other forms of nets with non-sequential computations may be presented uniformly in this style.

For each kind of generalized automata, we have tried to establish Galois connections between automata and net systems following two alternative approaches: one categorical, in which types of nets are viewed as schizophrenic objects (living both in the category of transition systems and in the category of nets), and the other order theoretical. Both approaches were used in earlier works to obtain Galois connections between ordinary automata and net systems. The strong point of the categorical approach is the complete symmetry of the dual constructions of state graphs and synthesized nets. The strong point of the order theoretical approach is mathematical simplicity.

Both methods are carried successfully from ordinary automata and nets to Δ-parametric automata and nets, if one defines morphisms of Δ-automata with sufficient care. Thus in some sense, making transitions functional does not deeply modify the relationships between automata and nets. The situation is different when labels of transitions are replaced with complex actions. On the one hand, the order theoretic Galois connection may be carried from ordinary automata and nets to (Σ, E)-transition systems and nets with minor adaptations, on the other hand, we did not succeed to construct a dual adjunction between (Σ, E)-transition systems and nets based on a schizophrenic object. The most robust approach to establishing net representation theorems for automata seems thus to combine regions as morphisms with order theoretic Galois connections.

Let us indicate directions for future work. One direction is trying to come up with a joint generalization of automata which would encompass Δ-parameterized transitions *and* complex actions, thus allowing to combine control and true concurrency. We have spent some time on this, without success so far. A second direction is to consider Δ-automata parameterized on sets Δ with structure. For instance, our framework is general enough to incorporate hybrid nets, but it seems that a thorough treatment of hybrid nets should take into account considerations of continuity (see Droste and Shortt's work presented in this volume). A third direction is to investigate the semantics of Petri nets with complex transitions. We have only touched the surface of problems in this respect. Last but not least, in order to make the theory useful, the algorithmic aspects of the constructions put forward need be considered for specific types of nets.

References

1. Badouel, E., Splitting of Actions, Higher-Dimensional Automata and Net Synthesis. Inria-RR 3013 (1996)
2. Badouel, E., Darondeau, P. Dualities between Nets and Automata Induced by Schizophrenic Objects. Proc. Sixth Int. Conf. on CTCS, LNCS 953 (1995) 24–43
3. Badouel, E., Darondeau, P. Theory of Regions. Lectures on Petri Nets I: Basic Models, Advances in Petri Nets, Reisig and Rozenberg eds., LNCS 1491 (1998) 529–586
4. David, R., Alla, H. Continuous Petri Nets. Proc. Eighth European Workshop on Applications and Theory of Petri Nets, Zaragoza (1987) 275–294
5. David, R., Alla, H., Modeling of Hybrid Systems Using Continuous and Hybrid Petri Nets. Proc. PNPM'97 (1997) 47–58
6. Davey, B.A., Priestley, H.A., *Introduction to Lattices and Order*, Cambridge University Press (1990).
7. Droste, M., Shortt, R.M., Petri Nets and Automata with Concurrency Relations - an Adjunction. Semantics of Programming Languages and Model Theory, Droste and Gurevitch eds. (1993) 69–87

8. Droste, M., Shortt, R.M., Continuous Petri Nets and Transition Systems. (in this volume)
9. Ehrenfeucht, A., Rozenberg, G., Partial 2-structures; *Part I*: Basic Notions and the Representation Problem;n *Part II*: State Spaces of Concurrent Systems. *Acta Informatica*, 27, 315–342 & 343–368 (1990)
10. Feuzeu, T., Synthèse de réseaux de Petri purs à partir de produits synchrones d'automates. Mémoire de fin d'études, ENSP-Yaoundé (2000)
11. Isbell, J. R.. General Functorial Semantics, I. *American Journal of Mathematics*, vol. 94, pp.: 535-596, 1972.
12. Johnstone, P.T., *Stone spaces*. Cambridge University Press (1982).
13. Juhas, G., Reasoning about Algebraic Generalisation of Petri Nets. Proc. ATPN'99, LNCS 1639 (1999) 324–343
14. Meseguer, J., Montanari, U., Petri Nets are Monoids. *Information and Computation*, 88-2 (1990) 105–155
15. Mukund, M., Petri Nets and Step Transition Systems. *IJFCS*, 3-4 (1992) 443–478
16. Porst, H.E., Tholen, W., Concrete Dualities. *Category Theory at Work*, Herrlich and Porst eds., Heldermann-Verlag (1991) 11–136

On Concurrent Realization of Reactive Systems and Their Morphisms[*]

Marek A. Bednarczyk and Andrzej M. Borzyszkowski

Institute of Computer Science, Gdańsk Branch, Polish Acad. of Sc.
Abrahama 18, 81-825 Sopot, Poland
http://www.ipipan.gda.pl

Abstract. The paper introduces the notion of *concurrent* realization of reactive systems. A framework is also presented in which labelled safe Petri nets as concurrent realizations of concrete asynchronous systems are constructed. The construction is uniform in the sense that it extends to a realization of arbitrary commuting diagrams. We discuss applicability of the framework to construct *maximally concurrent* realizations of reactive systems.

1 Introduction

The intuitive idea of *reactive system* admits various formalizations. This paper accepts it as a starting point that the most abstract among them is the one provided by the notion of *transition system*. Another formalization, and the one that offers most intuitive description of concurrent behavior, is provided by the notion of Petri net. And indeed, given a Petri net as a model of a reactive system one automatically is also given a transition system model in the form of the *case graph* of the net. Each Petri net can thus be seen as a concurrent realization of its case graph. So, a **fundamental problem** of Petri net theory is the following.

> Given a transition system S, find a Petri net N such that its case graph $Cg(N)$ is, essentially, equal to S.

The first satisfactory answer to the problem was offered by Ehrenfeucht and Rozenberg, cf. [14]. They characterized a class of transition systems which can be seen as case graphs of a subclass of nets called *elementary nets*. Moreover, they provided a construction which for any transition system from this class *synthesizes* a required net. Our paper, together with several other papers included in this volume, see [2, 12], attempts to achieve further unification of the Petri net theory, by extending the boundaries of synthesis to larger classes of transition systems and nets.

The fundamental problem can be seen as an instance of a general task of relating different classes of models. A fruitful, and well-established method of solving this type of problems is provided by category theory, cf. [16]. Firstly, one

---

[*] Partially supported by State Committee for Scientific Research grant 8 T11C 037 16.

H. Ehrig et al. (Eds.): Unifying Petri Nets, LNCS 2128, pp. 346–379, 2001.

should, for each class of models, come up with a suitable notion of *morphism*, and thereby turn the class into a *category*. Then, the aim is to establish a *functorial* translation of objects and morphisms from one category to the other, and back. It is desirable that the functors obtained in this way are *adjoint*. This would, for instance, provide means of transportation of categorical constructions between the classes. Ideally, one of the classes turns out to be more abstract than the other. This, formally, takes the form of the adjunction being a *(co-)reflection*. Existence of a (co-)reflection boils down to the observation that translating an abstract object to the less abstract category, and back, gives, essentially, the same abstract object again. To put it formally, the object thus obtained should be *isomorphic* to the original — the property often imposed in mathematically satisfactory formalizations of the fundamental problem of Petri net theory.

The line of research described above was initiated in the 80s, and led by Winskel. As a result many models of distributed and concurrent computations were inter-related. In fact it was Winskel who first established a coreflection in the form required by the fundamental problem. But his solution worked only for *unfoldings* of concurrent behaviors on both sides, i.e., a coreflection was established between prime event structures and occurrence nets.

In this paper we address the problem of finding a (maximally) concurrent realization of a reactive system within *a* categorical framework. Until now, all categorically motivated realization procedures looked for an adjunction between a category of abstract behaviors and a category of Petri nets, see [20, 11, 30, 2]. As a result, in order to guarantee universality of the construction, the Petri nets constructed by the adjunctions tended to be huge, literally *saturated* with places. In practical applications this price for the universality seems too high. Thus, in order to work with smaller nets, the original regional construction [14] was ramified. The idea was to construct nets of the desired kind, and with a small or minimal number of places, see [1, 9, 10]. To our knowledge, none of these constructions has been shown to have a categorical underpinning.

Several novel ideas are put forward in this paper.

First, a notion of *concurrent realization* of a reactive system is proposed. The idea is that the concurrency present in a Petri net taken as a realization should implement concurrency encoded in, or admissible for the given reactive system.

Second, concurrent realizations of a given behavior are sought in the category of *labelled* Petri nets. Originally, see [14], the problem of synthesis was to find a Petri net the transitions of which would be the actions of the synthesized transition system. This is a severe restriction — even simple classes of nets, like *safe nets*, when equipped with labellings can model behaviors which are beyond the scope of unlabelled synthesis presented in [1, 11, 20, 30]. Note, though, that with labelling allowed one could end up with a trivial solution: each step being synthesized as a separate transition. Thus, it is the demand that the Petri nets taken as realizations, should also implement the concurrency present in the abstract behaviors that makes the problem non-trivial.

Third, we show that thus generalized synthesis problem can be solved uniformly for a large class of concurrent behaviors represented by *concrete* asynchronous systems. In [19] Morin characterized concrete asynchronous systems as products of the usual sequential transition systems in a suitable category. Our idea is to utilize this characterization, and conduct synthesis of sequential systems only. Then, to obtain a realization of a concrete asynchronous system it is enough take the product of the sequential nets realizing its sequential component. The justification comes from a result that states that the notion of realization is consistent with products in both categories.

As a by-product, we offer a discussion of *general morphisms* of Petri nets, see [5]. The class, introduced independently and much earlier by Vogler in his studies of processes of nets ([26]), deserves to be known as much as the classes of morphisms introduced and studied by Winskel ([28]), and Nielsen et al. ([20]). General morphisms, we argue, are indispensable in the context of categorical synthesis of small nets.

The paper is organized as follows.

Sect. 2 explains why should we care about morphisms at all. Then, Sect. 3 discusses categories of Petri nets, and their morphisms in particular. Here, we explain in what sense the morphisms introduced by Vogler are general, and what does it mean that a labelled Petri net realizes a reactive system. Sect. 4 recalls the aforementioned work [19] of Morin on asynchronous systems decomposable as products of sequential ones. We also show how one can build realizations of concrete asynchronous systems and their morphisms. Finally, Sect. 5 provides some examples and indicates how one can go about finding concrete behaviors of reactive systems, and discusses our plans for further research in this area.

Acknowledgments. We would like to acknowledge stimulating discussions with Philippe Darondeau, Wiesław Pawłowski, Rafał Somla and Andrzej Tarlecki. Remarks of the anonymous referees helped us improve the presentation.

2 Why Should We Care about Morphisms

2.1 Reactive Systems, Their Categories and Properties

A *transition system* S is a triple $S = \langle S, A, T \rangle$ where S and A are sets, while $T \subseteq S \times A \times S$. Thus, a reactive system represented by a transition system S, has *states*—the elements of S. It is capable to react to some external actions—the collection of all potential *actions* of S is A, also called the *alphabet* of S. Finally, at each state the system is capable to accept/perform one of its actions and thereby change its state—this is captured by the *transition relation* T.

Let S be a transition system. Only finite transition systems are considered in this paper, therefore S and A are finite sets.

We let p, q, etc., to range over states, while a, b and so on range over actions. Usually, we write $p \xrightarrow{a} q$ whenever $\langle p, a, q \rangle \in T$, and call it a *a-step* in S. Then, notation $p \xrightarrow{a}$, $p \xrightarrow{a}\!\!\!\!\!/\;$ and $p \longrightarrow q$ is used to indicate that either $p \xrightarrow{a} q$ holds in S for some q, or for none, or for some a, respectively. Various sub- and/or super-scripts decorating transition systems are carried over to the components.

For instance, A_2' denotes the alphabet of S_2', and so on. This convention applies in the sequel to all structures and their components.

S is *deterministic* whenever $p \xrightarrow{a} q$ and $p \xrightarrow{a} r$ implies $q = r$. In the paper we consider only deterministic transition systems.

Transition systems are often considered together with a designated *initial* state $\hat{s} \in S$. The set R_S of *reachable states* of such a *pointed* or *initialized* transition system $\langle S, \hat{s} \rangle = \langle S, \hat{s}, A, T \rangle$ is then inductively defined as the least set such that $\hat{s} \in R_S$, and $q \in R_S$ whenever $p \in R_S$ and $p \longrightarrow q$ in S. A transition system S is *reachable* whenever all its states are reachable, that is $S = R_S$.

Given transition systems S_1 and S_2 their morphism f is a pair $f = \langle \sigma, \lambda \rangle$ where $\sigma : S_1 \to S_2$ is a total function, while $\lambda : A_1 \rightharpoonup A_2$ is a partial function, which together preserve the transition relation of S_1 in the following sense.

$$p \xrightarrow{a} q \quad and \quad \lambda a \Downarrow \quad implies \quad \sigma(p) \xrightarrow{\lambda a} \sigma(q)$$
$$p \xrightarrow{a} q \quad and \quad \lambda a \Uparrow \quad implies \quad \sigma(p) = \sigma(q)$$

Notation $\lambda a \Downarrow$, resp., $\lambda a \Uparrow$, above indicates that λa is defined, resp., undefined. When S_1 and S_2 are initialized, f also preserves the initial states, i.e., $\sigma(\hat{s}_1) = \hat{s}_2$.

This yields a category **TS** of transition systems, when composition of morphisms is defined componentwise and with pairs of identity functions as the identity morphisms. All transition systems considered in the sequel are assumed to be initialized. Subclasses of deterministic, reachable, and deterministic-and-reachable transition systems define important subcategories: $d\mathbf{TS}$, $r\mathbf{TS}$ and $dr\mathbf{TS}$, respectively. There is also an evident functor $\Re e : \mathbf{TS} \to r\mathbf{TS}$ that, given S, produces its reachable subsystem $\Re e(S) = \langle R_S, \hat{s}, A, T \cap (R_S \times A \times R_S) \rangle$. It cuts down to subcategories with deterministic systems.

Intuitively, a morphism $f : S \to S'$ explains how the dynamic activity in S is simulated in S'. The dynamic behavior of transition systems is often characterized by means of properties expressible in a modal logic. Then, the existence of a morphism like f, with $f = \langle \sigma, \lambda \rangle$, is a proof that the following hold.

- Modulo λ, the *liveness* properties valid in S also hold in S', i.e., what is possible for S, is also possible for S'.
- Modulo λ, the *safety* properties valid in S', are also valid in S. Thus, S cannot do what is forbidden for S', i.e., all inevitable properties of S' hold in S.

To substantiate the above claims let us consider a simple modal language.

2.2   A Simple Language of Modal Properties

The set $ML^{\Diamond \Box}(A)$ of formulae of the language, ranged over by φ and ψ, is defined by the following grammar.

$$\varphi ::= \mathbf{true} \mid \mathbf{false} \mid \varphi \wedge \psi \mid \varphi \vee \psi \mid \Diamond \varphi \mid \Box \varphi \mid \mathbf{accept}^a \mid \mathbf{refuse}^a$$

The construction of $ML^{\Diamond \Box}(A)$ is parameterized with the choice of alphabet A of actions in the sense that each action a in A induces atomic formulae $\mathbf{accept}^a$ and $\mathbf{refuse}^a$.

Each transition system S can be treated as a Kripke frame on which modal formulae can be interpreted, see [22]. The formal interpretation of $ML^{\Diamond\Box}(A)$ is presented in Table 1.

Table 1. The interpretation of $ML^{\Diamond\Box}(A)$

$s \models \mathbf{true}$	*always*	$s \models \varphi \wedge \psi$	*iff*	$s \models \varphi$ and $s \models \psi$
$s \models \mathbf{false}$	*never*	$s \models \varphi \vee \psi$	*iff*	$s \models \varphi$ or $s \models \psi$
$s \models \mathbf{accept}^a$	*iff* $s \xrightarrow{a}$	$s \models \mathbf{refuse}^a$	*iff*	$s \not\xrightarrow{a}$

$$s \models \Diamond\varphi \quad \textit{iff} \quad s \xrightarrow{*} t \quad \textit{and} \quad t \models \varphi \quad \text{for some } t$$
$$s \models \Box\varphi \quad \textit{iff} \quad s \xrightarrow{*} t \quad \textit{implies} \quad t \models \varphi \quad \text{for all } t$$

Where $s \xrightarrow{*} t$ means $s = s_0 \xrightarrow{a_1} \ldots \xrightarrow{a_n} s_n = t$ for a possibly empty sequence of actions.

Let $ML^{\Diamond}(A)$ be the subset of $ML^{\Diamond\Box}(A)$ obtained by removing all formulae which contain a sub-formula of the form $\Box\varphi$ or $\mathbf{refuse}^a$. Similarly, $ML^{\Box}(A)$ is obtained by removing formulae containing $\Diamond\varphi$ or $\mathbf{accept}^a$.

Formulae in $ML^{\Diamond}(A)$ are existential in nature. If a system satisfies a formula $\mathbf{accept}^a$ then it is capable of performing action a. Similarly, if the system satisfies $\Diamond\varphi$ then during its evolution it may reach a state in which φ will be satisfied. Conversely, formulae in $ML^{\Box}(A)$ are universal in nature. If the system satisfies formula $\Box\varphi$ it will always preserve φ during its evolution. If it satisfies $\mathbf{refuse}^a$ then there is no way in which it can perform action a.

Consequently, formulae in $ML^{\Diamond}(A)$ and $ML^{\Box}(A)$ are identified as *liveness* and *safety* properties, respectively.

Now, each partial function $\lambda : A \rightharpoonup A'$ induces a translation, also denoted λ, of formulae in $ML^{\Diamond\Box}(A)$ to formulae in $ML^{\Diamond\Box}(A')$ given in Table 2. Clearly, the translation cuts down to translations between liveness, resp., safety formulae.

Table 2. Lifting a partial function on actions to translation on formulae

$$\lambda\,\mathbf{true} = \mathbf{true} \qquad\qquad \lambda\,\mathbf{false} = \mathbf{false}$$
$$\lambda(\varphi \wedge \psi) = \lambda\varphi \wedge \lambda\psi \qquad\qquad \lambda(\varphi \vee \psi) = \lambda\varphi \vee \lambda\psi$$
$$\lambda(\Diamond\varphi) = \Diamond\lambda\varphi \qquad\qquad \lambda(\Box\varphi) = \Box\lambda\varphi$$
$$\lambda(\mathbf{accept}^a) = \begin{cases} \mathbf{accept}^{\lambda a} & \text{if } \lambda a \Downarrow \\ \mathbf{true} & \text{if } \lambda a \Uparrow \end{cases} \qquad \lambda(\mathbf{refuse}^a) = \begin{cases} \mathbf{refuse}^{\lambda a} & \text{if } \lambda a \Downarrow \\ \mathbf{false} & \text{if } \lambda a \Uparrow \end{cases}$$

With the machinery introduced above it is now possible to formally capture the intuitive explanation of the role of morphisms of transition systems with respect to the modal/temporal properties enjoyed by their sources and targets.

To strengthen the results let us introduce a class of special morphisms that generalizes Park's idea of *bisimulation* between transition systems.

Morphism $f : \mathcal{S} \to \mathcal{S}'$ is a *zig-zag* whenever λ is a bijection, and $\sigma(p) \xrightarrow{a'} r$ in $\mathcal{S}'$ implies the existence of q and a such that $\sigma q = r$, $\lambda a = a'$ and $p \xrightarrow{a} q$ in $\mathcal{S}$.

Proposition 1. *Let* $\langle \sigma, \lambda \rangle : \mathcal{S} \to \mathcal{S}'$. *Then, the following hold.*

$$
\begin{array}{llll}
s \models \varphi & \textit{implies} & \sigma s \models \lambda \varphi & \textit{for } \varphi \in ML^{\Diamond}(A) \\
s \models \varphi & \textit{whenever} & \sigma s \models \lambda \varphi & \textit{for } \varphi \in ML^{\Box}(A) \\
s \models \varphi & \textit{iff} & \sigma s \models \lambda \varphi & \textit{for } \varphi \in ML^{\Diamond\Box}(A), \quad \textit{provided } f \textit{ a zig-zag}
\end{array}
$$

The proof of the above is elementary, and in fact the result is stated here as an instance of a large family of similar results which can be found in the literature for each possible modal or temporal logic of properties of reactive systems, see for instance [22, Thm. 11.3].

The above proposition can be applied either to show that a given system $\mathcal{S}$ enjoys a liveness property, or that it satisfies a safety property. In the former case, one can try to find a morphism $f : \mathcal{T} \to \mathcal{S}$ from a simple 'test' system $\mathcal{T}$ that satisfies, modulo the translation, the required liveness property. In the latter case, one would look for a morphism $f : \mathcal{S} \to \mathcal{T}$ to a 'test' system satisfying, modulo the translation again, the interesting safety property.

In the sequel we turn to the problem of the concurrent realization of reactive systems. The above discussion, and Prop. 1 in particular, should justify our interest in the question of realization of morphisms as well.

3   Concurrent Realizations of Reactive Systems

In this section the notion of asynchronous system is recalled, and put forward as a formalization of the idea of concurrent behavior of reactive systems. We also recall the notion of labelled safe Petri net, to be used as the concurrent realizations. Then, a formal definition of the idea of a labelled Petri net *realizing* concurrent behavior of a reactive system represented as an asynchronous system is proposed. We discuss how difficult it is to *efficiently synthesize* concurrent behaviors of reactive systems as labelled Petri nets, and to synthesize their morphisms at the same time.

3.1   Asynchronous Systems

Asynchronous systems were introduced in 1986, independently by Shields and one of the authors, see [24, 3]. The idea was to generalize the notion of transition system so that *concurrency* gets reflected in the model.

An *asynchronous system* $\mathcal{A}$ is a pair $\mathcal{A} = \langle \mathcal{S}, \| \rangle$ where $\mathcal{S}$ is a deterministic transition system, and $\| \subseteq A \times A$ is an irreflexive and symmetric binary relation of *independence* defined on the alphabet of $\mathcal{S}$. The transition system underlying the asynchronous system should satisfy the following *swap* property with respect to the independence relation.

$$p \xrightarrow{a} q \xrightarrow{b} r \quad and \quad a \parallel b \quad implies \quad p \xrightarrow{b} s \xrightarrow{a} r \quad for \ some \ s \in S. \quad (1)$$

The swap property is intended to capture an idea that can be traced back to Mazurkiewicz, [17], that often *concurrent execution* of actions can be faithfully represented by independence of the actions. This, rather restrictive view, does not work well for *all* categories of concurrent devices. It does work, though, for an important class of safe Petri nets. As a result, *Mazurkiewicz traces* qualify as computations of asynchronous systems, cf. [3].

In the sequel only initialized asynchronous systems are considered, i.e., we assume that the underlying transition system of each asynchronous system is initialized. Given asynchronous systems $\mathcal{A}$ and $\mathcal{A}'$ their morphism $f : \mathcal{A} \to \mathcal{A}'$ is a morphism of the underlying initialized transition systems such that λ preserves independence in the following sense.

$$\lambda a \Downarrow \wedge \lambda b \Downarrow \wedge a \parallel b \quad implies \quad \lambda a \parallel' \lambda b \quad (2)$$

Asynchronous systems with morphisms defined above form a category **AS**.

Intuitively, the larger the independence relation, the more concurrent is the system. Consequently, asynchronous systems with empty independence relation may well be called *sequential systems*. Note that condition (2) is trivially satisfied by morphisms with a sequential system as the source. Therefore, the full subcategory of sequential systems is isomorphic to the category $d\mathbf{TS}$ of deterministic transition systems. We identify them in the sequel.

Clearly, a restriction of the independence relation of an asynchronous system preserves the validity of the swap condition. Thus, there is an evident forgetful functor that maps an asynchronous system to its underlying transition system. It is not difficult to verify that, under the identification, $d\mathbf{TS}$ becomes a coreflective subcategory of **AS**.

3.2 (Labelled) Petri Nets

Most of the material presented here and in the following sections is standard, cf. [23]. It is presented to make the paper self-contained and to fix the notation.

Let ω denote the set of natural numbers. Given a finite set X, a *multiset* on X is a function $M : X \to \omega$. In this paper only multisets over finite sets are considered. The set of all multisets on X is denoted μX. Relation $M' \leq M$, sum $M + M'$, and difference $M - M'$, are defined argumentwise, the latter is defined only under proviso $M' \leq M$. Subsets of X, qua characteristic functions, are identified with multisets in μX. Thus, elements of X, as singleton subsets of X, also live in μX, empty set $\emptyset$ is the empty multiset, finally, intersection $\cap$ and union $\cup$ of sets are conservatively extended to multisets as argumentwise minimum and maximum, respectively.

A *multirelation* $\beta : B \xrightarrow{r} B'$ is a multiset on $B \times B'$. By abuse of notation, a multirelation β is identified with a unique function $\beta : \mu B \to \mu B'$ defined by $(\beta M)b' = \Sigma_{b \in B} \beta \langle b, b' \rangle \cdot Mb$. As functions, multirelations are *monotone* and *additive*, i.e., satisfy $\beta(M + M') = \beta M + \beta M'$ and $\beta \emptyset = \emptyset$. In fact, every additive function comes this way. The *transpose* of β is a multirelation $\beta^{\mathsf{T}} : B' \xrightarrow{r} B$ defined by $\beta^{\mathsf{T}} b' b = \beta bb'$.

A finite *Petri net* $\mathcal{N}$ is a structure $\mathcal{N} = \langle B, E, F \rangle$ where B and E are finite disjoint sets of *places* and *events*, respectively, while $F : \mu(B \times E \cup E \times B)$ is a *flow multirelation*. A Petri net $\mathcal{N}$ is *marked* if, additionally, it is equipped with a marking $\hat{M} \in \mu B$ called the *initial* marking of $\mathcal{N}$.

We let e range over E, b range over B, M range over μB and γ range over μE. Notation $b^{\bullet}e$ for $F\langle b, e \rangle$ and $e^{\bullet}b$ for $F\langle e, b \rangle$ is often used. Also, taking the view of a multirelation as an additive function, we use $^{\bullet}e$ and $e^{\bullet}$ to denote the multisets of *preconditions* and *postconditions* of e, respectively. These, by additivity, are defined for arbitrary $\gamma \in \mu E$.

Firing relation $_{-} [\, _{-}\, \rangle \, _{-}$ on $\mu B \times \mu E \times \mu B$ is defined by: $M [\gamma \rangle M'$ iff $^{\bullet}\gamma \leq M$ and $M' = M - {^{\bullet}\gamma} + \gamma^{\bullet}$. We write $M [\gamma \rangle$, and say γ is *enabled* at M, if $^{\bullet}\gamma \leq M$.

The *case graph* of a Petri net $\mathcal{N}$ is a transition system $\mathcal{C}g(\mathcal{N}) = \langle \mu B, E, T \rangle$, where $M \xrightarrow{e} M'$ in $\mathcal{C}g(\mathcal{N})$ iff $M [e \rangle M'$. Thus, in the case graph all information about the concurrent execution of events in $\mathcal{N}$ is neglected. The case graph of a marked Petri net is defined as $\mathcal{C}g(\mathcal{N}, \hat{M}) = \mathcal{R}e(\mathcal{C}g(\mathcal{N}), \hat{M})$. Thus, only the reachable markings are considered as states. Let us recall that the set of reachable markings, denoted by $\mathcal{M}$, is the least set of markings such that $\hat{M} \in \mathcal{M}$ and $M_2 \in \mathcal{M}$ whenever $M_1 \in \mathcal{M}$ and $M_1 [e \rangle M_2$ for some e.

A marked Petri net $\mathcal{N}$ is *safe* if $\hat{M}$ is a set, and $e \in E$, M reachable and $M [\gamma \rangle$ imply that $^{\bullet}e$, $e^{\bullet}$, M and γ are sets. A *state machine* is a safe Petri net in which all these sets are singletons.

A *labelled* Petri net is a structure $\mathcal{L} = \langle \mathcal{N}, A, \ell \rangle$ where $\mathcal{N}$ is a Petri net, the *underlying* Petri net of $\mathcal{L}$, A is a set of *actions*, ranged over by a, and $\ell : E \to A$ is a *labelling* function. If $\mathcal{N}$ is marked then $\mathcal{L}$ is a marked labelled Petri net.

The case graph $\mathcal{C}g(\mathcal{L})$ of a labelled net $\mathcal{L}$ is a transition system $\langle \mu B, A, T \rangle$ where T is the least relation on $\mu B \times A \times \mu B$ that satisfies: $M [e \rangle M'$ implies $M \xrightarrow{\ell e} M'$. If $\mathcal{L}$ is marked, then $\mathcal{C}g(\mathcal{L}, \hat{M}) = \mathcal{R}e(\mathcal{C}g(\mathcal{L}), \hat{M})$.

3.3   Morphisms of (Labelled) Petri Nets

The notion of a morphism used in this paper was introduced by Vogler already in 1991, see [26]. His idea was to extend the notion of a process of safe nets to arbitrary P/T-nets. For that purpose it turned out that a new notion, more general than the the one earlier introduced by Winskel, is useful.

Unaware of this research we have rediscovered and studied the same notion in [5]. Our goal was to find a simple functorial realization of a sequential system as a labelled state machine, see Proposition 7. It turns out that this task cannot be achieved with the simpler notion of morphism introduced by Winskel.

Definition 1. *Let* N *and* N' *be Petri nets. A* general morphism $f : N \to N'$ *is a pair* $f = \langle \beta, \eta \rangle$ *where* $\eta : E \rightharpoonup E'$ *is a partial function and* $\beta : \mu B \to \mu B'$ *is a multirelation which together fulfill the following conditions.*

1. ${}^\bullet(\eta e) \le \beta({}^\bullet e)$.
2. $\beta(e^\bullet) = \beta({}^\bullet e) - {}^\bullet(\eta e) + (\eta e)^\bullet$.

If the nets are marked then, additionally, f should preserve the initial marking.

3. $\beta \hat{M} = \hat{M}'$.

In the definition above, the partial function η is lifted to a multirelation. Then, in particular, $\eta e = \emptyset$ whenever η is undefined on e. If this is the case, condition 2 simplifies to $\beta(e^\bullet) = \beta({}^\bullet e)$.

The terminology *general morphism* is justified by Proposition 3 which shows that this class of morphisms is in a sense optimal. First, though, let us verify that what we obtain is a category, denoted **PN** in the sequel.

Proposition 2. *Petri nets and general morphisms form a category.*

Proof. Obviously, the pair of identity functions (qua partial function and multirelation, respectively) form a morphism of Petri nets.

Conditions 1 and 2 can be checked by easy calculations. For marked nets, the composition of multirelations fulfills also condition 3.   □

Conditions 1 and 2 of Definition 1 together guarantee that general morphisms map steps in the source net into steps in the target net.

Lemma 1. *Let* $\langle \beta, \eta \rangle : N \to N'$ *be a morphism in* **PN**. *Then* $M_1 \, [e\rangle \, M_2$ *in* N *implies* $\beta M_1 \, [\eta e\rangle \, \beta M_2$ *in* N'.

Proof. Let $M_1 \, [e\rangle \, M_2$ in N, i.e., ${}^\bullet e \le M_1$ and $M_2 = M_1 - {}^\bullet e + e^\bullet$. Then, $\beta({}^\bullet e) \le \beta M_1$ and $\beta M_2 = \beta M_1 - \beta({}^\bullet e) + \beta(e^\bullet)$. By conditions 1 and 2 of Definition 1, ${}^\bullet(\eta e) \le \beta M_1$ and $\beta M_2 = \beta M_1 - {}^\bullet(\eta e) + (\eta e)^\bullet$, i.e., $\beta M_1 \, [\eta e\rangle \, \beta M_2$.   □

Interestingly, the converse to Lemma 1 also holds.

Proposition 3. *Let* N, N' *be Petri nets, while* $\beta : \mu B \to \mu B'$ *and* $\eta : E \rightharpoonup E'$ *satisfy the conclusion of Lemma 1. Then,* $\langle \beta, \eta \rangle : N \to N'$ *in* **PN**.

Proof. Clearly, ${}^\bullet e \, [e\rangle \, e^\bullet$ in N. Thus, $\beta({}^\bullet e) \, [\eta e\rangle \, \beta(e^\bullet)$ by the conclusion of Lemma 1. Now, conditions 1 and 2 follow by the definition of firing relation.   □

Another way of reading Lemma 1 is that the general morphisms of Petri nets give rise to morphisms between the corresponding case graphs.

Proposition 4. *The case graph construction is a functor* $Cg : $ **PN** $\to$ **TS**.

Proof. Immediate, by Lemma 1.   □

Thus, Proposition 3 says that **PN** is the largest category of Petri nets with morphisms given as pairs $f = \langle \beta, \eta \rangle$ such that the case graph construction maps f to a morphism of transition systems.

The definition of morphisms of labelled Petri nets is obtained by adding additional component to cope with the labels.

Definition 2. *A general morphism* $f : \mathcal{L} \to \mathcal{L}'$ *of labelled Petri nets is a triple* $f = \langle \beta, \eta, \lambda \rangle$ *where* $\langle \beta, \eta \rangle$ *is a general morphism of the underlying nets, and* $\lambda : A \rightharpoonup A'$ *is a partial function such that the following strong equality holds.*

4. $\ell; \lambda = \eta; \ell'$.

It is immediate to verify that labelled Petri nets with their morphisms constitute a category, denoted $\ell\mathbf{PN}$ in the sequel. The following generalization of Proposition 4 also holds.

Proposition 5. *The case graph construction on labelled nets, extended with* $Cg(\langle \beta, \eta, \lambda \rangle) = \langle \beta, \lambda \rangle$ *gives a functor* $Cg : \ell\mathbf{PN} \to \mathbf{TS}$.   □

3.4   Comparison with Winskel's Definition

It seems that Reisig was the first to view Petri nets as 2-sorted algebras. The sort of places lives in the category of multisets, with multirelations (additive functions) as morphisms. The sort of events lives in the category of partial functions. Operations are: pre- and post-conditions and, for marked nets, initial marking.

In [28, 29] Winskel proposed that morphisms of nets should be (presentations of) homomorphisms of such algebras. Consequently, a pair $\langle \beta, \eta \rangle$ as in Definition 1 is a *Winskel morphism* if it satisfies the following condition.

$- \ ^{\bullet}(\eta e) = \beta(^{\bullet}e) \text{ and } (\eta e)^{\bullet} = \beta(e^{\bullet}).$

Morphisms of marked nets preserve the initial marking as well: $\beta\hat{M} = \hat{M}'$.

It has been shown in [28] that Winskel morphisms satisfy Lemma 1. In fact, the following is immediate.

Proposition 6. *Every Winskel morphism is a general morphism.*   □

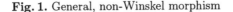

$$\mathcal{N}_{2e} \qquad\qquad\qquad \mathcal{N}_e$$

Fig. 1. General, non-Winskel morphism

The converse does not hold. To see this consider Fig. 1. It depicts a general morphism of marked Petri nets, let us call it $\langle \beta, \eta \rangle$, which is not a Winskel morphism since $\emptyset = {}^\bullet(\eta e_1) < \beta({}^\bullet e_1)$. In fact, it is easy to see that there is *no* Winskel morphism f such that $Cg(f) = Cg(\langle \beta, \eta \rangle)$.

Another, more complex example, is presented on Fig. 2. It demonstrates that the need to use general, non-Winskel morphisms does not depend on the partiality of the event part of morphisms. The morphism on Fig. 2 explains how the sequential execution of e_1's and e_2's in $\mathcal{N}_+$ can be simulated by a parallel execution of e_1 and e_2 in $\mathcal{N}_\parallel$. The multirelation component β of the morphisms is a relation. Clearly, the initial marking is preserved. The singleton initial marking of net $\mathcal{N}_+$ is shared as the precondition of both e_1' and e_2''. In the target net $\mathcal{N}_\parallel$, the images of e_1' and e_2'' do not have a common precondition. Thus, for example the inequality ${}^\bullet(\eta e_1') < \beta({}^\bullet e_1')$ is strict. In fact, all inequalities are strict in this case. Again, there is no such Winskel morphism from $\mathcal{N}_+$ to $\mathcal{N}_\parallel$.

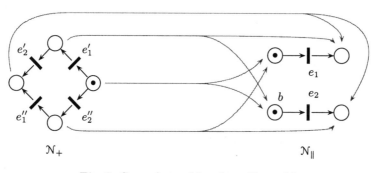

Fig. 2. General morphism from $\mathcal{N}_+$ to $\mathcal{N}_\parallel$

3.5   Properties of Case Graph Functor

Since nets are to serve as realizations of reactive systems, the latter represented as transition systems, it is important to explain what it means that a net $\mathcal{N}$ *realizes* a transition system $\mathcal{S}$. The answer put forward by Ehrenfeucht and Rozenberg in their seminal paper was that $\mathcal{S}$ should be, up to inessential renaming of components, equal to the case graph of $\mathcal{N}$. Formally, we demand equality up to an isomorphism of transition systems, notation $Cg(\mathcal{N}) \simeq \mathcal{S}$.

Thus, the problem of finding *a* realization of a reactive system could be cast as the problem of finding a construction Sn which, given a transition systems $\mathcal{S}$, synthesizes a net $Sn(\mathcal{S})$ such that $Cg(Sn(\mathcal{S})) \simeq \mathcal{S}$. Ideally, $Sn : \mathbf{TS} \to \mathbf{PN}$ would be a functor *inverse* to Cg. To see how hard is the task to find such a functor let us consider two properties that Cg could enjoy to make the task easier.

First, the case graph functor is not faithful. That is, given two Petri nets $\mathcal{N}_1$ and $\mathcal{N}_2$, and a morphism $f : Cg(\mathcal{N}_1) \to Cg(\mathcal{N}_2)$ between their case graphs, there can be many general Petri net morphisms $\langle \beta, \eta \rangle : \mathcal{N}_1 \to \mathcal{N}_2$ such that

$f = Cg(\langle \beta, \eta \rangle)$. To see an example consider a somewhat more saturated version $\mathcal{N}_+^{\parallel}$ of $\mathcal{N}_+$ and the morphism from $\mathcal{N}_+^{\parallel}$ to $\mathcal{N}_{\parallel}$, all described on Fig. 3.

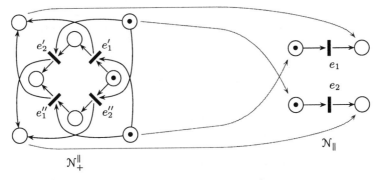

Fig. 3. General morphism from $\mathcal{N}_+^{\parallel}$ to $\mathcal{N}_{\parallel}$

There is a morphism from $\mathcal{N}_+^{\parallel}$ to $\mathcal{N}_+$ which simply erases from the source all places that are not present in the target. This morphism is mapped to an isomorphism of the corresponding case graphs. Composing it with the morphism presented on Fig. 2 yields another morphism from $\mathcal{N}_+^{\parallel}$ to $\mathcal{N}_{\parallel}$. Although different, both morphisms are glued by the case graph functor.

Since Cg is not faithful it follows that each inverse construction defined already on transition systems would have to choose one of possibly many morphisms. But the case graph functor is not full either, i.e., there are nets $\mathcal{N}_1$ and $\mathcal{N}_2$, and a morphism of their case graphs $f : Cg(\mathcal{N}_1) \to Cg(\mathcal{N}_2)$ such that there is *no* morphism $\langle \beta, \eta \rangle : \mathcal{N}_1 \to \mathcal{N}_2$ so that $f = Cg(\langle \beta, \eta \rangle)$. A counterexample follows. Hence, the initial choice of the nets realizing a transition system has to be wise.

In the example on Fig. 4 both Petri nets have isomorphic case graphs. Yet, no morphism from $\mathcal{N}'_+$ to $\mathcal{N}_+$ realizes the isomorphism between their case graphs.

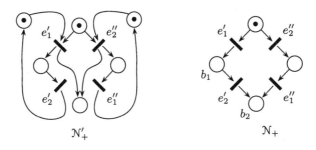

Fig. 4. Two realizations of $Cg(\mathcal{N}_+)$

Indeed, assume there is a morphism $\langle \beta, \eta \rangle : \mathcal{N}'_+ \to \mathcal{N}_+$ which maps e'_2 in $\mathcal{N}'_+$ to e'_2 in $\mathcal{N}_+$. Then this morphism relates the only postcondition of e'_2 in $\mathcal{N}'_+$ to place b_2 as the only postcondition of e'_2 in $\mathcal{N}_+$ as the following argument shows. Condition 2 of Definition 1 implies $\beta({}^\bullet e'_2) + b_2 = \beta(e'_2{}^\bullet) + b_1$. So $\beta(e'_2{}^\bullet)b_2 > 0$ follows from $b_1 \neq b_2$. The postcondition constituting $e'_2{}^\bullet$ in $\mathcal{N}'_+$ belongs to the initial marking. Thus, it may only be related to a place in the initial marking of $\mathcal{N}_+$, and we arrive at a contradiction.

Incidently, the isomorphism in the opposite direction from $Cg(\mathcal{N}_+)$ to $Cg(\mathcal{N}'_+)$ can be realized, cf. Sect. 4.4, but not as a Winskel morphism.

3.6   Labelled Petri Nets as Concurrent Realizations of Reactive Systems

We have started this section by arguing that to talk about the concurrent realization of reactive systems one has to be able to express concurrency in the underlying formal model. Subsequently, asynchronous systems have been suggested as a suitable extension of transition systems. With accord to the discussion in Sect. 3.5 we insist that a net $\mathcal{N}$ realizing given asynchronous system $\mathcal{A} = \langle \mathcal{S}, \| \rangle$ should have its case graph isomorphic to the transition system underlying the asynchronous system, i.e., $Cg(\mathcal{N}) \simeq \mathcal{S}$. However, in most applications one is neither interested in *all* markings of a Petri net, nor in *all* states of a transition system. Most of the time people use initialized transition systems and marked Petri nets. Consequently, we restrict the requirement to the sets of reachable markings and states, respectively. In fact, this requirement has already been built into the definition of the case graph of a marked Petri net.

Now, concurrency is represented on both sides. In a Petri net it even comes in two forms. Two events are *structurally* concurrent if their pre- and postconditions are disjoint, while *dynamic* concurrency is the ability to execute a multiset of events at a reachable marking. Concurrency in an asynchronous system is the ability to perform two or more independent actions one after another, and so in any order. Thus, in order for a net to be a *concurrent realization* of an asynchronous system it should be the case that the concurrency represented on both sides agree. The following definition intends to capture the above ideas and is central to further developments.

Definition 3. *A labelled marked net* $\mathcal{L} = \langle \mathcal{N}, \hat{M}, A, \ell \rangle$ *is called a* realization *of a reachable asynchronous system* $\mathcal{A} = \langle \mathcal{S}, \| \rangle$, *where* $\mathcal{S} = \langle S, \hat{s}, A, T \rangle$, *if the following hold.*

1. $Cg(\mathcal{L}) = \langle \mathcal{M}, \hat{M}, A, [\_ \rangle \rangle$ *equals* $\mathcal{S}$ *up to an isomorphism that is identity on labels.*
2. $M \in \mathcal{M}$ *and* $M [e_1 + e_2 \rangle$ *imply* $\ell e_1 \| \ell e_2$;
3. $\ell e_1 \| \ell e_2$ *implies* $({}^\bullet e_1 \cup e_1^\bullet) \cap ({}^\bullet e_2 \cup e_2^\bullet) = \emptyset$.

A morphisms $\langle \beta, \eta, \lambda \rangle : \mathcal{L} \to \mathcal{L}'$ *is a* realization *of a morphisms* $f : \mathcal{A} \to \mathcal{A}'$ *of asynchronous systems if* $Cg(\langle \beta, \eta, \lambda \rangle)$ *equals (up to the isomorphism)* f.

The first condition above generalizes the usually accepted requirement that the sequential case graph of a Petri net realization of a transition system should be isomorphic with this transition system, and that the alphabet of the transition system should play the role of the set of events in the net.

Condition 3(2) says that the dynamic concurrency present in the realization is always justified in the specification.

Finally, condition 3(3) ensures that all events labelled as potentially concurrent in the specification are structurally concurrent in the realization.

Without condition 3(3) the problem would admit a trivial solution. This is the case when one considers sequential systems. One can, for instance, disambiguate different occurrences of the same action by means of a labeling. Suppose that, together with the labeling, an elementary transition system is obtained in this way. Then one can apply the construction based on regions, cf. [20], and carry the labeling over to thus synthesized elementary net. The result would be a labelled safe Petri net saturated with places, and with reachable markings as the distinguished family.

Here, thanks to general morphisms of [26, 5], a simpler realization functor from sequential systems to labelled safe Petri nets is put forward. In practical applications, see [8], it is better to work with Petri nets which are as small as possible. Thus, many synthesis techniques have been proposed which aim at *minimizing* the number of places in the synthesized net, cf. [1, 9, 10]. None of these techniques, though, have been shown to be functorial, even for sequential systems. Our proposal is based on simple kind of sequential nets known as state machines.

Formally, we define functor Sm returning a state machine for *every* transition system. Thus, by convention, we can consider Sm to be defined on all sequential (asynchronous) systems.

Definition 4. *Let* $S = \langle S, \hat{s}, A, T \rangle$ *be a transition system. Then* $Sm(S)$ *is a labelled Petri net* $\langle S, T, F, \hat{s}, A, \ell \rangle$ *where* $^{\bullet}(p \xrightarrow{a} q) = p$, $(p \xrightarrow{a} q)^{\bullet} = q$ *and* $\ell(p \xrightarrow{a} q) = a$.

Let $\langle \sigma, \lambda \rangle : S \to S'$ *be a morphism of transition systems. Then the morphism of labelled Petri nets is defined as* $Sm(\langle \sigma, \lambda \rangle) = \langle \sigma, \eta, \lambda \rangle : Sm(S) \to Sm(S')$ *Here* $\eta : T \rightharpoonup T'$ *fulfills* $\eta(p \xrightarrow{a} q) = \sigma p \xrightarrow{\lambda a} \sigma q$ *if* λa *is defined and undefined otherwise.*

Proposition 7. Sm *is a realization functor for sequential asynchronous systems.*

Proof. Obviously, $Cg(Sm(\langle S, \emptyset \rangle)) = S$ and $Cg(Sm(\langle \sigma, \lambda \rangle)) = \langle \sigma, \lambda \rangle$.   □

The functoriality of the construction hinges on the use of general morphisms. In fact, general morphisms were introduced precisely to achieve this functoriality. If a morphism between transition systems is total on actions, i.e., *synchronous* in Winskel's terminology, then this morphism, qua a Petri net morphism, is a Winskel morphism.

Since [14], which has introduced the idea of *region*, the problem of synthesis of nets from transition systems has received a lot of attention. Here, [1] serves as a recent account on the developments in theoretical net synthesis based on the theory of regions. Yet, to our knowledge, all theoretical approaches to synthesis have always sought to reconstruct the actions of a transition system as the events of the synthesized net. As a result, the theoretical synthesis has limited applicability, i.e., not all transition systems can be synthesized. For instance the transition system S with transitions $\hat{s} \xrightarrow{a} p_1 \xrightarrow{b} p_2$ and $\hat{s} \xrightarrow{b} q_1 \xrightarrow{a} q_2$, and with $p_2 \neq q_2$, is not synthesizable within this framework.

This is in marked contrast to more practical approaches, see [8–10], which strive always to provide a solution, even at a price of resorting to heuristics. One of the techniques used in tools like Petrify ([8]), when the theoretical synthesis fails, is *action splitting* in the elaborated transition system. Then, the process of theoretical synthesis can be applied to the modified transition system with better chances of success. Action splitting is nothing else than introduction of an implicit labelling of events. From this perspective, introduction of the labelling into the definition of concurrent realization can be seen as a step towards reconciliation of the theoretical approach with the practical needs.

Our notion of concurrent realization is very general. In particular, it admits arbitrary labelled marked Petri nets as potential concurrent realizations of asynchronous systems. In fact, without too much work the notion can be generalized even further to allow classes of concurrent behaviors more general than asynchronous systems, see e.g., [11, 25].

On the other hand, the addition of labelling makes even a small class of safe nets very expressive. The category of safe Petri nets is attractive to work with since it admits many constructions, cf. [6, 28, 30]. Therefore, in the next sections of the paper we develop a theory in which labelled safe Petri nets serve as the pool in which concurrent realizations are sought.

4   Concurrent Realizations of Concrete Asynchronous Systems

In this section the notion of a *concrete* asynchronous system is recalled, and some recent refinements and results are presented. Then, we show that arbitrary finite diagrams in the full subcategory of concrete asynchronous systems admit systematic realization in the category of safe labelled Petri nets.

4.1   Concrete Asynchronous Systems

Recently, see [19], Morin has provided a characterization of those asynchronous systems which are isomorphic to the *mixed products* of sequential systems computed in a subcategory of reachable asynchronous systems.

It all starts with a choice of suitable morphisms. Recall that in Def. 3(1) we insisted that the morphism does not rename the actions. This property characterizes the notion of morphism used by Morin. Formally speaking, morphisms used in [19] do not constitute a subclass of morphisms studied in [3, 30], as recalled in Sect. 3.1, since their state part is often partial, defined on the reachable states only. The following modification rectifies this minor difference, but it should be kept in mind when we recall Morin's results. The terminology used follows [4].

A morphism $f : \mathcal{S} \to \mathcal{S}'$ of transition systems is *rigid* whenever $A \supseteq A'$ and $\lambda : A \rightharpoonup A'$ is the partial identity transposed to this inclusion, i.e., $\lambda a = a$ iff $a \in A'$, and undefined otherwise. A morphism of asynchronous systems is *rigid* iff it is a rigid morphism of the underlying transition systems.

One can easily verify that the class of rigid morphisms includes identities, and is closed under composition. Hence, asynchronous systems with rigid morphisms form a subcategory $\mathbf{AS}^r$ of $\mathbf{AS}$.

Note that transition systems $\mathcal{S}$ and $\mathcal{S}'$ are rigid isomorphic, notation $\mathcal{S} \overset{r}{\simeq} \mathcal{S}'$, iff both share the same alphabet and if there exist a bijection between their states which preserves and reflects the transition relation. In the sequel we often consider rigid isomorphisms *up to reachable parts*, notation $\mathcal{S} \overset{r}{\sim} \mathcal{S}'$, defined by $\mathcal{R}e(\mathcal{S}) \overset{r}{\simeq} \mathcal{R}e(\mathcal{S}')$. Thus, Def. 3(1) could now be restated as $\mathcal{C}g(\mathcal{L}) \overset{r}{\simeq} \mathcal{S}$.

Asynchronous systems $\mathcal{A}$ and $\mathcal{A}'$ are rigid isomorphic, notation $\mathcal{A} \overset{r}{\simeq} \mathcal{A}'$, iff $\mathcal{S} \overset{r}{\simeq} \mathcal{S}'$ and both share the same independence. Similar characterization applies to $\mathcal{A} \overset{r}{\sim} \mathcal{A}'$ defined by $\mathcal{R}e(\mathcal{A}) \overset{r}{\simeq} \mathcal{R}e(\mathcal{A}')$.

In [19] Morin has shown that the category of asynchronous systems with rigid morphisms admits products, called *mixed* for historical reasons, cf. [13].

The *mixed product* of a family $(\mathcal{A})_{i \in I}$ of asynchronous systems, denoted $\prod_{i \in I}^r \mathcal{A}_i$, is an asynchronous system $\langle \prod_{i \in I} S_i, (\hat{s}_i)_{i \in I}, \bigcup_{i \in I} A_i, T, \| \rangle$, where

- $(p_i)_{i \in I} \overset{a}{\longrightarrow} (q_i)_{i \in I}$ iff either $a \in A_i$ and $p_i \overset{a}{\longrightarrow} q_i$ or $a \notin A_i$ and $p_i = q_i$,
- $a \| b$ iff $a \|_i b$ whenever $\{a, b\} \subseteq A_i$.

Thus, an a-step in the product comes about by a *synchronous* execution of a-steps in all components with a in their alphabets, while the other components remain idle.

The reader should note that when all the components of a product are sequential the independence in the product is rather special. Namely, two actions are independent in the product iff they never occur in the same component.

The kth projection $\pi_k : \prod_{i \in I}^r \mathcal{A}_i \to \mathcal{A}_k$ consists of the kth projection on states, and the partial function transpose to the inclusion $A_k \subseteq \bigcup_{i \in I} A_i$. Now, we can quote the following result proved in [19, Lemma 1.3].

Proposition 8. *Mixed products with their projections are categorical products in the category of asynchronous systems with rigid morphisms.* □

An asynchronous system $\mathcal{A}$ is *concrete* if there exists a family of sequential asynchronous systems $(\mathcal{A}_i)_{i \in I}$ such that $\mathcal{A} \overset{r}{\sim} \prod_{i \in I}^r \mathcal{A}_i$.

As an example consider two sequential asynchronous systems, one with transitions $p \overset{a}{\longrightarrow} q$ and $p \overset{b}{\longrightarrow} q'$ and the other one with transitions $r \overset{c}{\longrightarrow} r' \overset{a}{\longrightarrow} r''$. Their mixed product contains unreachable states, see Fig. 5–6. Both, this product and its reachable part, are considered concrete.

Now, let A be a set and $\|$ an independence relation on A. We let $\mathbb{Q} = \mathbb{Q}(A, \|)$ be the family of *all* cliques of dependent actions, $\mathbb{Q} = \{\varDelta \subseteq A \mid (\varDelta \times \varDelta) \subseteq \nparallel\}$. The subfamily of *maximal* cliques is denoted by $\mathbb{X}$.

Let $\mathcal{A} = \langle S, \hat{s}, A, T, \| \rangle$ be an asynchronous system. Given $\varDelta \subseteq A$, one defines an equivalence relation on S, also denoted by $\varDelta$, as the least equivalence relation such that the following holds.

$- p \overset{a}{\longrightarrow} q$ and $a \notin \varDelta$ implies $p \varDelta q$.
$- p \varDelta r$ and $p \overset{a}{\longrightarrow} q$ and $r \overset{a}{\longrightarrow} s$ implies $q \varDelta s$.

The construction above allows to consider quotients of asynchronous systems.

Let $\varDelta \in \mathbb{Q}(A, \|)$. Then, a $\varDelta$-*quotient* of an asynchronous system $\mathcal{A} = \langle S, \hat{s}, A, T, \| \rangle$, is the quotient sequential system $\kappa_\varDelta(\mathcal{A}) = \langle S/\varDelta, [\hat{s}]_\varDelta, \varDelta, T_\varDelta, \emptyset \rangle$, where $T_\varDelta$ is the least relation such that $p \overset{a}{\longrightarrow} q$, $a \in \varDelta$, implies $[p]_\varDelta \overset{a}{\longrightarrow} [q]_\varDelta$.

Above, and in the sequel, $[s]_\varDelta$ stands for the class of states $\varDelta$-equivalent to s. Each quotient $\kappa_\varDelta(\mathcal{A})$ comes with a rigid *quotient* morphism $[\_] : \mathcal{A} \to \kappa_\varDelta(\mathcal{A})$ given on states by $s \mapsto [s]_\varDelta$. The assumption $\varDelta \in \mathbb{Q}(A, \|)$ is necessary and sufficient to ensure that the morphism preserves independence relation, cf. (2).

Morin has shown, cf. [19, Lemma 2.2], that (the reachable part of) a mixed product of sequential systems is isomorphic to (the reachable part of) the mixed product of its quotients.

Lemma 2. *Put* $\mathcal{A} = \prod_{i \in I}^r \mathcal{A}_i$, *and let* A_i *denote the alphabet of* $\mathcal{A}_i$, *for* $i \in I$. *Then* $\mathcal{A} \overset{r}{\sim} \prod_{i \in I}^r \kappa_{A_i}(\mathcal{A})$. $\square$

Let $\mathbb{M} \subseteq \mathbb{Q}$ be a family of cliques of the dependence relation $\nparallel$ of an asynchronous system $\mathcal{A}$. Consider the following properties.

state-state $\mathbb{M}$-*separation*	$p \neq q$	*implies*	$(\exists \varDelta \in \mathbb{M}) \, [p]_\varDelta \neq [q]_\varDelta$ (3)
state-action $\mathbb{M}$-*separation*	$p \overset{a}{\longmapsto}$	*implies*	$(\exists \varDelta \in \mathbb{M}, a \in \varDelta) \, [p]_\varDelta \overset{a}{\longmapsto}$ (4)
$\mathbb{M}$ *fully captures* $\|$	$a \nparallel b$	*implies*	$(\exists \varDelta \in \mathbb{M}) \, \{a, b\} \subseteq \varDelta$ (5)

Now, the two $\mathbb{M}$-separation axioms (3) and (4) provide a simple criterion for deciding when a reachable asynchronous system is concrete, cf. [19, Thm. 2.3].

Theorem 1 (Morin). *A reachable asynchronous system* $\mathcal{A}$ *is concrete iff there exists a family* $\mathbb{M} \subseteq \mathbb{Q}$ *such that axioms (3) and (4) are satisfied.* $\square$

Moreover, given $\mathbb{M} \subseteq \mathbb{Q}$ such that reachable $\mathcal{A}$ satisfies both $\mathbb{M}$-separation axioms, it follows that $\mathcal{A} \overset{r}{\cong} \mathcal{R}e\left(\prod_{\varDelta \in \mathbb{M}}^r \kappa_\varDelta(\mathcal{A})\right)$. If $\mathbb{M}' \subseteq \mathbb{Q}$ is another family of cliques such that $(\forall \varDelta \in \mathbb{M}) \, (\exists \varDelta' \in \mathbb{M}') \, \varDelta \subseteq \varDelta'$, then $\mathcal{A}$ satisfies $\mathbb{M}'$-separation axioms as well. Hence, in the quest for a family that ensures separation axioms it is enough to consider $\mathbb{M}$ such that $\mathbb{X} \subseteq \mathbb{M}$, e.g., $\mathbb{M} = \mathbb{X}$ or $\mathbb{M} = \mathbb{Q}$.

4.2   Concrete Asynchronous Systems Revisited

Let $\|$ be an independence relation on a set A. Assume that there is a construction that assigns to the family $\mathbb{Q} = \mathbb{Q}(A, \|)$ of all cliques a subfamily $\mathbb{M} \subseteq \mathbb{Q}$ which also covers A, i.e., $\bigcup \mathbb{M} = A$.

Then, given an asynchronous system $\mathcal{A}$ with actions A and independence $\|$, we associate with it a family $\mathcal{F}_{\mathbb{M}}(\mathcal{A}) = \{\kappa_\Delta(\mathcal{A}) \mid \Delta \in \mathbb{M}\}$ of quotients.

The family of the corresponding quotient morphisms gives a cone in the category of asynchronous systems with rigid morphisms. Thus, from the general nonsense it follows that there exists a unique rigid $\epsilon_\mathcal{A}^{\mathbb{M}} : \mathcal{A} \to \prod^r \mathcal{F}_{\mathbb{M}}(\mathcal{A})$.

Note that the products in full subcategories of reachable asynchronous systems are computed as reachable parts of the products in **AS**, resp., **AS**^{r}. Then, if $\mathcal{A} = \mathcal{R}e(\mathcal{A})$ is reachable we obtain $\epsilon_\mathcal{A}^{\mathbb{M}} : \mathcal{A} \to \mathcal{R}e(\prod^r \mathcal{F}_{\mathbb{M}}(\mathcal{A}))$ by taking appropriate corestriction on the state component.

Here comes an elementary characterization of $\epsilon_\mathcal{A}^{\mathbb{M}}$.

Lemma 3. $\epsilon_\mathcal{A}^{\mathbb{M}} : \mathcal{A} \to \mathcal{R}e(\prod^r \mathcal{F}_{\mathbb{M}}(\mathcal{A}))$ *is a rigid morphism* $\epsilon_\mathcal{A}^{\mathbb{M}} = \langle \sigma, \lambda \rangle$ *where* $(\sigma s) = ([s]_\Delta)_{\Delta \in \mathbb{M}}$ *and* $\lambda = \mathrm{id}_A$.

Moreover, λ *reflects the independence whenever* $\mathbb{M}$ *satisfies (5).*

Proof. Consider $\langle \sigma, \lambda \rangle$ as defined. λ can indeed be taken as the identity because of $\bigcup \mathbb{M} = A$. Given a transition $p \xrightarrow{a} q$ in $\mathcal{A}$, either $a \in \Delta$, then $[p]_\Delta \xrightarrow{a} [q]_\Delta$, or $a \notin \Delta$, then $[p]_\Delta = [q]_\Delta$. This is precisely the definition of an a-transition in the product.

Two independent actions cannot belong to the same clique, hence they are independent in the product too. The converse holds when $\mathbb{M}$ fully captures the independence relation, (5). □

Now, Theorem 1 can be reconstructed.

Proposition 9. *Let* $\mathcal{A}$ *be a reachable asynchronous system. Then, the following are equivalent.*

1. $\mathcal{A}$ *is concrete.*
2. *There exists a family* $\mathbb{M} \subseteq \mathbb{Q}$, $A = \bigcup \mathbb{M}$, *such that* $\epsilon_\mathcal{A}^{\mathbb{M}} : \mathcal{A} \to \mathcal{R}e(\prod^r \mathcal{F}_{\mathbb{M}}(\mathcal{A}))$ *is a rigid isomorphism.*
3. *There exists a family* $\mathbb{M} \subseteq \mathbb{Q}$, $A = \bigcup \mathbb{M}$, *such that* $\mathcal{A}$ *fulfills both* $\mathbb{M}$-*separation axioms (3) and (4), and axiom (5) too.*

Moreover, each $\mathcal{R}e(\prod^r \mathcal{F}_{\mathbb{M}}(\mathcal{A}))$ *satisfies* $\mathbb{M}$-*separation axioms.*

Proof. Suppose $\mathcal{A} \overset{r}{\simeq} \mathcal{R}e(\prod_{i \in I}^r \mathcal{A}_i)$ for some family of sequential systems. Define $\mathbb{M} = \{A_i \mid i \in I\}$. Clearly, $\mathbb{M} \subseteq \mathbb{Q}$ and $A = \bigcup \mathbb{M}$. To see that $\epsilon_\mathcal{A}^{\mathbb{M}}$ is an isomorphism let us show that every projection from $\mathcal{A}$ onto $\mathcal{A}_k$ factorizes through the equivalence relation defined by clique A_k. In the base case $(s_i)_{i \in I}$ is A_k-equivalent to $(s_i')_{i \in I}$ due to the existence of an a-transition, with $a \notin A_k$, from the former vector to the latter. But then $s_k = s_k'$. This equality is preserved in the inductive case. The above gives morphisms from $\kappa_{A_k}(\mathcal{A})$ to $\mathcal{A}_k$, and we easily check that, together, they give rise to an inverse to $\epsilon_\mathcal{A}^{\mathbb{M}}$.

The converse implication follows by definition.

The state-state separation axiom is equivalent to $\epsilon_\mathcal{A}^{\mathbb{M}}$ being a 1-1 mapping.

The image of $\mathcal{A}$ under $\epsilon_\mathcal{A}^{\mathbb{M}}$ consists of tuples $([s]_\Delta)_{\Delta \in \mathbb{M}}$, $s \in S$. The initial state in the product is of this shape. To see that all reachable states in the product belong to the image one has to check that given a transition $([s]_\Delta)_{\Delta \in \mathbb{M}} \xrightarrow{a}$ in a product we can find a single state s' such that $s' \xrightarrow{a}$ in $\mathcal{A}$ and $s \Delta s'$ for all Δ. In view of the state-state separation lemma, s is the only candidate for $s \xrightarrow{a}$. Hence, in the context of state-state separation axiom $\epsilon_{\mathcal{A}}^{\mathbb{M}}$ is onto iff the state-action separation axiom holds.

Finally, if $\mathbb{M}$ does not satisfy (5) it can be enlarged to a family of cliques that fully captures $\parallel$ without violating the $\mathbb{M}$-separation axioms.

The last statement of the theorem follows from the previous ones.   □

An asynchronous system that does not fulfill state-state separation axiom has states p, q and $\hat{s}$, transitions $\hat{s} \xrightarrow{a} p \xrightarrow{b} q$ and $\hat{s} \xrightarrow{b} q \xrightarrow{a} q$, with $a \parallel b$. Then the only cliques are singletons, and $\hat{s}$ and p are both $\{b\}$- and $\{a\}$-equivalent.

Consider now an asynchronous system with transitions $\hat{s} \xrightarrow{a} p$ and $\hat{s} \xrightarrow{a} q$, all states different, and with relation $a \parallel b$. Again, the only cliques are singletons, and $p \xrightarrow{b}$ while $\hat{s} \xrightarrow{b}$ and $\hat{s}$ is $\{b\}$-equivalent to p, i.e., $[p]_{\{b\}} \xrightarrow{b}$. The reader will check that the product of the two quotients of the system has an extra state allowing for a concurrent execution of a and b.

Consider a morphism $f = \langle \sigma, \lambda \rangle : \mathcal{A}' \to \mathcal{A}$ of asynchronous systems.

For each $\Delta \in \mathbb{Q}$ let $\lambda^{\leftarrow}(\Delta) = \{a' \in A' \mid \lambda a' \in \Delta\}$ be the *strict inverse image* of Δ via λ. Above, by convention, $\lambda a' \in \Delta$ holds only when $\lambda a'$ is defined.

Lemma 4. *Let $f : \mathcal{A}' \to \mathcal{A}$ be a morphism of asynchronous systems. Given $\Delta \in \mathbb{Q}$ put $\Delta' = \lambda^{\leftarrow}(\Delta)$. Then, $\Delta' \in \mathbb{Q}'$. Moreover, $p \Delta' q$ implies $\sigma p \Delta \sigma q$. Hence, $f_\Delta : \kappa_{\Delta'}(\mathcal{A}') \to \kappa_\Delta(\mathcal{A})$, where $\sigma_\Delta([s]_{\Delta'}) = [\sigma s]_\Delta$ and $\lambda_\Delta a = \lambda a$, is a morphism of (sequential) asynchronous systems.*

Proof. Let $a, b \in \Delta' = \lambda^{\leftarrow}(\Delta)$, where $\Delta \in \mathbb{Q}$. Then $\lambda a \not\parallel \lambda b$. So, $a \parallel' b$ is impossible since morphisms preserve independence. Thus, Δ' is a clique.

The next claim follows by induction on the derivation of $p \Delta' q$. Indeed, in the base case $p \Delta' q$ since $p \xrightarrow{a} q$ where $a \notin \lambda^{\leftarrow}(\Delta)$. Now, if λa defined, then surely $\lambda a \notin \Delta$. Thus, since $\sigma p \xrightarrow{\lambda a} \sigma q$, it follows that $\sigma p \Delta \sigma q$, as required. Same in case λa undefined, since then $\sigma p = \sigma q$. The inductive step is even easier.

The last claim is immediate.   □

The following proposition makes use of the lemma and allows to glue the morphisms described there.

Proposition 10. *Let $f : \mathcal{A}' \to \mathcal{A}$ in **AS**. Let $\mathbb{M}'$ and $\mathbb{M}$ be families of cliques on $\mathcal{A}'$ and $\mathcal{A}$ respectively. Assume $\Delta \in \mathbb{M}$ implies $\lambda^{\leftarrow}(\Delta) \in \mathbb{M}'$ and assume $\mathbb{M}'$ fully captures the independence relation in $\mathcal{A}'$.*

Then $f^\Pi : \prod^r \mathcal{F}_{\mathbb{M}'}(\mathcal{A}') \to \prod^r \mathcal{F}_{\mathbb{M}}(\mathcal{A})$ defined by $\sigma^\Pi([s_{\Delta'}]_{\Delta'})_{\Delta' \in \mathbb{M}'} = ([\sigma(s_{\lambda^{\leftarrow}(\Delta)})]_\Delta)_{\Delta \in \mathbb{M}}$ and $\lambda^\Pi = \lambda$ is a morphism of asynchronous systems and the following diagram commutes.

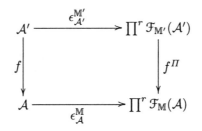

Proof. First we show that f^{Π} is a morphism of transition systems.

Indeed, by Lemma 4 both its parts are well defined.

Now, take a transition in $\prod^r \mathcal{F}_{M'}(\mathcal{A}')$. Without loss of generality the it has the form $([p_{\Delta'}]_{\Delta'})_{\Delta'} \xrightarrow{a} ([q_{\Delta'}]_{\Delta'})_{\Delta'}$, where for all $\Delta' \in M'$ the states $p_{\Delta'}$, $q_{\Delta'}$ are such that either $p_{\Delta'} \xrightarrow{a} q_{\Delta'}$ in $\mathcal{A}'$ where $a \in \Delta'$, or $p_{\Delta'} = q_{\Delta'}$ and $a \notin \Delta'$. Now, consider $\Delta \in M$ and let $\Delta' = \lambda^{-}(\Delta)$. Assume λa is defined. Then $\sigma(p_{\Delta'}) \xrightarrow{\lambda a} \sigma(q_{\Delta'})$ in $\mathcal{A}$. If $\lambda a \in \Delta$, then $[\sigma(p_{\Delta'})]_{\Delta} \xrightarrow{\lambda a} [\sigma(q_{\Delta'})]_{\Delta}$ in $\kappa_{\Delta}(\mathcal{A})$, otherwise, $[\sigma(p_{\Delta'})]_{\Delta} = [\sigma(q_{\Delta'})]_{\Delta}$. Therefore, $([\sigma(p_{\Delta'})]_{\Delta})_{\Delta} \xrightarrow{\lambda a} ([\sigma(q_{\Delta'})]_{\Delta})_{\Delta}$. If λa is undefined, then $\sigma(p_{\Delta'}) = \sigma(q_{\Delta'})$ for all Δ', and hence, σ^{Π} glues both states of the transition. Thus, f is a morphism of the underlying transition systems.

The diagram in question commutes, by construction of f^{Π}.

By Lemma 3, the independence in $\prod^r \mathcal{F}_{M'}(\mathcal{A}')$ is the same as in $\mathcal{A}'$. Surely, λ preserves it in $\mathcal{A}$, and this is included in the independence of $\prod^r \mathcal{F}_M(\mathcal{A})$. Hence, f^{Π} is a morphism of asynchronous systems too.   □

Consider $\mathbb{F}$, an arbitrary small commuting diagram in the category of all asynchronous systems. It may contain no morphisms, several or even all[1] morphisms in **AS**. Then Proposition 10 provides means to generate, for each asynchronous system that appears as an object in diagram $\mathbb{F}$, a family of cliques in a way which turns $\mathbb{F}$ into a commuting diagram: $\mathbb{F}^{\Pi} = \{f^{\Pi} \mid f \in \mathbb{F}\}$.

Proposition 11. *For each object $\mathcal{A}$ in $\mathbb{F}$ let $M = M(\mathcal{A})$ be a family of cliques on $\mathcal{A}$ that fully captures the independence. Then, there exists minimal extensions $M_{\mathbb{F}}$ of each M, such that $\mathbb{F}^{\Pi} = \{f^{\Pi} : \prod^r \mathcal{F}_{M_{\mathbb{F}}}(\mathcal{A}) \to \prod^r \mathcal{F}_{M'_{\mathbb{F}}}(\mathcal{A}') \mid f : \mathcal{A} \to \mathcal{A}' \in \mathbb{F}\}$ becomes a commuting diagram in **AS**.*

If, moreover, $\mathcal{A}$ is M-separated, then it satisfies $M_{\mathbb{F}}$-separation axioms.

Proof. It is easy to verify that the assignment $f \mapsto f^{\Pi}$ preserves identities. Moreover, it preserves the composition of morphisms whenever each of them satisfies the assumption of Proposition 10. Thus, the only problem is to fulfill the assumption by suitably extending each M.

$M_{\mathbb{F}}$'s are constructed in stages. First, for each $\mathcal{A}$ in $\mathbb{F}$ take $M_0 = M$. Then, given M_n put $M_{n+1} = M_n \cup \{\lambda^{-}(\Delta') \mid f : \mathcal{A} \to \mathcal{A}' \in \mathbb{F}, \Delta' \in M'_n\}$. Finally, put $M_{\mathbb{F}} = \bigcup_{n \geq 0} M_n$.

By construction, $\Delta' \in M'_{\mathbb{F}}$ implies $\lambda^{-}(\Delta') \in M_{\mathbb{F}}$ for each $f : \mathcal{A} \to \mathcal{A}'$ in $\mathbb{F}$.

---

[1] To avoid foundational issues we could consider, for instance, only transition systems in which all states and actions are subsets of a fixed set of large cardinality.

Finally, from $\mathbb{M} \subseteq \mathbb{M}_{\mathbb{F}}$ it follows that if $\mathcal{A}$ is $\mathbb{M}$-separated then it satisfies $\mathbb{M}_{\mathbb{F}}$-separation axioms. □

Proposition 11 can be applied, for example, to two universal choices of a family of cliques $\mathbb{M}$ for each $\mathcal{A}$ in $\mathbb{F}$: $\mathbb{M} = \mathbb{Q}$ and $\mathbb{M} = \mathbb{X}$. The first case is trivial in the sense that $\mathbb{Q}_{\mathbb{F}} = \mathbb{Q}$ whatever $\mathbb{F}$ is. In the second case the outcome depends on $\mathbb{F}$. If, for instance, we take $\mathbb{F} = \mathbf{AS}$, then $\mathbb{X}_{\mathbb{F}} = \mathbb{Q}$. But for finite $\mathbb{F}$ the resulting family $\mathbb{X}_{\mathbb{F}}$ is usually smaller.

In both cases, i.e., for $\mathbb{M} = \mathbb{X}$ or $\mathbb{M} = \mathbb{Q}$, the following is immediate.

Corollary 1. *If $\mathbb{F}$ is a diagram in the full subcategory of concrete asynchronous systems then $\mathcal{R}e(\mathbb{F}^{\varPi})$ is rigid isomorphic to $\mathbb{F}$.* □

4.3 Comparison with Elementary Transition Systems

The theory of concrete asynchronous systems subsumes, in the sense discussed below, the theory of elementary transition systems initiated by Ehrenfeucht and Rozenberg, see e.g. [14, 20]. The results presented below extend from elementary transition systems to *semi-elementary* transition systems of [21] and to asynchronous systems are obtained as the case graphs of (unlabelled) safe nets, see [30]. At the same time, as we shall see, there are simple examples of concrete asynchronous systems which fall beyond these classes.

A *region* R in a transition system $\mathcal{S} = \langle S, \hat{s}, A, T \rangle$ is a set of states, $R \subseteq S$, such that given a transition $p \xrightarrow{a} q$, the value of $\chi(p) - \chi(q)$, where χ stands for the characteristic function of R, does not depend on p and q, but is a function of a only. We write $R^{\bullet}a$ if $p \xrightarrow{a} q$ implies $\chi(p) - \chi(q) = 1$ and $a^{\bullet}R$ if $p \xrightarrow{a} q$ implies $\chi(p) - \chi(q) = -1$.

Deterministic and reachable transition system $\mathcal{S}$ is *elementary* if the following conditions are satisfied.

- *no loops:* $p \xrightarrow{a} q$ implies $p \neq q$,
- *no parallel arrows:* $p \xrightarrow{a} q$ and $p \xrightarrow{b} q$ imply $a = b$,
- *no junk:* every action can be enabled in some state, $p \xrightarrow{a}$,

and, additionally, regional separability axioms of Ehrenfeucht and Rozenberg:

- *state-state separation:* $p \neq q$ implies $p \in R$ and $q \notin R$, for some region R;
- *state-action separation:* $p \xrightarrow{a}$ implies $p \notin R$ and $R^{\bullet}a$ for some region R.

Given a transition system $\langle S, \hat{s}, A, T \rangle$ define a relation $\| \subseteq A \times A$ as follows.

$$a \parallel b \quad \textit{iff} \quad \xleftarrow{a} p \xrightarrow{b} \quad \textit{and} \quad \xrightarrow{a} q \xrightarrow{b} \quad \text{for some } p, q. \qquad (6)$$

In the sequel $\|$ defined by (6) is called the *canonical independence* induced by $\mathcal{S}$. The following result justifies this terminology.

Lemma 5. *An elementary transition system $\mathcal{S}$ together with its canonical independence relation makes an asynchronous system.*

Proof. To see that $\|$ is irreflexive suppose to the contrary $p \xrightarrow{a} q \xrightarrow{a}$ for some a, p, q. Then $p \neq q$ and, further, $p \in R$ and $q \notin R$, for some region R. Then we have both $R^\bullet a$ and q, a source of an a-arrow, is not in R—a contradiction.

Symmetry of $\|$ follows from the swap property. Indeed, suppose $p \xrightarrow{b} q \xrightarrow{a} r$ and $b \| a$. First we show that $p \xrightarrow{a}$. Indeed, otherwise there would be a region R such that $p \notin R$ and $R^\bullet a$. Then $q \in R$, hence $b^\bullet R$. But there exists a state being a source of both a- and b-arrows. Conditions $R^\bullet a$ and $b^\bullet R$ cannot be fulfilled simultaneously, hence $p \xrightarrow{a} q'$ for some q'. In a similar manner one can see that there exists r' such that $q' \xrightarrow{b} r'$. Now, given any region R and its characteristic function χ,

$$\chi(p) - \chi(q) = \chi(q') - \chi(r') \qquad b\text{-arrows}$$
$$\chi(p) - \chi(q') = \chi(q) - \chi(r) \qquad a\text{-arrows}$$

and hence $\chi(r) = \chi(r')$. By state-state separation condition, $r = r'$, which finishes the proof. □

In fact, only a minor modification of a proof of the above is required to show that also semi-elementary transition systems coincide with a subclass of asynchronous systems. The following provides some insight into the relationship between regions in an elementary transition system and quotients of its canonical asynchronous system.

Lemma 6. *Let R be a region in an elementary transition system S. Define a set of actions $\Delta = \Delta_R = \{a \in A \mid a^\bullet R \vee R^\bullet a\}$. Then Δ is a clique of dependent actions. Moreover, $p \Delta q$ implies $p \in R \Leftrightarrow q \in R$.*

Proof. Let $a \in \Delta$ and let $a \| b$. Then in the transition system there exist fragments like (6). Whichever of $a^\bullet R$ or $R^\bullet a$ is true, precisely one of the sources of b-arrows belongs to R. Hence, neither $b^\bullet R$ nor $R^\bullet b$ holds, consequently, $b \notin \Delta$.

Now, let $p \Delta q$ and let χ denote the characteristic function of R. The proof of $\chi(p) = \chi(q)$ goes by induction on the derivation of $p \Delta q$. The base case is $p \xrightarrow{a} q$ for some $a \notin \Delta$. By definition of Δ, neither $a^\bullet R$ nor $R^\bullet a$ holds, hence the claim. For the other case, let $r \xrightarrow{a} p$ and $s \xrightarrow{a} q$ be two a-arrows with $r \Delta s$. We have $\chi(p) - \chi(r) = \chi(q) - \chi(s)$ and $\chi(r) = \chi(s)$ by inductive hypotheses. Hence $\chi(p) = \chi(q)$, as required. □

Now, the following is immediate.

Proposition 12. *The canonical asynchronous system induced by an elementary transition system is concrete.*

Proof. Let us verify that with $\mathbb{M} = \{\Delta_R \mid R\text{ -a region}\}$ the canonical asynchronous system satisfies Morin's $\mathbb{M}$-separation axioms (3) and (4).

Suppose $p \neq q$. Then there exists a region which separates p and q. A clique Δ induced by this region as in Lemma 6 also separates the states.

Now, let $p \xrightarrow{a}$. Take a region R with $p \notin R$ and $R^\bullet a$. Then, the clique Δ induced by R contains a and separates p and a. Indeed, consider q, with $q \Delta p$. Then, by Lemma 6, $q \notin R$. Hence, $[p]_\Delta \xrightarrow{a}$. □

The converse to Proposition 12 is not true. For instance, each *sequential system* $\mathcal{A}$, i.e., each asynchronous system with $\|_{\mathcal{A}} = \emptyset$, does satisfy Morin's conditions. This is because the entire alphabet of $\mathcal{A}$ is a clique of dependent actions, and it generates the trivial equivalence. Hence, for instance, a sequential system that performs an action a twice in a row and then stops is concrete. Clearly, it is not elementary. In fact, it is neither semi-elementary nor does it satisfy the separation axiom 3 in [30]. Thus, the realization procedure described in this paper extends functorial realizability to cases not covered before. In fact, it will become apparent in the next section, it covers cases which cannot be realized by *any* unlabelled general Petri net.

But first, let us notice that the arguments presented above can be easily adopted to safe nets and their case graphs. Indeed, each net $\mathcal{N}$ induces structural independence $\|_{\mathcal{N}}$ on events defined by: $e_1 \|_{\mathcal{N}} e_2$ iff $({}^\bullet e_1 \cup e_1^\bullet) \cap ({}^\bullet e_2 \cup e_2^\bullet) = \emptyset$. Then, $\mathcal{A}_{\mathcal{N}} = \langle \mathcal{C}g(\mathcal{N}), \|_{\mathcal{N}} \rangle$ is an asynchronous system when $\mathcal{N}$ is a safe net, see [3].

Proposition 13. *Asynchronous system $\mathcal{A}_{\mathcal{N}}$ induced by a safe net $\mathcal{N}$ is concrete.*

Proof. Each place b of $\mathcal{N}$ induces a clique $\Delta_b = \{e \in E \mid b^\bullet e \vee e^\bullet b\}$ of structurally dependent actions. Moreover, one easily obtains an analogue of Lemma 6: $M \Delta_b M'$ implies $b \in M \Leftrightarrow b \in M'$. With it, the proof goes as in Prop. 12.   □

4.4   Labelled Safe Petri Nets and Their Products

The importance of categorical products as a fundamental tool to explain synchronization is well-known, cf. [28, 29]. Here, we investigate the Petri net counterpart of the mixed product ([13, 19]) in the context of nets with rigid morphism.

A *rigid morphism* of two labelled Petri nets $\langle \mathcal{N}, A, \ell \rangle$ and $\langle \mathcal{N}', A', \ell' \rangle$ with $A' \subseteq A$ is a general morphism $\langle \beta, \eta, \lambda \rangle$ of nets in which λ is the transpose partial function induced by the inclusion $A' \subseteq A$. Equivalently, we can keep λ implicit, and consider a morphism $\langle \beta, \eta \rangle$ of the underlying Petri nets which preserves the labelling in the following sense: for all $e \in E$, either ηe is defined, $\ell e \in A'$ and $\ell'(\eta e) = \ell e$ or ηe is undefined and $\ell e \notin A'$. Clearly, labelled safe Petri nets with rigid morphisms also form a category, denoted $\ell \mathbf{PN}^r$. Note that the case graph functor maps rigid morphisms of nets to rigid morphisms of transition systems.

The category of labelled (safe) Petri nets with rigid morphisms admits finite products, called *mixed* or *rigid* in the sequel. Essentially, it is the old Winskel construction, cf. [29], except that the synchronization of events is already incorporated into the product (due to the choice of rigid morphisms). Also, the presence of labelling requires some refinements of the construction.

Theorem 2. *Let $(\mathcal{L}_i)_{i \in I}$ be a finite family of labelled Petri nets. Their mixed product $\prod_{i \in I}^r \mathcal{L}_i = \langle B, E, F, \hat{M}, A, \ell \rangle$ is given by:*

- $B = \biguplus_{i \in I} B_i$, *the disjoint union of places,*
- $A = \bigcup_{i \in I} A_i$,

- $E \subseteq \prod_{i \in I}(E_i \cup \{\bot\})$ *is defined as follows.*

$$e \in E \quad \textit{iff} \quad (\exists\, a \in A)\ (\forall\, i \in I)\ (e_i \neq \bot \wedge \ell_i e_i = a) \vee (e_i = \bot \wedge a \notin A_i)$$

Here $\bot$ is a dummy event occurring in none of E_i,
- $F\langle b, e\rangle = F_i\langle b, e_i\rangle$ and $F\langle e, b\rangle = F_i\langle e_i, b\rangle$, *whenever $b \in B_i$. Here, by convention $F_i\langle b, \bot\rangle = 0$ and $F_i\langle\bot, b\rangle = 0$, for all $i \in I$.*
- $\hat{M} = \Sigma_{i \in I}\hat{M}_i$
- $\ell : E \to A$ *is determined uniquely due to the definitions of A and E.*

The place part of i-th projection is the relation transposed to the inclusion $B_i \subseteq B$. Its event part is a partial function which maps $\langle e_i\rangle_{i \in I}$ to e_i when $e_i \neq \bot$, and otherwise undefined. □

Proof. It is immediate that projections fulfill the conditions of Definition 1, and that they preserve labels. Moreover, they are Winskel morphisms.

Suppose a family of Petri net rigid morphisms $\langle \beta_i, \eta_i\rangle : \mathcal{L} \to \mathcal{L}_i$, $i \in I$ is given. In particular, $(\forall\, i)\ A_i \subseteq A$. Define $\langle \beta, \eta\rangle : \mathcal{L} \to \prod_{i \in I}^r \mathcal{L}_i$ as $\beta = \Sigma_{i \in I}\beta_i$ and $\eta e = (\eta_i e_i)_{i \in I}$, where the i's component equals $\bot$ whenever $\eta_i e_i$ is not defined. The requirement that morphisms preserve labelling ensures ηe is properly defined and the resulting morphism preserve labelling too. The definition of β ensures condition (3) of Definition 1 is fulfilled. Now $\beta({}^\bullet e) = \Sigma_{i \in I}\beta_i({}^\bullet e_i)$ while ${}^\bullet(\eta e) = \Sigma_{i \in I}{}^\bullet \eta_i e_i$ and similarly for $(\eta e)^\bullet$. Here summands are assumed to be zero if $\eta_i e_i$ is undefined. Hence conditions (1) and (2) of Definition 1 are fulfilled too.

It is clear from the construction that $\langle \beta, \eta\rangle$ composed with i-th projection equals $\langle \beta_i, \eta_i\rangle$ and that it is the only (rigid) morphism with this property. □

Rigid projections are Winskel morphisms. It can be shown that if a cone consists of Winskel morphisms only, then so is the mediating morphism. Thus, the constructions specializes to the subcategory with rigid Winskel morphisms.

In the rigid product two nets synchronize on shared labels. It may be instructive to consider an example of a rigid product and a canonical morphism.

Net $\mathcal{N}'_+$ in Fig. 4 is a rigid product of three state machines—the one with language $(e'_1 e'_2)^\star$, another with $(e''_2 e''_1)^\star$ and yet another with $e'_1 + e''_2$. Here, a synchronization does take place. There are morphisms from $\mathcal{N}_+$ to $\mathcal{N}'_+$ which yield projections on the level of case graphs, their product is a morphism from $\mathcal{N}_+$ to $\mathcal{N}'_+$ yielding identity on case graphs. As we have already mentioned, there is no morphism in the opposite direction.

The case graph functor preserves mixed products.

Proposition 14. *The case graph of the mixed product of labelled Petri nets is (rigid isomorphic to) the mixed product of their case graphs.*

Proof. Let $\mathcal{L}_i$, $i \in I$, be a finite family of Petri nets, and let $\mathcal{L} = \prod_{i \in I}^r \mathcal{L}_i$. A mapping $(M_i)_{i \in I} \mapsto \Sigma_{i \in I}M_i$ is a bijection between $\prod_{i \in I}\mu B_i$ and $\mu(\biguplus_{i \in I}B_i)$. Hence, $(\hat{M}_i)_{i \in I}$ is mapped into $\hat{M}$.

Let $(M_i)_{i \in I} \xrightarrow{a} (M'_i)_{i \in I}$ in the product of case graphs. It means that if $a \in A_i$ then $M_i\, [e_i\rangle\, M'_i$ for some e_i, $\ell_i e_i = a$, otherwise $M_i = M'_i$. If the former is the case, $M_i' = M_i - {}^\bullet e_i + e_i^\bullet$. Hence $\Sigma_{i \in I} M_i \, [e\rangle \, \Sigma_{i \in I} M_i'$, where $e = (e_i)_{i \in I}$, $\ell e = a$. Thus, the bijection is actually an isomorphism in **TS**. $\qquad\square$

While the mixed product of labelled Petri nets is a product in the subcategory with rigid morphisms, it is also true, that a cone of "consistent" non-rigid morphisms also gives rise to a not necessarily rigid morphism to the mixed product.

Proposition 15. *Consider a family of net morphisms* $\langle \beta_i, \eta_i, \lambda_i \rangle : \mathcal{L} \to \mathcal{L}_i$, *for* $i \in I$. *Assume all* λ_i *composed with inclusions* $\iota_i : A_i \subseteq \bigcup_{i \in I} A_i$ *agree, i.e., that there exists a function* $\lambda : A \to \bigcup_{i \in I} A_i$ *such that* $\lambda_i = \lambda; \iota_i^\mathsf{T}$. *Then there exists (a unique) morphism* $\langle \beta, \eta, \lambda \rangle : \mathcal{L} \to \prod_{i \in I}^r \mathcal{L}_i$ *of labelled Petri nets such that its compositions with rigid projections to* $\mathcal{L}_i$ *gives back the original morphisms.*

Proof. Define $\lambda : A \to \bigcup_{i \in I} A_i$ as the common relabelling. The construction of $\langle \beta, \eta \rangle$ given in the proof of Theorem 2 is applicable here too, the assumption of the common relabelling makes the definition of η correct. Together with λ they form the desired morphism. $\qquad\square$

4.5   Towards a Functorial Realization of Concrete Asynchronous Systems by Means of Labelled Safe Petri Nets

The notion of realization of a concurrent behavior of a reactive system has been introduced in Sect. 3.6, and formalized as a relation between reachable asynchronous systems and labelled marked Petri nets. Now, we are ready to show that this realization relation commutes with rigid products.

Theorem 3. *Let* $(\mathcal{A}_i)_{i \in I}$ *be a finite family of asynchronous systems and let* $(\mathcal{L}_i)_{i \in I}$ *be a family of their realizations. Then the mixed product of nets,* $\prod_{i \in I}^r \mathcal{L}_i$, *realizes the mixed product of asynchronous systems,* $\prod_{i \in I}^r \mathcal{A}_i$.

Proof. Let $\prod_{i \in I}^r \mathcal{L}_i = \langle B, E, F, \hat{M}, A, \ell \rangle$ be a mixed product of labelled nets and let $\prod_{i \in I}^r \mathcal{A}_i = \langle \prod_{i \in I}^r \mathcal{S}_i, \| \rangle$ be a product of asynchronous systems.

1. We have already established, see Proposition 14, that $\mathcal{C}g(\prod_{i \in I}^r \mathcal{L}_i)$ is rigid isomorphic to the product of $\mathcal{C}g(\mathcal{L}_i)$, hence also to $\prod_{i \in I}^r \mathcal{S}_i$.
2. Let M be reachable, $M \, [e+e'\rangle$, where $e = (e_i)_{i \in I}$, $e' = (e_i')_{i \in I}$ and let $\ell e, \ell e' \in A_i$ for some $i \in I$. Then $M \cap B_i \in \mathcal{M}_i$ and $M \cap B_i \, [e_i + e_i'\rangle$ in $\mathcal{N}_i$, hence $\ell e = \ell_i e_i \|_i \ell_i e_i' = \ell e'$. The choice of $i \in I$ does not matter.
3. Finally, let $b \in ({}^\bullet e \cup e^\bullet) \cap ({}^\bullet e' \cup e'^\bullet)$. There is only one $i \in I$ such that $b \in B_i$, for this i both $e_i \neq \bot$ and $e_i' \neq \bot$ and $b \in ({}^\bullet e_i \cup e_i^\bullet) \cap ({}^\bullet e_i' \cup e_i'^\bullet)$. Then $\ell_i e_i \not\|_i \ell_i e_i'$, hence $\ell e \not\| \ell e'$. $\qquad\square$

The reader should note that the above result does not presuppose safety of realizations. This, in principle, allows to use arbitrary labelled Petri nets. If, however, all nets are safe, then so is their mixed product.

Consider a reachable concrete asynchronous systems $\mathcal{A}$, and let $\mathbb{M} \subseteq \mathbb{Q}(A, \|)$ be any family of cliques such that $\mathbb{M}$-separation axioms hold for $\mathcal{A}$, and $\mathbb{M}$ fully captures $\|$. A realization of $\mathcal{A}$ by means of a labelled Petri net is obtained by applying the following *parametric* procedure.

Factorize $\mathcal{A}$ into the family $\mathcal{F}_{\mathbb{M}}(\mathcal{A}) = \{\kappa_\Delta(\mathcal{A}) \mid \Delta \in \mathbb{M}\}$ of sequential systems.
Realize each quotient $\kappa_\Delta(\mathcal{A})$ in $\mathcal{F}_{\mathbb{M}}(\mathcal{A})$ as a net $\mathcal{L}_\Delta$.
Compute $\prod_{\Delta \in \mathbb{M}}^r \mathcal{L}_\Delta$.

The first step has been done by Morin, cf. [19] and Sect. 4.2.

The last step produces a realization of the mixed product of sequential systems, by Theorem 3.

Thus, to fully explain how the procedure works it is enough to present a realization of sequential systems. We have already seen an easy and functorial way to realize sequential systems as labelled state machines, see Prop. 7.

Suppose that $\mathcal{R}$ is *any* such realization functor from the category of sequential systems to the category of labelled Petri nets. Is it now possible to extend $\mathcal{R}$ as described above so that this realization becomes a functor? More precisely, given a diagram $\mathbb{F}$ in **AS** in which all objects are concrete, one would like to find a realization of all the objects and all the morphisms which preserves commuting diagrams. Indeed, this can be done thanks to Prop. 10–11.

Theorem 4. *Let $\mathbb{F}$ be a commuting diagram in the full subcategory of concrete asynchronous systems. Then there exists a diagram $\mathcal{R}(\mathbb{F})$ (of the same shape) in the category of labelled Petri nets whose objects realize the respective objects of $\mathbb{F}$ and whose morphisms realize the morphisms of $\mathbb{F}$.*

Proof. For each object $\mathcal{A}$ in $\mathbb{F}$ let $\mathbb{M}_\mathbb{F}$ be a family of cliques of $\mathcal{A}$ build as in Prop. 11, and such that it ensures separation axioms and fully captures independence in $\mathcal{A}$. Consider a family of quotients $\mathcal{F}_{\mathbb{M}_\mathbb{F}}(\mathcal{A}) = \{\kappa_\Delta(\mathcal{A}) \mid \Delta \in \mathbb{M}_\mathbb{F}\}$. Realize every member of this family using a realization functor $\mathcal{R}$. Define a labelled net $\mathcal{R}(\mathcal{A})$ as $\prod^r \{\mathcal{R}(\kappa_\Delta(\mathcal{A})) \mid \Delta \in \mathbb{M}_\mathbb{F}\}$. In this way all objects of $\mathbb{F}$ are realized as labelled Petri nets.

Now, let $f : \mathcal{A}' \to \mathcal{A} \in \mathbb{F}$ be a morphism of asynchronous systems. The families of cliques $\mathbb{M}'_\mathbb{F}$ and $\mathbb{M}_\mathbb{F}$ of $\mathcal{A}'$ and $\mathcal{A}$, respectively, fulfill the requirements of Prop. 10. Hence, a family of morphisms $f_\Delta : \kappa_{\lambda\leftarrow(\Delta)}(\mathcal{A}') \to \kappa_\Delta(\mathcal{A})$, $\Delta \in \mathbb{M}_\mathbb{F}$ as defined in Lemma 4 is consistent. Applying to f_Δ's the realization functor we arrive at a family of Petri net morphisms $\mathcal{R}(f_\Delta) : \mathcal{R}(\kappa_{\lambda\leftarrow(\Delta)}(\mathcal{A}')) \to \mathcal{R}(\kappa_\Delta(\mathcal{A}))$, $\Delta \in \mathbb{M}_\mathbb{F}$. Composed with projections from $\mathcal{R}(\mathcal{A}')$ they make a cone of consistent morphisms in the category of Petri nets. We can now apply Prop. 15 to obtain a morphism $\mathcal{R}(f) : \mathcal{R}(\mathcal{A}') \to \mathcal{R}(\mathcal{A})$ which realizes f.   □

Corollary 2. *There is a realization functor from the category of concrete asynchronous systems to the category of labelled Petri nets.*

Proof. Consider $\mathbb{Q}$ as the universal family of cliques and apply the construction to the functor $\mathcal{S}m$.   □

Fig. 5–7 demonstrate an example of two simple transition systems, their mixed product, and their realization by means of labelled safe Petri nets. Actually, the example is so simple that the labelling is not really required. Note that realizations may introduce indistinguishable places, viz.: postconditions of a.

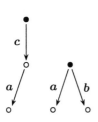 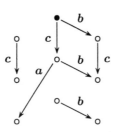 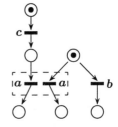

Fig. 5. Two simple transition systems

Fig. 6. The product of these transition systems

Fig. 7. Two state machines and their product realizing the transition systems

5   Examples, Conclusions and Further Work

Let us assume that a reactive system is given in the form of transition system $S = \langle S, \hat{s}, A, T \rangle$. We have shown how to find a concurrent realization of a reactive systems when its behavior is represented as a concrete asynchronous system $\langle S, \| \rangle$. The difficult question, though, is how to define a suitable $\|$.

Intuitively, the larger the independence relation, the more concurrent realization should be obtained. One could formally define a partial order on asynchronous systems over the same underlying transition system by saying that $\langle S, \|_1 \rangle$ is *more concurrent* than $\langle S, \|_2 \rangle$ whenever $\|_1 \supseteq \|_2$.

Now, if S elementary one can, by Prop. 12, derive from its structure the canonical independence relation which yields a concrete asynchronous system. So, one can ask whether what we obtain in this way is the most concurrent behavior of S? The answer is negative.

Indeed, consider transition system S_{abc} given by: $\hat{s} \xrightarrow{a} p \xrightarrow{b} q \xrightarrow{c} r$. This is an elementary transition system, and its canonical independence relation is empty. This is a way of saying that the system does not exhibit any concurrent activity. However, nothing prevents us from declaring a and c as independent. In fact, $\langle S_{abc}, \| \rangle$ with $a \| c$ and $a \nparallel b \nparallel c$ becomes a concrete asynchronous system! The corresponding two realizations of both systems are depicted on Fig. 8–9.

Fig. 8. Sequential realization of S_{abc} **Fig. 9.** "Concurrent" realization of S_{abc}

Interestingly, *both* nets are realizations of *both* concrete asynchronous systems which shows that the notion of realization is quite liberal. Other issues are also demonstrated by this simple example.

Firstly, the realization procedure that we have described strives to implement actions declared as independent by means of *structurally separated* transitions.

Secondly, the choice of decomposition of a behavior into a product of sequential components affects the size of the realization of the system. One of the parameters of the construction of realization presented in Sect. 4 was the choice of the family of cliques of the dependence relation. Thus, the net on Fig. 8 is determined by choosing $\mathbb{X} = \{\{a, b, c\}\}$, while the net on Fig. 9 is determined by choosing $\mathbb{M} = \{\{a, b\}, \{b, c\}\}$. In this case we could also choose $\mathbb{Q}$ which contains all subsets of $\{a, b, c\}$, and obtain a huge net, with many copies of indistinguishable places. This seems to indicate that when the independence is fixed, the most compact realizations would be obtained with $\mathbb{X}$.

Finally, on a related note, one should not declare two actions as independent unless they form a diamond somewhere.

This brings us back to the general question when $\mathcal{S}$ is not elementary, and yet one would like to find its non-trivial concurrent realization. One could look at this problem of finding a maximally concurrent realization of $\mathcal{S}$ as a kind of 'code optimization problem'. In fact, there is a way to turn a transition system $\mathcal{S}$ into maximally concurrent asynchronous system $\mathcal{A} = \langle \mathcal{S}, ||| \rangle$ where $|||$ is the largest symmetric and irreflexive relation that satisfies the following.

$$a \, ||| \, b \quad iff \quad p \xrightarrow{a} q \xrightarrow{b} r \quad implies \quad (\exists s) \, p \xrightarrow{b} s \xrightarrow{a} r \tag{7}$$

Sadly, the above construction is not 'functorial' in the sense, that a morphism $f : \mathcal{S} \to \mathcal{S}'$ in general does not map $|||$-independent actions to $|||'$-independent actions. More seriously, $\langle \mathcal{S}, ||| \rangle$ obtained in this way is not concrete in general. The problem are demonstrated in the next section with a well-known example.

5.1   Mandala

Suppose there are agents interested in using a resource. Simultaneous usage of the resource leads to its corruption. It is therefore imperative that some kind of scheduler is devised to control the access to the resource. The behavior of an individual agent A from the scheduler perspective could be defined as a simple CCS-like process. In the simple case described on Fig. 10 each agent would behave like $A \Leftarrow r.u.A$ where r and u stand for *request* and *use* phases, respectively. The more elaborate case of Fig. 11 arises when each phase of using the resource is split into two: e of *entering* the critical section, and phase ℓ of *leaving* the critical section and thereby releasing the resource.

In the simplest case, in which only two agents are considered, solutions take the form of *mandala*, see Fig. 10–11. The solutions are based on assumption[2] that the scheduler should accept the requests at any time, but the permission should be granted on the first-come-first-served basis. The reactive system on Fig. 11 is isomorphic to the flow-graph for mutual exclusion derived by Emerson and Clarke from a branching time temporal specification, see [15, Fig. 11].

A quick analysis of transition system $\mathcal{S}_2$ reveals that it is *not* elementary. First of all, states s and t reached from the initial state after performing $r_1 r_2$ and $r_2 r_1$

---

[2] This is questionable as it prohibits asynchronous reaction to request signals.

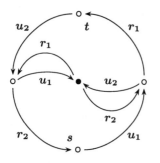

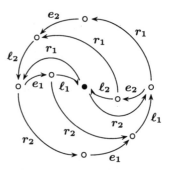

Fig. 10. Simple 2-agent scheduler S_2 **Fig. 11.** Emerson-Clarke scheduler S_3

are different, so state-state separation fails. Moreover, the transitions enabled in these two states are different, so state-event separation fails as well. Hence, the system cannot be realized using the regional construction of Ehrenfeucht-Rozenberg. In fact, there is no (unlabelled) net with S_2 as its case graph. The proof is simple: whenever $M \left[e_2\right\rangle M' \left[e_1\right\rangle M_1$ and $M \left[e_1\right\rangle M'' \left[e_2\right\rangle M_2$ hold in a Petri net, then $M_1 = M_2$ follows.

The above arguments work also for Emerson-Clarke scheduler S_3.

The only interesting concrete asynchronous system with S_2 as its underlying transition system is the sequential one. To see this assume to the contrary that, say, $r_1 \parallel u_2$. Then, s and u_2 could not be separated, whereas $s \xrightarrow{u_2}$. Indeed, consider any $\Delta \subseteq \{r_1, u_1, u_2, r_1\}$ with $u_2 \in \Delta$ where Δ is a clique of the dependence relation. Then $r_1 \notin \Delta$, by assumption, so by the construction of the quotient relation induced by Δ it follows that $s \Delta t$. Thus, $[s]_\Delta \xrightarrow{u_2}$.

A symmetric argument shows that also the other diamond in S_2 cannot be filled with concurrency, i.e., we cannot assume $r_2 \parallel u_1$. Similarly, one can show that $r_2 \nparallel e_1$ and $r_1 \nparallel e_2$ in any concrete asynchronous system built on S_3.

5.2   Realization of the Emerson-Clarke Scheduler

With mandala not much can be done within our framework. That is, the only implementation of mandala is the sequential one. In case of Emerson-Clarke scheduler, see Fig. 11, the situation is better. Namely, there exists maximal concrete independence relation: $r_1 \parallel \ell_2$ and $r_2 \parallel \ell_1$. The relation admits four maximal cliques.

The factorization of the resulting concrete asynchronous system into sequential systems is depicted on Fig. 12. The last of the four sequential systems is the notorious mandala. Taking the labelled state machine associated with each factor and computing their product in the category of labelled nets would give a safe net with 14 places and 14 transitions.

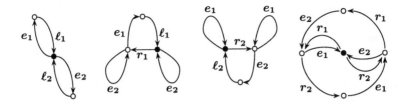

Fig. 12. The decomposition of the scheduler with $r_2 \not\parallel e_1$ and $r_1 \not\parallel e_2$

5.3 Zig-zag Morphism and Systematic Refinement of Concrete Asynchronous Systems

We do not know if there exists the biggest concurrent concrete asynchronous system over a given transition system. Nevertheless, assume that each transition system S in a diagram $\mathbb{F}$ is equipped with a concurrency relation $\parallel$ such that $\langle S, \parallel \rangle$ is a concrete asynchronous system. Then, we could iteratively keep restricting the independence relations until all f in $\mathbb{F}$ become morphisms of asynchronous systems. The resulting diagram in the full subcategory of concrete asynchronous systems can be then realized as a diagram of labelled safe Petri nets as described in Sect. 4. Thus, the problem would be to find *a* maximally concurrent independence such that $\langle S, \parallel \rangle$ is concrete.

Each concrete $\langle S, \parallel \rangle$ satisfies $\parallel \subseteq \parallel\!\parallel$ where $\parallel\!\parallel$ is the maximal independence admissible for S. Moreover, whenever $\parallel_1 \subseteq \parallel_2$ then the identity morphism on S becomes is a morphism from $\imath : \langle S, \parallel_1 \rangle \to \langle S, \parallel_2 \rangle$ in **AS**. This suggests that in search for a maximal concrete independence one could start with $\parallel\!\parallel$ and subsequently restrict the independence until one arrives at a concrete system. This process will surely terminate since each sequential system is concrete.

The above idea can be generalized by replacing $\imath : \langle S, \parallel_1 \rangle \to \langle S, \parallel_2 \rangle$ by more general $f : \langle S_1, \parallel_1 \rangle \to \langle S_2, \parallel_2 \rangle$ and continue working with S_1. To maintain the same behavior any such morphism should satisfy certain requirements. For instance, modulo λ, it should satisfy the same properties as S_2.

Assume that λ is total. We can view S_1 as a transition system S over A_2 defined by $S = S_1$, $A = A_2$ and $p \xrightarrow{a} q$ in S iff $\lambda b = a$ and $p \xrightarrow{b} q$ in S_1. In the light of Proposition 1, if the morphism $\langle \sigma, \mathrm{id}_{A_2} \rangle : S \to S_2$ is a zig-zag, then S and S_2 satisfy the same properties. Hence, if S_1 is realized as a labelled net, then extending the labelling of the net with λ renders a realization of S_2.

Let us turn the above requirement into definition.

Definition 5. *A* proto-zig-zag *is a morphism* $f : S \to S'$ *such that* λ *is total, and if* $\sigma s \xrightarrow{b} q$ *in* S' *then there is transition* $s \xrightarrow{a} p$ *in* S *such that* $\lambda a = b$ *and* $\sigma p = q$. *A* proto-zig-zag *between asynchronous systems is their morphism which is a proto-zig-zag of the underlying transition systems.*

Let A *and* A' *be asynchronous systems. Then* A implements A' via f, *and* f *is an* implementation *of* A', *whenever* $f : A \to A'$ *is a proto-zig-zag morphism.*

Proto-zig-zags are closed under composition, and contain identity morphisms. When the target of a proto-zig-zag is reachable then its components of are surjective functions. If $\mathcal{A} = \langle \mathcal{S}, \| \rangle$, then the pair of identities forms a proto-zig-zag from $\langle \mathcal{S}, \emptyset \rangle$ to $\mathcal{A}$. Thus, the sequential system $\langle \mathcal{S}, \emptyset \rangle$ is an implementation of $\mathcal{A}$.

Let $f' : \mathcal{A}' \to \mathcal{A}$ and $f'' : \mathcal{A}'' \to \mathcal{A}$ be two implementations. Then f' is *better* than f'', notation $f'' \preceq f'$ when there exists $f : \mathcal{A}'' \to \mathcal{A}'$ such that $f'' = f; f'$. Surely, any such f is necessarily a proto-zig-zag, too.

The idea, now, would be to iterate refinement steps, by analogy to what was proposed in [9, 10] and implemented in [8], so that with each refinement the new, transformed specification is *more concrete*. Each implementation preserves the behavior of the target in the source modulo the λ-part of the implementation, see Proposition 1. Thus, the realization obtained at the end, with labelling composed with all λ's parts of the implementations obtained along the way, is a realization of the original system.

Assume that one insists that the states are shared by the implementing and the implemented asynchronous systems. Then, there are basically two orthogonal ways in which an asynchronous system $\mathcal{A}'$ can implement $\mathcal{A}$ via $f : \mathcal{A}' \to \mathcal{A}$.

- *action splitting*: f glues some actions of $\mathcal{A}'$ into a single action of $\mathcal{A}$, independence in $\mathcal{A}$ is reflected in $\mathcal{A}'$;
- *concurrency reduction*: f does not glue any actions, but the independence relation in $\mathcal{A}'$ is smaller then the one in $\mathcal{A}$.

Then, the less splitting of actions, the smaller loss of independence, the better, i.e., more concurrent, an implementation.

Clearly, the implementation by means of the induced sequential system is of the second type. All other implementations can be seen as compositions of a split implementation followed by a concurrency reducing implementation.

5.4   Further Research

The framework proposed here subsumes those synthesis procedures which targeted subclasses of safe nets, cf. [20, 21]. It remains to be checked if Morin's separation axioms also imply those of [30].

More importantly, though, would be to relate our framework to other frameworks, especially those underlying tools used in practice.

It remains as an interesting challenge if one can push further the frontiers of uniform realizations. One possibility would be to target non-safe labelled nets and use more liberal abstract models of concurrent behaviors, like in [11, 25]. Indeed, by implementations involving splitting of some actions one can obtain from Emerson-Clarke scheduler a concrete asynchronous system that fulfills separation axioms of Droste-Shortt. In fact, there are several maximal, and hence incomparable ways to achieve this goal. For instance one can choose one of the lower diamonds, say $[r_2, e_1]$, and one of the upper ones, say $[r_1, \ell_2]$, and then introduce *two* actions for each of r_2, e_1, r_1 and ℓ_2, and adjust the independence relation accordingly. The corresponding implementation morphism is an identity on states, glues actions that got split, and preserves the independence relation.

Other option is to accept a more liberal notion of concurrent realization. This can be done by removing condition Def. 3(2). Intuitively, there is no sense to prohibit concurrent execution in the realization even when it is not present in the specification. Indeed, more compact realizations can be obtained this way.

And there is yet another option. Our notion of implementation has been based on strict morphisms, i.e., morphisms with total action relabelling part. In this way the implementation corresponds to the notion of strong bisimulation. We have grounds to believe that by allowing partial relabeling, and thereby *silent actions* in the implementing systems, one can achieve more concurrent behavior. In case of the Emerson-Clarke scheduler one can achieve the realization of maximally concurrent behavior easily, see Fig. 14 (to ease the comprehension we copied Fig. 11 for comparison). Many researchers are active in this field, see [27] for a recent attempt.

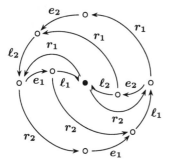
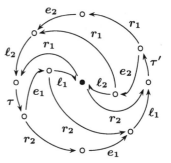

Fig. 13. Emerson-Clarke scheduler S_3 from Fig. 11

Fig. 14. Emerson-Clarke scheduler with silent moves

In this paper we have studied the notion of general morphism of Petri nets. General morphisms were shown to be the largest class of morphisms which are transformed by the case graph construction to morphisms of transition systems. This notion of general Petri net morphism can further be generalized along the lines of [18]. That is, one may consider richer structures on the event part of a Petri net and Petri net morphism. Indeed, Definition 1 works fine if one allows mapping an event in the source net of a morphism to a multiset of events in the target net.

However, as argued in [18], while it is natural to consider Petri nets with monoidal structure on events, in general the monoids need not be free. An example is Milner's synchronization: $\alpha + \bar{\alpha} = \tau$, where τ is a silent move. A good candidate could be finitely presentable monoids, i.e., monoids generated by a finite number of events and with a finite number of equalities. Then, a Petri net morphism should preserve these equalities.

Another line of generalization is to consider a richer class of morphisms of transition systems. A *simulation* between transition systems $S = \langle S, \hat{s}, A, \longrightarrow \rangle$

and $\mathcal{S}' = \langle S', \hat{s}', A', \longrightarrow \rangle$ is a pair consisting of a relation $\prec \subseteq S \times S'$ and a partial function $\eta : A \rightharpoonup A'$ such that initial states are related, $\hat{s} \prec \hat{s}'$, and

$$p \xrightarrow{a} q \text{ in } \mathcal{S}, \ \eta a\Downarrow, \ p \prec p' \quad implies \quad p' \xrightarrow{\eta a} q' \text{ in } \mathcal{S}' \text{ for some } q', q \prec q'$$
$$p \xrightarrow{a} q \text{ in } \mathcal{S}, \ \eta a\Uparrow, \ p \prec p' \quad implies \quad q \prec p'$$

Brown and Gurr [7] have defined the notion of a simulation of Petri nets in a somewhat restricted way. Their simulation of Petri nets, gives rise to the simulation of their case graphs, as expected. The general simulation of Petri nets would be a pair $\langle \beta, \eta \rangle$ which fulfill condition (1) of Definition 1 and, additionally, conditions

2'. $\beta(e^{\bullet}) \leq \beta(^{\bullet}e) - {}^{\bullet}(\eta e) + (\eta e)^{\bullet}$.
3'. $\beta \hat{M} \leq \hat{M}'$.

Again, simulations form a category, in fact, the largest one within the framework. The case graph is a functor—a simulation of the case graphs is a pair $\langle \prec_{\beta}, \eta \rangle$, where $M \prec_{\beta} M'$ iff $\beta M \leq M'$. Moreover, Winskel definition of products works. However, there is no obvious state machine construction at hand, hence the synthesis problem for simulations of transition systems requires further research.

References

1. E. BADOUEL and P. DARONDEAU. Theory of regions. In *Advances in Petri Nets*, vol. 1491 of *LNCS*, pp. 529–586. Springer, 1998.
2. E. BADOUEL, M. A. BEDNARCZYK and P. DARONDEAU. Generalized automata and their net representations. In this volume.
3. M. A. BEDNARCZYK. *Categories of Asynchronous Systems*. Ph. D. thesis, University of Sussex, England, 1988. CST 1-88.
4. M. A. BEDNARCZYK and A. M. BORZYSZKOWSKI. Concurrent realizations of reactive systems. In P. R. M. Hofmann, D. Pavlovic, ed., *Proc. Category Theory in Computer Science, 8th Conf., Edinburgh*, vol. 29 of *Electronic Notes in Theoretical Computer Science*, pp. 1–19. Elsevier, 1999.
5. M. A. BEDNARCZYK and A. M. BORZYSZKOWSKI. General morphisms of Petri nets; extended abstract. In M. N. Jiri Wiedermann, Peter van Emde Boas, ed., *Proc. Automata, Languages and Programming, 26th Intn'l Coll., Prague*, vol. 1644 of *LNCS*, pp. 190–199. Springer, 1999.
6. M. A. BEDNARCZYK, A. M. BORZYSZKOWSKI and R. SOMLA. Finite Completeness of Categories of Petri Nets. *Fundamenta Informaticæ*, vol. 43(1-4): pp. 21-48, 2000.
7. C. BROWN and D. GURR. Refinement and simulation of nets – a categorical characterization. In K. Jensen, ed., *Proc. Applications and Theory of Petri Nets*, vol. 616 of *LNCS*, pp. 76–92. Springer, 1992.
8. J. CORTADELLA, M. KISHINEVSKY, A. KONDRATYEV, L. LAVAGNO, and A. YAKOVLEV. Petrify: a tool for manipulating concurrent specifications and synthesis of asynchronous controllers. In *IEICE Trans. on Information & Systems*, vol. E80-D(3), pp. 315–325, 1997.
9. J. CORTADELLA, M. KISHINEVSKY, L. LAVAGNO, and A. YAKOVLEV. Synthesizing Petri nets from state-based models. In *Proc. International Conference on Computer Aided Design*, pp. 164–171, 1995.

10. J. CORTADELLA, M. KISHINEVSKY, L. LAVAGNO, and A. YAKOVLEV. Deriving Petri nets from finite transition systems. In *IEEE Transactions on Computers*, vol. 47(8), pp. 859–882, 1998.

11. M. DROSTE and R. M. SHORTT. Petri nets and automata with concurrency relation – an adjunction. In M. Droste and Y. Gurevich, eds., *Proc. Semantics of Programming Languages and Model Theory*, pp. 69–87, 1993.

12. M. DROSTE and R. M. SHORTT. Continuous Petri nets and transition systems. In this volume.

13. C. DUBOC. Mixed products and asynchronous automata. *Theoretical Computer Science*, vol. 48: pp. 183–199, 1986.

14. A. EHRENFEUCHT and G. ROZENBERG. Partial (set) 2-structures, part I and II. *Acta Informatica*, vol. 27(4): pp. 315–368, 1990.

15. E. A. EMERSON and E. M. CLARKE. Using branching time logic to synthesize synchronizations skeletons. *Science of Computer Programming*, vol. 2: pp. 241–266, 1982.

16. S. MACLANE. *Categories for the Working Mathematician*. Graduate Text in Mathematics. Springer, 1971.

17. A. MAZURKIEWICZ. Concurrent program schemes and their interpretations. DAIMI PB–78, Århus University, 1977.

18. J. MESEGUER and U. MONTANARI. Petri nets are monoids. *Information and Computation*, vol. 88: pp. 105–155, 1990.

19. R. MORIN. Decompositions of asynchronous systems. In *Proc. CONCUR'98*, *LNCS*, pp. 549–564. Springer, 1998.

20. M. NIELSEN, G. ROZENBERG, and P. S. THIAGARAJAN. Elementary transition systems. *Theoretical Computer Science*, vol. 96: pp. 3–33, 1992.

21. M. PIETKIEWICZ-KOUTNY and A. YAKOVLEV. Non-pure nets and their transition systems. TR. no. 528, Department of Computing Science, University of Newcastle upon Tyne, 1995.

22. S. POPKORN. *First Steps in Modal Logic*. Cambridge University Press, 1994.

23. W. REISIG. *Petri Nets*. EATCS Monographs in Theoretical Computer Science, vol. 4, Springer-Verlag, 1985.

24. M. W. SHIELDS. Multitraces, hipertraces and partial order semantics. *Formal Aspects of Computing*, vol. 4: pp. 649–672, 1992.

25. E. W. STARK. Compositional relational semantics for indeterminate dataflow networks. In *Proc. CTCS'89*, vol. 389 of *LNCS*, pp. 52–74. Springer, 1989.

26. W. VOGLER. Executions: a new partial-order semantics for Petri nets. *Theoretical Computer Science*, vol. 91: pp. 205–238, 1991.

27. W. VOGLER. Concurrent implementations of asynchronous transition systems. In *Proc. Application and Theory of Petri Nets 1999, ICATPN'99*, vol. 1630 of *LNCS*, pp. 284–303. Springer, 1999.

28. G. WINSKEL. Petri nets, algebras, morphisms and compositionality. *Information and Computation*, vol. 72: pp. 197–238, 1987.

29. G. WINSKEL. A category of labelled Petri nets and compositional proof system (extended abstract). In *Proc. Third IEEE Symposium on Logic in Computer Science*, pp. 142–154. IEEE, The Computer Society, Computer Society Press, 1988.

30. G. WINSKEL and M. NIELSEN. Models for concurrency. In S. Abramsky, D. M. Gabbay, and T. S. E. Maibaum, eds., *Handbook of Logic in Computer Science; Semantic Modeling*, vol. 4, pp. 1–148. Oxford University Press, 1994.

Transactions and Zero-Safe Nets*

Roberto Bruni and Ugo Montanari

Dipartimento di Informatica, Università di Pisa, Italia
{bruni,ugo}@di.unipi.it

Abstract When employing Petri nets to model distributed systems, one must be aware that the basic activities of each component can vary in duration and can involve smaller internal activities, i.e., that transitions are conceptually refined into *transactions*. We present an approach to the modeling of transactions based on *zero-safe nets*. They extend ordinary PT nets with a simple mechanism for transition synchronization. We show that the net theory developed under the two most widely adopted semantic interpretations (*collective token* and *individual token* philosophies) can be uniformly adapted to zero-safe nets. In particular, we show that each zero-safe net has two associated PT nets that represent the abstract counterparts of the modeled system according to these two philosophies. We show several applications of the framework, a distributed interpreter for ZS nets based on classical net unfolding (here extended with a *commit* rule) and discuss some extensions to other net flavours to show that the concept of *zero place* provides a unifying notion of transaction for several different kinds of Petri nets.

Introduction

A distributed system can be viewed as a collection of several components that evolve concurrently, by performing local actions, but that can also exchange information, e.g., according to suitable communication protocols. Operational models for distributed systems are often defined using suitable labeled transition systems. *Place/transition Petri nets* [41,43] (abbreviated as PT *nets*) can be viewed as particular structured transition systems, where the additional algebraic structure (i.e., monoidal composition of states and runs) offers a suitable basis for expressing the concurrency of local actions. In fact PT nets have been extensively used both as a foundational model for concurrent computations and as a specification language, due to their well studied theory, a simple graphical presentation and several supporting tools.

When designing large and complex systems via PT nets, the more convenient approach is to start by outlining a very abstract model and then to refine each transition (that might represent a complex activity of the system) into a net

---

* Research supported by CNR Integrated Project *Progettazione e Verifica di Sistemi Eterogenei Connessi mediante Reti*; by TMR Project *GETGRATS*; by Esprit Working Group *APPLIGRAPH*; and by MURST project *TOSCa: Tipi, Ordine Superiore e Concorrenza*.

H. Ehrig et al. (Eds.): Unifying Petri Nets, LNCS 2128, pp. 380–426, 2001.

that offers a more precise representation of the associated activity. For example, communication protocols for passing and retrieving values cannot ignore that agent synchronization is built on finer actions (e.g., for sending data requests and acknowledgments). Moreover, such actions must be executed according to certain local/global strategies that must be completed before the interaction is closed. Hence the abstract transition is seen, at the refined level, as a distributed computation (that we call a *transaction*) which succeeds only if all the involved component accomplish their tasks. In particular the *commit* of the transaction synchronizes all the terminal operations of local tasks. For the refinement to be correct, we must assume that the transaction is executed *atomically*, as if it were a transition. Thus, the execution strategy can be only partially distributed, since certain local choices must be globally coordinated. However, this is also the case in ordinary (non free-choice) PT nets. In fact, let us consider a generic interpreter for PT nets, where each transition synchronizes the consumption and production of its pre- and postset. This assumption requires that a local activity can influence the behavior of other transitions: Before executing any transition t, the interpreter must lock all the distributed resources that t will consume and this must be done atomically; otherwise a different transition t' could lock some of the resources needed by t. Therefore the interpreter can afford only a certain degree of distribution. This originates what can be called 'place synchronization.'

Several approaches have appeared in the literature that present different refinement techniques for top-down design of a concurrent system (e.g., *Petri Box calculus* [7,6] and *rule-based refinement* [39]). Many references to the subject can be found in [8,26]. Typically at each step a single transition (say t) of the actual net N is refined into a suitable subnet M, yielding the net $N[t \rightarrow M]$. This approach is somehow related to the notion of *general net morphism* proposed by Petri, that can be used to map the refined net into its abstract representative by collapsing the structure of M into the transition t. In general some constraints must be assumed on the net M for its behavior to be consistent with that of t, as e.g. in [49,48,47]. Our approach is slightly different, because all transitions of the abstract net are refined by runs of 'the same' *zero-safe net*.

Zero-safe nets (ZS nets) have been introduced in [13] to provide a basic synchronization mechanism for transitions as a built-in feature. In fact, PT nets allow for 'place synchronization' only, whereas 'transition synchronization' is an essential feature to write modular and expressive programs, and to model systems equipped with synchronization primitives (to achieve modularity in defining the net associated to the synchronous composition of two programs, the translations presented in the literature involve complex, and often ad hoc, constructions [50,28,38,22,30,7]).

Besides transitions and ordinary places (here called *stable* places), ZS nets include a distinguished set of *zero* places for modeling idealized resources that remain invisible to external observers, whilst *stable markings*, which just consist of tokens in stable places, define the observable states. Any operational step of a ZS net starts at some stable marking, evolves through hidden states (i.e., markings with some tokens in zero places, called *non-stable markings*) and eventually

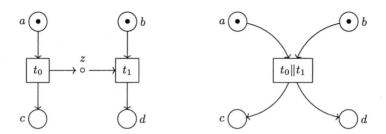

Fig. 1. A ZS net and its abstract counterpart.

leads to a stable marking. All the stable tokens produced during a certain step are released together only at the end of the step, i.e., they are 'frozen' until the commit is executed. The synchronization of transitions can thus be performed via zero tokens. The toy example in Figure 1 illustrates this basic mechanism. First, note that we extend the standard graphical representation for nets — in which boxes stand for transitions, circles for places, dots for tokens, and directed weighted arcs describe the flow relation with unary weights omitted — by using smaller circles to represent zero places. In the refined model (Figure 1, left), the initial marking $\{a, b\}$ is stable and enables the transition t_0 whose firing produces a 'frozen' stable token in c and a zero token in z. Hence, after the firing of t_0 we reach a non-stable marking. But now t_1 is enabled and its firing consumes the stable token in b and the zero token in z, and produces a frozen token in d. Since the reached marking is stable, the transaction is closed and frozen tokens are released. At the abstract level, we are not interested in observing the hidden intermediate state $\{b, c, z\}$. In fact we just consider stable places plus the atomic activity that takes $\{a, b\}$ and produces $\{c, d\}$ (Figure 1, right). Pursuing this view, a 'refined' ZS net and an 'abstract' PT net model the same system. The latter, where only stable places are considered, offers the synchronized view, which abstracts away the production and consumption of zero tokens.

In this paper we survey the operational and abstract semantics of ZS nets, together with several applications to the modeling of distributed systems. It is worth remarking that both the operational semantics of ZS nets and the construction of their abstract PT nets are characterized as two universal constructions, following the so-called '*Petri nets are monoids*' approach [32]. More precisely, the former can be characterized as an *adjunction* and the latter as a *coreflection*. The universal properties of the two constructions witness that they are the 'optimal' choices. In particular, by expressing the abstract semantics via coreflection we fully justify the choice of abstract PT nets as canonical representatives.

We stress that the synchronization mechanism of ZS nets can favor a uniform approach to concurrent language translation. For instance, in the case of CCS-like process algebras, the parallel composition of two nets modeling communicating processes involves the combinatorial analysis of all the admissible synchronizations, whereas we have shown in [17] that using zero places for modeling communication channels, the parallel composition of two nets can just merge the common channels. As an original contribution, here we show how to model distributed choices in a compositional way and discuss how the basic concept of 'zero place' can be exploited in other net flavours, still preserving some distinguishing features of the approach.

For what concerns ZS nets implementation, one has to specify the computational machinery for performing only correct transactions, recovering deadlocks and treating infinite low-level computations. We illustrate our proposal in [17] for equipping ZS nets with such a distributed operational tool and then extend it to deal with *read arcs* in contextual (zero-safe) nets.

Since the notion of *zero safe* place is to some extent orthogonal w.r.t. the different kinds of Petri nets considered in the literature (e.g., contextual, coloured, timed, probabilistic) we think that it can provide a unifying basis for developing a theory of concurrent transactions in Petri nets. Note that we employ the terminology 'transaction' with a meaning analogous to the one it finds in databases: a (sort of) program that when applied to a consistent state still leads to a consistent state, though not necessarily the consistency of the state is preserved by all steps in the program.

Origin and Structure of the Paper. The operational and abstract semantics of ZS nets according to the two more widely adopted net philosophies (called *collective token* and *individual token*) have been presented in [13,14] together with the associated universal constructions. A comparison between the two approaches has been discussed in [17], in the Ph.D. Thesis of the first author [10], and in the tutorial overview [15]. The distributed interpreter for ZS nets has been proposed in [16]. The modeling of distributed *don't know* choice and the extensions of the zero safe approach to other net flavours (e.g., read arcs) have not appeared elsewhere.

In Section 1 we recall PT nets and their semantics. Section 2 illustrates ZS nets and their operational and abstract semantics and uses two examples to motivate the usage of zero places. In Section 3 we give a compositional representation of a simple process algebra equipped with action prefix, parallel composition, restriction and *don't know* nondeterministic choice. The distributed interpreter for ZS nets is defined in Section 4. We conclude in Section 5 by extending the ZS net formalism to deal with read arcs. For detailed proofs of most results we refer to [17,16,10].

Acknowledgements. We thank Paolo Baldan for several interesting discussions on contextual nets and for his careful reading of a preliminary version of this work. We also thank José Meseguer and the anonymous referees for their suggestions that helped us in improving the presentation of the material.

1   Place/Transition Petri Nets

Definition 1 (Net). *A net is a triple $N = (S_N, T_N, F_N)$, where S_N is the set of places $a, a', \ldots, T_N$ is the set of transitions $t, t', \ldots$ (with $S_N \cap T_N = \varnothing$), and*

$F_N \subseteq (S_N \times T_N) \cup (T_N \times S_N)$ *is called the* flow relation. *The elements of the flow relation are called* arcs, *and we write $x \ F_N \ y$ for $(x, y) \in F_N$.*

We shall denote $S_N \cup T_N$ by N when no confusion can arise. Subscripts will be omitted if they are obvious from the context. For $x \in N$, the set $^\bullet x = \{y \in N \mid y \ F \ x\}$ is called the *preset* of x, and the set $x^\bullet = \{y \in N \mid x \ F \ y\}$ is called the *postset* of x. We only consider nets such that for any transition t, $^\bullet t \neq \varnothing$. Moreover, let $^\circ N = \{x \in N \mid {}^\bullet x = \varnothing\}$ and $N^\circ = \{x \in N \mid x^\bullet = \varnothing\}$ denote the sets of *initial* and *final* elements of N respectively. A place a is *isolated* if $^\bullet a \cup a^\bullet = \varnothing$.

Definition 2 (PT net). *A marked place/transition Petri net (*PT *net) is a tuple $N = (S, T, F, W, u_{\text{in}})$ such that (S, T, F) is a net, the function $W : F \to \mathbb{N}$ assigns a positive weight to each arc in F, and the finite multiset $u_{\text{in}} : S \to \mathbb{N}$ is the initial marking of N.*

We find convenient to view F as a function $F : (S \times T) \cup (T \times S) \to \{0, 1\}$, with $x \ F \ y \iff F(x, y) \neq 0$. Then, for PT nets we replace $\{0, 1\}$ by $\mathbb{N}$ and abandon W. Thus, the flow relation becomes a *multiset relation* $F : (S \times T) \cup (T \times S) \to \mathbb{N}$.

A *marking* $u : S \to \mathbb{N}$ is a finite multiset of places. It can be written either as $u = \{n_1 a_1, ..., n_k a_k\}$ where each n_i dictates the number of occurrences (*tokens*) of the place a_i in u, i.e., $n_i = u(a_i)$ (if $n_i = 0$ then the $n_i a_i$ is omitted), or as the formal sum $u = \bigoplus_{a_i \in S} n_i a_i$ denoting an element of the free commutative monoid $S^\oplus$ on the set of places S (the monoidal operation is defined by $(\bigoplus_i n_i a_i) \oplus (\bigoplus_i m_i a_i) = (\bigoplus_i (n_i + m_i) a_i)$ with 0 as the neutral element). The monoid $(\mu(S), \cup, \varnothing)$ of finite multisets on S (with multiset union as monoidal operation and the empty multiset as unit) is isomorphic to $S^\oplus$.

For any transition $t \in T$, let $\text{pre}(t)$ and $\text{post}(t)$ be the multisets over S such that $\text{pre}(t)(a) = F(a, t)$ and $\text{post}(t)(a) = F(t, a)$, for all $a \in S$. A PT net can be equivalently defined as the (marked) graph $(S^\oplus, T, \text{pre}, \text{post}, u_{\text{in}})$, with nodes in the monoid $S^\oplus$ and edges in T, where $\text{pre}(\_), \text{post}(\_) : T \to S^\oplus$ define the source and target of transitions, respectively. As usual we write $t : u \to v$ for a transition t with $\text{pre}(t) = u$ and $\text{post}(t) = v$. This definition emphasizes the algebraic structure of PT nets and allows us to define a category of nets by considering the obvious homomorphisms preserving such structure.

Definition 3 (Category Petri). *A PT net morphism $h : N \to N'$ is a pair of functions $h = (f : T \to T', g : S^\oplus \to S'^\oplus)$ with g a monoid homomorphism and with $g(\text{pre}(t)) = \text{pre}(f(t))$ and $g(\text{post}(t)) = \text{post}(f(t))$ for each $t \in T$. That is, h is a graph morphism whose node component g is a monoid homomorphism. (For marked nets, morphisms must also preserve initial markings, i.e., $g(u_{\text{in}}) = u'_{\text{in}}$.) The category* **Petri** *has (unmarked) PT nets as objects and PT net morphisms as arrows.*

Definition 4 (Firing). *Given a PT net N, let u and u' be markings of N. A transition $t \in T$ is enabled at u if $\text{pre}(t)(a) \leq u(a)$, for all $a \in S$. Moreover, we say that u evolves to u' under the firing of t, written $u \ [t\rangle \ u'$, if t is enabled at u and $u'(a) = u(a) - \text{pre}(t)(a) + \text{post}(t)(a)$, for all $a \in S$.*

Table 1. The inference rules for $\_ \Rightarrow_N \_$ and $\_ \Rightarrow_N^* \_$.

identities	generators	parallel composition	basic step	sequential composition
$u \in S^{\oplus}$	$t : u \to v \in T$	$u \Rightarrow_N v,\ u' \Rightarrow_N v'$	$u \Rightarrow_N v$	$u \Rightarrow_N^* v,\ v \Rightarrow_N w$
$u \Rightarrow_N u$	$u \Rightarrow_N v$	$u \oplus u' \Rightarrow_N v \oplus v'$	$u \Rightarrow_N^* v$	$u \Rightarrow_N^* w$

A *firing sequence* from u_0 to u_n is a sequence of markings and transitions such that $u_0 \ [t_1\rangle \ u_1...u_{n-1} \ [t_n\rangle \ u_n$. Besides firings and firing sequences, *steps* and *step sequences* are usually introduced.

Definition 5 (Step). *Given a* PT *net N, we say that a multiset $X : T \to \mathbb{N}$ is enabled at u if $\sum_{t \in T} X(t) \cdot \mathrm{pre}(t)(a) \leq u(a)$ for all $a \in S$.*

Moreover, we say that u evolves to u' under the step X, written $u \ [X\rangle \ u'$, if X is enabled at u and $u'(a) = u(a) + \sum_{t \in T} X(t) \cdot (\mathrm{post}(t)(a) - \mathrm{pre}(t)(a))$ for all $a \in S$.

Given a marking u of N, we denote by $[u\rangle$ the set of all the markings that are reachable from u via some firing sequence. The *reachable markings* of the net $N = (S, T, F, u_{\mathrm{in}})$ are the elements of the set $[u_{\mathrm{in}}\rangle$.

The dynamics of a net can be expressed by the one-step relation $\_ \Rightarrow_N \_$ defined by the three leftmost inference rules in Table 1: identities represent idle resources, generators represent the firing of a transition within the minimal marking that can enable it, and parallel composition provides concurrent execution of generators and idle steps. Then, it is obvious that $u \Rightarrow_N v \iff \exists (X : T \to \mathbb{N}).u \ [X\rangle \ v$.

The extension of this approach to computations $u_0 \Rightarrow u_1 \Rightarrow \cdots \Rightarrow u_n$ is not straightforward. Indeed, concurrent semantics must consider as equivalent all the computations where the same *concurrent* events are executed in different orders, and we cannot leave out of consideration the distinction between *collective* and *individual token philosophies* (noticed e.g., in [27], but see also [11,12]).

The simplest approach relies on the collective token philosophy (*CTph*), where semantics does not distinguish among tokens which are available at the same place, because any such token is regarded to be *operationally* equivalent to all the others. A major drawback of this approach is that it leaves out of consideration the fact that operationally equivalent resources may have different origins and histories, carrying different *causality* information. Instead, according to the individual token philosophy (*ITph*), causal dependencies are central to net dynamics. As a consequence, only the computations that refer to isomorphic *Goltz-Reisig processes* [29] can be identified, and causality information is fully maintained (the *CTph* relies instead on the *commutative processes* of Best and Devillers [5]). If one is simply interested in 'reachability' matters, then the distinction between the *CTph* and *ITph* is irrelevant, and the obvious two rightmost rules in Table 1 can be introduced (transitive closure). Otherwise, suitable *proof terms* for computations can be introduced and axiomatized to faithfully recover the two different philosophies. In this sense, Best-Devillers and Goltz-Reisig processes can be seen as *concurrent computation strategies* for CTph (resp. ITph) and can be shown to correspond to equivalence classes of proof terms modulo natural algebraic axiomatizations [23].

Commutative processes can be characterized by quotienting step sequences.

Definition 6 (Diamond transformation). *Given a* PT *net* N, *let*

$$s = u_0 \ [t_1\rangle \ u_1 \cdots u_{i-1} \ [t_i\rangle \ u_i \ [t_{i+1}\rangle \ u_{i+1} \cdots u_{n-1} \ [t_n\rangle \ u_n$$

be a step sequence of N, *where* t_i *and* t_{i+1} *are concurrently enabled by* u_{i-1}, *in the sense that* $(\mathrm{pre}(t_i) \cup \mathrm{pre}(t_{i+1}))(a) \leq u_{i-1}(a)$ *for any* $a \in S_N$. *Let* s' *be the firing sequence obtained from* s *by firing* t_i *and* t_{i+1} *in the reverse order, i.e.,*

$$s' = u_0 \ [t_1\rangle \ u_1 \cdots u_{i-1} \ [t_{i+1}\rangle \ u_i' \ [t_i\rangle \ u_{i+1} \cdots u_{n-1} \ [t_n\rangle \ u_n.$$

Then, the sequence s' *is called a* diamond transformation *of* s.

Since in step sequences transitions can be fired concurrently, we let the step sequence $u_0 \ [X_1\rangle \ u_1 \cdots u_{i-1} \ [X_i \cup X_{i+1}\rangle \ u_{i+1} \cdots u_{n-1} \ [X_n\rangle \ u_n$ *be a diamond transformation of* $u_0 \ [X_1\rangle \ u_1 \cdots u_{i-1} \ [X_i\rangle \ u_i \ [X_{i+1}\rangle \ u_{i+1} \cdots u_{n-1} \ [X_n\rangle \ u_n$ *with* X_i *and* X_{i+1} *concurrently enabled by* u_{i-1} *(and vice versa).*

Diamond transformations define a symmetric relation whose reflexive and transitive closure gives the right equivalence w.r.t. the CTph interpretation.

The notion of (causal) *process* is due to Goltz and Reisig [29] and gives a more precise account of causal dependencies between firings and tokens.

Definition 7 (Occurrence net). *A net* K *is a (deterministic) occurrence net if (1) for all* $a \in S_K$, $|{}^\bullet a| \leq 1 \wedge |a^\bullet| \leq 1$ *and (2)* F_K^* *is acyclic.*[1]

Definition 8 (Process). *A process for a* PT *net* N *is a net morphism* $P \colon K \to N$, *from an occurrence net* K *to* N, *such that* $P(S_K) \subseteq S_N$, $P(T_K) \subseteq T_N$, ${}^\circ K \subseteq S_K$, *and for all* $t \in T_K$, $a \in S_N$, $F_N(a, P(t)) = |P^{-1}(a) \cap {}^\bullet t|$ *and* $F_N(P(t), a) = |P^{-1}(a) \cap t^\bullet|$.

Two processes P and P' of N are *isomorphic* and thus identified if there exists a net isomorphism $\psi \colon K_P \to K_{P'}$ such that $\psi; P' = P$. As usual we denote the set of *origins* (i.e., minimal or initial places) and *destinations* (i.e., final or maximal places) by $O(K) = {}^\circ K$ and $D(K) = K^\circ \cap S_K$, respectively. For concatenating causal computations, the notion of *concatenable process* has been introduced in [23]. Concatenable processes are obtained from processes by imposing a total ordering on the origins that are instances of the same place and, similarly, on the destinations. The orderings are defined by means of label-indexed ordering functions. Given a set S with a labeling function $l \colon S \to S'$, a *label-indexed ordering function* for l is a family $\beta = \{\beta_a\}_{a \in S'}$ of bijections, where $\beta_a \colon l^{-1}(a) \to \{1, \ldots, |l^{-1}(a)|\}$. Thus, for $x, y \in l^{-1}(a)$ we let $x \sqsubseteq y \iff \beta_a(x) \leq \beta_a(y)$.

---

[1] F^* denotes the reflexive and transitive closure of relation F.

Table 2.

$u \in S^\oplus$	$t: u \to v \in T$	$\alpha: u \to v,\ \beta: u' \to v'$	$\alpha: u \to v,\ \beta: v \to w$
$id_u: u \to u$	$t: u \to v$	$\alpha \otimes \beta: u \oplus u' \to v \oplus v'$	$\alpha; \beta: u \to w$

Definition 9 (Concatenable process). *A concatenable process for a* PT *net
N is a triple $C = (P, {}^\circ\ell, \ell^\circ)$ where $P: K \to N$ is a process, and ${}^\circ\ell,\ \ell^\circ$ are label-
indexed ordering functions for the function P restricted to $O(K)$ and $D(K)$
respectively.*

Two concatenable processes C and C' are *isomorphic* if P_C and $P_{C'}$ are
isomorphic via a morphism that preserves all the orderings. A partial binary
operation $\_;\_$ (associative up to isomorphism and with identities) of concatenation of concatenable processes (whence their names) can be easily defined: we
take as source (target) the image through P of the initial (maximal) places of
K_P; then the composition of $C = (P, {}^\circ\ell, \ell^\circ)$ and $C' = (P', {}^\circ\ell', \ell'^\circ)$ is realized by
merging, when it is possible, the maximal places of K_P with the initial places of
$K_{P'}$ according to their labeling and ordering functions. Concatenable processes
admit also a monoidal *parallel* composition $\_ \otimes \_$ (commutative up to a natural
isomorphism), which can be represented by putting two processes side by side.
We refer the interested reader to [23] for the formal definitions of $C; C'$ and
$C \otimes C'$, which make the concatenable processes of a PT net N be the arrows of
a symmetric monoidal category $\mathcal{CP}(N)$ (whose objects are the markings of N).
The symmetries of $\mathcal{CP}(N)$ are given by concatenable processes with empty set
of transitions (token permutation is expressed by different orderings ${}^\circ\ell$ and ℓ°).

1.1 Petri Nets Are Monoids

Several interesting aspects of net theory can be profitably developed within
category theory, see, e.g., [52,32,9,24,40,35]. We focus on the so-called 'Petri
nets are monoids' approach initiated in [32] (see also [23,33,45,34,46,12]). The
idea is to extend (part of) the algebraic structure of states to the level of proof
terms associated to the rules in Table 1 in such a way to capture the basic laws of
concurrent and causal computations. The proof terms we consider are inductively
defined in Table 2. In [32,23] it is shown that axiomatic equivalences on such
proof terms can precisely characterize several standard semantic constructions.
In particular, commutative processes can be characterized by lifting the multiset
structure of states to the level of computations in a functorial way, yielding a
strictly symmetric monoidal category $\mathcal{T}(N)$ (it is called 'strictly symmetric'
because the monoidal operation is commutative). For each net N, the category
$\mathcal{T}(N)$ has markings of N as objects, and proof terms modulo the axioms in
Table 3 as arrows. Abusing the notation, in Table 3 the parallel composition
of arrows is denoted by $\oplus$, instead of $\otimes$, to emphasize that it is commutative
and can be viewed as multiset union. The functoriality law is the analogous
of diamond transformation. Denoting by **CMonCat** the category of strictly

Table 3.

neutral:	$id_\varnothing \oplus \alpha = \alpha,$	
commutativity:	$\alpha \oplus \beta = \beta \oplus \alpha,$	
associativity:	$(\alpha \oplus \beta) \oplus \alpha' = \alpha \oplus (\beta \oplus \alpha'),$	$(\alpha;\beta);\alpha' = \alpha;(\beta;\alpha'),$
identities:	$\alpha; id_u = id_v; \alpha = \alpha,$	$id_u \oplus id_v = id_{u \oplus v},$
functoriality:	$(\alpha;\beta) \oplus (\alpha';\beta') = (\alpha \oplus \alpha');(\beta \oplus \beta').$	

Table 4.

$$\frac{u, u' \in S^\oplus}{\gamma_{u,u'} : u \oplus u' \to u' \oplus u}$$

Table 5.

neutral:	$id_\varnothing \otimes \alpha = \alpha \otimes id_\varnothing = \alpha,$	
associativity:	$(\alpha \otimes \beta) \otimes \alpha' = \alpha \otimes (\beta \otimes \alpha'),$	$(\alpha;\beta);\alpha' = \alpha;(\beta;\alpha'),$
identities:	$\alpha; id_u = id_v; \alpha = \alpha,$	$id_u \otimes id_v = id_{u \oplus v},$
functoriality:	$(\alpha;\beta) \otimes (\alpha';\beta') = (\alpha \otimes \alpha');(\beta \otimes \beta'),$	
naturality:	$(\alpha \otimes \alpha'); \gamma_{v,v'} = \gamma_{u,u'}; (\alpha' \otimes \alpha),$	
coherence:	$\gamma_{u,v \oplus v'} = (\gamma_{u,v} \otimes id_{v'}); (id_v \otimes \gamma_{u,v'}),$	$\gamma_{u,v}; \gamma_{v,u} = id_{u \oplus v}.$

symmetric monoidal categories (as objects) and monoidal functors (as arrows), $\mathcal{T}(\_)$ extends to a functor from **Petri** to **CMonCat**.

Proposition 1 (cf. [32]). *The presentation of $\mathcal{T}(N)$ given above precisely characterizes the algebra of commutative processes of N, i.e., the arrows in $\mathcal{T}(N)$ are in bijection with the commutative processes of N.*

Under the *ITph*, for analogous results to hold, one must resort to symmetric monoidal categories, where parallel composition is commutative only up to a natural isomorphism. In fact, suitable auxiliary arrows called *symmetries* are present (see Table 4) that can model the possible reorganizations of minimal and maximal places of a process. We recall here the definition of the category $\mathcal{P}(N)$ introduced in [23] and finitely axiomatized in [45].

Definition 10. *Let N be a* PT *net. The category $\mathcal{P}(N)$ is the monoidal quotient of the free symmetric monoidal category $\mathcal{F}(N)$ generated by N, modulo the two axioms: (i) $\gamma_{a,b} = id_a \otimes id_b$, if $a, b \in S$, and $a \neq b$; and (ii) $s;t;s' = t$, if $t \in T$ and s, s' are symmetries (where $\gamma_{\_,\_}$, $id_\_$, $\_\otimes\_$, and $\_;\_$ are, resp., the symmetry isomorphism, the identities, the tensor product, and the composition of $\mathcal{F}(N)$).*

We remark that in $\mathcal{F}(N)$ the tensor product is not commutative and the symmetries satisfy the naturality axiom and the MacLane coherence axioms [31]. For the reader's convenience, the axioms of $\mathcal{F}(N)$ are recalled in Table 5.

Proposition 2. *The presentation of $\mathcal{P}(N)$ given above precisely characterizes the algebra of concatenable processes of the* PT *net N.*

The constructions $\mathcal{T}(N)$ and $\mathcal{P}(N)$ provide a useful syntax that can be used for denoting commutative processes and concatenable processes, respectively.

2   Zero-Safe Nets

We recall the notion of *safety* in PT nets.

Definition 11 (*n*-safe net). *A place is n-safe if it contains at most n tokens in any reachable marking. A net is n-safe if all its places are n-safe.*

Thus, the adjective '0-safe' for nets means that all places cannot contain any token in all reachable markings. We use the terminology *zero-safe net* — using the word 'zero' instead of the digit '0' — to mean that the net contains special places, called *zero places*, whose role is that of coordinating the atomic execution of several transitions, which, from an abstract viewpoint, will appear as synchronized. However no new interaction mechanism is needed, and the coordination of the transitions participating in a step is handled by the ordinary token-pushing rules of nets, assuming late delivery of stable tokens (postponed to the end of the transaction). These places are 'zero-safe' in the sense that they cannot contain any token in any *observable* state.

Definition 12 (ZS net). *A zero-safe net (ZS net) is a tuple* $B = (S_B, T_B, F_B, u_B, Z_B)$ *where* $N_B = (S_B, T_B, F_B, u_B)$ *is the* underlying PT net *of B and the set* $Z_B \subseteq S_B$ *is the set of* zero places. *The places in* $L_B = S_B \setminus Z_B$ *are called* stable places. *A stable marking is a multiset of stable places, and the initial marking* u_B *must be stable.*

Stable markings describe *observable* states, whereas the presence of one or more zero tokens in a given marking makes it be *unobservable*. We call *stable tokens* and *zero tokens* the tokens that respectively belong to stable places and to zero places. Since $S^\oplus$ is a free commutative monoid, it is isomorphic to the cartesian product $L^\oplus \times Z^\oplus$ and we can write $t \colon (u, x) \to (v, y)$ for a transition t with $\mathrm{pre}(t) = u \oplus x$ and $\mathrm{post}(t) = v \oplus y$, where u and v are stable multisets and x and y are multisets over Z. In a way similar to PT nets, ZS nets can also be seen as suitable graphs, yielding the following category.

Definition 13 (Category dZPetri). *A ZS net morphism between two ZS nets B and B′ is a PT net morphism* $(f, g) \colon N_B \to N_{B'}$ *where g preserves the partitioning of places (i.e.,* $g(a) \in L_{B'}^\oplus$ *if* $a \in L_B$ *and* $g(a) \in Z_{B'}^\oplus$ *if* $a \in Z_B$*) and satisfies the additional condition of mapping zero places into pairwise disjoint (nonempty) zero markings (i.e., for all* $z \neq z' \in Z_B$*, if* $g(z) = n_1 a_1 \oplus \cdots \oplus n_k a_k$ *and* $g(z') = m_1 b_1 \oplus \cdots \oplus m_l b_l$ *then we have that* $a_i \neq b_j$ *for* $i = 1, \ldots, k$ *and* $j = 1, \ldots, l$*), which is called the* disjoint image *property. The category* **dZPetri** *has ZS nets as objects and ZS net morphisms as arrows.*

Since $S^\oplus$ is equivalent to $L^\oplus \times Z^\oplus$, ZS net morphisms become triples of the form $h = (f, g_L, g_Z)$, where both g_L and g_Z are monoid homomorphisms on the free commutative monoids of stable and zero places, respectively.

Proposition 3. *The category* **Petri** *is a full subcategory of* **dZPetri**.

Table 6. The inference rules for $\_\Rightarrow_B \_$.

underlying	horizontal composition	commit
$u \oplus x \Rightarrow_{N_B} v \oplus y,\; u,v \in L^{\oplus},\; x,y \in Z^{\oplus}$	$(u,x) \rightrightarrows_B (v,y),\; (u',y) \rightrightarrows_B (v',y')$	$(u,0) \rightrightarrows_B (v,0)$
$(u,x) \rightrightarrows_B (v,y)$	$(u \oplus u', x) \rightrightarrows_B (v \oplus v', y')$	$u \Rightarrow_B v$

As for PT nets, we can define the behavior of ZS nets by means of a step relation $\_\Rightarrow_B \_$, defined by the inference rules in Table 6. An auxiliary relation $\_\rightrightarrows_B \_$ is introduced for modeling *transaction segments*. We can take advantage of the step relation $\_\Rightarrow_{N_B} \_$ of the underlying net for concurrently executing several transitions (rule **underlying**). The rule **horizontal composition** acts as parallel composition for stable resources and as sequential composition for zero places. We call it 'horizontal' because we prefer to view it as a synchronization mechanism rather than as the ordinary sequential composition of computations, which flows vertically from top to bottom. The rule **commit** selects the transaction segments that correspond to acceptable steps: They must start from a stable marking and end up in a stable marking. As a particular instance of the horizontal composition of two transaction segments $(u,0) \rightrightarrows_B (v,0)$ and $(u',0) \rightrightarrows_B (v',0)$, we can derive their parallel composition $(u \oplus u', 0) \rightrightarrows_B (v \oplus v', 0)$.

2.1   Introductory Example: Dining Philosophers

A simple example that illustrates the coordination role played by zero places relies on the modeling of the well-known 'dining philosophers' problem: There are n philosophers (with $n \geq 2$) sitting on a round table; each having a plate in front with some food on it; between each couple of plates there is a fork, with a total of n forks on the table; each philosopher cyclically thinks and eats, but to eat he needs both the fork on the left and that on the right of his plate; after eating a few mouthfuls, the philosopher puts the forks back on the table and starts thinking again.

The PT net for the case $n = 2$ is illustrated in Figure 2. A token in one of the places PhH_i, PhE_i, and PhT_i, for $1 \leq i \leq 2$, means that the ith philosopher is hungry, is eating, and is thinking, respectively. A token in the place Fk_i means that the ith fork is on the table. The transitions Take_i, Drop_i, and Hungry_i represent that the ith philosopher takes the forks and starts eating, finishes eating and drops the forks, feels his stomach hungry and prepares to eat, respectively. Note that Take_i requires both forks and thus cannot be fired if the other philosopher is eating. The initial marking of the net is $\{\mathrm{PhT}_1, \mathrm{PhT}_2, \mathrm{Fk}_1, \mathrm{Fk}_2\}$ (i.e., both philosophers are thinking and both forks are on the table).

Of course, this model does not tell how the philosophers access the 'resources' needed to eat, whereas the action Take_i is not trivial and requires some atomic mechanism for getting the forks. At a more refined level, for example, the strategy for executing the action Take_i could be specified as 'take the ith fork (if possible),

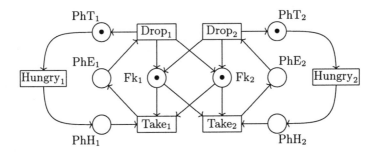

Fig. 2. An abstract view for (two) dining philosophers.

then the $((i \bmod 2) + 1)$th fork (if possible) and eat,' hence it is not difficult to imagine a deadlock where each philosopher takes one fork and cannot continue, since conflict arises. The fact is that the coordination mechanism is hidden inside transitions whose granularity is too coarse.

The situation is completely different if one wants to model the system using *free choice* nets,[2] where all decisions are local to each place. To see this, let us concentrate our attention to a subpart of the net in Figure 2, depicted in Figure 3(a), which will suffice to illustrate the point. We can translate any net into a free choice net by adding special transitions that perform the local decisions required. For example, the free choice net in Figure 3(b) corresponds to the net in Figure 3(a), but models a system where two decisions can take place independently: One decision concerns the assignment of the first fork either to the first or the second philosopher, the other decision concerns the assignment of the second fork. Then, it might happen that the first fork is assigned to the first philosopher ($Ch_{1,1}$) and the second fork is assigned to the second philosopher ($Ch_{2,2}$), and in such case the translated net deadlocks and none of the $Take_i$ actions can occur. Thus, the translated net admits computations not allowed in the abstract system of Figure 3(a).

Zero-safe nets overcome this deadlock problem by executing only certain atomic transactions, where tokens produced in low-level resources are also consumed. In our example, the invisible resources consist of places $Fk_{i,j}$ for $1 \le i, j \le 2$, that can be interpreted as zero-places. In this way the computation performing $Ch_{1,1}$ and $Ch_{2,2}$ is forbidden, because it stops in an invisible state, i.e., a state that contains zero tokens. Figure 4 represents the low-level model as a ZS net. (Recall that smaller circles stand for zero-places.)

---

[2] We recall that a net is *free choice* if for any transitions t_1 and t_2 whose presets are not disjoint, then the presets of t_1 and t_2 consist of exactly one place, or equivalently, a net is free choice if for any place s in the preset of two or more transitions, then the preset of any such transition is exactly $\{s\}$.

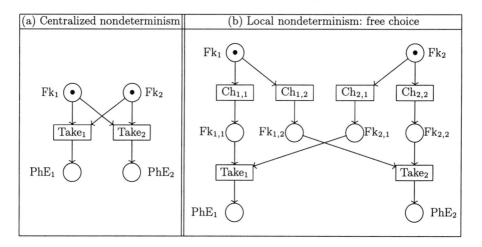

Fig. 3. Global vs (completely) local choices.

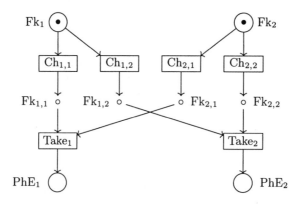

Fig. 4. Atomic free choice.

2.2   CTph vs. ITph: The Multicasting System Example

At an abstract level, the system modeled via a ZS net B can be equivalently de-scribed via a PT net $\mathcal{E}(B)$ such that $S_{\mathcal{E}(B)} = L_B = S_B \smallsetminus Z_B$ and $(\_ \Rightarrow_{\mathcal{E}(B)} \_) = (\_ \Rightarrow_B \_)$. Among the several PT nets that satisfy the above conditions we would like to choose the optimal one: Informally the transitions of such net should represent the proofs of transaction steps $u \Rightarrow_B v$ taken up to concurrent equiva-lence and such that they cannot be decomposed into smaller transaction proofs. When these two conditions are satisfied, the concurrent kernel of the possible behaviors has been identified, and all the steps can be generated by it.

We have seen in Section 1 that when dealing with concurrency, there is a real dichotomy between the *CTph* and the *ITph*. According to the *CTph*, all those fir-ing sequences obtained by repeatedly permuting pairs of (adjacent) concurrently

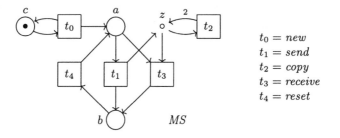

Fig. 5. The zs net *MS* representing a multicasting system.

enabled firings are identified. We call *abstract stable transactions* the resulting equivalence classes of zs net behaviours. However, acting in this way, causal dependencies on zero tokens are lost, and the class of computations captured by abstract nets may be too abstract for some applications. According to the *ITph*, instead, causal dependencies are a central aspect. As a consequence, only the transactions which refer to isomorphic Goltz-Reisig processes are identified, and we call *connected transactions* the induced equivalence classes. To illustrate these concepts, we recall the 'multicasting' example, taken from [14]. The zs net *MS* depicted in Figure 5 is designed to model a *multicasting system*: As in a broadcasting system, an agent can simultaneously send the same message to an unlimited number of receivers, but here the receivers are not necessarily all the remaining agents.

Each token in place a represents a different *active* agent that is ready to communicate, while tokens in b represent *inactive* agents. The zero place z models a buffer where tokens are messages (e.g., data, values). The transition *new* permits creating fresh agents. Each firing of *send* opens a one-to-many communication: A message is put in the buffer z and the agent which started the communication is frozen in b until the end of the current transaction. Each time the transition *copy* fires, a new copy of a message is created. To complete a transaction, as many firings of *receive* are needed as the number of copies created by *copy* plus one. Each firing of *receive* synchronizes an active agent with a copy of the message and then freezes the agent. At the end of a session, all the suspended agents are moved into place b. The transition *reset* activates an inactive agent. The graph corresponding to the zs net *MS* has the following set of arcs: $T_{MS} = \{t_0\colon (c, 0) \to (a \oplus c, 0), t_1\colon (a, 0) \to (b, z), t_2\colon (0, z) \to (0, 2z),$ $t_3\colon (a, z) \to (b, 0), t_4\colon (b, 0) \to (a, 0)\}$.

In Figure 6 we see the infinite abstract PT net A_{MS} for the refined zs net *MS*, according to the *CTph* (see Definition 17). As it will be explained later, the *abstract* net A_{MS} comes equipped with a *refinement morphism* ϵ_{MS}^C to the refined net *MS*. The refinement morphism maps each place of A_{MS} into the homonymous stable place of *MS* and defines a bijection between the transitions of A_{MS} and the abstract stable transactions of *MS*: The transition σ_n of A_{MS} represents a one-to-n transmission. By contrast, under the *ITph*, different copy

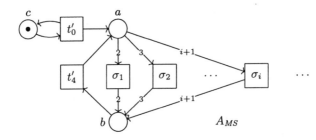

Fig. 6. The abstract net for the multicasting system under the *CTph*.

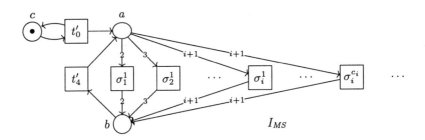

Fig. 7. The *causal* abstract net for the multicasting system under the *ITph*.

policies[3] for a one-to-n transmission may be distinguished. The infinite *causal abstract* PT net I_{MS} corresponding to the refined ZS net MS under the *ITph* (see Definition 24) is displayed in Figure 7. It comes equipped with a *causal refinement morphism* ϵ^I_{MS} to the refined net MS. Such morphism maps each place of I_{MS} into the homonymous stable place of MS, and defines a bijection between the transitions of I_{MS} and the connected transactions of MS. We assume that the generic transition σ^k_n corresponds to the one-to-n transmission that follows the k-th codified copy policy (we denote by c_n the number of different copy policies associated to the one-to-n transmission).

Zero places can be used to coordinate and synchronize in a single transaction any number of transitions of the refined net. Thus it may well happen that the refined net is finite while the abstract net is infinite. This is the case for this example, in which communication events can involve any number of receivers.

2.3 Collective Token Approach

Operational Semantics under the CTph. A *stable step* of a ZS net B may involve the execution of several transitions. At the beginning, the state must contain enough stable tokens to concurrently enable the stable presets of all

---

[3] We call *copy policy* any strategy (e.g., sequential, with maximal parallelism) for making copies of the messages in the buffer z.

these transitions. As the computation progresses, the firings can only consume the stable tokens that were also available at the beginning of the computation and the zero tokens that have been produced by some fired transition. A stable step whose intermediate markings are all nonstable and which consumes all the available stable tokens is called a *stable transaction*.

Definition 14 (Stable step and stable transaction). *Let B be a ZS net. A firing sequence $s = u_0\,[t_1\rangle\,u_1 \ldots u_{n-1}\,[t_n\rangle\,u_n$ of the underlying net N_B is a stable step of B if:*

- $\sum_{i=1}^n \mathrm{pre}(t_i)(a) \leq u_0(a)$ *for all $a \in S_B \smallsetminus Z_B$ (concurrent enabling);*
- u_0 *and u_n are stable markings of B (stable fairness).*

We write $u_0\{[s\rangle u_n$ to denote the stable step s, and $O(s)$ and $D(s)$ to denote the u_0 and u_n respectively. A stable step s is a stable transaction if in addition:

- *the markings $u_1, \ldots, u_{n-1}$ are not stable (atomicity);*
- $\sum_{i=1}^n \mathrm{pre}(t_i)(a) = u_0(a)$ *for all $a \in S_B \smallsetminus Z_B$ (perfect enabling).*

A *stable step sequence* is a sequence of stable steps $u_0\{[s_1\rangle u_1 \ldots u_{n-1}\{[s_n\rangle u_n$. We then say that u_n is *reachable* from u_0. We recall that stable tokens produced during the transaction become operative in the system only after the commit.

Example 1. Consider the ZS net *MS* of Figure 5.

The firing sequence $\{2a\}\,[t_1\rangle\,\{a, b, z\}\,[t_4\rangle\,\{2a, z\}\,[t_3\rangle\,\{a, b\}$ is not a stable step, because the concurrent enabling condition is not satisfied.

The sequence $\{4a\}\,[t_1\rangle\,\{3a, b, z\}\,[t_2\rangle\,\{3a, b, 2z\}\,[t_3\rangle\,\{2a, 2b, z\}\,[t_3\rangle\,\{a, 3b\}$ is a stable step but not a stable transaction, because the perfect enabling condition is not satisfied.

The firing sequence $s' = \{2a, b\}\,[t_1\rangle\,\{a, 2b, z\}\,[t_3\rangle\,\{3b\}\,[t_4\rangle\,\{a, 2b\}$ is a stable step but not a stable transaction, because the atomicity constraint is not satisfied.

The firing sequence $s'' = \{2a, b\}\,[t_1\rangle\,\{a, 2b, z\}\,[t_4\rangle\,\{2a, b, z\}\,[t_3\rangle\,\{a, 2b\}$ is a stable transaction (compare it with the first sequence of this example).

To obtain a more satisfactory notion of stable step (transaction) in the concurrent setting of *CTph*, we can then consider commutative processes.

Definition 15 (Abstract sequence). *Equivalence classes of sequences (w.r.t. diamond transformation) are called* abstract sequences *and are ranged over by σ. The abstract sequence of s is written $[\![s]\!]$. We also write $\mathrm{pre}([\![s]\!]) = O(s)$ and $\mathrm{post}([\![s]\!]) = D(s)$ to denote respectively the origins and the destinations of $[\![s]\!]$.*

Definition 16 (Abstract stable step and abstract transaction). *Given a ZS net B, an* abstract stable step *is an abstract sequence $[\![s]\!]$ of the underlying net N_B, where s is a stable step. An* abstract stable transaction *is an abstract sequence of N_B that contains only stable transactions of B. We denote by Υ_B the set of all abstract stable transactions of B.*

The equivalence induced by diamond transformation preserves stable steps (because the diamond transformation preserves the properties of concurrent enabling and of stable fairness required by Definition 14) but does not preserve stable transactions. Generally speaking, the problem is that two stable transactions that are concurrently enabled could be interleaved in such a way that the resulting sequence is a stable transaction. Of course, such transaction cannot be considered as a representative of an atomic activity of the system, because it can be expressed in terms of two concurrent sub-activities. Therefore, we take as representatives of abstract stable transactions all those stable transactions whose equivalence classes contain only stable transactions.

Abstract Semantics under the CTph. It is now possible to define abstract representatives of those systems modeled by ZS nets in terms of PT nets whose transitions are abstract stable transactions.

Definition 17 (Abstract net). *Given a* ZS *net* $B = (S_B, T_B, F_B, u_B, Z_B)$, *its abstract net is the net* $A_B = (S_B \smallsetminus Z_B, \Upsilon_B, F, u_B)$, *with* $F(a, \sigma) = \mathrm{pre}(\sigma)(a)$ *and* $F(\sigma, a) = \mathrm{post}(\sigma)(a)$ *for all* $a \in S_B \smallsetminus Z_B$ *and* $\sigma \in \Upsilon_B$.

Example 2. Consider the following firing sequences of the underlying net N_{MS} of the ZS net MS in Figure 5: $s_{new} = \{c\}\ [t_0\rangle\ \{a, c\}$, $s_{res} = \{b\}\ [t_4\rangle\ \{a\}$, $s_1 = \{2a\}\ [t_1\rangle\ \{a, b, z\}\ [t_3\rangle\ \{2b\}$, $s_2 = \{3a\}\ [t_1\rangle\ \{2a, b, z\}\ [t_2\rangle\ \{2a, b, 2z\}\ [t_3\rangle\ \{a, 2b, z\}\ [t_3\rangle\ \{3b\}, \ldots$,

$$s_i = \{(i+1)a\}\ [t_1\rangle\ \{ia, b, z\}\ \underbrace{[t_2\rangle\ \cdots\ [t_2\rangle}_{i-1}\{ia, b, iz\}\ \underbrace{[t_3\rangle\ \cdots\ [t_3\rangle}_{i}\ \{(i+1)b\}, \cdots$$

We have $\Upsilon_{MS} = \{t'_0, t'_4, \sigma_1, \ldots, \sigma_i, \ldots\}$ with $t'_0 = [\![s_{new}]\!]$, $t'_4 = [\![s_{res}]\!]$ and $\sigma_i = [\![s_i]\!]$, for $i \geq 1$. The abstract net A_{MS} of MS is shown in Figure 6. It consists of three places and infinitely many transitions: One transition for creating a new active process, one for reactivating a process after a synchronization, and one for each possible multicasting involving a different number of receivers.

Proposition 4. *The reachable markings of* A_B *and of* B *are the same.*

Universal Constructions in the CTph. We recast the operational and abstract (*CTph*) semantics of ZS nets in a categorical framework via two universal constructions. The first construction starts from the category **dZPetri** (where ZS nets are seen as programs) and exhibits an adjunction to a category **HCatZPetri** consisting of machines equipped with suitable operations on states and transitions (e.g., parallel composition and a special kind of sequential composition, called *horizontal*). This adjunction corresponds to the operational semantics of ZS nets, in the sense that the transitions of the machine $\mathscr{Z}[B]$ associated to a ZS net B are exactly the abstract stable steps of B. Moreover, abstract stable transactions can be characterized algebraically as special transitions of $\mathscr{Z}[B]$, called *prime arrows*. The second construction starts from a different category **ZSN** of ZS nets (strictly related to **HCatZPetri**), having the ordinary category **Petri** of PT nets as a subcategory, and yields a coreflection that recovers the abstract net construction in Definition 17. We remark that **ZSN** allows one to map transitions of a machine into prime arrows of another machine, yielding a very general notion of 'implementation morphism.'

Definition 18 (Category HCatZPetri). *A* ZS *graph*

$$H = ((L \cup Z)^{\oplus}, (T, \oplus, 0, id, \cdot), \text{pre}, \text{post})$$

is both a ZS *net and a reflexive Petri commutative monoid.*[4] *In addition, it is equipped with a partial function* _ · _, *called* horizontal composition, *such that:*

$$\frac{\alpha: (u, x) \to (v, y), \ \beta: (u', y) \to (v', z)}{\alpha \cdot \beta: (u \oplus u', x) \to (v \oplus v', z)}. \tag{1}$$

Horizontal composition is associative and has identities $id_{(0,x)}: (0, x) \to (0, x)$ *for any* $x \in Z^{\oplus}$. *Moreover, the commutative monoidal operator* _ $\oplus$ _ *is functorial w.r.t. horizontal composition. A* ZS *graph morphism* $h = (f, g_L, g_Z): H \to H'$ *between two* ZS *graphs* H *and* H' *is both a* ZS *net morphism and a reflexive Petri monoid morphism such that* $f(\alpha \cdot \beta) = f(\alpha) \cdot f(\beta)$. ZS *graphs (as objects) and their morphisms (as arrows) form the category* **HCatZPetri**.

Horizontal composition acts as a sequential composition on zero places and as a parallel composition on stable places.

Proposition 5. *If* $\alpha: (u, 0) \to (v, 0)$ *and* $\alpha': (u', 0) \to (v', 0)$ *are two transitions of a* ZS *graph then* $\alpha \cdot \alpha' = \alpha \oplus \alpha'$.

Theorem 1. *Let* $\mathscr{U}: \mathbf{HCatZPetri} \to \mathbf{dZPetri}$ *be the functor which forgets about the additional structure on transitions, i.e.,*

$$\mathscr{U}[((L \cup Z)^{\oplus}, (T, \oplus, 0, id, \cdot), \text{pre}, \text{post})] = (L^{\oplus} \times Z^{\oplus}, T, \text{pre}, \text{post}).$$

The functor $\mathscr{U}$ *has a left adjoint* $\mathscr{L}: \mathbf{dZPetri} \to \mathbf{HCatZPetri}$.

The functor $\mathscr{L}: \mathbf{dZPetri} \to \mathbf{HCatZPetri}$ maps a ZS net B into the ZS graph which is defined by the inference rules in Table 7 modulo suitable axioms: Transitions form a commutative monoid (with $\oplus$ and $id_{(0,0)}$); the horizontal composition _ · _ is associative and has identities $id_{(0,x)}$; finally, the monoidal operator _ $\oplus$ _ is functorial w.r.t. horizontal composition and identities.

Example 3. Let MS be the ZS net defined in Section 2.2. The arrow $t_1 \cdot t_3 \in \mathscr{L}[MS]$ has source $(2a, 0)$ and target $(2b, 0)$, while $(t_1 \oplus id_{(a,0)}) \cdot (id_{(b,0)} \oplus t_3)$ goes

---

[4] A *reflexive Petri commutative monoid* is a Petri net together with a function $id: S^{\oplus} \to T$, where the set of transitions is a commutative monoid $(T, \oplus, 0)$ and pre, post and id are monoid homomorphisms, with $\text{pre}(id(x)) = \text{post}(id(x)) = x$.

Table 7. Free construction of $\mathscr{Z}[B]$.

$$\frac{(u,x) \in L_B^{\oplus} \times Z_B^{\oplus}}{id_{(u,x)}:(u,x) \to (u,x) \in \mathscr{Z}[B]}$$

$$\frac{\alpha:(u,x) \to (v,y),\ \beta:(u',x') \to (v',y') \in \mathscr{Z}[B]}{\alpha \oplus \beta:(u \oplus u', x \oplus x') \to (v \oplus v', y \oplus y') \in \mathscr{Z}[B]}$$

$$\frac{t:(u,x) \to (v,y) \in T_B}{t:(u,x) \to (v,y) \in \mathscr{Z}[B]}$$

$$\frac{\alpha:(u,x) \to (v,y),\ \beta:(u',y) \to (v',z) \in \mathscr{Z}[B]}{\alpha \cdot \beta:(u \oplus u',x) \to (v \oplus v',z) \in \mathscr{Z}[B]}$$

from $(3a \oplus b, 0)$ to $(a \oplus 3b, 0)$. As another example, the following expressions all denote the same arrow (i.e., the one-to-three communication):

$$
\begin{aligned}
t_1 \cdot t_2 \cdot (t_2 \oplus t_3) \cdot (t_3 \oplus t_3) &= t_1 \cdot t_2 \cdot (t_2 \oplus id_{(0,z)}) \cdot (id_{(0,2z)} \oplus t_3) \cdot (t_3 \oplus t_3) \\
&= t_1 \cdot t_2 \cdot (t_2 \oplus id_{(0,z)}) \cdot (t_3 \oplus t_3 \oplus t_3) \\
&= t_1 \cdot t_2 \cdot (t_2 \oplus id_{(0,z)}) \cdot (t_3 \oplus id_{(0,2z)}) \\
&\quad \cdot (t_3 \oplus id_{(0,z)}) \cdot t_3.
\end{aligned}
$$

Definition 19 (Prime arrow). *An arrow* $\alpha:(u,0) \to (v,0)$ *of a* ZS *graph* H *is prime if it cannot be expressed as the monoidal composition of nontrivial arrows (i.e.,* $\alpha = \beta \otimes \gamma$ *implies that* $\beta = id_{(0,0)}$ *or* $\gamma = id_{(0,0)}$*).*

The following theorem defines the correspondence between the algebraic and operational semantics of ZS nets.

Theorem 2. *Given a* ZS *net* B*, there is a bijection between arrows* $\alpha:(u,0) \to (v,0)$ *in* $\mathscr{Z}[B]$ *and abstract stable steps of* B*. Moreover, if such an arrow is prime then the corresponding abstract stable step is an abstract stable transaction.*

Example 4. The prime arrows in $\mathscr{Z}[MS]$ are $\tau_0 = t_0$, $\tau_4 = t_4$, $\alpha_1 = t_1 \cdot t_3$, $\alpha_2 = t_1 \cdot t_2 \cdot (t_3 \oplus id_{(0,z)}) \cdot t_3$, $\ldots$, $\alpha_i = t_1 \cdot t_2 \cdot (t_2 \oplus t_3) \cdot \ldots \cdot (t_2 \oplus t_3) \cdot (t_3 \oplus t_3)$, and so on, where the expression $(t_2 \otimes t_3)$ appears exactly $i - 2$ times in α_i.

To characterize the abstract semantics, we introduce a category **ZSN** of ZS nets, where the morphisms may map a transition into a transaction.

Definition 20. *An* abstract transition *of a* ZS *net* B *is either a prime arrow of* $\mathscr{Z}[B]$ *or a transition of* B *(seen as an arrow in* $\mathscr{Z}[B]$*).*

Definition 21 (Category ZSN). *Given two* ZS *nets* B *and* B'*, a* refinement morphism $h: B \to B'$ *is a* ZS *net morphism* $(f, g_L, g_Z): B \to \mathscr{U}[\mathscr{Z}[B']]$ *such that the function* f *maps transitions into abstract transitions. The category* **ZSN** *has* ZS *nets as objects and refinement morphisms as arrows. The composition between two refinement morphisms* $h: B \to B'$ *and* $h': B' \to B''$ *is defined as the* ZS *net morphism* $\mathscr{U}[\tilde{h}'] \circ h: B \to \mathscr{U}[\mathscr{Z}[B'']]$*, where* $\tilde{h}': \mathscr{Z}[B'] \to \mathscr{Z}[B'']$ *is the unique extension of* h' *to a morphism in* **HCatZPetri**.

Theorem 3. *The Category* **Petri** *is embedded into* **ZSN** *fully and faithfully as a coreflective subcategory and the right adjoint functor* $\mathscr{A}[\_]$ *is such that* $\mathscr{A}[B] = A_B$ *for any* ZS *net* B. *Furthermore, the counit component* ϵ_B^C *maps transitions of the abstract net into appropriate abstract transactions.*

The universal property of the coreflection witnesses that A_B is the PT net that better approximates the abstract *CTph* behaviour of B.

2.4 Individual Token Approach

In this section, the basic activities of ZS nets are defined accordingly to the *ITph*. This choice has a great impact on the resulting notion of transaction.

Operational Semantics under the ITph. In the *ITph*, a marking can be thought of as an indexed (over the places) collection of ordered sequences of tokens and each firing must exactly specify *which* tokens are consumed.

In [14], inspired by [44], we presented a *stack based approach* to the implementation of *ITph* states. The idea was to choose a canonical interpretation of the tokens that have to be consumed and produced in a firing and to introduce *permutation firings* with the task of rearranging ordered tokens: A marking is represented as a collection of stacks, one for each place and thus the extraction and the insertion of tokens follow the LIFO policy. However, permutation firings can modify the token positions in the stacks. Causal firings were essentially introduced as a concrete means to describe the token flow, providing an intuitive grasp of the underlying mechanism. In this presentation, we prefer to resort to the more compact algebraic notation given in Section 1.1, that precisely denotes concatenable processes. For the interested reader, causal firing sequences can be thought of as arrows in $\mathcal{P}(N_B)$ having the form

$$\omega = s_0; (t_1 \otimes id_{u_1}); s_1; (t_2 \otimes id_{u_2}); s_2; \dots; (t_n \otimes id_{u_n}); s_n,$$

where the s_i are permutations, the t_i are transitions, and the u_i are suitable markings. In the following, we shall keep the terminology of causal firing sequences for such arrows. For ω a causal firing sequence, we denote by $pr(\omega)$ the concatenable process it represents (considered up to isomorphism).

Example 5. Let N_{MS} be the underlying net of the ZS net *MS* defined in Figure 5 (i.e., in N_{MS} we do not distinguish between stable and zero places). The concatenable processes associated to the sequences

$$\omega = (t_0 \otimes id_b); (t_4 \otimes id_{a\oplus c}); (t_1 \otimes id_{a\oplus c}); (t_3 \otimes id_{b\oplus c}): b \oplus c \to 2b$$
$$\omega' = (t_4 \otimes id_c); (t_0 \otimes id_a); (t_1 \otimes id_{a\oplus c}); (t_3 \otimes id_{b\oplus c}): b \oplus c \to 2b$$
$$\omega'' = (t_0 \otimes id_b); (t_4 \otimes id_{a\oplus c}); (\gamma_{a,a} \otimes id_c); (t_1 \otimes id_{a\oplus c}); (t_3 \otimes id_{b\oplus c}): b \oplus c \to 2b$$

are presented in Figure 8. We use the standard notation that labels the places and transitions of the occurrence net K with their images in N. A superscript for each initial place and a subscript for each final place denote, respectively, the value of $°\ell$ and $\ell°$.

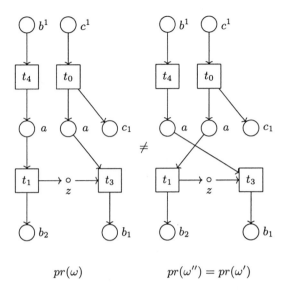

$$pr(\omega) \qquad\qquad pr(\omega'') = pr(\omega')$$

Fig. 8. The concatenable processes for ω, ω', and ω'' of Example 5.

Before continuing, let us introduce some terminology that will be used in defining the *lTph* semantics of ZS nets. A process is *full* if it does not contain idle (i.e., isolated) places. A process is *active* if it includes at least one transition, *inactive* otherwise. An active process is *decomposable into parallel activities* if it is the parallel composition of two (or more) active processes. If such a decomposition does not exist, then the process is called *connected*. A connected process may involve idle places, but it does not admit globally disjoint activities (i.e., the adjective refers to activities and not to states). Finally, the set of *evolution places* (that represent resources which are first produced and then consumed) of a process C is the set $E_C = \{P(a) \mid a \in K, \mid {}^\bullet a\mid = \mid a^\bullet\mid = 1\}$.

To forget about the ordering functions of origins and destinations we can quotient concatenable processes modulo the underlying Goltz-Reisig processes.

Definition 22. *Let N be a net. Two causal firing sequences ω and ω' are causally equivalent, written $\omega \approx \omega'$ if $pr(\omega) = (P\colon K \to N,\ {}^\circ\ell, \ell^\circ)$ and $pr(\omega') = (P'\colon K' \to N,\ {}^\circ\ell', \ell'^\circ)$ with process P isomorphic to P'. The equivalence class of ω is denoted by $[\![\omega]\!]_\approx$. We use ξ to range over equivalence classes. Since the relation $\approx$ respects the initial and final marking, we extend the notation letting $O(\xi) = O(\omega)$ and $D(\xi) = D(\omega)$, for $\xi = [\![\omega]\!]_\approx$.*

In the *lTph*, state changes are given in terms of *connected steps*, which may involve the concurrent execution and synchronization of several transitions. A connected transaction is a connected step such that no intermediate marking is stable, and which consumes all the available stable tokens of the starting state.

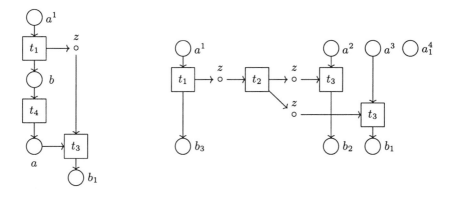

Fig. 9. The concatenable processes $pr(\omega_1)$ (left) and $pr(\omega_2)$ (right).

Definition 23 (Connected step and transaction). *Given a* ZS *net B, let ω be a causal firing sequence of the underlying* PT *net N_B. The equivalence class $\xi = [\![\omega]\!]_\approx$ is a connected step of B, written $O(\xi)[\![\xi\rangle D(\xi)$, if:*

- *$O(\omega)$ and $D(\omega)$ are stable markings (stable fairness);*
- *$E_{pr(\omega)} \subseteq Z_B$ (atomicity).*

Furthermore, the connected step ξ is a connected transaction of B if:

- *$pr(\omega)$ is connected;*
- *$pr(\omega)$ is full.*

We denote by Ξ_B (ranged by δ) the set of connected transactions of B.

A *connected step sequence* is a sequence $u_0[\![\xi_1\rangle u_1 \ldots u_{n-1}[\![\xi_n\rangle u_n$ of connected steps, and we then say that u_n is reachable from u_0. Connected steps differ from stable steps in that they allow for a finer causal relationship among events. Fullness ensures the absence of idle resources in connected transactions. Note that all conditions in Definition 23 impose constraints only over the Goltz-Reisig process associated with $pr(\omega)$.

Example 6. Let us consider the ZS net *MS* in Figure 5 and the causal firing sequences

$$\omega_1 = t_1; (t_4 \otimes id_z); t_3 : a \to b$$
$$\omega_2 = (t_1 \otimes id_{3a}); (t_2 \otimes id_{3a \oplus b}); (t_3 \otimes id_{2a \oplus b \oplus z}); (t_3 \otimes id_{a \oplus 2b}): 4a \to a \oplus 3b$$
$$\omega_3 = (t_1 \otimes id_{a \oplus c}); (t_3 \otimes id_{b \oplus c}); (t_0 \otimes id_{2b}): 2a \oplus c \to a \oplus 2b \oplus c.$$

The equivalence class $[\![\omega_1]\!]_\approx$ is not a connected step since the 'atomicity' requirement is not fulfilled (Figure 9, left). The equivalence class $[\![\omega_2]\!]_\approx$ is a connected step but not a connected transaction since the associated process is connected but not full (Figure 9, right). Likewise, $[\![\omega_3]\!]_\approx$ is a connected step but not a

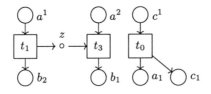

Fig. 10. The concatenable process $pr(\omega_3)$.

connected transaction since the associated process is not connected (Figure 10). The equivalence class of the causal firing sequence $(t_1 \otimes id_{4a}); (t_2 \otimes id_{4a \oplus b}); (t_2 \otimes id_{4a \oplus b \oplus z}); (t_2 \otimes id_{4a \oplus b \oplus 2z}); (t_3 \otimes id_{3a \oplus b \oplus 3z}); (t_3 \otimes id_{2a \oplus 2b \oplus 2z}); (t_3 \otimes id_{a \oplus 3b \oplus z}); (t_3 \otimes id_{4b}) : 5a \rightarrow 5b$ is a connected transaction.

Abstract Semantics under the ITph. In the *ITph* based approach it is also possible to define an abstract view of the systems modeled via ZS nets. Since transactions rewrite multisets of stable tokens, PT nets are again a natural candidate for the abstract representation.

Definition 24 (Causal abstract net). *Let B be a ZS net. The net $I_B = (S_B \setminus Z_B, \Xi_B, F, u_B)$, with $F(a, \delta) = \text{pre}(\delta)(a)$ and $F(\delta, a) = \text{post}(\delta)(a)$ for all $a \in S_B \setminus Z_B$ and $\delta \in \Xi_B$, is the* causal abstract net *of B.*

Proposition 6. *The reachable markings of I_B and of B are the same.*

Example 7. Let *MS* be the ZS net in Figure 5. Its causal abstract net I_{MS} is shown in Figure 7. Transition t'_0 is the basic activity which creates a new communicating process and it corresponds to $[\![t_0]\!]_\approx$. Similarly $t'_4 = [\![t_4]\!]_\approx$. Each σ_i^k describes a different one-to-i communication. The index k identifies the copy policy under consideration. For each i, we denote by c_i the number of different copy policies for the communication one-to-i and we have a bijective correspondence among copy policies and the *complete* binary trees[5] with exactly i leaves.

Universal Constructions in the ITph. In this section, analogously to what has been done for the *CTph*, we present the categorical constructions that characterize the operational and abstract semantics of ZS nets under the *ITph*. The first adjunction goes from **dZPetri** to a category **ZSCGraph** of more structured models, called ZS *causal graphs*, equipped not only with parallel and horizontal compositions as in **HCatZPetri**, but also with a family of *swappings* playing the role of zero token permutations. Again, the connected transactions are characterized as prime arrows of ZS causal graphs. The second construction starts

---

[5] We recall that a binary tree is *complete* if any internal node has exactly two children and we do not distinguish between *left* and *right* children.

from a category **ZSC** of zs nets and more complex morphisms, having the ordinary category **Petri** of PT nets as a subcategory, and yields a coreflection that recovers exactly the construction of the causal abstract net.

Definition 25 (Category ZSCGraph). *A* zs *causal graph*

$$E = ((L \cup Z)^{\oplus}, (T, \otimes, 0, id, *, e_{-,-}), \text{pre}, \text{post})$$

is both a zs *net and a reflexive Petri monoid. In addition, it comes equipped with a partial function* $_*\_$ *called* horizontal composition,

$$\frac{\alpha: (u, x) \to (v, y), \ \beta: (u', y) \to (v', y')}{\alpha * \beta: (u \oplus u', x) \to (v \oplus v', y')},$$

and a family of horizontal swappings, $\{e_{x,y}: (0, x \oplus y) \to (0, y \oplus x)\}_{x,y \in Z^{\oplus}}$. *Horizontal composition is associative and has identities* $id_{(0,x)}$ *for all* $x \in Z^{\oplus}$. *The monoidal operator* $_\otimes\_$ *is functorial w.r.t. horizontal composition. The (horizontal)* naturality *axiom,* $e_{x,x'} * (\beta \otimes \alpha) = (\alpha \otimes \beta) * e_{y,y'}$ *holds for any* $\alpha: (u, x) \to (v, y)$ *and* $\beta: (u', x') \to (v', y')$. *Furthermore, the* coherence *axioms* $e_{x,y} * e_{y,x} = id_{(0,x \oplus y)}$ *and* $e_{x,y \oplus y'} = (e_{x,y} \otimes id_{(0,y')}) * (id_{(0,y)} \otimes e_{x,y'})$ *must be satisfied. A morphism h between two* zs *causal graphs E and E' is a* zs *net monoidal morphism which in addition respects horizontal composition and swappings. This defines the category* **ZSCGraph**.

Again, horizontal composition is the key feature of the approach: It avoids the construction of steps which reuse stable tokens.

Proposition 7. *If* $\alpha: (u, 0) \to (v, 0)$ *and* $\alpha': (u', 0) \to (v', 0)$ *are two transitions of a* zs *causal graph then* $\alpha \otimes \alpha' = \alpha' \otimes \alpha$ *and* $\alpha * \alpha' = \alpha \otimes \alpha'$.

The next theorem defines the algebraic semantics of zs nets by means of a universal property.

Theorem 4. *The obvious forgetful functor* $\mathscr{U}$: **ZSCGraph** → **dZPetri** *has a left adjoint* $\mathscr{CG}$: **dZPetri** → **ZSCGraph**.

The functor $\mathscr{CG}$ maps a zs net B into the zs causal graph $\mathscr{CG}[B]$ whose arrows are generated by the inference rules in Table 8 modulo suitable axioms (see [17] for details). The zs causal graph $\mathscr{CG}[B]$ is still too concrete w.r.t. the operational (*ITph*) semantics of zs nets. More precisely we need two more axioms.

Definition 26. *Given a* zs *net B, we denote by* $\mathscr{CG}[B]/\Psi$ *the quotient of the free* zs *causal graph* $\mathscr{CG}[B]$ *generated by B in* **ZSCGraph** *modulo the axioms*

$$d_{z,z'} = id_{(0,z \oplus z')}, \text{ if } z \neq z' \in Z_B, \tag{2}$$

$$d * t * d' = t, \text{ if } t \in T_B \text{ and } d, d' \text{ are swappings}. \tag{3}$$

Table 8. Free construction of $\mathscr{CG}[B]$.

$$\frac{(u, x) \in L_B^{\oplus} \times Z_B^{\oplus}}{id_{(u,x)} \colon (u, x) \to (u, x) \in \mathscr{CG}[B]} \qquad \frac{z, z' \in Z_B}{d_{z,z'} \colon (0, z \oplus z') \to (0, z' \oplus z) \in \mathscr{CG}[B]}$$

$$\frac{t \colon (u, x) \to (v, y) \in T_B}{t \colon (u, x) \to (v, y) \in \mathscr{CG}[B]} \qquad \frac{\alpha \colon (u, x) \to (v, y), \ \beta \colon (u', y) \to (v', z) \in \mathscr{CG}[B]}{\alpha * \beta \colon (u \oplus u', x) \to (v \oplus v', z) \in \mathscr{CG}[B]}$$

$$\frac{\alpha \colon (u, x) \to (v, y), \ \beta \colon (u', x') \to (v', y') \in \mathscr{CG}[B]}{\alpha \otimes \beta \colon (u \oplus u', x \oplus x') \to (v \oplus v', y \oplus y') \in \mathscr{CG}[B]}$$

The quotient $\mathscr{CG}[B]/\Psi$ is such that for any $k \colon \mathscr{CG}[B] \to E \in \mathbf{ZSCGraph}$ which respects axioms (2) and (3), there exists a unique arrow $k_\Psi \colon \mathscr{CG}[B]/\Psi \to E$ such that $k_\Psi \circ Q_\Psi = k$ in $\mathbf{ZSCGraph}$, where $Q_\Psi \colon \mathscr{CG}[B] \to \mathscr{CG}[B]/\Psi$ is the obvious zs causal graph morphism associated to the (least) congruence generated by the imposed axiomatization.

Proposition 8. *For any morphism $h \colon B \to B'$ in $\mathbf{dZPetri}$ there is a unique extension $\hat{h} \colon \mathscr{CG}[B]/\Psi \to \mathscr{CG}[B']/\Psi$ of h in $\mathbf{ZSCGraph}$.*

Example 8. Let MS be the zs net defined in Section 2.2. The arrow $t_1 * t_3 \in \mathscr{CG}[MS]/\Psi$ has source $(2a, 0)$ and target $(2b, 0)$, while $(t_1 \otimes id_{(a,0)}) * (id_{(b,0)} \otimes t_3)$ goes from $(3a \oplus b, 0)$ to $(a \oplus 3b, 0)$. As another example, all the following expressions denote the same arrow:

$$
\begin{aligned}
t_1 * t_2 * (t_2 \otimes t_3) * (t_3 \otimes t_3) &= t_1 * t_2 * (t_2 \otimes id_{(0,z)}) * (t_3 \otimes t_3 \otimes t_3) \\
&= t_1 * t_2 * d_{z,z} * (t_2 \otimes id_{(0,z)}) * (t_3 \otimes t_3 \otimes t_3) \\
&= t_1 * t_2 * (id_{(0,z)} \otimes t_2) * (t_3 \otimes t_3 \otimes t_3) \\
&= t_1 * t_2 * (t_3 \otimes t_2) * (t_3 \otimes t_3).
\end{aligned}
$$

To give the expected correspondence between algebraic and operational semantics we reuse in the current setting the notion of prime arrows.

Theorem 5. *Given a zs net B, there is a one-to-one correspondence between arrows $\alpha \colon (u, 0) \to (v, 0) \in \mathscr{CG}[B]/\Psi$ and the connected steps of B. Moreover, if such an arrow is prime (and is not an identity) then the corresponding connected step is a connected transaction.*

Example 9. In our running example, some prime arrows of $\mathscr{CG}[MS]$ are t_0, $t_1 * t_3$, and $t_1 * t_2 * (t_2 \otimes t_2) * (t_3 \otimes t_2 \otimes t_3 \otimes t_3) * (t_3 \otimes t_3)$, while the arrow $(t_1 \otimes t_1) * d_{z,z} * (t_2 \otimes t_3) * (t_3 \otimes t_3)$ is not prime.

To recover the abstract semantics of zs nets in the *lTph*, we define a category $\mathbf{ZSC}$ whose objects are zs nets and whose morphisms allow for the refinement of a transition into an abstract connected transaction.

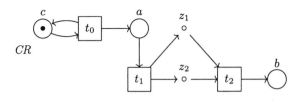

Fig. 11.

Definition 27. *Given a* ZS *net* B, *a causal abstract transition of* $\mathscr{CG}[B]/\Psi$ *is either a prime arrow of* $\mathscr{CG}[B]/\Psi$ *or a transition of* B *(seen as arrow).*

Definition 28 (Category ZSC). *Given two* ZS *net* B *and* B', *a causal refinement morphism* $h: B \to B'$ *is a* ZS *net morphism* $h = (f, g_L, g_Z)$ *from* B *to (the image through the forgetful functor of)* $\mathscr{CG}[B']/\Psi$, *such that function* f *maps transitions into causal abstract transitions. The category* **ZSC** *has* ZS *nets as objects and causal refinement morphisms as arrows, with composition defined similarly to that in* **ZSN**.

Theorem 6. *Category* **Petri** *is embedded in* **ZSC** *fully and faithfully as a coreflective subcategory and the right adjoint functor* $\mathscr{I}[\_]$ *is such that* $\mathscr{I}[B] = I_B$ *for any* ZS *net* B. *Furthermore, the counit component of the coreflection* ϵ_B^I *maps each transition of the abstract net into the appropriate connected transaction.*

2.5   One More Example

We remark that the impact of different philosophies on the modeled system is considerable. This has been already suggested by the multicasting system example, but there are many other examples where the dichotomy is immediate. Let us consider the ZS net CR in Figure 11. Then, according to the *CTph* the abstract net A_{CR} has only two transitions that correspond to t_0 and $t_1 \cdot t_2$, whereas, according to the *ITph* the causal abstract net I_{CR} has infinitely many transitions: t_0, $t_1 * t_2$, $(t_1 \otimes t_1) * d_{z,3z} * (t_2 \otimes t_2)$, and so on. Note the analogy between I_{CR} and the abstract net A_{MS} of the multicasting system example.

We end this section by observing that all PT nets can be constructed using ZS nets whose transitions have four fixed 'shapes.' The needed components are illustrated in Figure 12. We leave to the reader, as an easy task, to combine these shapes for building a generic transition, though of course several other sets of building blocks could have been chosen.

3   Translation of Languages with Synchronization Primitives

In this section we give some general hints for modeling CCS-like communication via ZS nets. The idea is to represent each channel by a pair of zero places, one

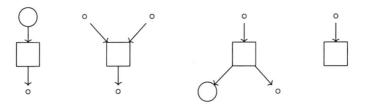

Fig. 12.

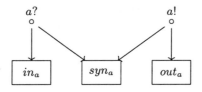

Fig. 13. The zs net Z_a.

for input and one for output, and to model each input (output) action on a
channel with a transition that produces a token on the input (output) zero place
associated to that channel. A special transition, also associated to the channel,
is enabled by a token in the input and a token in the output zero place. If the
channel is restricted, this is the only transition that can consume those tokens,
thus synchronizing the input and output actions that produced the two tokens. If
the channel is not restricted, two additional transitions can consume the tokens
separately. Thus, for every channel name a we define a zs net Z_a consisting
of two zero places $a!$ and $a?$, and three transitions in_a, syn_a, and out_a (see
Figure 13). For a set $\mathbf{A} = \{a_1, ..., a_n\}$ of channel names, we denote by $Z(\mathbf{A})$ the
zs net obtained as the disjoint union of $Z_{a_1}, \ldots, Z_{a_n}$.

Definition 29 (Interfaced net). *Given a set $\mathbf{A} = \{a_1, ..., a_n\}$ of channel
names, an $\mathbf{A}$-interfaced net is a triple $\langle B, \mathbf{A}, P \rangle$, where B is a zs net — in
our translation the initial marking will always be a set — and P is an injective
mapping from $Z(\mathbf{A})$ to B, which preserves the zs net structure. The set $\mathbf{A}$ is
called the interface of the net.*

Two $\mathbf{A}$-interfaced nets $\langle B, \mathbf{A}, P \rangle$ and $\langle B', \mathbf{A}, P' \rangle$ are *isomorphic* if there exists a zs net isomorphism ψ from B to B' that 'preserves interfaces' (in the sense
that it must preserve the injective images of $Z(\mathbf{A})$).

The *simple process algebra* (SPA) considered in [17] is equipped with the
operations of inaction nil, input and output action prefix $a.\_$ and $\bar{a}.\_$, parallel
composition $\_|\_$, and restriction $\_\backslash a$, whose associated SOS rules are given in
Table 9. (We let μ range over input (a), output ($\bar{a}$) and silent (τ) actions, and
let λ range over input/output actions.) We will show later how to deal with
distributed nondeterministic sum.

Table 9. SOS rules for the simple process algebra SPA.

$$\frac{}{\lambda.p \xrightarrow{\lambda} p} \qquad \frac{p \xrightarrow{\mu} q}{p|r \xrightarrow{\mu} q|r} \qquad \frac{p \xrightarrow{\lambda} q,\ p' \xrightarrow{\bar{\lambda}} q'}{p|p' \xrightarrow{\tau} q|q'} \qquad \frac{p \xrightarrow{\mu} q}{r|p \xrightarrow{\mu} r|q} \qquad \frac{p \xrightarrow{\mu} q \ \ \mu \notin \{a, \bar{a}\}}{p\backslash a \xrightarrow{\mu} q\backslash a}$$

Each agent p is modeled by an $fv(p)$-interfaced net $[\![p]\!]_{zs}$, where the set $fv(p)$ is the set of the free (i.e., non-restricted) channel names in p. The definition of $[\![p]\!]_{zs}$ is given by initiality (i.e., it is the unique SPA-algebra homomorphism from the term algebra), and thus it is enough to define the corresponding operations on interfaced nets.

Inaction. The inactive net *nil* is a $\varnothing$-interfaced net $\langle B, \varnothing, \varnothing \rangle$, where B consists of a single place that contains one token in the initial marking.

Action prefix. The interfaced net $a.\langle B, \mathbf{A} \cup \{a\}, P \rangle$ is given by adding a new stable place b and a new transition t to B. The initial marking consists of a token in b. The transition t takes a token in b and produces the initial marking of B plus a token in the zero place $P(a?)$. If the name a is not contained in the interface of the given net, then also a copy of Z_a has to be added, and the injective mapping P is extended in the obvious way. A similar construction is defined for an output action prefix $\bar{a}.p$ (we substitute $a!$ for $a?$ in the postset of the new transition t).

Parallel composition. We let $\langle B_1, \mathbf{A_1}, P_1 \rangle | \langle B_2, \mathbf{A_2}, P_2 \rangle = \langle B, \mathbf{A_1} \cup \mathbf{A_2}, P \rangle$, with B given by the union of B_1 and B_2 where only $P_1(Z(\mathbf{A_1} \cap \mathbf{A_2}))$ and $P_2(Z(\mathbf{A_1} \cap \mathbf{A_2}))$ are identified, and with the mapping P given by the union of P_1 and P_2. The initial marking of B is the union of the initial markings of B_1 and B_2.

Restriction. If a does not appear in the interface, then $\langle B, \mathbf{A}, P \rangle \backslash a = \langle B, \mathbf{A}, P \rangle$. Otherwise, $\langle B, \mathbf{A} \cup \{a\}, P \rangle \backslash a = \langle B', \mathbf{A}, P' \rangle$, with $B' = B \smallsetminus \{P(in_a), P(out_a)\}$ and P' is P restricted to $Z(\mathbf{A})$.

The image of $Z(fv(p))$ in B (via P) plays the role of the interface, since it is the only part of the net $[\![p]\!]_{zs}$ that is modified by the construction defined above: It can be increased (as in the case of action prefix), it can be merged with another interface (as in the case of parallel composition) and it can also be restricted (as in the case of the restriction operator). It is worth noting that for each agent p, with $[\![p]\!]_{zs} = \langle B, \mathbf{A}, P \rangle$, we have $A_B = I_B$.

The relation between SPA agents and their associated interfaced nets can be formalized by adding a labeling function ϕ from the transitions of abstract nets to the set of actions.

Definition 30 (Labels of transactions). *Let p be an agent. For each (connected) transaction ξ of $[\![p]\!]_{zs}$, we define*

Roberto Bruni and Ugo Montanari

Table 10. Two kinds of nondeterministic choice.

don't know		don't care	
$\dfrac{p \xrightarrow{\lambda} q}{p + r \xrightarrow{\lambda} q}$	$\dfrac{p \xrightarrow{\lambda} q}{r + p \xrightarrow{\lambda} q}$	$\dfrac{}{p + r \xrightarrow{\tau} p}$	$\dfrac{}{r + p \xrightarrow{\tau} p}$

$$\phi(\xi) \stackrel{\text{def}}{=} \begin{cases} a_i \text{ if } a_i \in fv(p) \text{ and } P(in_{a_i}) \text{ is fired in } \xi \\ \bar{a}_i \text{ if } a_i \in fv(p) \text{ and } P(out_{a_i}) \text{ is fired in } \xi \\ \tau \text{ otherwise} \end{cases}$$

The definition of labels is not ambiguous, because each transaction ξ of $[\![p]\!]_{zs}$ contains at most one firing of transitions in $P(Z(fv(p)))$.

Definition 31 (Bisimilarity between agents and markings). *Let p be an agent, let N be a net whose transitions are labeled by ϕ over the set of actions, and let u be a marking of N. We say that p is bisimilar to u in N if there exists a relation $\sim$ between agents and markings of N such that $p \sim u$, and: (1) for each transition $p \xrightarrow{\mu} p'$ there exists a firing $u \,[t\rangle\, u'$ of N such that $\phi(t) = \mu$ and $p' \sim u'$; (2) for each firing $u \,[t\rangle\, u'$ of N with $\phi(t) = \mu$ there exists a transition $p \xrightarrow{\mu} p'$ such that $p' \sim u'$.*

Proposition 9. *Let p be an agent, and $[\![p]\!]_{zs} = \langle B, \mathbf{A}, P \rangle$, then p is bisimilar to the initial marking of the abstract net A_B.*

A comparison with other net semantics presented in the literature for CCS-like algebras is out of the scope of this presentation. We just remark the linearity of our encoding and that it provides a reasonable concurrent semantics for SPA agents, as formalized by Proposition 9 above.

Restricted names have only local scopes, and agents that differ only for local names (i.e., agents that can be obtained one from the other by α-conversion) can be considered equivalent. We use the symbol $\_ \equiv_\alpha \_$ to denote such equivalence (e.g., we have $(a_1.a_2.nil|\bar{a}_2.nil)\backslash a_2 \equiv_\alpha (a_1.a_3.nil|\bar{a}_3.nil)\backslash a_3$ for any $a_3 \neq a_1$). It is worth noting that our translation supports α-conversion.

Proposition 10. *If $p \equiv_\alpha q$, then $[\![p]\!]_{zs}$ and $[\![q]\!]_{zs}$ are isomorphic.*

3.1 Distributed Sum

Usually, one can distinguish between two kinds of nondeterminism: *don't know* and *don't care*. In the former, an alternative is selected via a sort of 'lookahead' (e.g., only if the associated subprocess can move), whereas the latter is 'blind.' The difference between the two is evident just by looking at the SOS rules in Table 10, as 'don't care choice' is modeled via axioms. In the next we rely on don't know choice, which is more complicate to deal with.

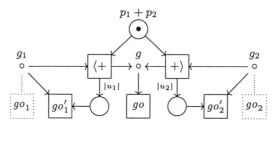

Fig. 14.

To model don't know choice in distributed implementations of CCS-like languages, the classical approach is to make a cross product of the initial markings of the subcomponents in such a way that when one thread r in one component moves, then all the threads in the other component will never be enabled since r consumes some of their premises. Of course this is an expensive construction that adds a lot of auxiliary structure causing state explosion. Using ZS nets the 'interface' approach described above can be exploited for accommodating a more compact solution (though the classical one is still possible).

The idea is that besides channel (zero) places, also a 'generic action' zero place, say g, is added that receives tokens from all transitions generated by the action prefix construction. Using this place we can establish whether or not a thread can evolve. Normally, a transition go can consume exactly one token from g and produce nothing. (Both g and go are part of any interface). Since we are interested in catching 'top level' actions, every time a process is prefixed by an action we remove from the interface g and go and add new instances of them which become connected to the prefixed action only. The general situation is that we have two such interfaced nets and we want to model their nondeterministic sum. The first step is to replace the two go transitions by transitions that are in some way controlled by the choice-point. The relevant part of the construction is illustrated in Figure 14; for the rest we assume that the two 'argument' nets are put in parallel, merging their interfaces except that for g and go components (denoted by g_i and go_i in figure) that are carried out of the interface, while fresh g and go are inserted in the composed net. We call u_1 and u_2 the markings of the two argument nets, that form, together with a token in the place '$p_1 + p_2$' the initial marking of the composed net. To see how the construction behaves, suppose that p_1 can perform a certain action, then a zero token appears in g_1 that can be consumed only by firing '$\langle+$' since go_1 has been deleted and go'_1 is not yet enabled. The firing of '$\langle+$' consumes the only stable token in $p_1 + p_2$ and hence, all threads in the second net cannot complete any transaction (because tokens in g_2 can never be consumed). We have decided to put $|u_1|$ tokens in the stable place that rule go'_1, so that when other top level threads of p_1 will be able to fire, then enough tokens will be available to close all transactions asynchronously. Propositions 9 and 10 are still valid for this extended framework.

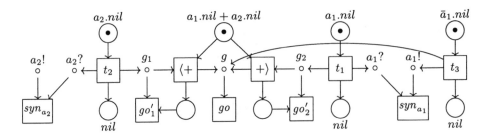

Fig. 15. The interfaced net for the process $((a_1.nil + a_2.nil)|\bar{a}_1.nil)\backslash a_1 \backslash a_2$.

Example 10. The interfaced net for the agent $((a_1.nil + a_2.nil)|\bar{a}_1.nil)\backslash a_1 \backslash a_2$ is presented in Figure 15. Note that the presence of the arc from t_3 to g is motivated by the fact that g (and go) are part of the interfaces of the two nets associated with $a_1.nil + a_2.nil$ and $\bar{a}_1.nil$ and thus g becomes shared when the two agents are composed in parallel. If t_2 tries to fire, then the zero token in a_2? cannot be consumed. If t_1 tries to fire, then the only possibility is that also t_3 fires producing a token in a_1! so that also syn_{a_1} is enabled and the token in a_1? can be consumed. If this is the case, then one has still to consume the zero tokens in g and g_2 produced by the firings of t_3 and t_1, respectively. Thus, $+\rangle$ must be fired that produces another zero token in g. Then, transition go can be fired twice to conclude the transaction. This transaction corresponds to the transition $((a_1.nil + a_2.nil)|\bar{a}_1.nil)\backslash a_1 \backslash a_2 \xrightarrow{\tau} (nil|nil)\backslash a_1 \backslash a_2$, and is the only possible one.

Note that since '$\langle+$' and '$+\rangle$' take one token only from g_1 and g_2 respectively, then τ-moves cannot force the choice (because synchronizations produce two tokens in g_1 or g_2). This was also the reason for writing λ and not μ in the rules for don't know choice in Table 10. To deal with this possibility, it suffices to add two variants of '$\langle+$' and '$+\rangle$' that consume two tokens from g_1 and g_2 respectively. More generally, one might want to synchronize any number of threads as triggers of the same left/right choice. This can be easily done by augmenting the composed nets with one additional zero place and three transitions, as illustrated in Figure 16. The transitions dc_i have the duty of sequentially decrementing the number of tokens in g_i (takes two and puts one back) until only one token is left that can be consumed by go'_i, thus synchronizing the choice with all threads of the ith component that tried to move. Since only one token is present in the place $p_1 + p_2$, then only one token can be produced in z and the dc_i preserve this invariant under firing. Therefore, only one transition between $\langle+$ and $+\rangle$ can fire in the transaction. As a consequence, it is not possible that both dc_1 and dc_2 are fired in the same transaction, as otherwise a zero token would remain in one of the zero places g_i. Note that the *CTph* and the *ITph* can yield different abstract nets, as the latter distinguishes between the different ways for dc_i to consume the tokens in g_i. One may argue that there is a centralized choice point and that

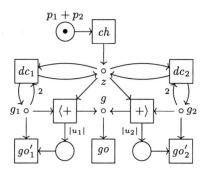

$$p_1 + p_2$$

Fig. 16.

therefore two threads of p_1 cannot force concurrently the 'left choice' but must be synchronized on it. We think that this is by no means a limitation of the approach, as e.g. the classical solution requires a 'token synchronization' between each thread of one component and all other threads of the other components, via a conflict resolution similar to the one illustrated in the dining philosophers example (here the tokens in the cross product of the two initial markings of subnets are 'forks' and the threads are the 'philosophers' that want to eat).

4   An Operational Definition for Transactions

The issues that we want to address in this section regard ZS nets implementation. The problem is that the operational semantics relies on some sort of meta-definition, where one computes on the underlying net, builds transaction segments, and then can discard 'bad' behaviors and accept the 'good' ones, acting as a filter. This means that there are important questions which can be asked for any actual interpreter — Is backtracking necessary? Is the implementation correct? And complete? Does a more efficient implementation exist? We try to answer these questions (see the Conclusions) by defining a machinery for computing on ZS nets. The idea is to adapt the classical net unfolding to pursue concurrently all the nondeterministic runs of the ZS net under inspection, in such a way that 'commit' stable states are recognized and generated.

Whether one is interested in distinguishing between different concurrent proofs or is just interested in the step relation $\_ \Rightarrow_B \_$ is an important issue. In particular, given a ZS net B and a stable marking u we address the problem of computing in a distributed and efficient fashion the set of markings that can be reached from u via an atomic transaction step, i.e., the set $\{v \mid u \Rightarrow_B v\}$ (that is invariant under the *CTph* and *ITph*). The solution relies on a modification of the interpreter for unfolding PT nets [37,51,33], which is extended with a commit rule enforcing the synchronous termination of transactions.

4.1 PT Net Unfolding

The unfolding of a net gives a constructive way to generate all its possible computations, offering a satisfactory mathematical description of the interplay between nondeterminism and causality (and concurrency). In fact the unfolding construction allows for bridging the gap between PT nets and *prime algebraic domains*. The obvious references to this approach are [37,51,33], but we suggest also the interesting overview [25]. It is worth remarking that our presentation is slightly different from usual ones (but reminiscent of [51]), since it is presented as the least net generated by suitable inference rules, rather than by making explicit the chain of finite nets that approximate it.

The construction provides a distributed interpreter for PT nets. We remark that the unfolding applies only to marked nets, i.e., it requires an initial marking.

Starting from a net N, the unfolding produces a nondeterministic occurrence net $\mathcal{U}(N)$ (an acyclic net, where transition pre- and post-markings are sets instead of multisets and where each place has at most one entering arc), together with a mapping from $\mathcal{U}(N)$ to N that tells which places and transitions of the unfolding are instances of the same element of N. Hence the places of $\mathcal{U}(N)$ represent the tokens and the transitions (called *events*) the occurrences of transitions in all possible runs. For this kind of nets the notions of *causally dependent*, of *conflicting* and of *concurrent* elements can be straightforwardly defined and are represented by the binary relations $\_ \preceq \_$, $\_\#\_$ and $\mathbf{co}(\_,\_)$, respectively. Formally, the relation $\_ \preceq \_$ is the transitive and reflexive closure of the *immediate precedence* relation $\_ \prec_0 \_$ defined as $\prec_0 \overset{\text{def}}{=} \{(a,t) \mid a \in \text{pre}(t)\} \cup \{(t,a) \mid a \in \text{post}(t)\}$, while the binary conflict relation is defined as the minimal symmetric relation that contains $\_\#_0\_$ (defined by $t_1 \#_0 t_2 \overset{\text{def}}{\Leftrightarrow} t_1 \neq t_2 \ \wedge \ \text{pre}(t_1) \cap \text{pre}(t_2) \neq \varnothing$), and that is hereditary with respect to $\_ \preceq \_$. Since the conflict relation must be irreflexive, then $\_ \preceq \_$ and $\_\#\_$ have empty intersection. The concurrency relation is defined by letting $\mathbf{co}(o_1, o_2)$ if it is not the case that ($o_1 \prec o_2$ or $o_2 \prec o_1$ or $o_1 \# o_2$). In particular, the relation $\mathbf{co}$ is usually extended to sets of elements by writing $\mathbf{co}(X)$ if for all $o_1, o_2 \in X$ we have $\mathbf{co}(o_1, o_2)$.

More concretely, the places of $\mathcal{U}(N)$ have the form $\langle a, n, H \rangle$, where $a \in S_N$, n is a positive natural number that is used to distinguish different tokens with the same history, and H is the history of the place under inspection and therefore either consists of just one event (the one that produced the token) or is empty (if the token is in the initial marking). Analogously, a generic transition of $\mathcal{U}(N)$ has the form $\langle t, H \rangle$ with $t \in T_N$, since each transition is completely identified by its history H, which in this case consists of the set of consumed tokens. The set H cannot be empty since transitions with empty preset are not allowed. The net $\mathcal{U}(N)$ is defined as the minimal net generated by the rules in Table 11.

We now give a computational interpretation of such rules. The first rule defines the initial marking of $\mathcal{U}(N)$. The second rule is the core of the unfolding: It searches for a set Θ of concurrent tokens that enables a transition t of N, atomically locks them, fires the event e (that is an occurrence of t), and produces some fresh tokens Υ according to post(t). Notice that the condition $\mathbf{co}(\Theta)$ depends only on the histories H_i for $i \in I$, and therefore cannot be altered by successive

Table 11. The unfolding $\mathcal{U}(N)$.

$$\frac{u_{\mathrm{in}}(a) = n,\ 1 \le k \le n}{\langle a, k, \varnothing \rangle \in S_{\mathcal{U}(N)}}$$

$$\frac{t\colon u \to \bigoplus_{j \in J} n_j b_j \in T_N,\ \Theta = \{\langle a_i, k_i, H_i \rangle \mid i \in I\} \subseteq S_{\mathcal{U}(N)},\ \mathbf{co}(\Theta),\ u = \bigoplus_{i \in I} a_i}{e = \langle t, \Theta \rangle \in T_{\mathcal{U}(N)},\ \varUpsilon = \{\langle b_j, m, \{e\}\rangle \mid j \in J,\ 1 \le m \le n_j\} \subseteq S_{\mathcal{U}(N)},\ \mathrm{pre}(e) = \Theta,\ \mathrm{post}(e) = \varUpsilon}$$

Table 12. The unfolding $\mathcal{U}(B)$.

$$\frac{u(a) = n,\ 1 \le k \le n}{\langle a, k, \varnothing \rangle \in S_{\mathcal{U}(B)}}$$

$$\frac{t\colon (u, x) \to (v, \bigoplus_{j \in J} n_j z_j) \in T_B,\ \Theta = \{\langle s_i, k_i, H_i \rangle \mid i \in I\} \subseteq S_{\mathcal{U}(B)},\ \mathbf{co}(\Theta),\ u \oplus x = \bigoplus_{i \in I} s_i}{e = \langle t, \Theta \rangle \in T_{\mathcal{U}(B)},\ \varUpsilon = \{\langle z_j, m, \{e\}\rangle \mid j \in J,\ 1 \le m \le n_j\} \subseteq S_{\mathcal{U}(B)},\ \mathrm{pre}(e) = \Theta,\ \mathrm{post}(e) = \varUpsilon}$$

$$\frac{u \in \mathscr{R}(B, u)}{} \qquad \frac{\varGamma \subseteq T_{\mathcal{U}(B)},\ \mathbf{co}(\varGamma),\ \mathbf{ZProd}(\varGamma) = \mathbf{ZCons}(\varGamma)}{u \ominus \mathbf{SCons}(\varGamma) \oplus \mathbf{SProd}(\varGamma) \in \mathscr{R}(B, u)}$$

firings. In fact, as in *memoizing* for logic programming, or more generally in *dynamic programming*, the history is completely encoded in the tokens, so that it is not necessary to compute it at every firing. Also, note that histories retain concurrent information rather than just sequential, therefore each token/event is generated exactly once (though it can be later referred to many times). Moreover, several occurrences of the second rule can be applied concurrently and therefore the unfolding can be implemented as a distributed algorithm.

4.2 zs Net Unfolding

The unfolding of the underlying net N_B does not yield a faithful representation of the behavior of B. In fact, we must forbid the consumption of stable resources that were not inserted in the starting marking. Moreover, we must be able to apply the commit when the transaction step has consumed all the zero tokens produced so far.

The net $\mathcal{U}(B)$ is defined as the minimal net generated by the rules in Table 12. Together with the unfolding net we compute a set of (reachable) stable markings $\mathscr{R}(B, u)$ for the initial (stable) marking u of the unfolding.

The first two rules define the unfolding, which remains similar to the classical algorithm, except for the fact that stable tokens in the postset of the fired transition are not released to the system. In fact, while the set Θ must contain enough tokens to provide both the stable and the zero resources needed by t (as expressed by the condition $u \oplus x = \bigoplus_{i \in I} s_i$), the tokens that are produced by the occurrence of t applied to Θ (i.e., tokens in the set $\varUpsilon = \mathrm{post}(e)$) just match the zero place component $\bigoplus_{j \in J} n_j z_j$ of $\mathrm{post}(t)$ and not the stable place component v (it is not released until a commit related to e will occur). The third rule is obvious. The fourth rule defines the commit of a transaction step.

To shorten the notation, we introduce the following functions that, given an event e, return the set of zero tokens respectively consumed and produced by the ancestors of e (and by e itself), i.e., we let

$$\mathbf{ZCons}(e) \stackrel{\text{def}}{=} \bigcup_{\langle t, \Theta \rangle \preceq e} \{\langle z, k, H \rangle \in \Theta \mid z \in Z_B\}, \quad \mathbf{ZProd}(e) \stackrel{\text{def}}{=} \bigcup_{e' \preceq e} \mathrm{post}(e'),$$

where $\mathbf{ZCons}(e)$ is the set of zero tokens that have been consumed by some $e' \preceq e$; similarly $\mathbf{ZProd}(e)$ represents the set of zero tokens that have been produced by some $e' \preceq e$ (note that for any $\langle z, k, H \rangle \in \mathbf{ZCons}(e)$ we have $\langle z, k, H \rangle \preceq e$, while $\mathbf{ZProd}(e)$ can also contain tokens that are concurrent with e or produced by e). We remark that $\mathbf{ZCons}(e) \subseteq \mathbf{ZProd}(e)$, because the marking u is stable and therefore it does not contain zero tokens with empty histories. For stable places the situation is different, since we are just interested in knowing how many tokens have been consumed and will be produced for each place by the antecedents of e, thus:

$$\mathbf{SCons}(e) \stackrel{\text{def}}{=} \bigoplus_{\langle t:(u,x) \to (v,y), \Theta \rangle \preceq e} u, \qquad \mathbf{SProd}(e) \stackrel{\text{def}}{=} \bigoplus_{\langle t:(u,x) \to (v,y), \Theta \rangle \preceq e} v.$$

The four functions that we have defined are extended to sets of events in the obvious way. We remark that while $\mathbf{ZCons}(\_)$ and $\mathbf{ZProd}(\_)$ return sets (of zero places in the unfolding net), the functions $\mathbf{SCons}(\_)$ and $\mathbf{SProd}(\_)$ return multisets (of stable places in the original net).

The fourth rule takes a set Γ of concurrent events and checks that any zero token produced by their antecedents is consumed by an antecedent of some event in Γ. The latter condition can be conveniently expressed as the equality $\mathbf{ZProd}(\Gamma) = \mathbf{ZCons}(\Gamma)$. In fact, if a certain token o is in $\mathbf{ZProd}(\Gamma)$, then the condition states that there exists at least an event $e \in \Gamma$ and a unique[6] $e' \preceq e$ such that $o \in \mathrm{pre}(e)$. If these premises are satisfied, then the rule extends $\mathcal{R}(B, u)$ with the multiset obtained by subtracting from u the stable resources consumed by all the antecedents of events in Γ, but adding those that would have been produced during the step. This rule defines a commit, since it synchronizes local commits, as the following result shows.

Proposition 11. *If $\Gamma \subseteq T_{\mathcal{U}(B)}$ such that $\mathbf{co}(\Gamma)$ and $\mathbf{ZProd}(\Gamma) = \mathbf{ZCons}(\Gamma)$, then for any $e = \langle t, \Theta \rangle \in \Gamma$ we have that t does not produce any zero token.*

Note the analogy between the 'commit' rule that takes a set of concurrent events and the 'unfolding' rule that takes a set of concurrent tokens: This is to some extent related to our view of ZS nets as a formalism for expressing transition synchronization rather that just token synchronization.

The resulting algorithm is as much distributed as the classical one when applied to the abstract net of B. In fact all the useful relations are defined by just looking at the history of the elements in the premises, which, under the atomicity assumption reduce to the stable preset of the abstract step. To improve efficiency, the sets $\mathbf{ZProd}(e)$, $\mathbf{ZCons}(e)$, $\mathbf{SProd}(e)$ and $\mathbf{SCons}(e)$ could be

---

[6] Otherwise a conflict would arise.

also encoded in e more directly, although they can be easily calculated from the history component. The main result can be formulated as the following theorem.

Theorem 7. $\mathscr{R}(B, u) = \{v \mid u \Rightarrow_B v\}.$

Since the unfolding encodes the proof of the transaction step (via the history components), it is possible to use the same scheme for computing the abstract net (whatever philosophy is preferred). However, for doing this efficiently, we must be able to recognize isomorphic processes. For example, note that given a certain computed process, any renaming of the stable tokens in the initial marking (i.e., any permutation of tokens in the same place) yields a different but isomorphic process that is also calculated during the unfolding. To solve this problem, we can either try to avoid having several isomorphic processes in the unfolding by some clever construction, or check at commit-time if the freshly computed transaction is isomorphic to some transaction already computed.

Since a ZS net B can contain cycles that produce an unbounded number of zero tokens, the unfolding can become infinite. So an important question concerns the decidability of $\mathscr{R}(B, u)$. In a private communication to the authors [20], Nadia Busi proved that such set is indeed decidable. Roughly speaking the idea is to simulate the behaviors of B by a PT net with exactly one inhibitor arc,[7] for which the reachability problem has been solved in [42]. The set $\mathscr{R}(B, u)$ is recursively enumerated by the inference system, and if it is infinite we cannot do any improvement. But when $\mathscr{R}(B, u)$ is finite, it would be desirable to find some condition for halting the execution of the algorithm, that otherwise could continue computing transaction segments that cannot be completed. Finding some general condition for halting the unfolding of ZS is an open problem that we leave for future investigations.

5   Zero-Safe Nets and Read Arcs

We now show how to extend the zero-safe net paradigm with read arcs, in the style of contextual nets [36]. The idea is to model transitions that can read certain tokens without consuming them, so that multiple readings on the same token can take place concurrently (using ordinary PT nets, the naive way of modeling readings via self-loops[8] is not appropriate because the accesses to read tokens are sequentialized).

---

[7] Nets with inhibitor arcs, also called with negative arcs, have been introduced in [1,2] for modeling systems where the presence of certain resources can inhibit the firing of some transitions. We recall that the reachability problem is undecidable for the class of nets with two or more inhibitor arcs.

[8] Given a net N, a self-loop consists of two arcs $(a, t), (t, a) \in F_N$ for a place $a \in S_N$ and a transition $t \in T_N$.

Table 13. The inference rules for $\_ \Longrightarrow_{\mathsf{N}} \_$.

identities	generators	parallel composition
$u \in S^{\oplus}$	$t \colon u \xrightarrow{w} v \in T$	$u_1 \oplus w \xRightarrow{w_1 \oplus w}_{\mathsf{N}} v_1 \oplus w, \ u_2 \oplus w \xRightarrow{w_2 \oplus w}_{\mathsf{N}} v_2 \oplus w$
$u \xRightarrow{u}_{\mathsf{N}} u$	$u \oplus \lfloor w \rfloor \xRightarrow{\lfloor w \rfloor}_{\mathsf{N}} v \oplus \lfloor w \rfloor$	$u_1 \oplus u_2 \oplus w \xRightarrow{w_1 \oplus w_2 \oplus w}_{\mathsf{N}} v_1 \oplus v_2 \oplus w$

5.1 Contextual Nets

Definition 32. *A marked contextual net (c-net) is a tuple* $\mathsf{N} = (S, T, F, \mathsf{C}, u_{\mathrm{in}})$, *where* $N_{\mathsf{N}} = (S, T, F, u_{\mathrm{in}})$ *is the underlying* PT *net and* $\mathsf{C} \colon S \times T \to \mathbb{N}$ *is the context relation.*

We denote by $\mathrm{ctx}(t)$ the multiset of places defined by $\mathrm{ctx}(t)(a) = \mathsf{C}(a, t)$ for all $a \in S$ and by $\lfloor u \rfloor$ the underlying set of places of a multiset u (i.e., $\lfloor u \rfloor = \{a \mid u(a) > 0\}$). Informally, the minimum amount of resources that a transition t requires to be enabled is $\mathrm{pre}(t) \oplus \lfloor \mathrm{ctx}(t) \rfloor$: The tokens in $\mathrm{pre}(t)$ are fetched, while those in $\lfloor \mathrm{ctx}(t) \rfloor$ are just read, and other transitions can access them, concurrently with t. The minimum requirement involves $\lfloor \mathrm{ctx}(t) \rfloor$ and not $\mathrm{ctx}(t)$ because the same token can be read more than once. However, t can also read different tokens from the same place, up to the maximum established by $\mathrm{ctx}(t)$. For $t \in T$ with $\mathrm{pre}(t) = u$, $\mathrm{post}(t) = v$ and $\mathrm{ctx}(t) = w$, we write $t \colon u \xrightarrow{w} v$. In the following, we shall overload the symbol $\subseteq$ to denote multiset inclusion.

Definition 33. *Let u and v be markings of a c-net* N *and let X be a finite multiset of transitions of* N. *We say that X is enabled at u if* $\lfloor \bigoplus_{t \in T} \mathrm{ctx}(t) \rfloor \oplus \bigoplus_{t \in T} X(t) \cdot \mathrm{pre}(t) \subseteq u$. *Moreover, we say that u evolves to v via X, written* $u \, [X\rangle \, v$ *if X is enabled at u and $u \, [X\rangle \, v$ is a step of the underlying* PT *net* N_{N}.

Note that if u has enough tokens to satisfy also the 'context' of X, then v is obtained from u by removing $\bigoplus_{t \in T} X(t) \cdot \mathrm{pre}(t)$ and then adding $\bigoplus_{t \in T} X(t) \cdot \mathrm{post}(t)$. The step relation can be equivalently defined by the inference rules in Table 13, that carry also information about the context used in the step. The meaning of $u \xRightarrow{w}_{\mathsf{N}} v$ is that from the marking u there is a step that leads to v reading w — note that there must exist two markings u_1 and v_1 such that $u = u_1 \oplus w$ and $v = v_1 \oplus w$. Idle tokens are seen as part of the context of a step. Transitions yield basic steps, where only the minimal context is required. When building larger steps, any part of the contexts of the two substeps can be shared. For example, from $w \xRightarrow{w}_{\mathsf{N}} w$ and from the step $u \oplus \lfloor w \rfloor \xRightarrow{\lfloor w \rfloor}_{\mathsf{N}} v \oplus \lfloor w \rfloor$ associated to $t \colon u \xrightarrow{w} v$, we obtain $u \oplus w \xRightarrow{w}_{\mathsf{N}} v \oplus w$, because $\lfloor w \rfloor \subseteq w$, and therefore $\lfloor w \rfloor$ can be shared.

For sequential composition of steps we have several alternatives: (1) to forget about all the information on context; (2) to arbitrarily forget about part of the context; (3) to define the context of the composed sequence in a canonical way.

Table 14. Three sets of inference rules for $\_ \Longrightarrow^*_N \_$.

	basic step	sequential composition
(1)	$\dfrac{u \overset{w}{\Longrightarrow}_N v}{u \Longrightarrow^*_N v}$	$\dfrac{u \Longrightarrow^*_N v, \; v \overset{w}{\Longrightarrow}_N v'}{u \Longrightarrow^*_N v'}$
(2)	$\dfrac{u \overset{w_1 \oplus w}{\Longrightarrow}_N v}{u \overset{w}{\Longrightarrow}^*_N v}$	$\dfrac{u \overset{w}{\Longrightarrow}^*_N v, \; v \overset{w_1 \oplus w}{\Longrightarrow}_N v'}{u \overset{w}{\Longrightarrow}^*_N v'}$
(3)	$\dfrac{u \overset{w}{\Longrightarrow}_N v}{u \overset{w}{\Longrightarrow}^*_N v}$	$\dfrac{u_1 \overset{w_1}{\Longrightarrow}^*_N v_1, \; v_1 \overset{w_2}{\Longrightarrow}_N v_2, \; w = w_1 \cap w_2}{u_1 \overset{w}{\Longrightarrow}^*_N v_2}$

The three cases are illustrated in Table 14, where $u \cap v$ denotes the multiset of places such that $(u \cap v)(a) = \min(u(a), v(a))$, for all $a \in S$. In particular, the third set of rules keeps track of the maximal possible context of a sequence. The three definitions lead to the same set of reachable markings. Though (2) and (3) are similar, in principle the former is more appropriate for the *ITph* (because the shared context is not necessarily the maximal one), while the latter can deal well with the *CTph* (cf. the 'maximum sharing hypothesis' of [18,19]).

In a way analogous to PT nets, step sequences for c-nets can be considered up to diamond transformation, originating commutative contextual processes. Instead, for accommodating causal dependencies, (causal) contextual processes are introduced.

Definition 34. *A* deterministic occurrence c-net *is a finite, acyclic (w.r.t. the preorder in which t precedes t' if either $\mathrm{post}(t) \cap (\mathrm{pre}(t') \cup \mathrm{ctx}(t')) \neq \varnothing$ or $\mathrm{ctx}(t) \cap \mathrm{pre}(t') \neq \varnothing$) c-net* O *such that: (1) for all $t \in T$, $\mathrm{pre}(t)$ and $\mathrm{post}(t)$ are sets (not multisets) and (2) for all $t_0 \neq t_1 \in T$, $\mathrm{pre}(t_0) \cap \mathrm{pre}(t_1) = \mathrm{post}(t_0) \cap \mathrm{post}(t_1) = \varnothing$.*

The dependencies between events in an occurrence c-net can be of two kinds: 'causal' and 'temporal.' When $\mathrm{post}(t) \cap (\mathrm{pre}(t') \cup \mathrm{ctx}(t')) \neq \varnothing$, then t *causes* t', because t produces a token that is necessary for enabling t'. When $\mathrm{ctx}(t) \cap \mathrm{pre}(t') \neq \varnothing$, then t cannot happen after t', because t' consumes (part of) the context needed by t and since we are describing a deterministic computation where both t and t' must fire, we have that t temporally precedes t'. In [4,3] it is shown that these two notions are precisely characterized by a causal dependency relation $<$ and a relation $\nearrow$ called *asymmetric conflict*. The former is the transitive closure of the relation $\prec$ defined by: (i) if $s \in \mathrm{pre}(t)$, then $s \prec t$; (ii) if $s \in \mathrm{post}(t)$, then $t \prec s$; and (iii) if $\mathrm{post}(t) \cap \mathrm{ctx}(t') \neq \varnothing$, then $t \prec t'$. The asymmetric conflict relation is the union of the causal dependency relation together with the strict asymmetric conflict relation $\rightsquigarrow$ defined by letting $t \rightsquigarrow t'$ if $\mathrm{ctx}(t) \cap \mathrm{pre}(t') \neq \varnothing$ or $t \neq t' \; \wedge \; \mathrm{pre}(t) \cap \mathrm{pre}(t') \neq \varnothing$. The conflict relation #

Table 15. The two common inference rules for $\_ \stackrel{\_}{\Rightarrow}_B \_$.

underlying	commit
$u \oplus x \stackrel{w \oplus z}{\Longrightarrow}_{N_B} v \oplus y, \; u,v,w \in L^\oplus, \; x,y,z \in Z^\oplus$	$(u,0) \stackrel{w}{\Rightarrow}_B (v,0)$
$(u,x) \stackrel{w}{\Rightarrow}_B (v,y)$	$u \stackrel{w}{\Rightarrow}_B v$

is then induced by $\nearrow$ and $\leq$ (the reflexive closure of $<$) via the rules below:

$$\frac{t_0 \nearrow t_1 \nearrow \ldots \nearrow t_n \nearrow t_0}{\#\{t_0, t_1, \ldots, t_n\}} \qquad \frac{\#(A \cup \{t\}) \;\; t \leq t'}{\#(A \cup \{t'\})}$$

where A is a finite set of transitions.

Note that $\#$ relates finite sets of transitions and not just pairs of transitions. However, when read arcs are not present, then $\#$ is just the closure under set union of the ordinary binary conflict relation of PT nets. It follows that for each deterministic occurrence c-net O the relation $\nearrow_O$ is acyclic and thus the net is conflict free. The last relation we shall introduce regards place concurrency. A set U of places is called concurrent, written $\mathbf{co}(U)$, if: (1) for any $a, a' \in U$, it is not the case that $a < a'$; and (2) $\nearrow$ is acyclic when restricted to the set $\Uparrow (U) = \bigcup_{a \in U} \Uparrow (a)$, where $\Uparrow (x) = \{t \in T \mid t \leq x\}$ is the set of ancestors of x.

Definition 35. *A* contextual process *(c-process)* P *for a c-net* N *consists of a* deterministic occurrence c-net O *together with a pair of functions* $f_P : T_O \to T_N$ *and* $g_P : S_O \to S_N$ *that respect sources, targets and contexts of transitions.*

5.2   Zero-Safe Contextual Nets

We now merge the features of ZS nets with that of c-nets, by allowing the combined use of zero places and read arcs in our models.

Definition 36 (ZS c-net). *A* ZS c-net *is a tuple* $B = (S, T, F, C, Z, u_{in})$ *such that* $N_B = (S, T, F, C, u_{in})$ *is a c-net and* (S, T, F, Z, u_{in}) *is a ZS net.*

Note that zero places can be used as context, because $Z \subset S$. In defining the dynamics of ZS c-nets, we can follow two main alternatives. The crucial point is whether to forbid or not that a stable token is read (possibly many times) and then also fetched during the same transaction. While the rules **underlying** and **commit** are identical for both alternatives (see Table 15), the difference between the two is expressed by the rule for sequential composition. For simplicity, on zero tokens we consider the step relation that forgets about the information on contexts, as it is equivalent to the other possible choices from the 'reachability' point of view.

In order to allow consumption of previously read stable tokens in the same transaction, we need the complex rule below:

$$(\phi): \frac{(u_1 \oplus w_1 \oplus w, x) \stackrel{w_1 \oplus w}{\rightrightarrows}_{\mathsf{B}} (v_1 \oplus w_1 \oplus w, y), \ (u_2 \oplus w_1 \oplus w, y) \stackrel{w}{\rightrightarrows}_{\mathsf{B}} (v_2 \oplus w, y')}{(u_1 \oplus u_2 \oplus w_1 \oplus w, x) \stackrel{w}{\rightrightarrows}_{\mathsf{B}} (v_1 \oplus v_2 \oplus w, y')}.$$

The idea is that the second step can consume the tokens in w_1 that the first step reads. The context w is instead shared between the two steps.

For *not* allowing consumption of previously read stable tokens in the same transaction, it suffices to introduce the simpler rule

$$(\psi) \ \frac{(u_1 \oplus w, x) \stackrel{w}{\rightrightarrows}_{\mathsf{B}} (v_1 \oplus w, y), \ (u_2 \oplus w, y) \stackrel{w}{\rightrightarrows}_{\mathsf{B}} (v_2 \oplus w, y')}{(u_1 \oplus u_2 \oplus w, x) \stackrel{w}{\rightrightarrows}_{\mathsf{B}} (v_1 \oplus v_2 \oplus w, y')}.$$

The rule sequentializes on zero tokens, while composing in parallel on stable tokens (sharing the whole context w of the two substeps). We recall that w can contain idle tokens, and therefore, given any two steps $(u_1, x) \stackrel{w_1 \oplus w}{\rightrightarrows}_{\mathsf{B}} (v_1, y)$ and $(u_2, y) \stackrel{w_2 \oplus w}{\rightrightarrows}_{\mathsf{B}} (v_2, y')$ we can always 'complement' such steps with markings w_2 and w_1 respectively, to obtain $(u_1 \oplus w_2, x) \stackrel{w_1 \oplus w_2 \oplus w}{\rightrightarrows}_{\mathsf{B}} (v_1 \oplus w_2, y)$ and $(u_2 \oplus w_1, y) \stackrel{w_1 \oplus w_2 \oplus w}{\rightrightarrows}_{\mathsf{B}} (v_2 \oplus w_1, y')$, so that the rule for horizontal composition can be applied.

To distinguish between the two interpretations, we write either $u \stackrel{w}{\Rightarrow}_{\mathsf{B}, \phi} v$ or $u \stackrel{w}{\Rightarrow}_{\mathsf{B}, \psi} v$, depending on which rule among ϕ and ψ is considered. Of course $u \stackrel{w}{\Rightarrow}_{\mathsf{B}, \psi} v$ implies $u \stackrel{w}{\Rightarrow}_{\mathsf{B}, \phi} v$ but not vice versa.

Though the operational and abstract semantics can be defined either according to the *CTph* or to the *ITph*, we prefer to follow the latter interpretation only: in this way it is possible to distinguish the places that are used as context during the transaction and the abstract counterpart is still a c-net, whereas the *CTph* might introduce some confusion.

For the *ITph* the transactions correspond to full and connected deterministic c-processes of the underlying c-net such that the origins and destinations are stable and the evolution places are zero places. Thus, the context of a transaction is the set of places that are both minimal and maximal (transactions do not contain isolated places but can contain places that are simply read). The abstract net associated to a zs c-net B is a c-net that has the stable places of B as places, the transactions of B as transitions with preset, postset and context defined in the obvious way (if we allow to first read and then consume a stable token in the same transaction, then the token is put in the preset of the corresponding abstract transition, not in the context).

Example 11 (Multicasting revisited). To illustrate the use of read arcs in the zero-safe framework, we show an improved specification of the multicasting system. Let us consider the zs c-net *CMS* in Figure 17 (as usual, read arcs are depicted as undirected lines). The idea is that copying can be avoided, as all receivers can simply 'read' the same copy. Thus, a firing of t_1 opens the session producing

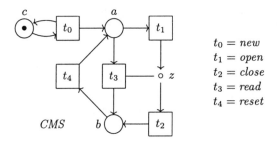

$$t_0 = new$$
$$t_1 = open$$
$$t_2 = close$$
$$t_3 = read$$
$$t_4 = reset$$

Fig. 17. The zs c-net *CMS* for multicasting.

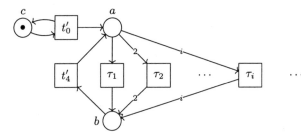

Fig. 18. The abstract c-net for the multicasting system *CMS*.

the message in the buffer z, then many receivers can concurrently read the information by firing t_3 and then the session is closed by a firing of t_2 that removes the message from the buffer. Of course, many multicasting sessions can take place concurrently. At the abstract level the system is then represented by the c-net in Figure 18. Though this time also the empty transmission τ_1 is possible, the analogy with the abstract net A_{MS} of the multicasting system presented in Section 2 is evident (τ_i represents the one-to-$(i-1)$ transmission).

5.3 A Distributed Contextual Interpreter

We conclude by showing that the interpreter of Section 4 can be modified to deal with contexts by considering the unfolding of zs nets proposed in [4]. As a matter of notation, we write $t: (u, x) \xrightarrow{(v,y)} (u', x')$ for a transition t with preset $u \oplus x$, context $v \oplus y$ and postset $u' \oplus x'$, with $u, v, u' \in (S \setminus Z)^{\oplus}$ and $x, y, x' \in Z^{\oplus}$. The rules for dealing with the case where stable tokens can be first read and then also consumed in the same transaction are illustrated in Table 16. Note that, to store the context accessed, events are encoded as triples rather than as couples.

The main differences w.r.t. the interpreter of Section 4 concern the firing rule and the commit rule. In fact, here the execution of a transition has to keep track of the context, differentiating it from the fetched tokens. For this purpose, we have supplied the set Φ. Note that we do not record multiplicities of readings, as they are not important to establish the correctness of a transaction; hence we

Table 16. The unfolding $\mathcal{U}(\mathrm{B})$ with commit according to (ϕ).

$$\frac{u(a) = n,\ 1 \le k \le n}{\langle a, k, \varnothing \rangle \in S_{\mathcal{U}(\mathrm{B})}}$$

$$\frac{t \colon (u, x) \xrightarrow{(v, y)} (w, \bigoplus_{j \in J} n_j z_j) \in T_{\mathrm{B}},\ \Theta = \{\langle s_i, k_i, H_i \rangle \mid i \in I\} \subseteq S_{\mathcal{U}(\mathrm{B})},\ \Phi = \{\langle s_i', k_i', H_i' \rangle \mid i \in I'\} \subseteq S_{\mathcal{U}(\mathrm{B})},\ \mathbf{co}(\Theta \cup \Phi),\ \Theta \cap \Phi = \varnothing,\ u \oplus x = \bigoplus_{i \in I} s_i,\ \lfloor v \oplus y \rfloor \subseteq \bigoplus_{i \in I'} s_i' \subseteq v \oplus y}{e = \langle t, \Theta, \Phi \rangle \in T_{\mathcal{U}(\mathrm{B})},\ \Upsilon = \{\langle z_j, m, \{e\} \rangle \mid j \in J,\ 1 \le m \le n_j\} \subseteq S_{\mathcal{U}(\mathrm{B})},\ \mathrm{pre}(e) = \Theta,\ \mathrm{post}(e) = \Upsilon,\ \mathrm{ctx}(e) = \Phi}$$

$$\frac{}{u \in R_\phi(\mathrm{B}, u)} \qquad \frac{\Gamma \subseteq T_{\mathcal{U}(\mathrm{B})},\ \nearrow_{\Uparrow(\Gamma)} \text{ is acyclic},\ <_\Gamma = \varnothing,\ \mathbf{ZProd}(\Gamma) = \mathbf{ZCons}(\Gamma)}{u \ominus \mathbf{SCons}(\Gamma) \oplus \mathbf{SProd}(\Gamma) \in R_\phi(\mathrm{B}, u)}$$

just check that the context contains enough tokens (more than $\lfloor v \oplus y \rfloor$), but less than the maximum allowed ($v \oplus y$). However, if one is interested in computing the associated processes, then also multiplicities should be considered: They should be assigned to the s_i' so as to exactly match $v \oplus y$. Of course, for the transition to fire, both the context and the preset must be concurrently available and disjoint. When the premises are satisfied, then the event e and the zero places in Υ are inserted in the unfolding. For the commit, we cannot just assume that transitions in Γ are concurrent, as the following example demonstrates.

Example 12. Let us consider the zs c-net in Figure 19 (reminiscent of *CMS* in Figure 17). There are two admissible transactions: The first is given by a firing of t_1 followed by a firing of t_2 that consumes the token produced in z by t_1. The second consists of a firing of t_1, followed by a firing of t_3 that reads the token in z, and then by a firing of t_2. The unfolding progressively introduces the following tokens and events:

- $s_1 = \langle a, 1, \varnothing \rangle$, $s_2 = \langle a, 2, \varnothing \rangle$;
- $e_1 = \langle t_1, \{s_1\}, \varnothing \rangle$, $z_1 = \langle z, 1, \{e_1\} \rangle$;
- $e_2 = \langle t_1, \{s_2\}, \varnothing \rangle$, $z_2 = \langle z, 1, \{e_2\} \rangle$;
- $e_3 = \langle t_3, \{s_1\}, \{z_2\} \rangle$ and $e_4 = \langle t_3, \{s_2\}, \{z_1\} \rangle$ (note that, e.g., $\langle t_3, \{s_1\}, \{z_1\} \rangle$ cannot be introduced because $s_1 < z_1$);
- $e_5 = \langle t_2, \{z_1\}, \varnothing \rangle$ and $e_6 = \langle t_2, \{z_2\}, \varnothing \rangle$.

If we require that the commit is given by concurrent transitions only, then the only admissible Γ are $\{e_5\}$, $\{e_6\}$ and $\{e_5, e_6\}$. From these sets we cannot derive any information about the occurrence of e_3 and e_4. However, there are asymmetric conflicts between e_3 and e_6 and between e_4 and e_5, and therefore these events are not completely unrelated. So the question is, e.g., 'how can we distinguish between the deterministic c-process that involves e_1 and e_5 only from the one that involves also e_4?' Our answer amounts to take Γ as consisting of 'compatible' (but not necessarily concurrent) events that are not causally dependent. This is expressed by requiring the asymmetric conflict relation to be acyclic (when restricted to the ancestors of events in Γ) and the intersection between $<$ and $\Gamma \times \Gamma$ to be empty. Under these assumptions, the commit can happen under

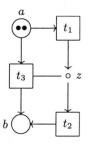

Fig. 19.

Table 17. The commit of the unfolding according to (ψ).

$$\frac{\Gamma \subseteq T_{\mathcal{U}(\mathsf{B})}, \ \nearrow_{\Uparrow(\Gamma)} \text{ acyclic}, \ <_\Gamma = \varnothing, \ \mathbf{ZProd}(\Gamma) = \mathbf{ZCons}(\Gamma), \ \mathbf{Spre}(\Gamma) \cap \mathbf{Sctx}(\Gamma) = \varnothing}{u \in R_\psi(\mathsf{B}, u) \qquad \qquad u \ominus \mathbf{SCons}(\Gamma) \oplus \mathbf{SProd}(\Gamma) \in R_\psi(\mathsf{B}, u)}$$

any of the following Γ: $\Gamma_1 = \{e_5\}$, $\Gamma_2 = \{e_6\}$, $\Gamma_3 = \{e_3, e_6\}$, $\Gamma_4 = \{e_4, e_5\}$, and $\Gamma_5 = \{e_5, e_6\}$.

Since stable contexts are left unchanged by the transaction, then the marking inserted after the commit just computes the tokens consumed and those produced. Note that all conditions can be verified 'locally', just by looking at the encoded history of the chosen premises (Θ and Φ for the firing rule and Γ for the commit rule). It might happen that a stable token is first read and also consumed before the end of the transaction; this makes clear the difference between read arcs and self-loops, as in the second case the stable token cannot be reused in the same transaction. It can be verified that the set $\mathscr{R}_\phi(\mathsf{B}, u)$ computed in this way is exactly the set of markings that are reachable in one step from u.

Theorem 8. $\mathscr{R}_\phi(\mathsf{B}, u) = \{v \mid u \stackrel{w}{\Rightarrow}_{\mathsf{B}, \phi} v\}$.

The obvious alternative is to forbid stable tokens to be read and consumed in the same transaction, according to the rule (ψ). In this case, the rules for computing $\mathscr{R}_\phi(\mathsf{B}, u)$ must be changed as shown in Table 17, where $\mathbf{Spre}(\_)$ and $\mathbf{Sctx}(\_)$ are the pointwise extensions of the functions:

$$\mathbf{Spre}(e) \stackrel{\text{def}}{=} \bigcup_{\langle t, \Theta, \Phi \rangle \preceq e} \{\langle a, k, H \rangle \in \Theta \mid a \in L_\mathsf{B}\},$$

$$\mathbf{Sctx}(e) \stackrel{\text{def}}{=} \bigcup_{\langle t, \Theta, \Phi \rangle \preceq e} \{\langle a, k, H \rangle \in \Phi \mid a \in L_\mathsf{B}\}.$$

Note that $\mathbf{Spre}(\_)$ and $\mathbf{Sctx}(\_)$ return sets of stable places in the unfolding, and we have $\mathbf{SCons}(e) = \bigoplus_{\langle a, k, H \rangle \in \mathbf{Spre}(e)} a$.

Theorem 9. $\mathscr{R}_\psi(\mathsf{B}, u) = \{v \mid u \stackrel{w}{\Rightarrow}_{\mathsf{B}, \psi} v\}$.

Since both interpreters are based on the same unfolding net $\mathcal{U}(\mathsf{B})$, it is evident from the two different commit rules that $\mathscr{R}_\psi(\mathsf{B}, u) \subseteq \mathscr{R}_\phi(\mathsf{B}, u)$, as $\mathscr{R}_\psi(\mathsf{B}, u)$ has an additional premise.

$$\text{HCatZPetri} \underset{u_C}{\overset{\mathscr{X}}{\underset{\longrightarrow}{\overset{\longleftarrow}{\bot}}}} \text{dZPetri} \underset{u_I}{\overset{c\mathcal{G}}{\underset{\longrightarrow}{\overset{\longleftarrow}{\bot}}}} \text{ZSCGraph} \qquad \text{ZSN} \underset{\mathcal{A}}{\overset{\longleftarrow}{\underset{\longrightarrow}{\bot}}} \text{Petri} \underset{\mathcal{I}}{\overset{\longleftarrow}{\underset{\longrightarrow}{\bot}}} \text{ZSC}$$

Fig. 20. Operational and abstract semantics of zero-safe nets.

Conclusions

We have proposed the framework of zero-safe nets as a basis for modeling and implementing distributed transactions. In fact, ZS nets can provide both the refined view of the systems where actions have finer grain and an abstract view where transactions are seen just as transitions of an ordinary PT net. Working at the level of ZS nets allows one to keep smaller the size of the system description (for example the abstract net can have an infinite number of transitions also when the refined net is finite).

After surveying the operational and abstract semantics of the framework, we have shown how to encode many features of concurrent systems, as e.g. distributed choice, in a compositional manner and how to combine zero places with read arcs. It is worth remarking that the construction of transactions can be defined, in the language of category theory, as an adjunction, i.e., it is a free construction and thus preserves several net composition operations (defined as colimits in the category of ZS nets). The construction of the abstract net defines a coreflection, whose universal properties confirm that it is the optimal such construction. These constructions can be pursued according to either the *CTph* or the *ITph*. The two approaches yield the same step relation but different abstract nets. The categorical semantics (summarized by the four adjunctions in Figure 20) recovers the operational and abstract semantics of ZS nets, introducing an algebraic characterization of the whole framework.

We have also illustrated a distributed interpreter for computing on (ordinary and contextual) ZS nets. We want to remark that the resulting implementation does not violate the locality assumptions, since it is completely analogous to the widely accepted implementation for PT nets. This interpreter satisfactorily answers the questions formulated at the beginning of Section 4: Backtracking is not necessary, correctness is given by Theorem 7, and completeness is ensured by our inference system. We are confident that formal halting criteria can be found for expressive classes of ZS nets.

As future work, we plan to extend the concept of 'zero place' to other net flavours, as e.g., coloured and timed nets, nets with inhibitor arcs, and probabilistic nets. In fact we conjecture that our basic mechanism for expressing 'transition synchronization' can be helpful also in these richer models for a compositional modeling of systems and for describing execution protocols. Another ongoing line of research concerns the study of hierarchical ZS nets, where one can have different levels of abstraction. Finally, it would be interesting to apply contextual ZS nets to the modeling and study of serializability in transaction systems, in a way which is somehow analogous to the research conducted in [44,21].

References

1. T. Agerwala. A complete model for representing the coordination of asynchronous processes. Hopkins Computer Research Report 32, John Hopkins University, 1974.
2. T. Agerwala and M. Flynn. Comments on capabilities, limitations and "correctness" of Petri nets. *Computer Architecture News*, 4(2):81–86, 1973.
3. P. Baldan. *Modelling concurrent computations: From contextual Petri nets to graph grammars*. PhD thesis, Computer Science Department, University of Pisa, 2000. Published as Technical Report TD-1/00.
4. P. Baldan, A. Corradini, and U. Montanari. An event structure semantics for P/T contextual nets: Asymmetric event structures. In M. Nivat, editor, *Proceedings of FoSSaCS'98, 1st International Conference on Foundations of Software Science and Computation Structures*, volume 1378 of *Lect. Notes in Comput. Sci.*, pages 63–80. Springer Verlag, 1999.
5. E. Best and R. Devillers. Sequential and concurrent behaviour in Petri net theory. *Theoret. Comput. Sci.*, 55:87–136, 1987.
6. E. Best, R. Devillers, and J. Esparza. General refinement and recursion for the Petri Box calculus. In P. Enjalbert, A. Finkel, and K.W. Wagner, editors, *Proceedings STACS'93*, volume 665 of *Lect. Notes in Comput. Sci.*, pages 130–140. Springer Verlag, 1993.
7. E. Best, R. Devillers, and J. Hall. The Box calculus: A new causal algebra with multi-label communication. In G. Rozenberg, editor, *Advances in Petri Nets'92*, volume 609 of *Lect. Notes in Comput. Sci.*, pages 21–69. Springer Verlag, 1992.
8. W. Brauer, R. Gold, and W. Vogler. A survey of behaviour and equivalence preserving refinements of Petri nets. In G. Rozenberg, editor, *Advances in Petri Nets'90*, volume 483 of *Lect. Notes in Comput. Sci.*, pages 1–46. Springer Verlag, 1991.
9. C. Brown and D. Gurr. A categorical linear framework for Petri nets. In *Proceedings of LICS'90, 5th Annual IEEE Symposium on Logic in Computer Science*, pages 208–218. IEEE Computer Society Press, 1990.
10. R. Bruni. *Tile Logic for Synchronized Rewriting of Concurrent Systems*. PhD thesis, Computer Science Department, University of Pisa, 1999. Published as Technical Report TD-1/99.
11. R. Bruni, J. Meseguer, U. Montanari, and V. Sassone. A comparison of petri net semantics under the collective token philosophy. In J. Hsiang and A. Ohori, editors, *Proceedings of ASIAN'98, 4th Asian Computing Science Conference*, volume 1538 of *Lect. Notes in Comput. Sci.*, pages 225–244. Springer Verlag, 1998.
12. R. Bruni, J. Meseguer, U. Montanari, and V. Sassone. Functorial semantics for petri nets under the individual token philosophy. In M. Hofmann, G. Rosolini, and D. Pavlovic, editors, *Proceedings of CTCS'99, 8th Category Theory and Computer Science*, volume 29 of *Elect. Notes in Th. Comput. Sci.* Elsevier Science, 1999.
13. R. Bruni and U. Montanari. Zero-safe nets, or transition synchronization made simple. In C. Palamidessi and J. Parrow, editors, *Proceedings EXPRESS'97, 4th workshop on Expressiveness in Concurrency*, volume 7 of *Elect. Notes in Th. Comput. Sci.* Elsevier Science, 1997.
14. R. Bruni and U. Montanari. Zero-safe nets: The individual token approach. In F. Parisi-Presicce, editor, *WADT'97, 12th workshop on Recent Trends in Algebraic Development Techniques*, volume 1376 of *Lect. Notes in Comput. Sci.*, pages 122–140. Springer Verlag, 1998.
15. R. Bruni and U. Montanari. Zero-safe nets: Composing nets via transition synchronization. In H. Weber, H. Ehrig, and W. Reisig, editors, *Int. Colloquium on*

Petri Net Technologies for Modelling Communication Based Systems, pages 43–80. Fraunhofer Gesellschaft ISST, 1999.

16. R. Bruni and U. Montanari. Executing transactions in zero-safe nets. In M. Nielsen and D. Simpson, editors, *Proceedings of ICATPN2000, 21st Int. Conf. on Application and Theory of Petri Nets*, volume 1825 of *Lect. Notes in Comput. Sci.*, pages 83–102. Springer Verlag, 2000.

17. R. Bruni and U. Montanari. Zero-safe nets: Comparing the collective and individual token approaches. *Inform. and Comput.*, 156:46–89, 2000.

18. R. Bruni and V. Sassone. Algebraic models for contextual nets. In U. Montanari, J.D.P. Rolim, and E. Welzl, editors, *Proceedings of ICALP2000, 27th Int. Coll. on Automata, Languages and Programming*, volume 1853 of *Lect. Notes in Comput. Sci.*, pages 175–186. Springer Verlag, 2000.

19. R. Bruni and V. Sassone. Two algebraic process semantics for contextual nets. This Volume.

20. N. Busi. On zero safe nets, April 1999. Private communication.

21. N. De Francesco, U. Montanari, and G. Ristori. Modeling concurrent accesses to shared data via Petri nets. In E.-R. Olderog, editor, *Programming Concepts, Methods and Calculi*, IFIP Transactions A-56, pages 403–422. North Holland, 1994.

22. P. Degano, R. De Nicola, and U. Montanari. A distributed operational semantics for CCS based on condition/event systems. *Acta Inform.*, 26(1-2):59–91, 1988.

23. P. Degano, J. Meseguer, and U. Montanari. Axiomatizing the algebra of net computations and processes. *Acta Inform.*, 33(7):641–667, 1996.

24. H. Ehrig and J. Padberg. Uniform approach to Petri nets. In C. Freska, M. Jantzen, and R. Valk, editors, *Proceedings Foundations of Computer Science: Potential-Theory-Cognition*, volume 1337 of *Lect. Notes in Comput. Sci.*, pages 219–231. Springer Verlag, 1997.

25. R.J. van Glabbeek. Petri nets, configuration structures and higher dimensional automata. In J.C.M. Baeten and S. Mauw, editors, *Proceedings CONCUR'99*, volume 1664 of *Lect. Notes in Comput. Sci.*, pages 21–27. Springer Verlag, 1999.

26. R.J. van Glabbeek and U. Goltz. Refinement of actions and equivalence notions for concurrent systems. Hildesheimer Informatik Bericht 6/98, Institut fuer Informatik, Universitaet Hildesheim, 1998.

27. R.J. van Glabbeek and G.D. Plotkin. Configuration structures. In D. Kozen, editor, *Proceedings of LICS'95, 10th Annual IEEE Symposium on Logics in Computer Science*, pages 199–209. IEEE Computer Society Press, 1995.

28. R.J. van Glabbeek and F. Vaandrager. Petri net models for algebraic theories of concurrency. In J.W. de Bakker, A.J. Nijman, and P.C. Treleaven, editors, *Proceedings PARLE*, volume 259 of *Lect. Notes in Comput. Sci.*, pages 224–242. Springer Verlag, 1987.

29. U. Goltz and W. Reisig. The non-sequential behaviour of Petri nets. *Inform. and Comput.*, 57:125–147, 1983.

30. R. Gorrieri and U. Montanari. On the implementation of concurrent calculi into net calculi: Two case studies. *Theoret. Comput. Sci.*, 141(1-2):195–252, 1995.

31. S. MacLane. *Categories for the Working Mathematician*. Springer Verlag, 1971.

32. J. Meseguer and U. Montanari. Petri nets are monoids. *Inform. and Comput.*, 88(2):105–155, 1990.

33. J. Meseguer, U. Montanari, and V. Sassone. Process versus unfolding semantics for place/transition Petri nets. *Theoret. Comput. Sci.*, 153(1-2):171–210, 1996.

34. J. Meseguer, U. Montanari, and V. Sassone. Representation theorems for Petri nets. In Ch. Freksa, M. Jantzen, and R. Valk, editors, *Foundations of Computer*

Science: Potential - Theory - Cognition, to Wilfried Brauer on the occasion of his sixtieth birthday, volume 1337 of *Lect. Notes in Comput. Sci.*, pages 239–249. Springer Verlag, 1997.

35. J. Meseguer, P.C. Ölveczky, and M.-O. Stehr. Rewriting logic as a unifying framework for Petri nets. This Volume.

36. U. Montanari and F. Rossi. Contextual nets. *Acta Inform.*, 32:545–596, 1995.

37. M. Nielsen, G. Plotkin, and G. Winskel. Petri nets, event structures and domains, part I. *Theoret. Comput. Sci.*, 13:85–108, 1981.

38. E.R. Olderog. Operational Petri net semantics for CCSP. In G. Rozenberg, editor, *Advances in Petri Nets'87*, volume 266 of *Lect. Notes in Comput. Sci.*, pages 196–223. Springer Verlag, 1987.

39. J. Padberg. *Abstract Petri nets: Uniform approach and rule-based refinement.* PhD thesis, Technische Universität Berlin, 1996.

40. J. Padberg. Classification of Petri nets using adjoint functors. In *EATCS Bulletin*, volume 66, pages 85–91. European Association for Theoretical Computer Science, 1998.

41. C.A. Petri. *Kommunikation mit Automaten.* PhD thesis, Institut für Instrumentelle Mathematik, Bonn, 1962.

42. K. Reinhardt. Reachability in Petri nets with inhibitor arcs. Technical Report WSI-96-30, Wilhelm Schickard Institut für Informatik, Universität Tübingen, 1996.

43. W. Reisig. *Petri Nets: An Introduction.* EATCS Monographs on Theoretical Computer Science. Springer Verlag, 1985.

44. G. Ristori. *Modelling Systems with Shared Resources via Petri Nets.* PhD thesis, Computer Science Department, University of Pisa, 1994. Published as Technical Report TD-5/94.

45. V. Sassone. An axiomatization of the algebra of Petri net concatenable processes. *Theoret. Comput. Sci.*, 170(1-2):277–296, 1996.

46. V. Sassone. An axiomatization of the category of Petri net computations. *Math. Struct. in Comput. Sci.*, 8(2):117–151, 1998.

47. I. Suzuki and T. Murata. A method for stepwise refinement and abstraction of Petri nets. *J. Comput. and System Sci.*, 27:51–76, 1983.

48. R. Valette. Analysis of Petri nets by stepwise refinement. *J. Comput. and System Sci.*, 18:35–46, 1979.

49. W. Vogler. Behaviour preserving refinements of Petri nets. In G. Tinhofer and G. Schmidt, editors, *Proceedings 12th International Workshop on Graph-Theoretic Concepts in Computer Science*, volume 246 of *Lect. Notes in Comput. Sci.*, pages 82–93. Springer Verlag, 1987.

50. G. Winskel. Event structure semantics of CCS and related languages. In M. Nielsen and E. Meineche Schmidt, editors, *Proceedings ICALP'82*, volume 140 of *Lect. Notes in Comput. Sci.*, pages 561–567. Springer Verlag, 1982.

51. G. Winskel. Event structures. In W. Brauer, editor, *Proceedings Advanced Course on Petri Nets*, volume 255 of *Lect. Notes in Comput. Sci.*, pages 325–392. Springer Verlag, 1987.

52. G. Winskel. Petri nets, algebras, morphisms and compositionality. *Inform. and Comput.*, 72:197–238, 1987.

Two Algebraic Process Semantics
for Contextual Nets

Roberto Bruni[1] and Vladimiro Sassone[2]

[1] Dipartimento di Informatica, Università di Pisa, Italia
bruni@di.unipi.it
[2] Dipartimento di Matematica e Informatica, Università di Catania, Italia
vs@dmi.unict.it

Abstract We show that the so-called 'Petri nets are monoids' approach initiated by Meseguer and Montanari can be extended from ordinary place/transition Petri nets to *contextual* nets by considering suitable non-free monoids of places. The algebraic characterizations of net concurrent computations we provide cover both the *collective* and the *individual* token philosophy, *uniformly* along the two interpretations, and coincide with the classical proposals for place/transition Petri nets in the absence of read-arcs.

Introduction

The basic features common to any 'flavour' of Petri net [28] essentially are that states are (multi)sets of distributed, abstract resources, and that actions only involve the coordination of local parts of the state, as they can consume some of the resources available and release fresh resources. Accordingly, a computation can be described abstractly as a partial order of events in which any two events are either causally dependent – when one could not have been executed without a resource provided by the other – or concurrent – when they could have happened in any order, because they affect independent subsystems. These features make net models suitable for representing in a satisfactory way *concurrent* and *distributed systems* in many interdisciplinary applications.

Meseguer and Montanari in [23,24] (and successively in [12,13,31,32,6,14] several authors) have recast these facts in algebraic terms to unveil properties of net computations and, especially, of the intrinsic concurrency of the net model. The underlying idea of the so-called 'Petri nets are monoids' approach is to lift the algebraic structure of states to the level of computations, so that the distribution of the resources is reflected on the performed actions, analogously to what happens in *rewriting logic* [21,22], in *structured transition systems* [11] and in *tile logic* [17,4]. In the case of ordinary place/transition Petri nets (PT nets),

Research supported by CNR Integrated Project *Metodi e Strumenti per la Progettazione e la Verifica di Sistemi Eterogenei Connessi mediante Reti di Comunicazione*; by Esprit Working Groups *CONFER2* and *COORDINA*; and by MURST project *TOSCa: Tipi, Ordine Superiore e Concorrenza*.

H. Ehrig et al. (Eds.): Unifying Petri Nets, LNCS 2128, pp. 427–456, 2001.

states are multisets of places, or equivalently, elements of the free commutative monoid over the set of places. Moreover, a computation can be obviously composed with any computation that originates from the same state in which the first ends, yielding a computation that is the concatenation of the two. Hence, computations possess by nature an intrinsic (partial) operation of 'sequential' composition that gives rise to a *category* – arrows are computations, identities representing unused tokens. Lifting the monoidal structure of states to the category of computations results in a *monoidal category of computations*, where the functoriality law of the monoidal tensor product expresses a basic fact about the *true concurrency* of the model. Namely, that in any computation the relative order in which two concurrent actions are executed is always immaterial. In fact, if α_1 and α_2 are computations such that α_i, for $i = 1, 2$, originates in u_i and leads to v_i (written $\alpha_i : u_i \to v_i$), then

$$(\alpha_1 \oplus id_{u_2}); (id_{v_1} \oplus \alpha_2) = \alpha_1 \oplus \alpha_2 = (id_{u_1} \oplus \alpha_2); (\alpha_1 \oplus id_{v_2}),$$

where $\oplus$ is the tensor product (modeling concurrent composition of computations) originated from multiset union on states, $\_;\_$ is the operation of sequential composition, and the id_{u_i}, id_{v_i} are idle components of computations, with e.g., $id_{u_1}; \alpha_1 = \alpha_1 = \alpha_1; id_{v_1}$.

The extensive use of PT nets has given rise to different schools of thought concerning their semantic interpretation. In particular, the main distinction is drawn between *collective* and *individual token philosophies* (see e.g. [18]). According to the collective token philosophy (*CTph*), one is not interested in distinguishing among different *tokens* in the same place (i.e., among instances of the same resource), because all such tokens are operationally equivalent. However, tokens may have different origins and histories, carrying different *causality* information and hence consuming one instance rather than another, can make the difference from being causally dependent or not on some previous event. The point of view of the individual token philosophy (*ITph*) is that these causal dependencies may well form an essential information that should not be discarded when, e.g., flow analysis is concerned. Of course, causal dependencies may influence the degree of concurrency in abstract computations, and therefore *CTph* and *ITph* lead to quite different concurrent semantics.

For ordinary PT nets the algebraic approach has been pursued under both philosophies, characterizing different kinds of *net processes*, ranging from Best and Devillers commutative processes [3] (that support the *CTph*) to concatenable processes [13,31] and strongly concatenable processes [32] (that support the *ITph*). Note that the *ITph* relies on a tensor product which can be commutative only *up to* a monoidal natural isomorphism. Therefore, the algebraic approach requires some special mechanism in order to accommodate the lifting of the (commutative) monoidal structure of states. It is worth mentioning that the algebraic approach under the *ITph* is completely straightforward for the recent proposal of *pre-nets* [6] whose states are based on strings rather than multisets. From this point of view, the approach initiated by Meseguer and Montanari is

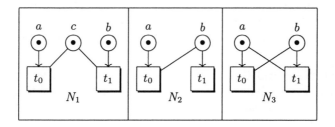

Fig. 1.

completely general and can be applied to more general net models where, e.g., tokens are some kind of more complex data [33,15].

Several extensions of the basic PT net paradigm have been considered in the literature that either increase the expressive power or give a better representation of existing phenomena. This paper focuses on extending the 'Petri nets are monoids' approach to *contextual nets*, also known as nets with *read-arcs*, or *condition-arcs*, or *test-arcs* [10,27,20,35]. The motivating idea behind 'read-arcs' is that of *reading* resources *without consuming* them, thus providing a way of modeling multiple concurrent accesses to the same resource. Using ordinary PT nets such readings must be rendered as *self-loops*, and this imposes an unwanted sequentialization of concurrent readings. On the contrary, with contextual nets, besides pre and post-sets, transitions also have '*contexts*', that is resources that are necessary for the enabling but are *not* affected by the firing. Contextual nets have found applications e.g., to transaction serializability in databases [30], concurrent constraint programming [26], and asynchronous systems [34].

Independently of *CTph* and *ITph*, for contextual nets several different approaches have been proposed that differ in the way in which contexts are read. For example, let us consider the nets N_1, N_2 and N_3 in Figure 1, taken from [35]. (As usual, places are represented by circles, tokens by black bullets, transitions by boxes, pre- and post-sets by directed weighted arcs, and contexts by undirected weighted arcs, with unary weights always omitted.) According to the semantic interpretation of [27], the transitions t_0 and t_1 can fire concurrently in N_1, but neither in N_2 nor in N_3, since the basic assumption is that a token cannot be read and consumed in the same step. In [20], instead, the concurrent step is allowed for all three nets, the basic assumption being that t_0 and t_1 can both start together, read the context tokens, and need them *not* while the actions take place. Besides its possible merits, we find this interpretation not fully convincing as, for instance, in N_3 we would end up in a state that cannot be reached by any firing sequence. Thus, to some extent, the firing steps of [20] allow certain transition occurrences to synchronize. The basic assumption of [35] that firings have duration leads to consider ST-traces, where explicit *transition-starts* and *transition-ends* events are fired. Hence N_2 can start t_0 and then t_1 before t_0 completes, allowing the concurrent step $\{t_0, t_1\}$ (with the hypothesis that t_0

starts first). On the contrary, in N_3 if either t_0 or t_1 starts, then the context for the other transition is consumed and the concurrent step is forbidden. We follow the interpretation of [27] that fits better our understanding of contexts.

Contextual Nets and Collective Token Philosophy. The algebraic theory for PT nets developed under the *CTph* is well consolidated, and the relationships between its *computational, algebraic* and *logical* interpretations are by now very clear [5]. Starting with the classical 'token-game' semantics, many computational models for Petri nets have been proposed that follow the *CTph*. In particular, the *commutative processes* of Best and Devillers [3] reconcile the 'diamond' equivalence (cf. § 1.1) on firing and step sequences, and express very nicely the concurrency of the model. They also admit an exact algebraic representation by means of the universal construction $\mathcal{T}(\_)$ that yields *strictly symmetric strict monoidal categories* from the category of PT nets. More precisely, given a PT net N, the objects of $\mathcal{T}(N)$ are markings and its arrows correspond to the commutative processes of N [24,13].

Surprisingly, *CTph* semantics for contextual nets have received poor attention in the literature, not only for what concerns the algebraic treatment. Whether because the problem has been underestimated, or simply because the *ITph* is more fascinating, we cannot tell. In any case, we think that it is useful to address this discrepancy with the semantics of ordinary PT nets. Moreover, although one can easily extend the diamond equivalence to firing sequences on contextual nets, the formalization of a good *algebraic* model is not at all straightforward. Inspired by a suggestion made by Meseguer in [22], we give here a satisfactory treatment of this issue. The idea is to consider monoidal categories with a commutative tensor product taken – differently from the case of PT nets – over a *non-free* monoid of places. In particular, we regard each token a as an *atom* (for lack of a better analogy) that can emit 'negative' particles a^- (*electrons*) while keeping track of their number, i.e., as in [22], we assume that for all $k \in \mathbb{N}$, $a = a^k \oplus k \cdot a^-$, where a^k represents an atom that has released exactly k particles to the environment.

Replacing context arcs on a with self-loop arcs on a^-, we are able to give an axiomatic construction of a monoidal category whose arrows between standard markings (i.e., containing no negative particles) are (isomorphic to) the concurrent computations of the net according to the *CTph*. A key ingredient for this result to hold is the so-called *maximum sharing hypothesis*, an axiom expressing that concurrent readings can *always* be seen as sharing the same token, a fundamental idea in *CTph*.

Contextual Nets and Individual Token Philosophy. Building on the notion of *process* introduced by Goltz and Reisig in [19], several authors have shown that the semantics of nets in the *ITph* can still be understood in terms of *symmetric monoidal categories*, where the tensor product, this time denoted by $\_ \otimes \_$, is commutative only up to a monoidal natural isomorphism γ called *symmetry* (for strictly symmetric monoidal categories the transformation γ is just the identity). In particular, a simple variation of Goltz-Reisig processes called *concatenable processes* is introduced in [13] (see also [31]), which admits sequential composition and yields a symmetric monoidal category $\mathcal{P}(N)$ for each net N. Note that $\otimes$ is commutative on the objects of $\mathcal{P}(N)$. A refined version of concatenable processes is given by strongly concatenable processes [32] where origins and destinations are totally ordered (as opposed to the orderings of origins and destinations of concatenable processes that are indexed by the places). Also several unfolding semantics (see e.g. [36,25]) have been proposed that give a denotational interpretation of the interplay between concurrency, causality and nondeterminism.

For contextual nets both the process and the unfolding approaches have been studied [27,8,9,2,1], giving a satisfactory understanding of the computational model via the introduction of *asymmetric event structures*. The algebraic approach, however, has been pursued only in a recent paper by Gadducci and Montanari [16] using *match-share* categories. Their basic idea is that, together with symmetries, two additional auxiliary constructors must be present: one for *duplicating* tokens and one for *matching* them. Formally, for each place a the auxiliary arrows $\nabla_a: a \to a \otimes a$ and $\Delta_a: a \otimes a \to a$ are added to the computational model (and suitably axiomatized, by letting e.g., $\nabla_a; \Delta_a = id_a$ and $\nabla_a; \gamma_{a,a} = \nabla_a$ with id_a the identity arrow on a and $\gamma_{a,a}$ the symmetry that swaps two tokens in a). Read-arcs can then be replaced by self-loops (i.e., if the transition t consumes u, reads v and produces w, then one considers a derived transition $t_v: u \otimes v \to w \otimes v$), and reading without consuming modeled by duplicating the context, firing the transition concurrently with an idle copy of the context, and then matching the idle copy with the corresponding produced tokens (i.e., by considering the arrow $\widehat{t_v} = (id_u \otimes \nabla_v); (t_v \otimes id_v); (id_w \otimes \Delta_v)$ illustrated in Figure 2(a)). Multiple concurrent access is achieved by producing via duplication – and then absorbing via matching – enough copies of the context. In [16], a suitable axiomatization of duplicators and matchers is introduced and proved to represent faithfully the basic fact about concurrent access: steps sharing the same context, but otherwise disjointly enabled, can execute concurrently or in any interleaved order with *no* noticeable difference (e.g., using the notation above, the term $(id_{u'} \otimes \widehat{t_v}); (\gamma_{u',w} \otimes id_v); (id_w \otimes \widehat{t'_v}); (\gamma_{w',w} \otimes id_v)$, illustrated in Figure 2(b), for t' that consumes u', reads v and produces w', is equivalent to $(\gamma_{u',u} \otimes id_v); (id_u \otimes \widehat{t'_v}); (\gamma_{u,w'} \otimes id_u); (id_{w'} \otimes \widehat{t_v})$ in Figure 2(c), and both admit a normal form where the subterms t_v and t'_v are executed concurrently, as illustrated in Figure 2(d)).

The main drawback of this approach is that the initial model contains *too many* arrows and, therefore, in order to obtain a bijection with contextual processes one has to carve a suitable subcategory. Although the arrows of this subcategory can be characterized by inspecting their structure, the lack of a global correspondence somehow weakens the framework. We aim at improving the approach of [16] starting from the observation that the unwanted arrows are due to *redundant* information in the model. In fact, once a context token is read by a transition we know the 'real' token it is connected to: the one duplication was applied to. Hence, the match operation, needed for express-

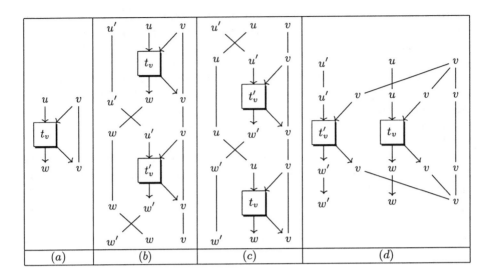

Fig. 2.

ing concurrent readings, does not add any further information and may introduce inconsistent behaviors. For example, given two tokens in the place a, one can first duplicate both and then match each copy of the first token with a copy of the second token: it should be evident that the resulting arrow (written $(\nabla_a \otimes \nabla_a); (id_a \otimes \gamma_{a,a} \otimes id_a); (\Delta_a \otimes \Delta_a))$ is meaningless from the computational viewpoint, unless the two tokens represent the same context. We overcome this problem by extending to the *ITph* the approach proposed for the *CTph* in the first part of the paper.

The key of our proposal is to *regulate* the use of symmetries on the markings so to *forbid* the swapping of a a^k and an adjacent a^-. This prevents the migration of electrons from atom to atom, as it might happen in the *CTph* and in [16]. The absence of electron migration represents, in the *ITph*, a sort of dual to the maximum sharing hypothesis, that we call *exact sharing hypothesis*. Most notably, the restriction is imposed simply by omitting the corresponding symmetries from the model. And reintroducing them would in fact result in a redundant framework perfectly analogous to the one provided by match-share categories. Observe that this yields a monoidal category that, formally speaking, is *not* symmetric anymore: we allow only selected commutations by explicitly including selected symmetries. These will include, of course, all the symmetries between standard markings (i.e., those in which tokens have released no particles), and will exclude all those that may lead to confuse the causal histories of tokens. Our main result is that, again, the arrows between standard markings are in bijection with a slight refinement of contextual processes, called *concatenable contextual processes*. In this, it is crucial that the model be able to treat particles in different ways depending on the context. On the one hand, according to the

ITph, we need to distinguish between a^- released by different atoms, but on the other hand, similarly to the *CTph*, we want to identify those particles generated by the same a. This is the precise content of our exact sharing hypothesis, as formalized by a new axiom that we call (Δ).

Origin and structure of the paper. This paper builds on the work reported in [7]. Besides extending *loc. cit.* by detailed examples and proofs of the main results, we improve its treatment of the *ITph* in many respects. In particular, in [7] we relied on a distinction between forward and backward contexts, realized through a second kind of electron, a_-, in addition to a^-. Moreover, differently from here, our representation result was phrased in terms of *strongly* concatenable contextual processes. Axiom (Δ) is instrumental in these improvements, and is first introduced here.

In Section 1 we recall some basics about contextual nets and the algebraic semantics of PT nets. In Sections 2 and 3 we define algebraic semantics for contextual nets under both the *CTph* and the *ITph*, providing original characterization results for commutative and concatenable contextual processes. We remark that in the absence of read-arcs, our semantics coincide with the classical ones.

Acknowledgements. We would like to thank José Meseguer and Paolo Baldan for some interesting discussion on the topic and also Matteo Coccia for his reading of a preliminary version of our work. We are also grateful to the anonymous referees for their careful reading of the manuscript (they spot several well-hidden typos) and their helpful comments.

1   Preliminaries

1.1   Contextual Nets

Contextual nets were introduced for extending PT nets with the 'read without consume' operation [10,27,20,35]. The states of contextual nets are called *markings* and represent distributions of resources (*tokens*) in typed repositories (*places*). Given the set of places S, markings can be seen as finite multisets $u: S \to \mathbb{N}$, where $u(a)$ denotes the number of tokens that place a carries in u. The set of finite multiset on S is the free commutative monoid on S. We denote it by $S^\oplus$, and indicate multiset inclusion, union and difference by $\subseteq$, $\oplus$ and $\ominus$, respectively, with $u \ominus v$ defined only for $v \subseteq u$. For k a natural number and u a multiset, $k \cdot u$ is the multiset such that $(k \cdot u)(a) = k \cdot u(a)$ for all a. We denote by $\lfloor u \rfloor$ the underlying set of u, that can be seen as the multiset such that $\lfloor u \rfloor(a) = 1$ if $u(a) > 0$ and $\lfloor u \rfloor(a) = 0$ otherwise. If $u = \lfloor u \rfloor$ and $v = \lfloor v \rfloor$ we use the standard set notation $u \cup v$ and $u \cap v$ to denote, respectively, the union and intersection of u and v. Since we consider finite multisets only, the reader should not get confused if in the following the adjective 'finite' is sometimes omitted.

Definition 1. *A contextual net N is a tuple $(S, T, \partial_0, \partial_1, \varsigma)$, where S is the set of places, T is the set of transitions, $\partial_0, \partial_1: T \to S^\oplus$ are the pre and post-set* *functions, and* $\varsigma: T \to S^{\oplus}$ *is the context function. Besides the usual assumption that* $\varsigma(t)$ *and* $\partial_0(t) \oplus \partial_1(t)$ *are disjoint for each transition* t, *we assume that* $\varsigma(t)$ *is a set.*

Informally, $\partial_0(t) \oplus \varsigma(t)$ is the minimum amount of resources that t requires to be enabled. Of these resources, those in $\partial_0(t)$ are retrieved and consumed, while those in $\varsigma(t)$ are just read and left on their repositories. When t has accomplished its task, it returns $\partial_1(t)$ fresh tokens and releases the context. Only at this point other transitions will be able to consume the tokens in $\varsigma(t)$, whereas they can use the same context concurrently with t.

Definition 2. *Let* u *and* v *be markings, and* X *a finite multiset of transitions of a contextual net* $N = (S, T, \partial_0, \partial_1, \varsigma)$. *We say that* u *evolves to* v *under the step* X, *in symbols* $u \; [X\rangle \; v$, *if the transitions in* X *are concurrently enabled at* u, *i.e.,* $\left\lfloor \bigoplus_{t \in T} \varsigma(t) \right\rfloor \oplus \bigoplus_{t \in T} X(t) \cdot \partial_0(t) \subseteq u$, *and*

$$v = u \ominus \left(\bigoplus_{t \in T} X(t) \cdot \partial_0(t) \right) \oplus \bigoplus_{t \in T} X(t) \cdot \partial_1(t).$$

A step sequence *from* u_0 *to* u_n *is a sequence* $u_0 \; [X_1\rangle \; u_1 \ldots u_{n-1} \; [X_n\rangle \; u_n$.

Thus the execution of the step X requires that the marking u contains at least *all* the tokens in the preconditions $\partial_0(t)$ of transitions $t \in X$ plus at least *one* token for each place that is used as context by some transition in X. This matches the intuition that a token can be used as context by many transitions at the same time. From the point of view of concurrency, the fact that transitions in X are executed in a step means that they can be equivalently executed in any order. Thus, likewise ordinary PT nets, step sequences for contextual nets can be considered up to the equivalence induced by the *diamond transformation* relation $\_ \diamond \_$ defined by $u \; [X \oplus Y\rangle \; v \diamond u \; [X\rangle \; u_1 \; [Y\rangle \; v$ for any step $u \; [X \oplus Y\rangle \; v$ (and suitable u_1). The *diamond equivalence* is the reflexive, symmetric, transitive and sequences concatenation closure of the relation $\_ \diamond \_$.

Definition 3. *Given a contextual net* N, *the strictly symmetric strict monoidal category (cf. § 1.2) of* contextual commutative processes $\mathcal{CT}(N)$ *has the markings of* N *as objects, its step sequences, taken modulo the diamond equivalence, as arrows, and composition is given by sequence concatenation.*

In the *ITph*, computations are commonly described in terms of structures representing the causal relationships between event occurrences. In the case of nets, this is fruitfully formalized through the following notion of process. We remark that these notions are conservative extension of the corresponding notions for ordinary PT nets, to which they reduce in the absence of read-arcs. The relation $\_ \nearrow \_$ referred to in the definition below is the least preorder in which t precedes t', written $t \nearrow t'$, if either $\partial_1(t) \cap (\partial_0(t') \cup \varsigma(t')) \neq \emptyset$, see Figures 3(a) and 3(b), or $\varsigma(t) \cap \partial_0(t') \neq \emptyset$, see Figure 3(c). (Relation $\_ \nearrow \_$ is used in [2,1] for nondeterministic contextual processes; note however that we deal with deterministic processes only.)

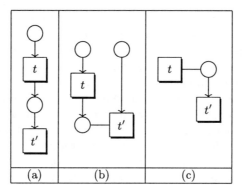

Fig. 3. Three situations in which t (immediately) precedes t'.

Definition 4. *A (deterministic) contextual process net is a finite, acyclic w.r.t.* $\nearrow$, *contextual net* Θ *such that*

1. *for all* $t \in T_\Theta$, $\partial_0(t)$ *and* $\partial_1(t)$ *are sets (as opposed to multisets), and*
2. *for all pairs* $t_0 \neq t_1 \in T_\Theta$, $\partial_i(t_0) \cap \partial_i(t_1) = \varnothing$, *for* $i = 0, 1$.

Remark 1. One could argue that in the contextual process net illustrated in Figure 3(c) the transition t' might also fire before t, inhibiting it. In fact, this cannot be the case. Since the net is a process, i.e., the description of a deterministic run, both t and t' must be fired, and the only possible interpretation is that t must execute before t'. There is however no causal dependence between the two events, but only a temporal one. Therefore $t \nearrow t'$ means that t precedes t', either causally or just temporally.

Two transitions t and t' in a deterministic occurrence net are called *concurrent* if they are not related by $\nearrow$ (i.e., if there is none of the two transitions that causally or temporally dependends on the other). We remark that the same definition does not apply to nondeterministic processes, where the concurrency relation must be defined on arbitrary sets of transitions and not just on pairs.

Definition 5. *A contextual process* π *of a contextual net* N *is a contextual process net* Θ *together with a pair of functions* $\langle \pi_T, \pi_S \rangle$, *where* $\pi_T: T_\Theta \to T_N$ *and* $\pi_S: S_\Theta \to S_N$, *that respect source, target and context, i.e., such that* $\partial_{Ni} \circ \pi_T = \pi_S \circ \partial_{\Theta i}$, *for* $i = 0, 1$, *and* $\varsigma_N \circ \pi_T = \pi_S \circ \varsigma_\Theta$, *where the symbol* $\circ$ *denotes the ordinary composition of functions. Contextual processes are considered up to isomorphism.*

If no confusion can arise, we denote the components π_T and π_S just by π.

1.2   Petri Nets Are Monoids

The paper [24] built on the monoidal structure of markings to provide an algebraic characterization of the concurrent computations of nets. The basic idea was to lift the structure of states to the level of transitions, providing an algebraic representation of concurrent firing. In turn, these 'algebraic' steps can be sequentially concatenated in order to express more complex computations. While sequential composition endows computations with a categorical structure – markings are objects, computations are arrows, and idle tokens are identities – the parallel composition yields a tensor product. The interplay of parallel and sequential composition, regulated by functoriality of tensor products, models a basic fact about concurrency, namely that concurrent transitions can occur in any relative order. Under the $CTph$ the tensor product can simply be commutative. Then, each PT net N freely generates a *strictly* symmetric strict monoidal category $\mathcal{T}(N)$ whose arrows are in bijection with the *commutative processes* of N [3].

Under the $ITph$ the situation is more complex. To be able to model causal dependencies, multisets of transitions are not enough. Degano, Meseguer and Montanari proposed to keep simple markings as objects, but to consider a tensor product *non* commutative on the arrows, together with a collection of arrows that may be used to explicitly change the order in which transitions fetch and produce tokens [13]. Such arrows, collected together as the components of a natural isomorphism, turn out to be the classical notion of *symmetry* in category theory, thus leading to the construction of a (non strictly) symmetric strict monoidal category $\mathcal{P}(N)$ for each net N, whose arrows define the *concatenable processes* of N. A more concrete construction, $\mathcal{Q}(N)$, was introduced in [32] in order to remove some deficiencies of the previous approach. The main feature of $\mathcal{Q}(N)$, which captures the so-called *strongly concatenable processes*, is that its objects are strings rather than multisets of tokens.

For the reader's convenience, we briefly recall the definition of monoidal categories and related concepts. As usual, for $\mathcal{C}$ a category, we denote the identity arrow on the object u by $id_u : u \to u$ and the composition of two arrows $f : u \to v$ and $g : v \to w$ by $f; g : u \to w$ (i.e., the operation $\_; \_$ composes in the diagrammatic order). In what follows we let $O_{\mathcal{C}}$ and $A_{\mathcal{C}}$ denote respectively the objects and the arrows of $\mathcal{C}$ and let $\times$ denote the ordinary cartesian product of categories.

Definition 6. *A* strict monoidal category *is a triple* $\langle \mathcal{C}, \otimes, e \rangle$, *where* $\mathcal{C}$ *is the underlying category, the functor* $\otimes : \mathcal{C} \times \mathcal{C} \to \mathcal{C}$ *is called* tensor product *and the object* $e \in O_{\mathcal{C}}$ *is called the* unit. *Moreover, the tensor product satisfies the associativity law* $f \otimes (g \otimes h) = (f \otimes g) \otimes h$ *for all* $f, g, h \in A_{\mathcal{C}}$ *and has the constant functor associated to* e *as neutral element, i.e.,* $id_e \otimes f = f = f \otimes id_e$, *for all* $f \in A_{\mathcal{C}}$.

For non-strict monoidal categories, the associativity and unit laws are satisfied only up to suitable natural isomorphisms. Since we shall always consider strict monoidal categories, the adjective 'strict' can be omitted to simplify the terminology. When the tensor product is commutative up to a suitable natural isomorphism, the monoidal category is called 'symmetric'.

Definition 7. *A* symmetric monoidal category *is a 4-tuple* $\langle \mathcal{C}, \otimes, e, \gamma \rangle$, *where* $\langle \mathcal{C}, \otimes, e \rangle$ *is a monoidal category and* $\gamma : \_{-1} \otimes \_{-2} \Rightarrow \_{-2} \otimes \_{-1}$ *is a natural isomor-*

phism satisfying the Kelly-MacLane coherence axioms expressed by the following equations:

$$\gamma_{u,v}; \gamma_{v,u} = id_u \otimes id_v$$
$$\gamma_{u,v\otimes w} = (\gamma_{u,v} \otimes id_w); (id_v \otimes \gamma_{u,w})$$

for all objects $u, v, w \in O_\mathcal{C}$.

Note that the equality $\gamma_{u,e} = id_u$ follows from the fact that $u \otimes e = u$ together with axioms above. When γ is the identity natural transformation, then the tensor product is commutative and the category is called 'strictly symmetric'. Commutative products are often denoted by the additive symbol $\oplus$ instead of $\otimes$. The arrows of a symmetric monoidal category that can be obtained as the sequential and parallel composition of identities and symmetries are called *permutations* and ranged by σ, σ', σ_1, and so on.

Definition 8. *Let $\langle \mathcal{C}, \otimes, e \rangle$ and $\langle \mathcal{C}', \otimes', e' \rangle$ be monoidal categories. A functor $F: \mathcal{C} \to \mathcal{C}'$ is called strict monoidal if $F(e) = e'$ and $F(f \otimes g) = F(f) \otimes' F(g)$ for all $f, g \in A_\mathcal{C}$.*

Again, we shall omit the term 'strict', since all monoidal functors that we consider are so. The category of monoidal categories and monoidal functors is commonly indicated by **MonCat**. Moreover, we denote by **CMonCat** the full subcategory of strictly symmetric monoidal categories, and use **CMonCat**$^\oplus$ for the full subcategory of **CMonCat** consisting of categories whose sets of objects are freely generated commutative monoids. In particular, we have that both $\mathcal{T}(N)$ and $\mathcal{CT}(N)$ belong to **CMonCat**$^\oplus$.

Definition 9. *Let $\langle \mathcal{C}, \otimes, e, \gamma \rangle$ and $\langle \mathcal{C}', \otimes', e', \gamma' \rangle$ be symmetric monoidal categories. A monoidal functor $F: \langle \mathcal{C}, \otimes, e \rangle \to \langle \mathcal{C}', \otimes', e' \rangle$ is called symmetric if $F(\gamma_{u,v}) = \gamma'_{F(u),F(v)}$.*

We denote by **SSMC** the subcategory of **MonCat** whose objects are symmetric monoidal categories and whose arrows are symmetric monoidal functors. Let **SSMC**$^\otimes$ (resp. **SSMC**$^\oplus$) be the full subcategory of **SSMC** consisting of monoidal categories whose sets of objects are freely generated monoids (resp. commutative monoids). Note that the tensor products of categories in **SSMC**$^\oplus$ are not necessarily commutative: the superscript $\oplus$ refers to commutative monoidal composition of objects only, not of arrows. We have $\mathcal{P}(N) \in$ **SSMC**$^\oplus$ and $\mathcal{Q}(N) \in$ **SSMC**$^\otimes$.

2   Collective Contexts

In [22], Meseguer suggested to represent contexts in rewriting logic theories by considering two kinds of entities for each term: 'counters' and 'copies'. Given a term, one can release as many copies of it as needed, while recording the number of such copies in the corresponding counter. Copies can only be accessed

(1)	$(r)^+ \oplus (r)^- = r$		(5)	$((r)^-)^- = \varnothing$
(2)	$((r)^-)^+ = (r)^-$		(6)	$((r)^+)^- = (r)^-$
(3)	$(r \oplus s)^+ = (r)^+ \oplus (s)^+$		(7)	$(r \oplus s)^- = (r)^- \oplus (s)^-$
(4)	$(\varnothing)^+ = \varnothing$		(8)	$(\varnothing)^- = \varnothing$
(unit)	$r \oplus \varnothing = \varnothing$		(ass)	$r \oplus (s \oplus r') = (r \oplus s) \oplus r'$
(comm)	$r \oplus s = s \oplus r$			

Fig. 4.

as contexts. On the contrary, when rewriting a term, one has to retrieve the counter and as many copies as indicated by the counter. That is, all the copies ever released. Formally, in the case of contextual nets, the data type of places is modified as follows:

$$a = \langle a, 0 \rangle$$
$$\langle a, n \rangle = \langle a, n+1 \rangle \oplus [a].$$

The terms having the form $\langle a, n \rangle$ (for a a place and n a natural number) are counters, and the $[a]$ are copies, with $a = \langle a, n \rangle \oplus n \cdot [a]$. Then, a transition with precondition a, context b and postcondition c becomes a rewrite rule $a \oplus [b] \Rightarrow c \oplus [b]$ with a self-loop on a copy of b. However, this fits well with the *CTph* approach only.

We tried to characterize the algebraic structure that gives the basis for Meseguer's encoding and have come out successfully with a representation that can be extended to deal with the *ITph* as well. As explained in the Introduction, we build the algebraic theory over a non-free monoid of places. In particular, apart from the commutative monoidal operation $\_ \oplus \_$ with unit $\varnothing$, we consider other two operations $(\_)^+$ and $(\_)^-$ that are axiomatized as in Figure 4, where we also included the ordinary unit, associativity and commutativity axioms for $\_ \oplus \_$. Quite simply, these mean that $(\_)^+$ and $(\_)^-$ are monoid homomorphisms – laws (3), (4), (7) and (8) – such that $(\_)^+ \oplus (\_)^- = id$, $(\_)^+ \circ (\_)^- = (\_)^-$, and $(\_)^- \circ (\_)^- = \varnothing$. Observe that (6) actually follows from (1), (7) and (5). We call the elements of this algebra *molecules*, ranged over by r, s, ... Given a set S, we let $\mu(S)$ denote the set of molecules generated by S, i.e., $\mu(S)$ is the quotient term algebra generated by S over the signature with $\varnothing$, $\oplus$, $(\_)^+$ and $(\_)^-$ (modulo the axioms in Figure 4).

By these laws we can always eliminate consecutive applications of $(\_)^+$ and $(\_)^-$, except for sequences of $(\_)^+$. We shall write r^k as a shorthand for $(\_)^+$ applied k times to r and omit the parentheses. We assume $r^0 = r$, but we remark that in general $r^+ = r^1 \neq r$.

Lemma 1. *For each molecule $r \in \mu(S)$ and each $k \in \mathbb{N}$, we have $(r^k)^- = r^-$.*

Proof. By induction on k, applying law (6). ◆

Proposition 1. *For each molecule $r \in \mu(S)$ and each $k \in \mathbb{N}$, we have $r^k = r^{k+1} \oplus r^-$.*

$$\frac{r \in \mu(S)}{id_r : r \to r} \qquad \frac{t \in T}{t : \varsigma(t)^{\cdot} \oplus \partial_0(t) \to \varsigma(t)^{\cdot} \oplus \partial_1(t)}$$

$$\frac{\alpha : r \to s, \ \beta : s \to s'}{\alpha ; \beta : r \to s'} \qquad \frac{\alpha : r \to s, \ \beta : r' \to s'}{\alpha \oplus \beta : r \oplus r' \to s \oplus s'}$$

Fig. 5.

$$\alpha ; (\beta ; \delta) = (\alpha ; \beta) ; \delta \qquad \qquad \alpha ; id_s = id_r ; \alpha = \alpha$$
$$\alpha \oplus (\beta \oplus \delta) = (\alpha \oplus \beta) \oplus \delta \qquad \alpha \oplus \beta = \beta \oplus \alpha \qquad \qquad \alpha \oplus id_{\varnothing} = \alpha$$
$$(\alpha ; \beta) \oplus (\delta ; \eta) = (\alpha \oplus \delta) ; (\beta \oplus \eta) \qquad id_{r \oplus s} = id_r \oplus id_s$$

Fig. 6.

Proof. By law (1), we have $r^{k} = (r^{k})^{+} \oplus (r^{k})^{-}$, and $(r^{k})^{-} = r^{-}$ by Lemma 1. ◆

Corollary 1. *For each molecule $r \in \mu(S)$ and each $k \in \mathbb{N}$, we have $r = r^{k} \oplus k \cdot r^{-}$.*

Of course we are interested in molecules generated from places, which can be of two forms: either a^{k} or a^{-}. From the computational point of view, the a^{-} are the basic contexts and carry very little information, since the nucleus a^{k} can produce as many of them as needed. To appreciate the point, we can think of the tokens as *ticket rolls* with unbounded number of tickets available. Readers just take a ticket and return it after use for recycle, whereas consumers must retrieve the entire roll, including all used tickets.

Definition 10. *For $N = (S, T, \partial_0, \partial_1, \varsigma)$ a contextual net, define $\mathcal{M}(N)$ as the category in* **CMonCat** *with objects the molecules on S, and arrows generated from the rules in Figure 5, modulo the axioms of strictly symmetric strict monoidal categories in Figure 6.*

We can now characterize contextual commutative processes algebraically.

Theorem 1. *The category $\mathcal{CT}(N)$ is isomorphic (via a monoidal functor) to the full subcategory of $\mathcal{M}(N)$ whose objects are $S_N^{\oplus}$.*

A very important property needed in the proof is what we call the *maximum sharing hypothesis*, that can be expressed as in the proposition below. This contains the core of the *CTph* for contextual nets, since it shows that whenever two or more tokens in the same place a are used as contexts, we can always find an equivalent computation where only one token in a is used (twice or more) as a context. In other words, tokens in the same place are completely interchangeable in contexts.

Proposition 2. *For each molecule $r \in \mu(S)$ and $k, n \in \mathbb{N}$, we have $r^{n} \oplus r^{k} = r^{n+k} \oplus r$.*

Proof. By Corollary 1, we have $r^{n+k} \oplus r = r^{n+k} \oplus r^k \oplus k \cdot r^-$. By commutativity (and associativity) of $\_ \oplus \_$ we get $r^{n+k} \oplus r = r^{n+k} \oplus k \cdot r^- \oplus r^k$. By applying k times Proposition 1 we have the result.   ◆

Before proving Theorem 1 we need some other technical lemmata.

Lemma 2. *Each molecule* $r \in \mu(S)$ *factorizes uniquely as* $u \oplus r_e \oplus r_n$ *where*[1]

▷ $u \in S^\oplus$;
▷ $r_e = k_1 \cdot a_1^- \oplus \ldots \oplus k_n \cdot a_n^-$ *with* $n \geq 0$ *and* $k_i > 0$, *for* $i = 1, \ldots, n$;
▷ $r_n = b_1^{h_1} \oplus \ldots \oplus b_m^{h_m}$ *with* $m \geq 0$ *and* $h_j > 0$, *for* $j = 1, \ldots, m$;

where all the a_i *and* b_j *are distinct places.*

Proof. The normal form representation follows by observing that $(\_)^+$ and $(\_)^-$ are monoid homomorphisms and, therefore, distribute over $\oplus$. Then, by laws (2), (5) and (6), we can reduce the molecule to the 'sum' of places a, electrons a^- and nuclei a^h. Then, by Proposition 2, we can simplify the expression to a form where at most one nucleus a^h with $h > 0$ is present for each a. Finally, if both a^h and $k \cdot a^-$ are present in the expression, we can simplify the expression according to the following three possibilities, until all the nuclei and electrons refer to different places.

$(h > k)$: then $a^h = (a^{h-k})^k$ and, by Lemma 1, $k \cdot a^- = k \cdot (a^{h-k})^-$, hence $a^h \oplus k \cdot a^- = (a^{h-k})^k \oplus k \cdot (a^{h-k})^- = a^{h-k}$ by Corollary 1;
$(h = k)$: then $a^k \oplus k \cdot a^- = a$ by Corollary 1;
$(h < k)$: then $a^h \oplus k \cdot a^- = a^h \oplus h \cdot a^- \oplus (k-h) \cdot a^- = a \oplus (k-h) \cdot a^-$ by applying Corollary 1 to $a^h \oplus h \cdot a^-$.   ◆

Lemma 3. *If the source of an arrow* $\alpha \in \mathcal{M}(N)$ *factorizes according to Lemma 2 as* $u \oplus r_e \oplus r_n$, *then* $\alpha : u \oplus r_e \oplus r_n \to v \oplus r_e \oplus r_n$ *for some* $v \in S^\oplus$.

Proof. It is straightforward to observe that r_e and r_n are invariants of the generation rules in Figure 5.   ◆

Lemma 4. *Each arrow* $\delta : r \to s$ *in* $\mathcal{M}(N)$ *can be decomposed as*

$$(t_1 \oplus id_{r_1}); (t_2 \oplus id_{r_2}); \ldots; (t_k \oplus id_{r_k}),$$

for some $k \geq 0$, *where all the* t_i *are transitions.*

Proof. By structural induction on the expression denoting δ. The complex case is when $\delta = \alpha \oplus \beta$. We can then apply the functoriality of $\oplus$ to get $\delta = (\alpha \oplus id_{r'}); (\beta \oplus id_{s'})$ where r' is the source of β and s' is the target of α. Then we apply the inductive hypothesis to α and β.   ◆

We are now ready to prove the main theorem.

---

[1] We choose the subscripts 'e' and 'n' as abbreviations for 'electron' and '(uncomplete) nucleus', respectively.

Proof. (of Theorem 1). We start by defining the functor $\mathsf{F}\colon \mathcal{CT}(N) \to \mathcal{M}(N)$. Given a generic step sequence $u_0 \; [X_1\rangle \; u_1 \ldots u_{n-1} \; [X_n\rangle \; u_n$ with length n (representing a generic arrow in $\mathcal{CT}(N)$), we let

$$\mathsf{F}(u_0 \; [X_1\rangle \; u_1 \ldots u_{n-1} \; [X_n\rangle \; u_n) = \mathsf{F}(u_0 \; [X_1\rangle \; u_1); \ldots ; \mathsf{F}(u_{n-1} \; [X_n\rangle \; u_n),$$

with $\mathsf{F}(u \; [X\rangle \; v)$ as defined below. Let

$\triangleright \; u_X = \bigoplus_{t \in \lfloor X \rfloor} X(t) \cdot \partial_0(t);$
$\triangleright \; v_X = \bigoplus_{t \in \lfloor X \rfloor} X(t) \cdot \partial_1(t);$
$\triangleright \; w_X = \bigoplus_{t \in \lfloor X \rfloor} X(t) \cdot \varsigma(t).$

We can assume that

$$w_X = k_1 \cdot a_1 \oplus k_2 \cdot a_2 \oplus \ldots \oplus k_m \cdot a_m$$

with $m \geq 0$, $k_i > 0$, for $i = 1, \ldots, m$ and all the a_i different places. Since the step X is enabled at u, then $u = u' \oplus u_X \oplus \lfloor w_X \rfloor$ for some $u' \in S^\oplus$. Hence $v = u' \oplus v_X \oplus \lfloor w_X \rfloor$. With this notation fixed, let

$$\mathsf{F}(u \; [X\rangle \; v) = id_{u'} \oplus X \oplus id_{a_1^{k_1}} \oplus id_{a_2^{k_2}} \oplus \ldots \oplus id_{a_m^{k_m}}.$$

Note that with this definition, the k_i tokens needed as context relatively to place a_i yield an idle nucleus $a_i^{k_i}$, for $i = 1, \ldots, m$. Also notice that when $X = \varnothing$ the result is just the identity on u.

To show that the mapping F is well-defined we must show that it respects the diamond equivalence, i.e., that when $u \; [X \oplus Y\rangle \; v$ is defined, then

$$\mathsf{F}(u \; [X \oplus Y\rangle \; v) = \mathsf{F}(u \; [X\rangle \; u_1); \mathsf{F}(u_1 \; [Y\rangle \; v)$$

with u_1 uniquely determined by u and X. This follows easily by definition of F and by the functoriality of the tensor product.

To show that F is faithful it suffices to observe that the only axiom that potentially may break this property (i.e., that could induce too many equalities on terms) is the functoriality of tensor product which, on the other hand, corresponds precisely to the diamond equivalence.

Finally, to show that F is full (on the full subcategory of $\mathcal{M}(N)$ whose objects are markings), we take a generic arrow $\alpha\colon u \to v \in \mathcal{M}(N)$ with $u, v \in S^\oplus$ and show that there exists a step sequence in $\mathcal{CT}(N)$ that is mapped to α by F. In fact, by Lemma 4 we take a 'linearization' of α (i.e., a sequential composition of transitions in parallel with identities) and show that the obvious firing sequence associated to it can be executed in N. In doing this we employ Lemma 3 and the fact that $u \in S^\oplus$. Observe that this construction defines the inverse to F. ◆

Example 1. Let us consider the net N in Figure 7. In $\mathcal{M}(N)$ we have three basic arrows

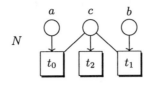

Fig. 7.

▷ $t_0: a \oplus c^- \to c^-$,
▷ $t_1: b \oplus c^- \to c^-$ and
▷ $t_2: c \to \varnothing$,

but neither t_0, nor t_1 can represent a commutative contextual process, since their sources and targets are not elements of $S^\oplus$. To remedy this, we must put t_0 and t_1 in an environment where the c^- become instances of a 'complete' token, as $id_{c^+} \oplus t_0: a \oplus c \to c$ and $id_{c^+} \oplus t_1: b \oplus c \to c$. The concurrent execution of t_0 and t_1 with shared context is instead written as $id_{c^2} \oplus t_0 \oplus t_1 : a \oplus b \oplus c \to c$: since two electrons are needed the idle nucleus has 'degree' 2. By the functoriality of $\_ \oplus \_$, we have that

$$id_{c^2} \oplus t_0 \oplus t_1 = (id_{c^+} \oplus t_0 \oplus id_b); (id_{c^+} \oplus t_1) = (id_{c^+} \oplus t_1 \oplus id_a); (id_{c^+} \oplus t_0),$$

(recall that $id_{c^2} \oplus id_{c^-} = id_{c^+}$), i.e., t_0 and t_1 can execute in any order. Also interesting is to observe that

$$(id_{c^+} \oplus t_0) \oplus ((id_{c^+} \oplus t_1); t_2) = ((id_{c^+} \oplus t_0); t_2) \oplus (id_{c^+} \oplus t_1),$$

i.e., we have no causal information about the token consumed by t_2: is it the one read by t_0, or the one read by t_1? In fact by $id_{c^+} \oplus t_0 = (id_{c^+} \oplus t_0); id_c$ and applying the functoriality of $\oplus$ we have:

$$(id_{c^+} \oplus t_0) \oplus ((id_{c^+} \oplus t_1); t_2) = (id_{c^+} \oplus t_0 \oplus id_{c^+} \oplus t_1); (id_c \oplus t_2).$$

Then, $id_c \oplus t_2 = t_2 \oplus id_c$ by commutativity of $\oplus$ and by applying the functoriality (in the opposite direction than before) we get the equality. Furthermore,

$$id_{c^+} \oplus t_0 \oplus id_{c^+} \oplus t_1 = id_{c^2} \oplus t_0 \oplus t_1 \oplus id_c$$

(by Proposition 2 and the commutativity of $\oplus$), and thus

$$(id_{c^+} \oplus t_0) \oplus ((id_{c^+} \oplus t_1); t_2) = (id_{c^2} \oplus t_0 \oplus t_1 \oplus id_c); (id_c \oplus t_2) = id_{c^2} \oplus t_0 \oplus t_1 \oplus t_2$$

i.e., t_0, t_1, and t_2 can be executed in a concurrent fashion without the possibility of distinguishing this case from those in which t_2 causally depends on t_0 or t_1.

Since a PT net N can always be seen as a contextual net with no read-arcs, in which case the commutative contextual processes of N are just the ordinary commutative processes (cf. § 1.2 and [24]), then by Theorem 1 we obtain the following corollary.

Corollary 2. *If N is a* PT *net, then $\mathcal{T}(N)$ is a full subcategory of $\mathcal{M}(N)$.*

We remark that the constructions we have shown can be easily extended to deal with multiplicities on read-arcs (i.e., to the case in which $\varsigma(t)$ is a multiset rather than a set).

3   Individual Contexts

The maximum sharing hypothesis creates obvious problems when dealing with the *ITph*, whose entire point is to be able to recognize which electrons are emitted from each token. For ordinary PT nets, the information about causality is recovered in the algebraic setting by using (non strictly) symmetric strict monoidal categories, i.e., by introducing *symmetries* to control rearrangements of tokens in process sequential composition. While at the level of states one can still view standard markings as indexed collections of ordered tokens (rather than resorting to take as states the elements of the free monoid on places, i.e., strings of places), at the level of computations (arrows), however, the tensor product is *not* commutative anymore, so that one is able to interpret in a canonical way the correct flow of causality through token histories. Thus, the first attempt to a uniform extension of the *CTph* treatment of the previous section to the *ITph* view is to introduce symmetries on molecules.

There is however another problem to solve. Since the context $\varsigma(t)$ is modeled by a self-loop on $\varsigma(t)^-$, two transitions with the same context can be concatenated on it, as if one depended on the execution of the other. This spurious causal dependency is to be avoided, as it gives rise to a wrong semantic model. To some extent, one would like to follow the *ITph* on 'complete' molecules (standard markings), and the *CTph* on electrons of the same nucleus, so that one has no information about which electron is consumed by a firing, but only about which molecule it comes from. We therefore need a canonical interpretation of molecules that respects this intuition. To fix the ideas, we take initially a non commutative monoidal operation $\otimes$ on molecules. Let us consider the molecule $a^2 \otimes a^+ \otimes a^- \otimes a^- \otimes a^-$. We would like to view it under an interpretation that connects each of the three electrons with one the two nuclei, and that is *invariant* not only under all possible computations that can originate from the state, but also under composition of the molecule to form larger states. Our idea is to associate an electron to the *first* incomplete nucleus (*ion*) that *precedes* it. In the present case, for instance, the first electron is associated to the second ion (a^+), while the second and third electrons (the two rightmost in the expression) are interpreted as electrons released by the leftmost ion a^2. A good way to explain the mechanism, is to view ions as open parentheses and electrons as closing parentheses, where of course an ion a^k opens several parentheses, namely k, at once. Clearly, we are mainly interested in balanced expressions, but unbalanced expressions must exist, and can always be completed by parallel composition to yield balanced terms. To complete the picture, consider now that the order in which atoms, nuclei and electrons of different kinds – i.e., coming from different

$$
\begin{array}{llll}
(9) & a^+ \otimes a^- = a & (13) & (p^-)^- = \varnothing \\
(10) & (p^-)^+ = p^- & (14) & (p^+)^- = p^- \\
(11) & (p \otimes q)^\chi = p^\chi \otimes q^\chi & (15) & \varnothing^\chi = \varnothing \\
(12) & a^\delta \otimes b^\epsilon = b^\epsilon \otimes a^\delta & (16) & a \otimes a^- = a^- \otimes a \\
(l.unit) & \varnothing \otimes p = p & (ass) & p \otimes (q \otimes p') = (p \otimes q) \otimes p' \\
(r.unit) & p \otimes \varnothing = p & &
\end{array}
$$

Fig. 8. Axioms for bimolecules (with $a \neq b \in S$, $\delta, \epsilon \in \mathbb{N} \cup \{^-\}$ and $\chi \in \{^+, ^-\}$).

places – appear in an expression is not relevant. Hence, the monoidal operation $\otimes$ better be commutative in such situations. In other words, we have:

$$
a^+ \otimes b^+ \otimes a^- \otimes b^- = a^+ \otimes b^+ \otimes b^- \otimes a^- = a^+ \otimes b \otimes a^- = a^+ \otimes a^- \otimes b = a \otimes b,
$$

but we definitely want that $a^+ \otimes a^+ \otimes a^- \neq a^+ \otimes a^- \otimes a^+$, because the particle a^- in the two terms is associated to different nuclei and, therefore, the two states may give rise to different causal histories when a transition reads that particle.

We call *bimolecules*, ranged over by $p, q, \ldots$, the (generalized) markings of the algebra illustrated above. It includes a set of axioms almost identical to those in Figure 4, plus some extra axioms to deal with restricted commutativity. Given a set S, we write $\nu(S)$ for the set of bimolecules on S. The complete axiomatization of bimolecules is shown in Figure 8. Note that law (9) – the analogous to (1) for molecules – on bimolecules applies only when a nucleus is immediately on the left of an electron, i.e., $a^- \otimes a^+ \neq a$. Furthermore, while law (1) applies to generic molecules, law (9) deals with a single atom (place) a.

The final and key ingredient in our construction is to *abandon* the symmetry of the monoidal categories involved. With a step similar to the one that led from strictly symmetric to symmetric categories, we choose (*non symmetric*) monoidal categories to which we adjoin *exactly* and *only* the symmetries we need. In this way, we are able to omit those symmetries that would cause migration of electrons from atom to atom. In the following we shall build on a construction somehow intermediate between $\mathcal{P}(N)$ and $\mathcal{Q}(N)$ for PT nets [12,32] and, therefore, take a non commutative monoid of objects: it is commutative only on some objects, in particular on the markings. We use the symbol $\otimes$ for the monoidal operation and denote the free monoid on the set S by $S^\otimes$.

Definition 11. *For $N = (S, T, \partial_0, \partial_1, \varsigma)$ a contextual net, $\mathcal{B}(N)$ is the monoidal category with objects the bimolecules on S, and arrows generated from the rules in Figure 9, together with the symmetries*

$$
\gamma_{a^\delta, b^\epsilon} : a^\delta \otimes b^\epsilon \to b^\epsilon \otimes a^\delta, \quad \text{for } a \neq b \in S \text{ and } \delta, \epsilon \in \mathbb{N} \cup \{^-\},
$$

$$
\gamma_{a, a^-} : a \otimes a^- \to a^- \otimes a,
$$

$$
\gamma_{a^-, a} : a^- \otimes a \to a \otimes a^-.
$$

$$\frac{p \in \nu(S)}{id_p \colon p \to p} \qquad \frac{t \in T}{t \colon \varsigma(t)^{\char`\-} \otimes \partial_0(t) \to \varsigma(t)^{\char`\-} \otimes \partial_1(t)} \qquad \frac{u,v \in S^{\otimes}}{\gamma_{u,v} \colon u \otimes v \to v \otimes u}$$

$$\frac{\alpha \colon p \to q, \ \beta \colon q \to r}{\alpha; \beta \colon p \to r} \qquad \frac{\alpha \colon p \to q, \ \beta \colon p' \to q'}{\alpha \otimes \beta \colon p \otimes p' \to q \otimes q'}$$

Fig. 9.

$$
\begin{aligned}
&\alpha; (\beta; \sigma) = (\alpha; \beta); \sigma &&\alpha; id_q = id_p; \alpha = \alpha &&(\alpha; \beta) \otimes (\alpha'; \beta') = (\alpha \otimes \alpha'); (\beta \otimes \beta')\\
&\alpha \otimes (\beta \otimes \sigma) = (\alpha \otimes \beta) \otimes \sigma &&\alpha \otimes id_{\varnothing} = id_{\varnothing} \otimes \alpha = \alpha &&id_{p \otimes q} = id_p \otimes id_q\\
&(\alpha \otimes \beta); \gamma_{q,q'} = \gamma_{p,p'}; (\beta \otimes \alpha) &&\gamma_{p,q}; \gamma_{q,p} = id_p \otimes id_q &&\gamma_{p,q \otimes r} = (\gamma_{p,q} \otimes id_r); (id_q \otimes \gamma_{p,r})
\end{aligned}
$$

Fig. 10.

The arrows are taken modulo the axioms of strict monoidal categories in Figure 10 (whenever the γ's are defined) and the laws:

$$\sigma; t; \sigma' = t \tag{17}$$

$$\gamma_{a^\delta, b^\epsilon} = id_{a^\delta \otimes b^\epsilon}, \quad \text{for } a \neq b \in S \text{ and } \delta, \epsilon \in \mathbb{N} \cup \{^{\char`\-}\} \tag{18}$$

$$\gamma_{a, a^{\char`\-}} = id_{a \otimes a^{\char`\-}} \tag{19}$$

$$id_{a^k} \otimes t = id_{a^{k+1}} \otimes t \otimes id_{a^{\char`\-}} \tag{Δ}$$

for all transitions $t \colon p \to q$, permutations $\sigma \colon p \to p$, $\sigma' \colon q \to q$, and $k > 0$.

Since $\gamma_{a^{\char`\-}, a}$ is inverse to $\gamma_{a, a^{\char`\-}} = id_{a \otimes a^{\char`\-}}$, it follows that $\gamma_{a^{\char`\-}, a} = id_{a \otimes a^{\char`\-}}$. Note that we *do not* introduce symmetries such as $\gamma_{a^{\char`\-}, a^{\char`\-}}$, $\gamma_{a^k, a^{\char`\-}}$, and γ_{a^k, a^n}, for $k, n \geq 1$, that would allow the particles to flow from a nucleus to a different one. For example, starting from $a \otimes a^+ = a^+ \otimes a^{\char`\-} \otimes a^+$ and applying an hypothetical arrow $id_{a^+} \otimes \gamma_{a^{\char`\-}, a^+}$, we would reach $a^+ \otimes a^+ \otimes a^{\char`\-} = a^+ \otimes a$, allowing the nuclei to exchange electrons, which is problematic. Another *non*-example would be applying the arrow $id_{a^+ \otimes a^+} \otimes \gamma_{a^{\char`\-}, a^{\char`\-}}$ to $a \otimes a = a^+ \otimes a \otimes a^{\char`\-} = a^+ \otimes a^+ \otimes a^{\char`\-} \otimes a^{\char`\-}$ because, after the exchange, the token of the first and second nucleus get confused. By forcing $\gamma_{a^{\char`\-}, a^{\char`\-}}$ to be the identity we would confuse the electrons of two different nuclei, because of the naturality axiom, and by leaving it free we would allow again for electrons migration. In fact, our *representation invariant* is that the electrons associated to a certain nucleus a^k in a bimolecule q are determined by following the discipline of proper nesting of open and closed parentheses. The absence of those symmetries maintains this invariant for us.

Laws (17) and (18) are classical laws for the $\mathcal{P}(N)$ construction; here they have a slightly more general role, because they also deal with nuclei and electrons. In particular, law (17) is the analogous of axiom (Ψ) for PT nets (cf. [12]). Law (19) says that electrons can be freely moved around 'complete atoms' of the same kind. Law (Δ) is original and really central to our development. In fact, even though the symmetries $\gamma_{a^{\char`\-}, a^{\char`\-}}$ are not allowed, we certainly do not want to distinguish between electrons of the same nucleus (the first released, the second, ...), as otherwise we would obtain a notion of computation very concrete and far from our target, that is to capture algebraically the notion of contextual process. Axiom (Δ) takes care of identifying such particles, as the Example 2 below illustrates.

Example 2. Let N be the contextual net in Figure 7. Then we have three basic arrows in $\mathcal{B}(N)$ associated to the transitions of N:

▷ $t_0: c^- \otimes a \to c^-$;
▷ $t_1: c^- \otimes b \to c^-$;
▷ $t_2: c \to \varnothing$.

Then, new arrows can be built by composing (sequentially and in parallel) these three arrows with identities and symmetries. For example, the arrow $id_{c^2} \otimes t_0 \otimes t_1$ goes from $c^2 \otimes c^- \otimes a \otimes c^- \otimes b = c \otimes a \otimes b$ to $c^2 \otimes c^- \otimes c^- = c$. Analogously we have the arrow $id_{c^2} \otimes t_1 \otimes t_0: c \otimes b \otimes a \to c$. Then, it is possible to prove that these two arrows are identified in $\mathcal{B}(N)$. In fact, we have:

$$id_{c^2} \otimes t_0 \otimes t_1 = (id_{c^2} \otimes t_0 \otimes id_{c^- \otimes b}); (id_{c^2} \otimes id_{c^-} \otimes t_1) \text{ (by functoriality)}$$
$$= (id_{c^+} \otimes t_0 \otimes id_b); (id_{c^+} \otimes t_1) \text{ (by law } \Delta).$$

Then, by naturality, we have $t_0 \otimes id_b = \gamma_{c^- \otimes a, b}; (id_b \otimes t_0); \gamma_{b, c^-}$, but these symmetries are just identities and therefore $t_0 \otimes id_b$ can be replaced by $id_b \otimes t_0$ in the expression above.

$$id_{c^2} \otimes t_0 \otimes t_1 = (id_{c^+} \otimes id_b \otimes t_0); (id_{c^+} \otimes t_1)$$
$$= (id_{c^2 \otimes c^-} \otimes id_b \otimes t_0); (id_{c^2} \otimes t_1 \otimes id_{c^-}) \text{ (by law } \Delta)$$
$$= id_{c^2} \otimes t_1 \otimes t_0 \text{ (by functoriality)}$$

Notice that, as formalised by the following Definition 12, there is only one concatenable contextual process that starts from $a \otimes b \otimes c$ and involves exactly one firing of t_0 and one firing of t_1. By repeatedly applying law (Δ) we then have, e.g.,

$$id_{c^2} \otimes t_0 \otimes t_1 = id_{c^{n+m+2}} \otimes t_1 \otimes id_{n \cdot c^-} \otimes t_0 \otimes id_{m \cdot c^-}$$

for all $n, m \in \mathbb{N}$. This means that the order in which the electrons are read is not important provided that they originated from the same nucleus.

To establish our representation result we need to refine contextual processes in order to be able to concatenate them. As for similar cases in the literature, this leads to the introduction of an *ordering* on the tokens in the source and target of the process net, yielding the notion of *concatenable contextual processes*.

Definition 12. *For N a contextual net, a* concatenable contextual process *is a tuple $(\pi, \Theta, \prec_0, \prec_1)$, where π is a contextual process with underlying contextual process net Θ, $\prec_0$ and $\prec_1$ are partial orders on the minimal and maximal places of Θ, respectively, such that: (1) $x \prec_i y$ implies that $\pi(x) = \pi(y)$; and (2) if $x \neq y$ are minimal places (respectively maximal places) such that $\pi(x) = \pi(y)$, then either $x \prec_0 y$ or $y \prec_0 x$ (respectively, $x \prec_1 y$ or $y \prec_1 x$).*

As usual, concatenable processes are taken up to isomorphism. The two conditions imposed in the definition above ensure that we order only places of Θ that are instances of the same place of N, and that on such places the ordering is total.

Likewise concatenable processes of PT nets, a partial operation of *sequential composition* can be defined. Provided the target of process π coincides with the source of process π', it merges the *maximal* places of π with the *minimal* places of π' according to the orders $\prec_1$ and $\prec'_0$.

Definition 13. *Let* $(\pi', \Theta', \prec'_0, \prec'_1)$ *and* $(\pi'', \Theta'', \prec''_0, \prec''_1)$ *be two concatenable contextual processes of a contextual net* N, *where* $T_{\Theta'} \cap T_{\Theta''} = \varnothing$ *and* $S_{\Theta'} \cap S_{\Theta''}$ *is both the set of maximal places for* Θ' *and the set of minimal places for* Θ'', *with* $\pi'(x) = \pi''(x)$ *for any* $x \in S_{\Theta'} \cap S_{\Theta''}$, *and* $x \prec'_1 y$ *iff* $x \prec''_0 y$ *for all* $x, y \in S_{\Theta'} \cap S_{\Theta''}$. *Then, their concatenation* $(\pi, \Theta' \cup \Theta'', \prec'_0, \prec''_1) = (\pi', \Theta', \prec'_0, \prec'_1); (\pi'', \Theta'', \prec''_0, \prec''_1)$ *is well defined, where* π *is the componentwise union of* π' *and* π'' *(i.e.,* $\pi(x) = \pi'(x)$ *if* $x \in \Theta'$ *and* $\pi(x) = \pi''(x)$ *if* $x \in \Theta''$*).*

The composition is well defined because by hypothesis we have $\pi'(x) = \pi''(x)$ for all $x \in \Theta' \cap \Theta'' = S_{\Theta'} \cap S_{\Theta''}$, i.e., merged places have the same names.

The *parallel composition* of two processes consists of taking their disjoint union and extending the orders on minimal and maximal places by $x \prec_i y$ whenever x belongs to the first process, y to the second, and $\pi(x) = \pi(y)$.

Definition 14. *Let* $(\pi', \Theta', \prec'_0, \prec'_1)$ *and* $(\pi'', \Theta'', \prec''_0, \prec''_1)$ *be two concatenable contextual processes of a contextual net* N, *where* $T_{\Theta'} \cap T_{\Theta''} = \varnothing$ *and* $S_{\Theta'} \cap S_{\Theta''} = \varnothing$. *Let* S'_0 *and* S''_0 *be the set of minimal places of* Θ' *and* Θ'', *respectively. Likewise, let* S'_1 *and* S''_1 *be the set of maximal places of* Θ' *and* Θ'', *respectively. Then, the parallel composition* $(\pi, \Theta' \cup \Theta'', \prec_0, \prec_1) = (\pi', \Theta', \prec'_0, \prec'_1) \otimes (\pi'', \Theta'', \prec''_0, \prec''_1)$ *is well defined, where*

▷ π *is the componentwise union of* π' *and* π''; *and*
▷ $x \prec_i y$ *iff* $(x, y \in S'_i \land x \prec'_i y) \lor (x, y \in S''_i \land x \prec''_i y) \lor (x \in S'_i \land y \in S''_i \land \pi'(x) = \pi''(y))$.

It can be shown that with these two operations the concatenable contextual processes of N form the arrows of a *strict monoidal category* $\mathcal{CP}(N)$. *Symmetries* can be defined by taking a process that contains just places (no transitions) with suitable orderings $\prec_0$ and $\prec_1$. Each place is both minimal and maximal. These symmetries make $\mathcal{CP}(N)$ a symmetric monoidal category in **SSMC**$^\oplus$.

Definition 15. *A concatenable contextual process is called* elementary *if it contains at most one transition.*

Definition 16. *Given a contextual net* N *and a transition* $t \in T_N$, *the elementary concatenable contextual process* $[t] = (\pi, \Theta, \prec_0, \prec_1)$ *associated to* t *is given by*

▷ $S_\Theta = \{\langle a, 0, n \rangle \mid a \in \lfloor \partial_0(t) \rfloor, \ 1 \leq n \leq \partial_0(t)(a)\} \cup$
$\{\langle a, 1, n \rangle \mid a \in \lfloor \partial_1(t) \rfloor, \ 1 \leq n \leq \partial_1(t)(a)\} \cup \{\langle a, 2, 1 \rangle \mid a \in \lfloor \varsigma(t) \rfloor\}$

▷ $T_\Theta = \{\langle t \rangle\}$;
▷ $\partial_0(\langle t \rangle) = \{\langle a, 0, n \rangle \mid a \in \lfloor \partial_0(t) \rfloor, \ 1 \le n \le \partial_0(t)(a)\}$;
▷ $\partial_1(\langle t \rangle) = \{\langle a, 1, n \rangle \mid a \in \lfloor \partial_1(t) \rfloor, \ 1 \le n \le \partial_1(t)(a)\}$;
▷ $\varsigma(\langle t \rangle) = \{\langle a, 2, 1 \rangle \mid a \in \lfloor \varsigma(t) \rfloor\}$;
▷ $\pi(\langle a, j, n \rangle) = a \quad and \quad \pi(\langle t \rangle) = t$;
▷ $\langle a, i, k \rangle \prec_0 \langle b, j, h \rangle \quad iff \quad a = b \land i = j = 0 \land k < h.$
▷ $\langle a, i, k \rangle \prec_1 \langle b, j, h \rangle \quad iff \quad a = b \land i = j = 1 \land k < h.$

Note that the places in $\{\langle a, 2, 1 \rangle \mid a \in \lfloor \varsigma(t) \rfloor\}$ are both minimal and maximal. Only the trivial (empty) order is needed on them, because we rely on the basic assumptions that $\varsigma(t)$ is a set and that $\varsigma(t) \cap \lfloor \partial_0(t) \cup \partial_1(t) \rfloor = \varnothing$, for any $t \in T_N$.

Proposition 3. *Each elementary concatenable contextual process $(\pi, \Theta, \prec_0, \prec_1)$ that contains exactly one transition, say x, can be obtained as $\sigma_1; ([\pi(x)] \otimes \sigma_2); \sigma_3$ for suitable elementary concatenable contextual processes σ_1, σ_2 and σ_3 that contains no transition.*

Proposition 4. *The concatenable contextual processes of a contextual net N can be obtained as the sequential composition of elementary concatenable contextual processes.*

Proof. Likewise the analogous statement for ordinary PT nets, the proof is by induction on the number of transitions in the process net (exploiting Proposition 3). ◆

Theorem 2. *The category $\mathcal{CP}(N)$ is isomorphic (via a symmetric monoidal functor) to the full subcategory of $\mathcal{B}(N)$ whose objects are the elements of $S^\otimes$ (which is symmetric).*

Before proving the main representation theorem above, we need some technical lemmata that state useful properties of the arrows in $\mathcal{B}(N)$. We start by extending some of the properties of molecules to the framework of bimolecules.

Lemma 5. *For each bimolecule p and each $k \in \mathbb{N}$, we have $(p^k)^- = p^-$.*

Proposition 5. *For each place a and each $k \in \mathbb{N}$, we have $a^k = a^{k+1} \otimes a^-$.*

Proof. The proof proceeds by induction on k. For the base case ($k = 0$) we get $a = a^+ \otimes a^-$ directly by law (9). For the inductive case, we assume the property to be valid for $k = n$ and prove it for $k = n + 1$. Then,

$$
\begin{aligned}
a^{n+1} &= (a^n)^+ \text{ (by definition)} \\
&= (a^{n+1} \otimes a^-)^+ \text{ (by inductive hypothesis)} \\
&= (a^{n+1})^+ \otimes (a^-)^+ \text{ (by law 11)} \\
&= a^{n+2} \otimes a^- \text{ (by law 10)}.
\end{aligned}
$$

This concludes the proof. ◆

Corollary 3. *For each place a and each $k \in \mathbb{N}$, we have $a = a^k \otimes k \cdot a^-$.*

Lemma 6. *Each bimolecule p can be decomposed as $p = p_1 \otimes p_2 \otimes \ldots \otimes p_n$, where each p_i has the form $k_{i,0} \cdot a_i^- \otimes a_i^{k_{i,1}} \otimes a_i^{k_{i,2}} \otimes \ldots \otimes a_i^{k_{i,n_i}}$ with $a_i \neq a_j$, for $i \neq j$.*

Lemma 7. *If $p \otimes q \in S^\otimes$, then for each $u \in S^\otimes$ we have $p \otimes u \otimes q \in S^\otimes$.*

Proof. It suffices to prove the property for $u \in S$, which can be done via a simple case analysis, exploiting the representation of p and q provided by Lemma 6 and applying law (16). ◆

Note that in the previous lemma, p and q are generic bimolecules and not necessarily markings, in fact $p \otimes q \in S^\otimes$ does not imply that $p \in S^\otimes \wedge q \in S^\otimes$. We can now state some invariant and decomposition properties for the arrows in $\mathcal{B}(N)$.

Lemma 8. *If $\alpha = id_p \otimes \gamma_{x,y} \otimes id_q$ with $p \otimes x \otimes y \otimes q \in S^\otimes$, then $p \otimes y \otimes x \otimes q \in S^\otimes$*

Proof. By a simple case analysis: all symmetries are collapsed to identities, except when $x = y = a$ for some $a \in S$. ◆

Lemma 9. *If $\alpha = id_p \otimes t \otimes id_q$ and $p \otimes \varsigma(t)^- \otimes \partial_0(t) \otimes q \in S^\otimes$, then $p \otimes \varsigma(t)^- \otimes \partial_1(t) \otimes q \in S^\otimes$.*

Proof. Follows from Lemma 7. ◆

Proposition 6. *Each $\alpha \in \mathcal{B}(N)$ can be decomposed as*

$$\alpha = \sigma_0; (id_{p_1} \otimes t_1 \otimes id_{q_1}); \sigma_1; (id_{p_2} \otimes t_2 \otimes id_{q_2}); \sigma_2; \ldots; (id_{p_n} \otimes t_n \otimes id_{q_n}); \sigma_n,$$

where the σ_i are permutations (i.e., sequential and parallel compositions of symmetries and identities) and the t_i are transitions.

Proof. By structural induction. The complex case is for $\alpha = \alpha_1 \otimes \alpha_2$ for some $\alpha_1 : r_1 \to r_1'$ and $\alpha_2 : r_2 \to r_2'$. But then, by functoriality we have $\alpha = (\alpha_1 \otimes id_{r_2}); (id_{r_1'} \otimes \alpha_2)$ and by inductive hypothesis

$$\alpha_1 = \sigma_0'; (id_{p_1'} \otimes t_1' \otimes id_{q_1'}); \sigma_1'; (id_{p_2'} \otimes t_2' \otimes id_{q_2'}); \sigma_2'; \ldots;$$
$$(id_{p_{n'}'} \otimes t_{n'}' \otimes id_{q_{n'}'}); \sigma_{n'}',$$
$$\alpha_2 = \sigma_0''; (id_{p_1''} \otimes t_1'' \otimes id_{q_1''}); \sigma_1''; (id_{p_2''} \otimes t_2'' \otimes id_{q_2''}); \sigma_2''; \ldots;$$
$$(id_{p_{n''}''} \otimes t_{n''}'' \otimes id_{q_{n''}''}); \sigma_{n''}'',$$

Then, by functoriality:

$$\alpha_1 \otimes id_{r_2} = (\sigma_0' \otimes id_{r_2}); (id_{p_1'} \otimes t_1' \otimes id_{q_1' \otimes r_2}); (\sigma_1' \otimes id_{r_2}); (id_{p_2'} \otimes t_2' \otimes id_{q_2' \otimes r_2});$$
$$(\sigma_2' \otimes id_{r_2}); \ldots; (id_{p_{n'}'} \otimes t_{n'}' \otimes id_{q_{n'}' \otimes r_2}); (\sigma_{n'}' \otimes id_{r_2}),$$
$$id_{r_1'} \otimes \alpha_2 = (id_{r_1'} \otimes \sigma_0''); (id_{r_1' \otimes p_1''} \otimes t_1'' \otimes id_{q_1''}); (id_{r_1'} \otimes \sigma_1''); (id_{r_1' \otimes p_2''} \otimes t_2'' \otimes id_{q_2''});$$
$$(id_{r_1'} \otimes \sigma_2''); \ldots; (id_{r_1' \otimes p_{n''}''} \otimes t_{n''}'' \otimes id_{q_{n''}''}); (id_{r_1'} \otimes \sigma_{n''}''),$$

From which the hypothesis follows trivially – $\sigma_i = \sigma_i' \otimes id_{r_2}$, for $i = 0, \ldots, n' - 1$, $\sigma_{n'} = (\sigma_{n'}' \otimes id_{r_2}); (id_{r_1'} \otimes \sigma_0'')$, and $\sigma_{n'+i} = id_{r_1'} \otimes \sigma_i''$, for $i = 1, \ldots, n''$. ◆

The main law (Δ) can then be extended to generic arrows whenever we know that the rightmost electron belongs to the nucleus that precedes the arrow.

Corollary 4. *For each $\alpha\colon p \to q \in \mathcal{B}(N)$, $a \in S$ and $k > 0$ such that $a^{k+1} \otimes p \otimes a^- = a^k \otimes p$ we have $id_{a^{k+1}} \otimes \alpha \otimes id_{a^-} = id_{a^k} \otimes \alpha$.*

Corollary 5. *If $\alpha\colon u \to q \in \mathcal{B}(N)$ with $u \in S^\otimes$, then $q \in S^\otimes$.*

Proof. Consequence of Proposition 6 and Lemmata 8 and 9.   ◆

Lemma 10. *If $\alpha = id_p \otimes t \otimes id_q \in \mathcal{B}(N)$ with t a transition and $p \otimes \varsigma(t)^- \otimes \partial_0(t) \otimes q \in S^\otimes$, then $\alpha = \sigma; (id_{\varsigma(t)^+} \otimes t \otimes id_u); \sigma'$ for some permutations σ and σ' and some marking $u \in S^\otimes$.*

Proof. By the decomposition of Lemma 6, and by the fact that the source of α is a marking, it follows that $p = \bigotimes_{a\in S} a^{k_{a,1}} \otimes \ldots \otimes a^{k_{a,n_a}}$ and $q = \bigotimes_{a\in S} h_a \cdot a^- \otimes h_a' \cdot a$. It follows that each electron a^- in $\varsigma(t)^-$ belongs to the closest ion on the left of the electron (namely, the i_ath nucleus of type a in p with i_a the greatest index in $1 \leq i_a \leq n_a$ such that $k_{a,i_a} > 0$). Moreover, the h_a electrons of type a in q can be attached to their corresponding nuclei in p, by applying law (Δ). Therefore we have $\alpha = id_{p'} \otimes t \otimes id_v$ where $p' = \bigotimes_{a\in S} (i_a - 1) \cdot a \otimes a^+ \otimes (n_a - i_a) \cdot a$ (if a is not read by t then the corresponding argument in the sum is just $n_a \cdot a$), and $v = \bigotimes_{a\in S} h_a' \cdot a$. Then, by naturality of symmetries, we have:

$$\alpha = (id_{p''} \otimes \gamma_{v',\varsigma(t)^- \otimes \partial_0(t)} \otimes id_v); (id_{p''} \otimes t \otimes id_{v'\otimes v}); (id_{p''} \otimes \gamma_{\varsigma(t)^- \otimes \partial_1(t),v'} \otimes id_v),$$

where $p'' = \bigotimes_{a\in S} (i_a - 1) \cdot a \otimes a^+$ and $v' = \bigotimes_{a\in S} (n_a - i_a) \cdot a$. In fact the symmetries that we have used in the expression are defined since they involve the swappings of 'complete' tokens with either 'complete' tokens, or electrons. By naturality we have also:

$$id_{p''} \otimes t \otimes id_{v'\otimes v} = (\gamma_{v'',\varsigma(t)\otimes\partial_0(t)} \otimes id_{v'\otimes v}); (id_{\varsigma(t)^+} \otimes t \otimes id_{v''\otimes v'\otimes v});$$
$$(\gamma_{\varsigma(t)\otimes\partial_1(t),v''} \otimes id_{v'\otimes v})$$

where $v'' = \bigotimes_{a\in S} (i_a - 1) \cdot a$. By taking

$$\sigma = (id_{p''} \otimes \gamma_{v',\varsigma(t)^-\otimes\partial_0(t)} \otimes id_v); (\gamma_{v'',\varsigma(t)\otimes\partial_0(t)} \otimes id_{v'\otimes v})$$
$$\sigma' = (\gamma_{\varsigma(t)\otimes\partial_1(t),v''} \otimes id_{v'\otimes v}); (id_{p''} \otimes \gamma_{\varsigma(t)^-\otimes\partial_1(t),v'} \otimes id_v)$$
$$u = v'' \otimes v' \otimes v$$

we have the thesis.   ◆

Proposition 7. *Each $\alpha\colon u \to q \in \mathcal{B}(N)$ with $u \in S^\otimes$ can be decomposed as*

$$\alpha = \sigma_0; (id_{\varsigma(t_1)^+} \otimes t_1 \otimes id_{u_1}); \sigma_1; (id_{\varsigma(t_2)^+} \otimes t_2 \otimes id_{u_2}); \sigma_2; \ldots;$$
$$(id_{\varsigma(t_n)^+} \otimes t_n \otimes id_{u_n}); \sigma_n,$$

where the σ_i are permutations, the t_i are transitions and $u_i \in S^\otimes$, for $i = 1, \ldots, n$.

Proof. The proof exploits the decomposition provided by Proposition 6 and then applies n times the result of Lemma 10. ♦

We are now ready to prove the main representation result of this section.

Proof. (of Theorem 2). We start by defining the monoidal functor $\mathsf{G} \colon \mathcal{CP}(N) \to \mathcal{B}(N)$, which is the identity on objects. By Proposition 3, the functor is completely determined by defining the mapping of elementary processes, since then $\mathsf{G}(\alpha; \beta) = \mathsf{G}(\alpha); \mathsf{G}(\beta)$ and $\mathsf{G}(\alpha \otimes \beta) = \mathsf{G}(\alpha) \otimes \mathsf{G}(\beta)$. For symmetries, the mapping is the classical one (see e.g. [31]). For the elementary process $[t]$ associated to the transition $t \in T_N$, we let $\mathsf{G}([t]) = id_{\varsigma(t)^+} \otimes t$. It remains to prove that:

1. G is well defined;
2. G is full (on the full subcategory of $\mathcal{B}(N)$ whose objects are markings);
3. G is faithful.

The fact that G is well defined means that different decompositions of the same process in terms of elementary processes are mapped to the same arrow. This corresponds to show that different orderings of the events in a process $\sigma = (\pi, \Theta, \prec_0, \prec_1)$ that are consistent with the ordering of events $\nearrow_\Theta$ yield the same arrow in $\mathcal{B}(N)$. To see this, it suffices to show that given a decomposition of the process σ, and taken any two concurrent events that are executed consecutively according to the order imposed by the fixed decomposition, then the decomposition in which the two concurrent events are executed in the reverse order is mapped to the same arrow of σ. The proof is easy (by functoriality of the tensor product) if the two events do not share a context. Otherwise, axiom (Δ) must be employed, as we did in Example 2. Formally, we consider the process $P = P_1; ([t_1] \otimes \sigma_1); \sigma; ([t_2] \otimes \sigma_2); P_2$ where σ_1 is the identity process on the marking $u_2 \oplus \partial_0(t_2) \oplus v$, σ is the process associated to the permutation $id_u \otimes \gamma_{u_1 \otimes \partial_1(t_1), u_2 \oplus \partial_0(t_2)} \otimes id_v$, and σ_2 is the identity process on the marking $u_1 \oplus \partial_1(t_2) \oplus v$, i.e., $\varsigma(t_1) = u \oplus u_1$, $\varsigma(t_2) = u \oplus u_2$, and the two occurrences share the context u (note that while u_1 and u_2 are not necessarily disjoint, the corresponding sets of tokens read by t_1 and t_2 in the process P are disjoint). Then, we have also $P = P_1; \sigma'; ([t_2] \otimes \sigma_2'); \sigma''; ([t_1] \otimes \sigma_1'); \sigma'''; P_2$, for suitable permutation processes:

σ' associated to $\qquad\qquad id_u \otimes \gamma_{u_1 \oplus \partial_0(t_1), u_2 \oplus \partial_0(t_2)} \otimes id_v,$

σ_2' idle process associated to $u_1 \oplus \partial_0(t_1) \oplus v,$

σ'' associated to $\qquad\qquad id_u \otimes \gamma_{u_2 \oplus \partial_1(t_2), u_1 \oplus \partial_0(t_1)} \otimes id_v,$

σ_1' idle process associated to $u_2 \oplus \partial_1(t_2) \oplus v,$

σ''' associated to $\qquad\qquad id_u \otimes \gamma_{u_1 \oplus \partial_1(t_1), u_2 \oplus \partial_1(t_2)} \otimes id_v.$

Hence we want to prove that the two decompositions are mapped to the same arrow in $\mathcal{B}(N)$. More precisely, we show that

$$\mathsf{G}(([t_1] \otimes \sigma_1); \sigma; ([t_2] \otimes \sigma_2)) = \mathsf{G}(\sigma'; ([t_2] \otimes \sigma_2'); \sigma''; ([t_1] \otimes \sigma_1'); \sigma''').$$

The complete proof is shown in Figure 11. We briefly comment the critical steps:

Step 20: we have exploited axiom (Δ) and then the fact that symmetries on electrons and tokens are identities to transform the second subexpression;

Step 21: we have applied the naturality of symmetries to the first and second subexpressions – in order to match source and target of t_1 with the components of the symmetries, observe that $u_1 \otimes u^- = u_1^+ \otimes u^- \otimes u_1^-$ since u and u_1 are disjoint;

Step 22: we have used axiom (Δ) to transform the second and third subexpressions;

Step 23: we have applied the functoriality of the tensor product to the second and third subexpressions;

Step 24: we have applied the functoriality of the tensor product to the second subexpressions to reverse the order in which t_2 and t_1 appear in the previous expressions;

Step 25: we have used axiom (Δ) to reduce the second and third subexpressions;

Step 26: we have applied the naturality of symmetries twice to expand the third subexpression;

Step 27: we have used axiom (Δ) and then the fact that symmetries on electrons are identities to transform the first, third and fifth subexpressions.

The fact that G is full follows from Propositions 4 and 7, since $\mathsf{G}([t]) = id_{\varsigma(t)^+} \otimes t$.

Finally, regarding faithfulness, let P_0 and P_1 be such that $\mathsf{G}(P_0) = \mathsf{G}(P_1)$, and let α be a term representing $\mathsf{G}(P_0)$. Observe, by simply inspecting the axioms that define $\mathcal{B}(N)$, that all the possible choices for α have the same number of transitions. More precisely, exactly same transitions occur in each term obtained by rewriting α according to such axioms. Moreover, by definition of G, these are in one-to-one correspondence with the transitions of P_0 and with those of P_1. We can therefore proceed by induction on the number n of transitions of α (and P_0 and P_1) to prove that P_0 and P_1 are isomorphic processes.

The base case, where n equals zero, is obvious, as α is simply a permutation. For the induction case, let fix any decomposition of α according to Proposition 7, say

$$\alpha = \sigma_0; (id_{\varsigma(t_1)^+} \otimes t_1 \otimes id_{u_1}); \sigma_1; (id_{\varsigma(t_2)^+} \otimes t_2 \otimes id_{u_2}); \sigma_2; \ldots;$$
$$(id_{\varsigma(t_n)^+} \otimes t_n \otimes id_{u_n}); \sigma_n,$$

An argument similar to the one employed to establish the well-definedness of G, but working in the opposite direction, proves that all the steps needed to transform α in the normal form selected above can be mimicked both on P_0 and P_1. It then follows that P_i, for $i = 0, 1$, can be written as $P_i = P_i'; \sigma_i; ([t_n] \otimes \sigma_i'); \sigma_i''$, where $\mathsf{G}(P_0') = \mathsf{G}(P_1')$. Then, by induction hypothesis, we can conclude that P_0' and P_1' are isomorphic processes. It is then easy to prove that so are P_0 and P_1.   ◆

Besides the fact that *all* the arrows of $\mathcal{B}(N)$ have a meaningful computational interpretation, a further advantage of the present approach with respect

$$G(([t_1] \otimes \sigma_1); \sigma; ([t_2] \otimes \sigma_2)) =$$

$$= (id_{u^+ \otimes u_1^+} \otimes t_1 \otimes id_{u_2 \otimes \partial_0(t_2) \otimes v}); (id_u \otimes \gamma_{u_1 \otimes \partial_1(t_1), u_2 \otimes \partial_0(t_2)} \otimes id_v);$$

$$(id_{u^+ \otimes u_2^+} \otimes t_2 \otimes id_{u_1 \otimes \partial_1(t_1) \otimes v})$$

$$= (id_{u^+ \otimes u_1^+} \otimes t_1 \otimes id_{u_2 \otimes \partial_0(t_2) \otimes v}); (id_{u^+} \otimes \gamma_{u_1 \otimes u^- \otimes \partial_1(t_1), u_2 \otimes \partial_0(t_2)} \otimes id_v);$$

$$(id_{u^+ \otimes u_2^+} \otimes t_2 \otimes id_{u_1 \otimes \partial_1(t_1) \otimes v}) \tag{20}$$

$$= (id_{u^+} \otimes \gamma_{u_1 \otimes u^- \otimes \partial_0(t_1), u_2 \otimes \partial_0(t_2)} \otimes id_v); (id_{u^+ \otimes u_2 \otimes \partial_0(t_2) \otimes u_1^+} \otimes t_1 \otimes id_v);$$

$$(id_{u^+ \otimes u_2^+} \otimes t_2 \otimes id_{u_1 \otimes \partial_1(t_1) \otimes v}) \tag{21}$$

$$= (id_{u^+} \otimes \gamma_{u_1 \otimes u^- \otimes \partial_0(t_1), u_2 \otimes \partial_0(t_2)} \otimes id_v); (id_{u^2 \otimes u_2^+ \otimes u^- \otimes u_2^- \otimes \partial_0(t_2) \otimes u_1^+} \otimes t_1 \otimes id_v);$$

$$(id_{u^2 \otimes u_2^+} \otimes t_2 \otimes id_{u_1^+ \otimes u^- \otimes u_1^- \otimes \partial_1(t_1) \otimes v}) \tag{22}$$

$$= (id_{u^+} \otimes \gamma_{u_1 \otimes u^- \otimes \partial_0(t_1), u_2 \otimes \partial_0(t_2)} \otimes id_v); (id_{u^2 \otimes u_2^+} \otimes t_2 \otimes id_{u_1^+} \otimes t_1 \otimes id_v); \tag{23}$$

$$= (id_{u^+} \otimes \gamma_{u_1 \otimes u^- \otimes \partial_0(t_1), u_2 \otimes \partial_0(t_2)} \otimes id_v); (id_{u^2 \otimes u_2^+} \otimes t_2 \otimes id_{u_1^+ \otimes u^- \otimes u_1^- \otimes \partial_0(t_1) \otimes v});$$

$$(id_{u^2 \otimes u_2^+ \otimes u^- \otimes u_2^- \otimes \partial_1(t_2) \otimes u_1^+} \otimes t_1 \otimes id_v); \tag{24}$$

$$= (id_{u^+} \otimes \gamma_{u_1 \otimes u^- \otimes \partial_0(t_1), u_2 \otimes \partial_0(t_2)} \otimes id_v); (id_{u^+ \otimes u_2^+} \otimes t_2 \otimes id_{u_1 \otimes \partial_0(t_1) \otimes v});$$

$$(id_{u^+ \otimes u_2 \otimes \partial_1(t_2) \otimes u_1^+} \otimes t_1 \otimes id_v); \tag{25}$$

$$= (id_{u^+} \otimes \gamma_{u_1 \otimes u^- \otimes \partial_0(t_1), u_2 \otimes \partial_0(t_2)} \otimes id_v); (id_{u^+ \otimes u_2^+} \otimes t_2 \otimes id_{u_1 \otimes \partial_0(t_1) \otimes v});$$

$$(id_{u^+} \otimes \gamma_{u_2 \otimes \partial_1(t_2), u_1 \otimes u^- \otimes \partial_0(t_1)} \otimes id_v); (id_{u^+ \otimes u_1^+} \otimes t_1 \otimes id_{u_2 \otimes \partial_1(t_2) \otimes v});$$

$$(id_{u^+} \otimes \gamma_{u_1 \otimes u^- \otimes \partial_1(t_1), u_2 \otimes \partial_1(t_2)} \otimes id_v); \tag{26}$$

$$= (id_u \otimes \gamma_{u_1 \otimes \partial_0(t_1), u_2 \otimes \partial_0(t_2)} \otimes id_v); (id_{u^+ \otimes u_2^+} \otimes t_2 \otimes id_{u_1 \otimes \partial_0(t_1) \otimes v});$$

$$(id_u \otimes \gamma_{u_2 \otimes \partial_1(t_2), u_1 \otimes \partial_0(t_1)} \otimes id_v); (id_{u^+ \otimes u_1^+} \otimes t_1 \otimes id_{u_2 \otimes \partial_1(t_2) \otimes v});$$

$$(id_u \otimes \gamma_{u_1 \otimes \partial_1(t_1), u_2 \otimes \partial_1(t_2)} \otimes id_v); \tag{27}$$

$$= G(\sigma'; ([t_2] \otimes \sigma_2'); \sigma''; ([t_1] \otimes \sigma_1'))$$

Fig. 11. The proof of $G(([t_1] \otimes \sigma_1); \sigma; ([t_2] \otimes \sigma_2)) = G(\sigma'; ([t_2] \otimes \sigma_2'); \sigma''; ([t_1] \otimes \sigma_1'))$.

to the match-share categories of [16] is that the arrows of the model category corresponding to pure concatenable process can be distinguished *just* by looking at their sources and targets, rather than by inspecting their construction. And as for the CTph, our proposal is a conservative extension of the ordinary concatenable process semantics (cf. § 1.2 and [12,31]).

Corollary 6. *If N is a PT net, then $\mathcal{P}(N)$ is a full subcategory of $\mathcal{B}(N)$.*

Moreover, the present axiomatics of $\mathcal{B}(N)$ improves sensibly the construction presented in [7]. In particular, the monoid of objects is here 'morally' commutative, thus making redundant the idea of instances of transitions and the related axioms [32,7]. Moreover, the exact sharing hypothesis has found a mature, satisfactory formulation in terms of law (Δ) which, among other things, allowed us to dispense with the particles a_-.

Concluding Remarks and Future Work

Building on an important suggestion of Meseguer in [22], we have shown a way to extend the algebraic semantics of PT nets proposed in [24] to contextual nets, both in the collective token and the individual token interpretation. The constructions rely on the choice of a non-free monoid of objects, whose elements we called molecules and bimolecules. In the case of the collective token philosophy, our work extends Meseguer's by identifying the *maximum sharing hypothesis* as the fundamental law of collective contextual processes. The key to transport these ideas to the individual token philosophy was to renounce to the symmetry of the monoidal category, being thus able to select only the symmetries consistent with our computational interpretation in terms of concatenable contextual processes. The axioms of *exact sharing* provided us with a way to regulate the interplay between all the different ingredients.

Although we have worked only at the level of single nets, we believe that our approach can be extended to constructions between categories of nets and models, with restrictions analogous to those well-known in the literature [31,32].

As one of the anonymous referees suggested, it would be interesting to apply our algebraic approach to high level Petri nets. In fact, these are often used for modeling programming languages where expressions can involve several variables *read but not modified*, so that in the computational analysis of the associated nets it would be important to understand the maximum degree of parallelism allowed in complex steps. Since the definition of high level nets has algebraic foundations, we think that our approach could be extended to that framework, but this is outside the scope of the present paper and left for future work.

References

1. P. Baldan. Modelling concurrent computations: From contextual Petri nets to graph grammars. Ph.D. thesis, TD-1/00, Dipartimento di Informatica, Università di Pisa, 2000.
2. P. Baldan, A. Corradini, and U. Montanari. An event structure semantics for P/T contextual nets: Asymmetric event structures. In *Proc. FoSSaCS'98, Foundations of Software Science and Computation Structures*, (M. Nivat, Ed.), vol. 1378 of *Lect. Notes in Comput. Sci.*, pp. 63–80. Springer, 1998.
3. E. Best and R. Devillers. Sequential and concurrent behaviour in Petri net theory. *Theoretical Computer Science*, 55:87–136, 1987.
4. R. Bruni. Tile Logic for Synchronized Rewriting of Concurrent Systems. Ph.D. thesis, TD-1/99, Dipartimento di Informatica, Università di Pisa, 1999.
5. R. Bruni, J. Meseguer, U. Montanari, and V. Sassone. A comparison of Petri net semantics under the collective token philosophy. In *Proc. ASIAN'98, 4th Asian Computing Science Conference*, (J. Hsiang, A. Ohori, Eds.), vol. 1538 of *Lect. Notes in Comput. Sci.*, pp. 225–244. Springer, 1998.

6. R. Bruni, J. Meseguer, U. Montanari, and V. Sassone. Functorial semantics for Petri nets under the individual token philosophy. In *Proc. CTCS'99, 8th conference on Category Theory and Computer Science*, (M. Hofmann, G. Rosolini, D. Pavlovic, Eds.), vol. 29 of *Elect. Notes in Comput. Sci.*, 19 pages. Elsevier Science, 1999.

7. R. Bruni and V. Sassone. Algebraic models for contextual nets. In *ICALP2000, 27th Int. Coll. on Automata, Languages and Programming*, (U. Montanari, J. Rolim, E. Welzl, Eds.), vol. 1853 of *Lect. Notes in Comput. Sci.*, pp. 175–186. Springer, 2000.

8. N. Busi and M. Pinna. Non sequential semantics for contextual P/T nets. In *Application and Theory of Petri Nets*, vol. 1091 of *Lect. Notes in Comput. Sci.*, pp. 113–132. Springer, 1996.

9. N. Busi. Petri nets with inhibitor and read arcs: Semantics, analysis and application to process calculi. Ph.D. thesis, TD-1/99, Dipartimento di Informatica, Università di Siena, 1998.

10. S. Christensen and N.D. Hansen. Coloured Petri nets extended with place capacities, test arcs and inhibitor arcs. In *ICATPN'93, 14th Int. Conf. Applications and Theory of Petri Nets*, (M.A. Marsan, Ed.), vol. 691 of *Lect. Notes in Comput. Sci.*, pp. 186–205. Springer, 1993.

11. A. Corradini and U. Montanari. An algebraic semantics for structured transition systems and its application to logic programs. *Theoretical Computer Science*, 103:51–106, 1992.

12. P. Degano, J. Meseguer, and U. Montanari. Axiomatizing net computations and processes. In *Proc. LICS'89, 4th Symposium on Logic in Computer Science*, pp. 175–185. IEEE Computer Society Press, 1989.

13. P. Degano, J. Meseguer, and U. Montanari. Axiomatizing the algebra of net computations and processes. *Acta Inform.*, 33(7):641–667, 1996.

14. J. Desel, G. Juhás, and R. Lorenz. Process semantics of Petri nets over partial algebra. In *Proc. ICATPN 2000, 21st Int. Conf. on Application and Theory of Petri Nets*, (M. Nielsen, D. Simpson, Eds.), vol. 1825 of *Lect. Notes in Comput. Sci.*, pp. 146–165. Springer, 2000.

15. H. Ehrig and J. Padberg. Uniform approach to Petri nets. In *Proc. Foundations of Computer Science: Potential-Theory-Cognition*, (C. Freska, M. Jantzen, R. Valk, Eds.), vol. 1337 of *Lect. Notes in Comput. Sci.*, pp. 219–231. Springer, 1997.

16. F. Gadducci and U. Montanari. Axioms for contextual net processes. In *Proc. ICALP'98, 25th International Colloquium on Automata, Languages, and Programming*, (K.G. Larsen, S. Skyum, G. Winskel, Eds.), vol. 1443 of *Lect. Notes in Comput. Sci.*, pp. 296–308. Springer, 1998.

17. F. Gadducci and U. Montanari. The tile model. In *Proof, Language and Interaction: Essays in Honour of Robin Milner*, (G. Plotkin, C. Stirling, and M. Tofte, Eds.). MIT Press, 2000.

18. R.J. van Glabbeek and G.D. Plotkin. Configuration structures. In *Proc. LICS'95, 10th Symposium on Logic in Computer Science*, pp. 199–209. IEEE Computer Society Press, 1995.

19. U. Goltz and W. Reisig. The non-sequential behaviour of Petri nets. *Inform. and Comput.*, 57:125–147, 1983.

20. R. Janicki and M. Koutny. Semantics of inhibitor nets. *Inform. and Comput.*, 123:1–16, 1995.

21. J. Meseguer, Conditional rewriting logic as a unified model of concurrency, *Theoretical Computer Science*, 96:73–155, 1992.

22. J. Meseguer. Rewriting logic as a semantic framework for concurrency: A progress report. In *Proc. CONCUR'96, 7th International Conference on Concurrency Theory*, (U. Montanari, V. Sassone, Eds.), vol. 1119 of *Lect. Notes in Comput. Sci.*, pp. 331–372. Springer, 1996.
23. J. Meseguer and U. Montanari. Petri nets are monoids: A new algebraic foundation for net theory. In *Proc. LICS'89, 3rd Symposium on Logic in Computer Science*, pp. 155–164. IEEE Computer Society Press, 1988.
24. J. Meseguer and U. Montanari. Petri nets are monoids. *Inform. and Comput.*, 88(2):105–155, 1990.
25. J. Meseguer, U. Montanari, and V. Sassone. On the semantics of place/transition Petri nets. *Math. Struct. in Computer Science*, 7:359–397, 1997.
26. U. Montanari and F. Rossi. Contextual occurrence nets and concurrent constraint programming. In *Graph Transformations in Computer Science*, vol. 776 of *Lect. Notes in Comput. Sci.*, pp. 280–285. Springer, 1994.
27. U. Montanari and F. Rossi. Contextual nets. *Acta Inform.*, 32:545–596, 1995.
28. C.A. Petri. *Kommunikation mit Automaten*. Ph.D. thesis, Institut für Instrumentelle Mathematik, Bonn, 1962.
29. W. Reisig. *Petri Nets: An Introduction*. EACTS Monographs on Theoretical Computer Science. Springer, 1985.
30. G. Ristori. *Modelling Systems with Shared Resources via Petri Nets*. Ph.D. thesis, TD-5/94, Dipartimento di Informatica, Università di Pisa, 1994.
31. V. Sassone. An axiomatization of the algebra of Petri net concatenable processes. *Theoretical Computer Science*, 170:277–296, 1996.
32. V. Sassone. An axiomatization of the category of Petri net computations. *Math. Struct. in Computer Science*, 8:117–151, 1998.
33. M.-O. Stehr, J. Meseguer, and P.C. Ölveczky. Rewriting logic as a unifying framework for Petri nets. This Volume.
34. W. Vogler. Efficiency of asynchronous systems and read arcs in Petri nets. In *Proc. ICALP'97, 24th International Colloquium on Automata, Languages, and Programming*, vol. 1256 of *Lect. Notes in Comput. Sci.*, pp. 538–548. Springer, 1997.
35. W. Vogler. Partial order semantics and read arcs. In *Proc. MFCS'97, 22nd Int. Symp. on Mathematical Foundations of Computer Science*, (P. Degano, R. Gorrieri, A. Marchetti-Spaccamela, Eds.), vol. 1295 of *Lect. Notes in Comput. Sci.*, pp. 508–517. Springer, 1997.
36. G. Winskel. Event structures. In *Proc. of Advanced Course on Petri Nets*, vol. 255 of *Lect. Notes in Comput. Sci.*, pp. 325–392. Springer, 1986.

Continuous Petri Nets and Transition Systems

Manfred Droste[1] and R.M. Shortt[†2]

[1] Institut für Algebra, Technische Universität Dresden, 01062 Dresden, Germany
droste@math.tu-dresden.de
[2] Department of Mathematics, Wesleyan University, Middletown, CT 06459, USA

Abstract. In many systems, the values of finitely many parameters can be influenced in a continuous way by controls acting with possibly varying strength over intervals of time. For this, we present general models of continuous Petri nets and of continuous transition systems with situation-dependent concurrency. With a suitable concept of morphisms, we obtain a categorial adjunction between these two models, and often even a coreflection. This shows that the concept of regions is also applicable in this continuous setting. Finally, we prove that our categories of continuous Petri nets and of continuous automata with concurrency have products and conditional coproducts.

1 Introduction

Petri nets have been a successful model for analyzing the behaviour of a variety of systems. Typically, these are described by finitely many parameters (conditions) which can be influenced and acted upon in finitely many ways. Whereas mostly only discrete values for the parameters and strength of the actions are considered, there is also widespread and growing interest in models with continuous parameter values or actions taking place over intervals of time instead of instantaneously, cf., e.g., [6] for continuous Petri nets with varying maximal speeds. Moreover, hybrid models containing both discrete and continuous components have been investigated, see, e.g., [7] for differential Petri nets, or [1] for decidability results on linear hybrid systems. For surveys, many practical examples and much research on such continuous and timed Petri nets, we refer the reader to [5, 2] and the present Petri net series.

It is the first goal of this paper to present a very general model for systems where the parameters can take on continuous values and the strength of the actions can be continuously distributed over intervals of time, both individually and in consort (concurrently). In examples, often there are finitely many controls which can influence the finitely many parameters of the system in such a continuous way, but here we admit also infinitely many parameters and controls. The static structure of the Petri net itself will be the same as for discrete place/transitions systems. The distribution of the strength of a control over an interval of time will be given by a control path, for which we admit piecewise continuous and even just measurable functions. The total effect of this control path on a parameter is obtained by a simple integration process. The whole dynamic

H. Ehrig et al. (Eds.): Unifying Petri Nets, LNCS 2128, pp. 457–484, 2001.
© Springer-Verlag Berlin Heidelberg 2001

behaviour of the system, i.e. the continuous change of the markings under the influence of control paths, leads to the model of a continuous transition system with concurrency. Here we only consider binary concurrency of control paths, but we allow auto-concurrency.

The main goal of this paper is to investigate the relationship between these continuous Petri nets and properties of their dynamic behaviour, as given by the associated continuous transition system. Indeed, we wish to show that techniques from the discrete case also apply here. This is the theory of regions, first used for describing the relationship between elementary Petri nets and classes of transition systems by Ehrenfeucht and Rozenberg [11], Nielsen, Rozenberg and Thiagarajan [13], Badouel and Darondeau [3] and Winskel and Nielsen [15] and for more general nets in Mukund [12] and in [9, 10]; for a survey, see [4].

Now we give an outline of this paper. First we develop basic results on the control paths involved (this requires only very basic integration theory of measurable functions). Then we present the model of continuous Petri nets, their dynamics and the associated automata with concurrency. Concurrency aspects will be illustrated by a simple example from engineering. We will define suitable morphisms for our continuous Petri nets and continuous automata with concurrency, and we will show that these two categories are related by an adjunction. We will characterize the continuous automata arising from given continuous Petri nets algebraically, using the concept of regions. We will also show that the subcategory of such automata is related to the continuous Petri nets by a coreflection. Finally, we use this coreflection result to show that these categories have products and conditional coproducts. We close with a comparison of our continuous model to related results for discrete Petri nets and automata with concurrency relations.

Acknowledgement. Rae Shortt died on July 11, 1999. I appreciated and miss him both as a friend and co-worker. M.D.

2   Control Paths

Let E be a non-empty set and let $\ell(E)$ be the linear subspace of $\mathbb{R}^E$ comprising all vectors whose components form an absolutely summable sequence. If $e \in E$ and $u \in \mathbb{R}^E$ then the eth component of u is denoted $\langle u, e \rangle$. Thus $\ell(E)$ is the set of all $u \in \mathbb{R}^E$ such that the norm

$$\|u\| = \sum_e |\langle u, e \rangle|$$

is finite. We note that in order for this sum to be finite, we must have $\langle u, e \rangle = 0$ for all but countably many elements $e \in E$ (since any uncountable sum of strictly positive reals is infinite).

The space of all bounded real functions on E is denoted $\ell^\infty(E)$. Thus $\ell^\infty(E)$ is the set of all $u \in \mathbb{R}^E$ such that

$$\|u\|_\infty = \sup_e |\langle u, e\rangle|$$

is finite. Then $\ell(E) \subseteq \ell^\infty(E)$. Also, we define the positive cone $\ell_+^\infty(E) = \{u \in \ell^\infty(E); \langle u, e\rangle \geq 0 \text{ for all } e \in E\}$ and $\ell_+(E) = \ell(E) \cap \ell_+^\infty(E)$.

We now let $\langle ., .\rangle : \ell^\infty(E) \times \ell(E) \longrightarrow \mathbb{R}$ be the usual bilinear pairing, that is $\langle u, v\rangle = \sum_e u(e) \cdot v(e)$. If now $e \in E$, then we define $\bar{e} \in \ell(E)$ to be the function given by

$$\bar{e}(e') = \begin{cases} 1 \text{ if } e' = e \\ 0 \text{ if } e' \neq e \end{cases}$$

Then, by abuse of notation, we write simply e for both e and $\bar{e}$. Thus, if $u \in \ell^\infty(E)$ and $v \in \ell(E)$, then

$$\langle u, v\rangle = \sum_e \langle u, e\rangle \langle v, e\rangle ,$$

which is consistent with the notation $\langle u, e\rangle$ introduced above. Hölder's Inequality implies that this sum is finite.

Given sets E and E', we see that every function $\eta : E \to E'$ induces a function $\bar{\eta} : \ell(E) \to \ell(E')$ defined by

$$\langle \bar{\eta}(u), e'\rangle = \sideset{}{'}\sum \langle u, e\rangle , \tag{1}$$

where the sum is taken over all $e \in E$ such that $\eta(e) = e'$ and is zero if there is no such e. It is easy to see that $\bar{\eta}$ is a linear transformation such that $\|\bar{\eta}(u)\| \leq \|u\|$ for all $u \in \ell(E)$. If $u \in \ell_+(E)$, then $\bar{\eta}(u) \in \ell_+(E')$ and $\|\bar{\eta}(u)\| = \|u\|$. Also, $\bar{\eta}(e) = \eta(e) \in \ell_+(E')$ for any $e \in E$. We also remark that if $\eta' : E' \to E''$ is another function, then $\overline{\eta' \circ \eta} = \bar{\eta}' \circ \bar{\eta}$, as is easy to check.

We now remark that if $\eta : E \to E'$, $u \in \ell^\infty(E')$ and $v \in \ell(E)$, then

$$\langle u, \bar{\eta}(v)\rangle = \langle u \circ \eta, v\rangle . \tag{2}$$

For the application we consider, E will represent a set of controls operating on a (physical or information-theoretic) system. These controls can be applied either singly or in consort, and each control can act with an adjustable intensity: thus, if $e_1, e_2, \ldots, e_n$ are elements of E, then the linear combination

$$t_1 e_1 + t_2 e_2 + \ldots + t_n e_n \quad (t_1 \geq 0, \ldots, t_n \geq 0)$$

represents the joint, simultaneous application of the controls $e_1, \ldots, e_n$ with respective intensities $t_1, \ldots, t_n$. Certain infinite bounded linear combinations of controls are also allowed: in fact, every element $u \in \ell(E)$ with $\langle u, e\rangle \geq 0$ for all $e \in E$ corresponds to a particular disposition of the controls, and conversely, each such "control profile" is represented by a positive function $u \in \ell(E)$.

The state of a system will change with time, and the control profile can likewise be altered over time. The changing control profiles are given by the "control paths". A function $c : \mathbb{R} \to \ell(E)$ is a *control path* if

(i) $c(t) \geq 0$ for $t \geq 0$, and $c(t) = 0$ for $t < 0$;
(ii) the quantity $x(c) = \sup \{t : c(t) \neq 0\}$ is finite
(iii) there is some $K \geq 0$ such that $\|c(t)\| \leq K$ for all $t \geq 0$;
(iv) the function $c : \mathbb{R} \to \ell_+(E)$ is weakly measurable, i.e. for each $e \in E$, the real-valued function on $\mathbb{R}$ given by $t \to \langle c(t), e \rangle$ is (Lebesgue) measurable: all continuous or piecewise continuous functions satisfy this mild condition.

We denote by $\mathcal{C}(E)$ the set of all control paths for E. It is assumed that each control path accomplishes its "task" (i.e. its effect upon the system) in a finite amount of time: thus $c(t) = 0$ for each t outside the interval $[0, x(c)]$ (condition (ii)) during which c acts. Condition (iii) means that the control profiles $c(t)$ are uniformly bounded for $t \in \mathbb{R}$. If $c : \mathbb{R} \to \ell(E)$ is a control path, then $w(c) : \mathbb{R} \to \ell(E)$ is the *weight* function defined by

$$\langle w(c)(t), e \rangle = \int_0^t \langle c(u), e \rangle \, du.$$

Using the boundedness condition (iii) for control paths, we check that

$$\sum_e |\langle w(c)(t), e \rangle| = \sum_e \left| \int_0^t \langle c(u), e \rangle \, du \right|$$
$$\leq \int_0^{x(c)} \sum_e |\langle c(u), e \rangle| \, du$$
$$= \int_0^{x(c)} \|c(u)\| \, du \leq K x(c)$$

so that $\|w(c)(t)\| \leq K x(c)$ for all $t \geq 0$. The function $w(c)(t)$ measures the cumulative effect of the control path c over the time interval $[0, t]$. Again, $w(c)(t) \in \ell_+(E)$. We define the *total weight* of c to be the vector $W(c) = \lim_{t \to \infty} w(c)(t)$ in $\ell_+(E)$.

The following result on the transformation of control paths will be essential for our morphisms both of continuous Petri nets and of continuous automata with concurrency.

Proposition 2.1. *Let $\eta : E \to E'$ be a function which is countable-to-one (i.e. $\eta^{-1}(e')$ is countable for each $e' \in E'$). Let $c \in \mathcal{C}(E)$ be a control path for E. Then $\bar{\eta} \circ c$ is a control path for E', and $w(\bar{\eta} \circ c) = \bar{\eta} \circ w(c)$.*

Proof. For each $t \geq 0$ and $e' \in E'$, we have, by (1)

$$\langle \bar{\eta}(c(t)), e' \rangle = \sum{}' \langle c(t), e \rangle,$$

where the sum $\sum'$ is taken over all $e \in E$ such that $\eta(e) = e'$. Since η is countable-to-one, this sum is countable. Since c is a control path, each of the functions

$t \to \langle c(t), e \rangle$ is measurable, and therefore so, too, is the function $t \to \langle \bar{\eta}(c(t)), e' \rangle$. The other requirements for a control path are easily verified; in particular, we have $\|\bar{\eta}(c(t))\| = \|c(t)\| \leq K$.

Next we verify the equality $w(\bar{\eta} \circ c) = \bar{\eta} \circ w(c)$. For each $e' \in E'$ and $t \geq 0$, we have

$$
\begin{aligned}
\langle w(\bar{\eta} \circ c)(t), e' \rangle &= \int_0^t \langle \bar{\eta}(c(s)), e' \rangle \, ds \\
&= \int_0^t {\sum}' \langle c(s), e \rangle \, ds \quad \text{(formula (1))} \\
&= {\sum}' \int_0^t \langle c(s), e \rangle \, ds \quad \text{(Beppo-Levi)} \\
&= {\sum}' \langle w(c)(t), e \rangle \\
&= \langle \bar{\eta}(w(c)(t)), e' \rangle,
\end{aligned}
$$

so that $w(\bar{\eta} \circ c) = \bar{\eta} \circ (w(c))$, as asserted. Again we note that since η is countable-to-one, each of the sums $\sum'$ is also countable: this allowed the exchange of summation and integral above.   □

Trivially, the requirement that $\eta : E \to E'$ be countable-to-one is satisfied if, for instance, E is countable. If $c : \mathbb{R} \to \ell(E)$ is a control path and $s \geq 0$, then $c_s : \mathbb{R} \to \ell(E)$ is the control path defined by $c_s(t) = c(t - s)$. Thus, c_s is the control path c "delayed" by s units of time. Also, $c^t : \mathbb{R} \to \ell(E)$ is the control path defined by

$$
c^t(u) = \begin{cases} c(u) & \text{for } u \leq t \\ 0 & \text{for } u > t; \end{cases}
$$

Thus, c^t is the control path c "cut off" after t units of time.

If c and c' are control paths in $\mathcal{C}(E)$, then so is their point-wise sum $c + c'$. If $r \geq 0$, then rc is again a control path. We see that $\mathcal{C}(E)$ is actually the positive cone of a partially ordered linear space. (We shall not have occasion to use this fact.)

For each $e \in E$, we define a control path c^e by

$$
c^e(t) = \begin{cases} e & \text{if } 0 \leq t \leq 1 \\ 0 & \text{if } t > 1 \text{ or } t < 0 \end{cases}
$$

The control path c^e corresponds to a "pure" application of the control e within the unit interval $[0, 1]$ with all other controls set at 0. Recalling the identification of e with $\bar{e} \in l(E)$, we have then

$$
\begin{aligned}
w(c^e)(t) &= \begin{cases} te & \text{if } 0 \leq t \leq 1 \\ e & \text{if } t > 1 \end{cases} \\
W(c^e) &= e.
\end{aligned}
$$

3   Continuous Petri Nets and Concurrency

Classically, a *Petri net* (*PN*) is defined as a quartuple (P, E, F, M_{in}) such that

1. P and E are disjoint sets;
2. $F = (F_1, F_2)$, where $F_i : P \times E \to \mathbb{N}$ are arbitary functions $(i = 1, 2)$;
3. $M_{in} : P \to \mathbb{N}$ is an arbitrary function.

Graphically, the elements of P and E are pictured as circles and rectangles, respectively. Each function $M : P \to \mathbb{N}$, called a *marking*, represents a possible distribution of tokens or marks on the "places" in P: there are $M(p)$-many tokens on circle p. The marking M_{in} gives an initial distribution of tokens. The number $F_1(p, e)$ [resp. $F_2(p, e)$] indicates the number of tokens taken from [resp. added, to] circle p when the event $e \in E$ occurs.

More precisely, an event $e \in E$ can *occur* or *fire* at a marking M and lead to the marking $M' = Me$, also denoted by $M \xrightarrow{e} M'$, if $M(p) \geq F_1(p, e)$ and $M'(p) = M(p) - F_1(p, e) + F_2(p, e)$ for each $p \in P$.

Thus, the initial distribution M_{in} of tokens is altered by a sequence of events $e_1, e_2, \ldots$ drawn from E. In this, the discrete, classical model, the quantities involved are integral, and the evolution of the system sequential.

Usually, in place/transition systems, the letter T is used in place of E, which is employed in condition/event systems. However, here we use the letter T later for the transitions of an automaton. We turn now to a continuous model of such nets.

Definition 3.1. A tuple $\mathcal{N} = (P, E, F, M_{in})$ is a *continuous Petri net* (CPN for short) if

1. P and E are disjoint, non-empty sets;
2. $F = (F_1, F_2)$, with functions $F_i : P \to \ell_+^\infty(E)$ $(i = 1, 2)$; we also write $F(p) = (F_1(p), F_2(p))$ for all $p \in P$;
3. $M_{in} : P \to \mathbb{R}_+$ is a given non-negative function.

The elements of P represent the various "substances" that the system keeps track of, e.g. the various chemicals present in an organic cell whose concentrations collectively describe the state of the cell at a given time. In a classical Petri net, these are the "conditions" or "places".

The numbers $\langle F_1(p), e \rangle$ [resp. $\langle F_2(p), e \rangle$] represent the amount by which the substance p is decreased [resp. increased] through action of the control e over a unit of time. In the cellular model, these are given by concentrations of chemicals p decreased [resp. increased] through e. In the classical Petri net, $\langle F_1(p), e \rangle$ [resp. $\langle F_2(p), e \rangle$] is the number of tokens to be removed from [resp. added to] the circle p when "event" e "fires": each e is represented as a rectangle, and the numbers $\langle F_i(p), e \rangle$ are attached to arrows between circle p and rectangle e, cf. Fig. 1.

If $\mathcal{N}$ is a CPN as above, then a non-negative function $M : P \to \mathbb{R}_+$ is called a *marking* of $\mathcal{N}$. Each marking represents a possible state of the system where the amount of substance p (concentration of chemical p, tokens on circle p ...) is

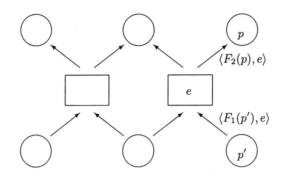

Fig. 1. Continuous Petri net

$M(p)$. The function M_{in} is the *initial marking*, i.e. the initial state of the system. We let

$$\mathcal{M} = \{M : P \to \mathbb{R} : M(p) \geq 0 \text{ for all } p \in P\}$$

be the set of all markings of $\mathcal{N}$. We now examine how a control path operates so as to influence and change a marking of the net. For markings M and M' of $\mathcal{N}$ and a control path $c : \mathbb{R} \to \ell_+(E)$, we write $M \xrightarrow{c} M'$ and also $M' = Mc$ if

(a) for each $t \geq 0$, the function $M_t : P \to \mathbb{R}$ defined by

$$M_t(p) = M(p) - \langle F_1(p), w(c)(t) \rangle + \langle F_2(p), w(c)(t) \rangle \qquad (3)$$

is a marking of $\mathcal{N}$. (We note that this condition need only be checked for $t \leq x(c)$; for $t \geq x(c)$ the function $w(c)(t)$ is constant - the work of c is finished at time $t = x(c)$.) As c operates over time, the marking M is continually transformed, and M_t is the intermediate marking at time t.

(b) $M_t = M'(= Mc)$ for all $t \geq x(c)$. (The marking M' is the end result obtained by applying the control path c to M.)

(c) For each $p \in P$ and $t \geq 0$, there is some $\triangle t > 0$ such that

$$\langle F_1(p), w(c)(t + \triangle t) - w(c)(t) \rangle \leq M_t(p). \qquad (4)$$

(If the process governed by c is to begin removal of substance p at time t, then some sufficient quantities of this substance must already be present. This is similar to the requirement in the discrete case that $F_1(p, e) \leq M(p)$ in order that e can fire at marking M.)

We say that control paths c and c' are *concurrent at a marking* $M \in \mathcal{M}$ and write $c \|_M c'$ if for each $s \geq 0$, the markings $M(c + c'_s)$ and $M(c_s + c')$ exist. Taking $s \geq x(c)$ [resp. $s \geq x(c')$] then yields the existence of the marking Mcc' [resp. $Mc'c$]. Linearity of the map $c \to \langle c(t), e \rangle$ then yields the equalities $Mcc' = Mc'c = M(c + c'_s) = M(c_s + c')$ for all $s \geq 0$.

The existence of Mcc' indicates that the "substances" $p \in P$ are present in M in sufficient quantity so as to allow the application of c and then are likewise present in Mc so as to allow subsequent application of c'. A similar formulation applies for $Mc'c$. If c and c' are concurrent at M, then both c and c' may be applied in either order or simultaneously (as in $M(c + c')$) or one applied and then, after a delay, the other (as in $M(c+c_s')$ or $M(c_s+c')$), with the same result in each case. We note that in our formulation, auto-concurrency is allowed: it can happen that $c \|_M c$ for certain control paths c.

Given a control $e \in E$, we define a marking $M^e \in \mathcal{M}$, putting $M^e(p) = \langle F_1(p), e \rangle$. It is easily verified that the marking $M^e c^e$ exists. Thus, each "pure" control path c^e can operate non-trivially on some state of the system.

It is useful to view the continuous Petri nets from the standpoint of category theory. We now describe the corresponding morphisms. Let $\mathcal{N} = (P, E, F, M_{in})$ and $\mathcal{N}' = (P', E', F', M_{in}')$ be CPN's. A CPN-*morphism* from $\mathcal{N}$ to $\mathcal{N}'$ is a pair (π, η), where $\pi : P' \to P$ and $\eta : E \to E'$ are functions with η countable-to-one such that $M_{in}' = M_{in} \circ \pi$ and $F_i'(p') \circ \eta = F_i \circ \pi(p')$ for all $p' \in P'$ and $i = 1, 2$; i.e. $\langle F_i'(p'), \eta(e) \rangle = \langle F_i(\pi(p')), e \rangle$ for all $p' \in P'$, $e \in E$ and $i = 1, 2$. By formula (2), we obtain $\langle F_i'(p'), \overline{\eta}(v) \rangle = \langle F_i(\pi(p')), v \rangle$ for any $p' \in P'$, $v \in \ell(E)$. (Cf. Fig. 2)

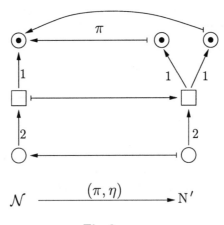

Fig. 2.

The requirement that η be countable-to-one is due to Proposition 2.1 which will be essential here and later on. The identity morphism on $\mathcal{N}$ is the pair (id_P, id_E), where, for any set Z, $id_Z : Z \to Z$ is the identity function $id_Z(z) = z$ on Z.

We note that formally these are precisely the same kind of morphisms as defined in [10] (and in [9]) for the classical discrete Petri nets. We will show that they are also a suitable concept for continuous Petri nets.

It is easy to verify the

Proposition 3.2. *Continuous Petri nets and their morphisms form a category in which the composition of two morphisms*

$$\mathcal{N} \xrightarrow{(\pi',\eta')} \mathcal{N}' \xrightarrow{(\pi'',\eta'')} \mathcal{N}''$$

is the morphism $\mathcal{N} \xrightarrow{(\pi,\eta)} \mathcal{N}''$, *where* $\pi = \pi' \circ \pi''$ *and* $\eta = \eta'' \circ \eta'$.

We let *CPN* be the category of continuous Petri nets and their morphisms. Next, we show that these morphisms preserve the dynamic behaviour of continuous Petri nets.

Proposition 3.3. *Let* $(\pi,\eta) : \mathcal{N} \to \mathcal{N}'$ *be a morphism between CPN's with* $\eta : E \to E'$ *countable-to-one. Then*

(a) *if* M *is a marking of* $\mathcal{N}$, *then* $M \circ \pi$ *is a marking of* $\mathcal{N}'$;
(b) *if* M *is a marking of* $\mathcal{N}$ *and* $c \in \mathcal{C}(E)$ *is such that* $M \xrightarrow{c} Mc$, *then*
   $$M \circ \pi \xrightarrow{\overline{\eta} \circ c} (M \circ \pi)\overline{\eta} \circ c = (Mc) \circ \pi;$$
(c) *if* $c \parallel_M c'$ *in* $\mathcal{N}$, *then* $\overline{\eta} \circ c \parallel_{M \circ \pi} \overline{\eta} \circ c'$ *in* $\mathcal{N}'$.

Proof. (a) Trivial.

(b) Note that $\overline{\eta} \circ c$ is a control path for $\mathcal{N}'$ due to Proposition 2.1. Also, $w(\overline{\eta} \circ c) = \overline{\eta} \circ w(c)$. Now consider the function $(M \circ \pi)_t : P' \to \mathbb{R}$ whose value at $p' \in P'$ is

$$
\begin{aligned}
(M \circ \pi)_t(p') &= M(\pi(p')) - \langle F_1'(p'), w(\overline{\eta} \circ c)(t) \rangle + \langle F_2'(p'), w(\overline{\eta} \circ c)(t) \rangle \\
&= M(\pi(p')) - \langle F_1'(p'), \overline{\eta}(w(c)(t)) \rangle + \langle F_2'(p'), \overline{\eta}(w(c)(t)) \rangle \\
&= M(\pi(p')) - \langle F_1'(\pi(p')), w(c)(t) \rangle + \langle F_2'(\pi(p')), w(c)(t) \rangle \\
&= M_t(\pi(p'));
\end{aligned}
$$

we have proved that $(M \circ \pi)_t = M_t \circ \pi$, so that $(M \circ \pi)_t$ is a marking of $\mathcal{N}'$ for each $t \geq 0$. Taking $t \geq x(c)$, we obtain $(M \circ \pi)(\overline{\eta} \circ c) = (Mc) \circ \pi$.

It remains for us to check that for each $t \in \mathbb{R}$, $p' \in P'$ and $\Delta t > 0$,

$$
\begin{aligned}
&\langle F_1'(p'), w(\overline{\eta} \circ c)(t + \Delta t) - w(\overline{\eta} \circ c)(t) \rangle \\
&= \langle F_1'(p'), \overline{\eta}(w(c)(t + \Delta t) - w(c)(t)) \rangle \\
&= \langle F_1(\pi(p')), w(c)(t + \Delta t) - w(c)(t) \rangle \, ;
\end{aligned}
$$

since $M \xrightarrow{c} Mc$, we have that there exists some $\Delta t > 0$ so that this quantity is at most $M_t(\pi(p')) = (M \circ \pi)_t(p')$, as required.

(c) Suppose that M is a marking for $\mathcal{N}$ and that c and c' are control paths of $\mathcal{N}$ such that $c \parallel_M c'$. Using parts (a) and (b) already established, we note the existence of the markings

$$
\begin{aligned}
(M \circ \pi)(\overline{\eta} \circ (c + c_s')) &= (M \circ \pi)((\overline{\eta} \circ c) + (\overline{\eta} \circ c_s')) \\
(M \circ \pi)(\overline{\eta} \circ (c_s + c')) &= (M \circ \pi)((\overline{\eta} \circ c_s) + (\overline{\eta} \circ c'))
\end{aligned}
$$

for all $s \geq 0$.

$\square$

We now consider the connection between transitions $M \xrightarrow{e} M'$ for markings in a classical Petri net and the control path transitions $M \xrightarrow{c} M'$ described above. To each transition event e we associate the "pure" control path c^e defined in section 2. Then for $0 \leq t \leq 1$, we have

$$M_t(p) = M(p) - \langle F_1(p), w(c^e)(t) \rangle + \langle F_2(p), w(c^e)(t) \rangle$$
$$= M(p) - \langle F_1(p), te \rangle + \langle F_2(p), te \rangle$$
$$= M(p) - t \langle F_1(p), e \rangle + t \langle F_2(p), e \rangle \, ;$$

putting $t \geq 1$ yields $M_1(p) = M'(p)$, as in the classical case. Since $M \xrightarrow{e} M'$ in the classical net, we have $\langle F_1(p), e \rangle \leq M(p)$; it follows that for $\Delta t = 1 - t > 0$, we have

$$\langle F_1(p), w(c^e)(t + \Delta t) - w(c^e)(t) \rangle$$
$$= \Delta t \langle F_1(p), e \rangle = \langle F_1(p), e \rangle - t \langle F_1(p), e \rangle$$
$$\leq M(p) - t \langle F_1(p), e \rangle \leq M_t(p),$$

this yields $M \xrightarrow{c^e} Mc^e = M'$.

In this way, we see how classical Petri net transitions can be modelled by CPN transitions $M \xrightarrow{c} M'$. The latter transitions have the advantage that the quantities $M_t(p)$ are always continuous functions of t. After executing them for one unit of time, we have obtained the same marking as in the classical discrete case. We will come back to this in the conclusion.

Other kinds of Petri nets can also be expressed using CPN's. This is in particular the case for so-called "timed Petri nets". These come in several varieties. Each models a system in which there is a certain delay between the start and completion. We consider two types (see [5], p. 189).

1. *P-timed nets*: Here, a timing delay d_p is assigned to each "place" $p \in P$: classically said, a "token must remain at p for time d_p before it can be moved". A continuous formulation of this is obtained by replacing the formula (3) with

$$M_t(p) = M(p) - \langle F_1(p), w(c)(t - d_p) \rangle + \langle F_2(p), w(c)(t) \rangle.$$

2. *T-timed nets*: In this model, a time log d_e is given to each "event" $e \in P$. When e is called upon to act, d_e units of time must pass before e begins to take effect. In a CPN, this feature can be included by restricting attention to those control paths c such that $\langle c(t), e \rangle = 0$ for $0 \leq t \leq d_e$ and $e \in E$ or by replacing each control path $c \in \mathcal{P}(E)$ with $\hat{c}$ defined by $\langle \hat{c}(t), e \rangle = \langle c(t - d_e), e \rangle$ for all $e \in E$.

4   Continuous Automata with Concurrency; the Functors *an* and *na*

After a brief discussion of classical automata, we introduce the notion of a continuous automaton with concurrency (CAC) as an abstract, time-sensitive transition system. Each CPN is seen to give rise to a corresponding CAC *via* the functor na; we then give a functor an that synthesizes a concrete CPN for each given abstract CAC.

Following [8], we define an *automaton with concurrency* as a tuple $\mathcal{A} = (S, E, T, s_0, \|)$, where

1. S is an arbitrary set containing a distinguished element s_0. (The elements of S are the *states* of the system; s_0 is the start state); E is a non-empty set disjoint from S;
2. T is a subset of $S \times E \times S$ forming a partially defined function $S \times E \to S$, i.e. if $(s, e, s') \in T$ and $(s, e, s'') \in T$ then always $s' = s''$; an element $(s, e, s') \in T$ is called a *transition*; we write $s \xrightarrow{e} s'$;
3. $\| = (\|_s)_{s \in S}$ is a family of symmetric binary relations on E such that if $e_1 \|_s e_2$, then there are states $s', s'', s''' \in S$ so that (s, e_1, s'), (s, e_2, s''), (s', e_2, s'''), and (s'', e_1, s''') are transitions in T: thus, from state s, state s''' can be reached by successive application of e_1 and e_2 in either order $e_1 e_2$ or $e_2 e_1$. (Here we do not require the relations $\|_s$ to be irreflexive; that is, we allow autoconcurrency.)

Next, we introduce a continuous model of such automata. In these, the actions can act continuously over intervals of time to transform the states. This is modelled using control paths.

Definition 4.1. A *continuous automaton with concurrency* (CAC for short) is a tuple $\mathcal{A} = (S, E, T, s_0, \|)$, where

1. S is a set containing a distinguished element s_0. (The elements of S are the *states* of the automaton; s_0 is the *start state*); E is a set disjoint from S;
2. T is a function defined on a subset of $S \times \mathbb{R}_+ \times \mathcal{C}(E)$ taking values in S and such that
   (i) if $c, c' \in \mathcal{C}(E)$, $s \in S$, and $t, t' \in \mathbb{R}_+$ are such that $s' = T(s, t, c)$ and $s'' = T(s', t', c')$ exist, then $s'' = T(s, t + t', c^t + c'_t)$;
   (ii) if $T(s, t, c)$ exists, then so does $T(s, u, c)$ for $0 \leq u \leq t$;
   (iii) if $T(s, t, c)$ exists, then $T(s, t, c) = T(s, t, c')$ whenever c' is a control path such that $c'(u) = c(u)$ for all $u \leq t$;
   (iv) $T(s, 0, c) = s$ for all $s \in S$ and $c \in \mathcal{C}(E)$;
3. $\| = (\|_s)_{s \in S}$ is a family of symmetric binary relations $\|_s$ on $\mathcal{C}$; it is required that whenever $c \|_s c'$, then, for all $t, u \geq 0$, we have that

$$T(s, t, c + c'_u) \text{ and } T(s, t, c_u + c')$$

exist; we note that putting $u = 0$ yields the existence of $T(s, t, c + c')$.

Here, the existence of $T(s, t, c)$ means that in state s we can execute the control path c over a length of time t. Clearly, we can then execute c for any shorter time $u \leq t$ (condition 4.1.2(ii)) and the state we obtain depends only on the values of $c(u)$ for $u \in [0, t]$ (condition 4.1.2(iii)). If $t = 0$, we remain in the same state (condition 4.1.2(iv)). If, furthermore, in state $T(s, t, c)$, we can execute another control path c' for a duration t', then we can also execute in state s the control path $c^t + c'_t$, which consists of c cut off after t units of time and c' delayed by t units of time, for the total duration $t + t'$, obtaining the same result; this is condition 4.1.2(i). Two control paths c, c' can be executed concurrently at state s, if we can execute them one after the other (in any order) or simultaneously with partial overlap; cf. condition 4.1.3.

We now illustrate the idea of a continuous automaton and the concept of concurrency in such structures with an example. We are thankful to Oksana Arnold for providing us with this example (see Fig. 3).

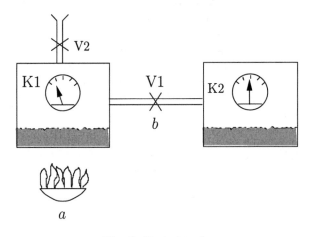

Fig. 3. Heated tanks

Consider two air-tight tanks K1 and K2, each of which is partly filled with fluid. The tanks are connected by a pipe with a valve V1. Tank K1 has an additional valve V2 as depicted in Figure 3. The vapor pressure in K2 is greater than in K1. The collection S of states of this system comprises all tuples (p_1, p_2, v_1, v_2), where p_1 and p_2 are the vapor pressures in K1 and K2, respectively, and v_1 [resp. v_2] is 0 or 1 according as valve V1 [resp. V2] is fully open or completely closed. We consider two controls or "events" a and b in E. Control a serves to heat tank K1: when V1 and V2 are closed vapor pressure p_1 rises to a very high level $q > p_2$. Control b opens V1 for a time, then closes it.

Let $s_1 = (p_1, p_2, 1, 1)$ and $s_2 = (p_1, p_2, 1, 0)$ be states of the system with $p_1 < p_2$. In s_1, both valves are closed; in s_2, valve V1 is closed but V2 is open. We assume that p_1 is the ambient atmospheric pressure. In state s_1, the sequential actions ab and ba affect the system differently. We have

$$s_1 \xrightarrow{a} (q, p_2, 1, 1) \xrightarrow{b} (r, r, 1, 1)$$

with $p_2 < r < q$, and

$$s_1 \xrightarrow{b} (u, u, 1, 1) \xrightarrow{a} (u', u, 1, 1)$$

with $p_1 < u < p_2$ and $u < u'$. However, starting from state s_2, the end results for ab and ba are the same. We have

$$s_2 \xrightarrow{a} (p_1, p_2, 1, 0) \xrightarrow{b} (p_1, p_1, 1, 0),$$
$$s_2 \xrightarrow{b} (p_1, p_1, 1, 0) \xrightarrow{a} (p_1, p_1, 1, 0).$$

We are thus justified in calling a and b concurrent in state s_2 ($a\|_{s_2}b$) but not in state s_1.

We now define the concept of a morphism of continuous automata. Let $\mathcal{A} = (S, E, T, s_0, \|)$ and $\mathcal{A}' = (S', E', T', s_0', \|')$ be two CAC's. A *morphism* from $\mathcal{A}$ to $\mathcal{A}'$ is a pair (σ, η) of functions $\sigma : S \to S'$, $\eta : E \to E'$ such that η is countable-to-one and the following hold:

1. if $T(s, t, c)$ is defined, then so is $T'(\sigma(s), t, \overline{\eta} \circ c)$, and
   $\sigma(T(s, t, c)) = T'(\sigma(s), t, \overline{\eta} \circ c)$;
2. $\sigma(s_0) = s_0'$;
3. $c \|_s c'$ in $\mathcal{A}$ implies that $\overline{\eta} \circ c \|_{\sigma(s)}' \overline{\eta} \circ c'$ in $\mathcal{A}'$.

Again, we require that η be countable-to-one in order that $\overline{\eta} \circ c$ is a control path for any $c \in \mathcal{C}(E)$. This concept of morphism is quite natural from universal algebra considerations: morphisms should preserve the structure of the underlying automaton (transitions, concurrency) and the initial state. The identity morphism on $\mathcal{A}$ is the pair (id_S, id_E).

Proposition 4.2. *The class of all CAC's and their morphisms form a category* <u>CAC</u>, *where the composition of two morphisms* $\mathcal{A} \xrightarrow{(\sigma', \eta')} \mathcal{A}' \xrightarrow{(\sigma'', \eta'')} \mathcal{A}''$ *is* $\mathcal{A} \xrightarrow{(\sigma, \eta)} \mathcal{A}''$, *with* $\sigma = \sigma'' \circ \sigma'$ *and* $\eta = \eta'' \circ \eta'$.

Next we indicate how each continuous Petri net gives rise to an abstract transition system, i.e. an automaton in which the notion of concurrency is present. We then show how to associate a continuous Petri net with each such abstract automaton.

We first construct a functor na from <u>CPN</u> to <u>CAC</u>. Let $\mathcal{N} = (P, E, F, M_{in})$ be a continuous Petri net. We define a corresponding CAC $na(\mathcal{N}) = (\mathcal{M}, E, T, M_{in}, \|)$ as follows:

1. $\mathcal{M} = \{M : P \to \mathbb{R}_+ : M(p) \geq 0 \text{ for all } p \in P\}$ is the set of all markings of $\mathcal{N}$, now to be understood as the states of $na(\mathcal{N})$;
2. the event set E is the same as in $\mathcal{N}$;
3. $T(M, t, c)$ is defined just in case Mc^t exists, and then $T(M, t, c) = Mc^t$;
4. $s_0 = M_{in}$, the initial marking of $\mathcal{N}$;
5. $c \|_M c'$ is as previously defined.

The following result is readily verified using the definitions of section 3.

Lemma 4.3. *If $\mathcal{N}$ is a CPN, then the system $na(\mathcal{N})$ defined above is a CAC.*

Let $\mathcal{N}$ and $\mathcal{N}'$ be CPN's and let (π, η) be a morphism from $\mathcal{N}$ to $\mathcal{N}'$. Let $na(\mathcal{N})$ and $na(\mathcal{N}')$ be the corresponding CAC's with state sets $\mathcal{M}$ and $\mathcal{M}'$, respectively. We define a function $\sigma : \mathcal{M} \to \mathcal{M}'$ by $\sigma(M) = M \circ \pi$, which is, by Proposition 3.3, a marking of $\mathcal{N}'$. The same proposition guarantees that $na(\pi, \eta) = (\sigma, \eta)$ is a morphism from $na(\mathcal{N})$ to $na(\mathcal{N}')$. Clearly, na respects composition of morphisms and identities. Thus we have

Proposition 4.4. *The correspondence na is a functor from* CPN *to* CAC.

We now define a functor an from $\underline{CAC}$ to $\underline{CPN}$. We note that if $\mathcal{N}$ is a *CPN*, then each condition $p \in P$ induces a real function $M \to M(p)$ from $\mathcal{M}$ to $\mathbb{R}_+$, where $\mathcal{M}$ is the set of markings of $\mathcal{N}$. Also, p induces the pair $(F_1(p), F_2(p))$ in $\ell_+^\infty(E) \times \ell_+^\infty(E)$. Essential properties of these induced functions motivate our definition of the set of *places* of $\mathcal{A}$.

Construction 4.5. Let $\mathcal{A} = (S, E, T, s_0, \|)$ be a CAC. We build a CPN $an(\mathcal{A}) = (P, E, F, M_{in})$ derived from $\mathcal{A}$. The elements of P (the *places*) are pairs $\varphi = (\varphi_S, \varphi_E)$, where $\varphi_S : S \to \mathbb{R}_+$ is a function, and $\varphi_E = ((\varphi_E)_1, (\varphi_E)_2)$ is a pair in $\ell_+^\infty(E) \times \ell_+^\infty(E)$. In order for φ to be a place, the functions $\varphi_S, (\varphi_E)_1, (\varphi_E)_2$ must satisfy the following hypotheses:
If $T(s, t, c)$ exists for some $(s, t, c) \in S \times \mathbb{R}_+ \times \mathcal{P}$, then

(I) $\varphi_S(T(s, t, c)) = \varphi_S(s) - \langle (\varphi_E)_1, w(c)(t) \rangle + \langle (\varphi_E)_2, w(c)(t) \rangle$

and, for each $u < t$, there is some $\Delta u > 0$ such that

(II) $\langle (\varphi_E)_1, w(c)(u + \Delta u) - w(c)(u) \rangle \leq \varphi_S(T(s, u, c))$.

We define $P = \{\varphi : \varphi \text{ is a place of } \mathcal{A}\}$.
The construction of $an(\mathcal{A})$ continues with the definitions of $F = (F_1, F_2)$ and $M_{in} : P \to \mathbb{R}_+ :$

$$F_i(\varphi) = (\varphi_E)_i \quad (i = 1, 2)$$

and $M_{in} = M_{s_0}$, where $M_s(\varphi) = \varphi_S(s)$ for $s \in S$ and $\varphi \in P$.

Lemma 4.6. *The structure $an(\mathcal{A}) = (P, E, F, M_{in})$ is a CPN.*

We now note that if $\mathcal{A} = na(\mathcal{N})$ is the automaton associated with a CPN $\mathcal{N}$ as previously defined, then places of $\mathcal{A}$ exist in abundance. For each $p \in P$, we define a pair $|p| = (|p|_\mathcal{M}, |p|_E)$, the *extent* of p, where $|p|_\mathcal{M} : \mathcal{M} \to \mathbb{R}_+$ is the evaluation map $|p|_\mathcal{M}(M) = M(p)$, and $|p|_E = (F_1(p), F_2(p)) \in \ell_+^\infty(E) \times \ell_+^\infty(E)$.

Lemma 4.7. *With $\mathcal{N}$ as above and $p \in P$, the extent $|p|$ is a place of the CAC $na(\mathcal{N})$ according to Construction 4.5.*

Proof. Routine.

We proceed with the definition of the functor an. Let $\mathcal{A} = (S, E, T, s_0, \|)$ and $\mathcal{A}' = (S', E', T', s_0', \|')$ be CAC's and let (σ, η) be a morphism from $\mathcal{A}$ to $\mathcal{A}'$. Consider the associated CPN's $an(\mathcal{A}) = (P, E, F, M_{in})$ and $an(\mathcal{A}') = (P', E', F', M_{in}')$. We define a pair $an(\sigma, \eta) = (\pi, \eta)$ with $\pi : P' \to P$ the function defined by $\pi(\varphi') = (\pi(\varphi')_S, \pi(\varphi')_E)$, where $\pi(\varphi')_S : S \to \mathbb{R}_+$ is given by $\pi(\varphi')_S = \varphi'_{S'} \circ \sigma$, and $\pi(\varphi')_E = ((\varphi'_{E'})_1 \circ \eta, (\varphi'_{E'})_2 \circ \eta)$.

Lemma 4.8. (a) *For each $\varphi' \in P'$, the pair $\pi(\varphi') = (\pi(\varphi')_S, \pi(\varphi')_E)$ defined above is a place of $\mathcal{A}$.*
(b) *The pair $(\pi, \eta) = an(\sigma, \eta)$ is a CPN-morphism from $an(\mathcal{A})$ to $an(\mathcal{A}')$.*

Proof. (a) Assume that $T(s, t, c)$ is defined. Then

$$
\begin{aligned}
\pi(\varphi')_S(T(s, t, c)) &= \varphi'_{S'}(\sigma(T(s, t, c))) \\
&= \varphi'_{S'}(T'(\sigma(s), t, \overline{\eta} \circ c)) \\
&= \varphi'_{S'}(\sigma(s)) - \langle (\varphi'_{E'})_1, w(\overline{\eta} \circ c)(t) \rangle + \langle (\varphi'_{E'})_2, w(\overline{\eta} \circ c)(t) \rangle \\
&= \pi(\varphi')_S(s) - \langle (\varphi'_{E'})_1 \circ \eta, w(c)(t) \rangle + \langle (\varphi'_{E'})_2 \circ \eta, w(c)(t) \rangle
\end{aligned}
$$

as required for equation (I). We then have (for each $u < t$ and $\Delta u > 0$)

$$
\begin{aligned}
\pi(\varphi')_S(T(s, u, c)) &= \varphi'_{S'}(\sigma(T(s, u, c))) \\
&= \varphi'_{S'}(T'(\sigma(s), u, \overline{\eta} \circ c)).
\end{aligned}
$$

Since φ' is a place, and $T'(\sigma(s), t, \overline{\eta} \circ c)$ is defined, there is some $\Delta u > 0$ such that the above quantity is at least

$$
\langle (\varphi'_{E'})_1, w(\overline{\eta} \circ c)(u + \Delta u) - w(\overline{\eta} \circ c)(u) \rangle = \langle (\varphi'_{E'})_1 \circ \eta, w(c)(u + \Delta u) - w(c)(u) \rangle
$$

as required for (II).

(b) By (a), π is well-defined. Next, we note that for every $s \in S$ and $\varphi' \in P'$, $M_s(\pi(\varphi')) = \pi(\varphi')(s) = \varphi'(\sigma(s)) = M'_{\sigma(s)}(\varphi')$, so that $M'_{\sigma(s)} = M_s \circ \pi$. In particular, $M'_{in} = M'_{s_0'} = M'_{\sigma(s_0)} = M_{s_0} \circ \pi = M_{in} \circ \pi$, as required. Now suppose that $\varphi' \in P'$ and $e \in E$. For $i = 1, 2$ we have

$$
\begin{aligned}
\langle F_i'(\varphi'), \eta(e) \rangle &= \langle (\varphi'_{E'})_i, \eta(e) \rangle && \text{(def. } F') \\
&= \langle (\varphi'_{E'})_i \circ \eta, e \rangle \\
&= \langle (\pi(\varphi')_E)_i, e \rangle && \text{(def. } \pi) \\
&= \langle F_i(\pi(\varphi')), e \rangle && \text{(def. } F).
\end{aligned}
$$

$\square$

We summarize these results in the

Lemma 4.9. *The correspondence an is a functor from $\underline{CAC}$ to $\underline{CPN}$.*

The lemma is simple: we note that an preserves identities and compositions.

5 An Adjunction

In this section, we will show that the functor $an : CAC \to CPN$ actually is an adjunction with left adjoint na.

Let $\mathcal{A} = (S, E, T, s_0, \|)$ be a CAC and let $an(\mathcal{A}) = (P, E, F, M_{in})$ be the corresponding continuous Petri net. Then let $na(an(\mathcal{A})) = (\mathcal{M}, E, T', s_0', \|')$ be the associated CAC. We define a function $\sigma_0 : S \to \mathcal{M}$ by $\sigma_0(s) = M_s$. (Recall that $M_s(\varphi) = \varphi_S(s)$ for all places $\varphi \in P$.) Let $id_E : E \to E$ be the identity function $id_E(e) = e$.

Lemma 5.1. *The pair* (σ_0, id_E) *is a CAC morphism from* $\mathcal{A}$ *to* $na(an(\mathcal{A}))$.

Proof. Suppose that $T(s, t, c)$ is defined. Then $\sigma_0(s) = M_s$ is non-negative and hence is a marking. We now assert the existence of $T'(\sigma_0(s), t, c) = T'(M_s, t, c)$. This is equivalent to the existence of $(M_s)c^t$, where

$$c^t(u) = \begin{cases} c(u) & \text{for } u \leq t \\ 0 & \text{for } u > t. \end{cases}$$

For all places $\varphi \in P$, all $0 \leq u \leq t$ and some $\Delta u > 0$, we have

$$
\begin{aligned}
& \langle F_1(\varphi), w(c)(u + \Delta u) - w(c)(u) \rangle && \\
&= \langle (\varphi_E)_1, w(c)(u + \Delta u) - w(c)(u) \rangle && (\text{def } F_1) \\
&\leq \varphi_S(T(s, u, c)) && (\text{place}) \\
&= \varphi_S(s) - \langle (\varphi_E)_1, w(c)(u) \rangle + \langle (\varphi_E)_2, w(c)(u) \rangle && (\text{place}) \\
&= M_s(\varphi) - \langle F_1(\varphi), w(c)(u) \rangle + \langle F_2(\varphi), w(c)(u) \rangle && \\
&= (M_s c)_u(\varphi), &&
\end{aligned}
$$

so that $M_s c^t$ exists. If $s' = T(s, t, c)$, then

$$
\begin{aligned}
\sigma_0(T(s, t, c)) = \sigma_0(s') = M_{s'} \quad &\text{and} \\
T'(\sigma_0(s), t, c) = T'(M_s, t, c) = M_s c^t.&
\end{aligned}
$$

As above, $M_s c^t(\varphi) = \varphi_S(T(s, t, c)) = \varphi_S(s') = M_{s'}(\varphi)$ for all $\varphi \in P$. As required, $\sigma_0(T(s, t, c)) = T'(\sigma_0(s), c, t)$.

Next, we see that $\sigma_0(s_0) = M_{s_0} = M_{in}$. Finally, we verify that if $c \|_s c'$ in $\mathcal{A}$, then also $c \|'_{M_s} c'$ in $na(an(\mathcal{A}))$. The relation $c \|_s c'$ implies that $T(s, t, c + c'_u)$ is defined for all $t, u \geq 0$. Then for each place φ of $\mathcal{A}$, as above we have

$$
\begin{aligned}
M_s(c + c'_u)^t(\varphi) &= \varphi_S(T(s, t, c + c'_u)) \\
&\geq \langle F_1(\varphi), w(c + c'_u)(t + \Delta t) - w(c + c'_u)(t) \rangle
\end{aligned}
$$

for some $\Delta t > 0$. Thus $M_s(c + c'_u)$ and, by symmetry, $M_s(c_u + c')$ exist, and $c \|'_{M_s} c'$. ☐

Let $\mathcal{N} = (P, E, F, M_{in})$ be a CPN and let $na(\mathcal{N}) = (\mathcal{M}, E, T, M_{in}, \|)$ be the corresponding CAC. Then consider $an(na(\mathcal{N})) = (P', E, F', M_{in}')$, the associated CPN. We recall that

- $P' = \{\varphi' : \varphi' = (\varphi'_{\mathcal{M}}, \varphi'_E)$ is a place of $na(\mathcal{N})\}$;
- $F'(\varphi') = ((\varphi_E)_1, (\varphi_E)_2)$;
- $M'_{in}(\varphi') = \varphi'(M_{in})$ for each $\varphi' \in P'$.

We now construct a morphism (π_0, id_E) from $an(na(\mathcal{N}))$ to $\mathcal{N}$, defining $\pi_0(p) = |p|$, the extent of p. By Lemma 4.7, $\pi_0 : P \to P'$ is well-defined.

Lemma 5.2. *The pair (π_0, id_E) is a CPN-morphism from $an(na(\mathcal{N}))$ to $\mathcal{N}$.*

Proof. Given any $p \in P$, we have $F(p) = |p|_E = F'(|p|) = F'(\pi_0(p))$. Next, we verify that $M_{in} = M'_{in} \circ \pi_0$: for each $p \in P$, we have $M'_{in}(\pi_0(p)) = M'_{in}(|p|) = |p|_{\mathcal{M}}(M_{in}) = M_{in}(p)$. $\square$

Theorem 5.3. *The functors $an : \underline{CAC} \to \underline{CPN}$ and $na : \underline{CPN} \to \underline{CAC}$ form an adjunction with an the left adjoint of na; the components of the units and co-units of this adjunction are the morphisms given in Lemma 5.1 and Lemma 5.2.*

Proof. Let $\mathcal{A} = (S, E, T, s_0, \|)$ be a CAC and let $an(\mathcal{A}) = (P, E, F, M_{in})$ be the associated CPN. Let $\mathcal{N} = (P', E', F', M'_{in})$ be a CPN and let $na(\mathcal{N}) = (\mathcal{M}', E', T', M'_{in}, \|')$ be the corresponding CAC.

First, given a CPN-morphism $an(\mathcal{A}) \xrightarrow{(\pi,\eta)} \mathcal{N}$ described by functions $\pi : P' \to P$ and $\eta : E \to E'$, we define a function $\sigma : S \to \mathcal{M}'$ by putting $\sigma(s) = M_s \circ \pi$. We assert that $\vartheta(\pi, \eta) = (\sigma, \eta)$ is a composition of CAC-morphisms

$$\mathcal{A} \xrightarrow{(\sigma_0, id_E)} na(an(\mathcal{A})) \xrightarrow{na(\pi,\eta)} na(\mathcal{N}),$$

where $\sigma_0(s) = M_s$ is the function considered in Lemma 5.1 Indeed, we have $na(\pi, \eta) = (\sigma', \eta)$ with $\sigma'(M) = M \circ \pi$, and $\sigma'(\sigma_0(s)) = \sigma_0(s) \circ \pi = M_s \circ \pi = \sigma(s)$ for all $s \in S$.

Secondly, suppose that $\sigma : S \to \mathcal{M}'$ and $\eta : E \to E'$ are given functions such that $\mathcal{A} \xrightarrow{(\sigma,\eta)} na(\mathcal{N})$ is a CAC-morphism. We define a function $\pi : P' \to P$ by setting $\pi(p') = (\pi(p')_S, \pi(p')_E)$, where $\pi(p')_S(s) = \sigma(s)(p')$ and $(\pi(p')_E)_i = F'_i(p') \circ \eta$. We will show that $\varphi(\sigma, \eta) := (\pi, \eta)$ is a composition of CPN-morphisms

$$an(\mathcal{A}) \xrightarrow{an(\sigma,\eta)} an(na(\mathcal{N})) \xrightarrow{(\pi_0, id_{E'})} \mathcal{N},$$

where $\pi_0(p') = |p'|$ is the function considered in Lemma 5.2: We have $an(\sigma, \eta) = (\pi', \eta)$, where $\pi'(\varphi')_S(s) = \varphi'_{\mathcal{M}'}(\sigma(s))$ and $\pi'(\varphi')_{E'} = \varphi'_{E'} \circ \eta$. Then

$$\pi'(\pi_0(p'))_S(s) = \pi_0(p')_{\mathcal{M}'}(\sigma(s)) = \sigma(s)(p') = \pi(p')_S(s) \quad \text{and}$$

$$\pi'(\pi_0(p'))_E = \pi_0(p')_{E'} \circ \eta = |p'|_{E'} \circ \eta = F'(p') \circ \eta = \pi(p')_E,$$

as required.

Finally, we prove that ϑ and φ are mutual inverses and establish a bijective correspondence between CPN-morphisms $an(\mathcal{A}) \to \mathcal{N}$ and CAC-morphisms $\mathcal{A} \to na(\mathcal{N})$.

First let $(\sigma, \eta) : \mathcal{A} \to na(\mathcal{N})$ be a morphism. We demonstrate that $\vartheta(\varphi(\sigma, \eta))$ $= (\sigma, \eta)$. Put $\varphi(\sigma, \eta) = (\pi, \eta)$ and $\vartheta(\pi, \eta) = (\sigma', \eta)$. For all $s \in S$ and $p' \in P'$, we have $\sigma'(s)(p') = M_s(\pi(p')) = \pi(p')(s) = \sigma(s)(p')$, so that $\sigma' = \sigma$.

Finally, let $(\pi, \eta) : an(\mathcal{A}) \to \mathcal{N}$ be a morphism. We show that $\varphi(\vartheta(\pi, \eta)) =$ (π, η). Put $\vartheta(\pi, \eta) = (\sigma, \eta)$, where $\sigma(s)(p') = M_s \circ \pi(p') = \pi(p')_S(s)$. Now for $\varphi(\vartheta(\pi, \eta)) = (\pi', \eta)$, we have $\pi'(p')_S(s) = \sigma(s)(p') = \pi(p')_S(s)$ for all $p' \in P'$ and $s \in S'$; also, $\pi'(p')_E = F'(p') \circ \eta = F(\pi(p')) = (\pi(p'))_E$. This proves that $\pi = \pi'$.   □

6 A Coreflection

In this section, we demonstrate the existence of a coreflection between $\underline{CPN}$ and a certain subcategory of $\underline{CAC}$. We will also obtain which continuous automata $\mathcal{A}$ (with only reachable states) are of the form $\mathcal{A} = na(\mathcal{N})$ for some continuous Petri net $\mathcal{N}$. Suppose that $\mathcal{A} = (S, E, T, s_0, \|) \in \underline{CAC}$, $s \in S$, $\Delta t \geq 0$, $t \geq 0$, $c \in \mathcal{C}(E)$ and φ is a place of $\mathcal{A}$. Then we define the quantities

$$N(\varphi, s, t, c) = \varphi_S(s) - \langle (\varphi_E)_1, w(c)(t) \rangle + \langle (\varphi_E)_2, w(c)(t) \rangle ,$$
$$K(\varphi, t, \Delta t, c) = \langle (\varphi_E)_1, w(c)(t + \Delta t) - w(c)(t) \rangle .$$

Definition 6.1. We will call a continuous automaton with concurrency $\mathcal{A} = (S, E, T, s_0, \|)$ *rich*, if the following conditions are satisfied:

1. the places separate the states of $\mathcal{A}$, i.e., whenever $u, v \in S$ are states such that $\varphi_S(u) = \varphi_S(v)$ for all places φ of $\mathcal{A}$, then $u = v$;
2. whenever $c \in \mathcal{C}(E)$, $t \geq 0$, and $s \in S$ such that for all places φ of $\mathcal{A}$, and for each $u < t$, there is some $\Delta u > 0$ with

$$K(\varphi, u, \Delta u, c) \leq N(\varphi, s, u, c),$$

then for all $u \leq t$, $s_u = T(s, u, c)$ is defined;
3. whenever $c, c' \in \mathcal{C}(E)$, $v \geq 0$, and $s \in S$ such that for all places φ, and for each $u \geq 0$, there is some $\Delta u > 0$ such that

$$K(\varphi, u, \Delta u, c + c'_v) \leq N(\varphi, s, u, c + c'_v)$$
$$K(\varphi, u, \Delta u, c_v + c') \leq N(\varphi, s, u, c_v + c')$$

then $c \parallel_s c'$.

Requirements (1)-(3) ensure that the structure of the given automaton $\mathcal{A}$ is directly correlated with properties of the places of $\mathcal{A}$. We see now that these conditions are satisfied by all CAC's arising from nets *via* the functor na.

Proposition 6.2. *Let $\mathcal{N}$ be a CPN. Then $\mathcal{A} = na(\mathcal{N})$ is rich.*

Proof. First, suppose that M and M' are states of $\mathcal{A}$, i.e. markings of $\mathcal{N}$, with $M \neq M'$. Then there is some $p \in P$ with $M(p) \neq M'(p)$. Now $|p|$ is, by Lemma

4.7, a place of $na(\mathcal{N})$ such that $|p|(M) \neq |p|(M')$. This establishes 6.1.1 for $\mathcal{A} = na(\mathcal{N})$.

To check 6.1.2 for $\mathcal{A}$, let $c \in \mathcal{C}(E)$, $t \geq 0$, M a marking such that for all places φ and for all $u < t$, there is some $\Delta u > 0$ such that

$$K(\varphi, u, \Delta u, c) \leq N(\varphi, M, u, c),$$

then putting $\varphi = |p|$ yields

$$\langle F_1(p), w(c)(u + \Delta u) - w(c)(u) \rangle \leq M_u(p),$$

so that $M_u = T(M, u, c)$ for all $u \leq t$, as required.

Next, we check condition 6.1.3 for $\mathcal{A}$. Let M be a marking of $\mathcal{N}$ and let $c, c' \in \mathcal{C}(E)$ be such that for all places φ and $u, v \geq 0$, there exists some $\Delta u > 0$ such that

$$K(\varphi, u, \Delta u, c + c'_v) \leq N(\varphi, M, u, c + c'_v)$$

$$K(\varphi, u, \Delta u, c_v + c') \leq N(\varphi, M, u, c_v + c').$$

Then taking $\varphi = |p|$ yields

$$\langle F_1(p), w(c + c'_v)(u + \Delta u) - w(c + c'_v)(u) \rangle \leq M(c + c'_v)_u(p)$$

$$\langle F_1(p), w(c_v + c')(u + \Delta u) - w(c_v + c')(u) \rangle \leq M(c_v + c')_u(p) .$$

The existence of $M(c + c'_v)$ and $M(c_v + c')$ follows, and $c \parallel_M c'$, as desired. $\square$

Next, we wish to show that for such rich continuous automata with concurrency, the morphism of Lemma 5.1 turns out to be an embedding. These are defined as follows.

Definition 6.3. Let $\mathcal{A} = (S, E, T, s_0, \parallel)$ and $\mathcal{A}' = (S', E', T', s'_0, \parallel')$ be two CAC's. A pair (σ, η) of functions $\sigma : S \to S', \eta : E \to E'$ is called an *embedding* of $\mathcal{A}$ into $\mathcal{A}'$, if the following conditions are satisfied:

1. σ and η are one-to-one, and $\sigma(s_0) = s'_0$;
2. $T(s, t, c)$ is defined if and only if $T'(\sigma(s), t, \overline{\eta} \circ c)$ is defined, and in this case $\sigma(T(s, t, c)) = T'(\sigma(s), t, \overline{\eta} \circ c)$, for any $s \in S, t \geq 0$ and $c \in \mathcal{C}(E)$;
3. $c \parallel_s c'$ in $\mathcal{A}$ if and only if $\overline{\eta} \circ c \parallel'_{\sigma(s)} \overline{\eta} \circ c'$ in $\mathcal{A}'$, for any $s \in S$ and $c, c' \in \mathcal{C}(E)$.

Furthermore, $\mathcal{A}$ is a *subautomaton* of $\mathcal{A}'$, if $S \subseteq S', E \subseteq E'$ and the pair (id_S, id_E) of identity mappings $id_S : S \to S', id_E : E \to E'$ is an embedding of $\mathcal{A}$ into $\mathcal{A}'$.

Again, this is the usual concept of embeddings from universal algebra considerations. Trivially, any embedding is a morphism. Now we show:

Theorem 6.4. *Let $\mathcal{A}$ be a rich CAC. Then the morphism $(\sigma_0, id_E) : \mathcal{A} \to na(an(\mathcal{A}))$ is an embedding.*

Proof. Let $\mathcal{A} = (S, E, T, s_0, \|)$. We form $an(\mathcal{A}) = (P, E, F, M_{in})$ according to Construction 4.5, and $\mathcal{A}' = na(an(\mathcal{A})) = (\mathcal{M}, E, T', M_0, \|')$ as before. Recall that $(\sigma_0, id_E) : \mathcal{A} \to \mathcal{A}'$ with $\sigma_0(s) = M_s (s \in S)$ is a morphism by Lemma 5.1. This already implies $\sigma_0(s_0) = M_0$ and one implication of conditions 6.3.2 and 6.3.3.

First suppose that $s, s' \in S$ with $M_s = M_{s'}$. Then for any place φ of $\mathcal{A}$, we have $\varphi(s) = M_s(\varphi) = M_{s'}(\varphi) = \varphi(s')$. By condition 6.1.1, $s = s'$, so that σ_0 is one-to-one.

Now let $s \in S, t \geq 0$ and $c \in \mathcal{C}(E)$, and assume that $T'(\sigma_0(s), t, c) =: M'$ is defined, i.e., $M_s \xrightarrow{c^t} M'$ in $na(an(\mathcal{A}))$. Then, for each place φ of $\mathcal{A}$ and all $u < t$, there exists some $\Delta u > 0$ such that

$$K(\varphi, u, \Delta u, c) = \langle F_1(\varphi), w(c)(u + \Delta u) - w(c)(u) \rangle$$
$$\leq (M_s)_u(\varphi)$$
$$= N(\varphi, M_s, u, c).$$

Invoking 6.1.2, we find that there exist states $s_u = T(s, u, c)$ for all $u \leq t$. Since (σ_0, id_E) is a morphism, we obtain $\sigma_0(T(s, t, c)) = M'$, proving condition 6.3.2.

Finally, suppose that $c, c' \in \mathcal{C}(E)$ and $s \in S$, with $c \|'_{M_s} c'$ in $\mathcal{A}'$. It remains only to show that $c\|_s c'$ in $\mathcal{A}$. For all $u, t \geq 0$ and each place φ of $\mathcal{A}$, there is some $\Delta t > 0$ such that

$$\langle F_1(\varphi), w(c + c'_u)(t + \Delta t) - w(c + c'_u)(t) \rangle \leq (M_s)_t(\varphi)$$
$$\langle F_1(\varphi), w(c_u + c')(t + \Delta t) - w(c_u + c')(t) \rangle \leq (M_s)_t(\varphi)$$

We re-write these inequalities as

$$K(\varphi, t, \Delta t, c + c'_u) \leq N(\varphi, M_s, t, c + c'_u)$$
$$K(\varphi, t, \Delta t, c_u + c') \leq N(\varphi, M_s, t, c_u + c')$$

and invoke 6.1.3 to conclude that $c \|_s c'$ as required. □

We also have

Lemma 6.5. *Let $\mathcal{A}'$ be a rich CAC and $\mathcal{A}$ a subautomaton of $\mathcal{A}'$. Then $\mathcal{A}$ is also rich.*

Proof. Let $\mathcal{A} = (S, E, T, s_0, \|)$ and $\mathcal{A}' = (S', E', T', s_0, \|')$. If $t \geq 0$ and $c \in \mathcal{C}(E)$, then $\overline{id_E} \circ c =: c' \in \mathcal{C}(E')$ satisfies $\langle c'(t), e \rangle = \langle c(t), e \rangle$ if $e \in E$, and $\langle c'(t), e' \rangle = 0$ if $e' \in E' \setminus E$; similarly for $w(c')(t), w(c)(t)$. Also, if $T(s, t, c)$ exists in $\mathcal{A}$, then $T'(s, t, c') = T(s, t, c)$. Using this, it is easy to see that if φ' is a place of $\mathcal{A}'$, then the restriction of φ' to S and E is a place of $\mathcal{A}$.

To check condition 6.1.1, let $u, v \in S$ such that $\varphi_S(u) = \varphi_S(v)$ for all places φ of $\mathcal{A}$. Then also $\varphi'_{S'}(u) = \varphi'_{S'}(v)$ for all places φ' of $\mathcal{A}'$ and hence $u = v$ since $\mathcal{A}'$ is rich. Similarly, conditions 6.1.2 and 6.1.3 follow, using also that the transitions and concurrency relations of $\mathcal{A}$ are restrictions of the ones of $\mathcal{A}'$. □

As a consequence, we have:

Corollary 6.6. *Let $\mathcal{A}$ be a CAC. Then $\mathcal{A}$ is rich if and only if $\mathcal{A}$ is isomorphic to a subautomaton of $na(\mathcal{N})$, for some CPN $\mathcal{N}$.*

Proof. Immediate by Theorem 6.4, Proposition 6.2 and Lemma 6.5.   □

Let $\mathcal{A} = (S, E, T, s_0, \|)$ be a CAC. A state $s \in S$ is called *reachable*, if $s = T(s_0, t, c)$ for some $t \geq 0$ and some control path c (note that in view of condition 4.1.2(i) there is no need to introduce the customary sequence of transitions leading from s_0 to s). Next we define

$$\mathcal{R}(\mathcal{A}) = (S', E, T', s_0, \|') \text{ ,where}$$
$$S' = \{s \in S; s \text{ is reachable}\},$$
$$T' = T \cap (S' \times \mathbb{R}^+ \times \mathcal{C}(E) \times S'),$$
$$\|' = (\|_s)_{s \in S'}.$$

It is easy to recognise that $R(\mathcal{A})$ is a CAC each state of which is reachable and that $R(\mathcal{A})$ is a subautomaton of $\mathcal{A}$. Now, if $(\sigma, \eta) : \mathcal{A} \to \mathcal{A}^*$ is a CAC-morphism, let $R(\sigma, \eta) = (\sigma', \eta)$, where σ' is the restriction of σ to the states of $R(\mathcal{A})$

Lemma 6.7. *Let $\mathcal{A}, \mathcal{A}'$ be two CAC's and let $(\sigma, \eta) : \mathcal{A} \to \mathcal{A}'$ be an embedding with η onto. Then $\mathcal{R}(\sigma, \eta) : \mathcal{R}(\mathcal{A}) \to \mathcal{R}(\mathcal{A}')$ is an isomorphism.*

Proof. Let $\mathcal{A} = (S, E, T, s_0, \|)$ and $\mathcal{A}' = (S', E', T', s_0', \|')$. Clearly, $\mathcal{R}(\sigma, \eta)$ is an embedding. We only have to show that σ maps the reachable states of $\mathcal{A}$ *onto* the reachable states of $\mathcal{A}'$. Let $s' = T'(s_0', t, c')$ for some $t \geq 0$ and some control path $c' \in \mathcal{C}(E')$ of $\mathcal{A}'$. Since η is bijective, so is $\overline{\eta}$, and $c = \overline{\eta}^{-1} \circ c'$ is a control path of $\mathcal{A}$ with $c' = \overline{\eta} \circ c$. Hence $s = T(s_0, t, c)$ exists in $\mathcal{A}$ and $\sigma(s) = s'$.   □

Corollary 6.8. *Let $\mathcal{A}$ be a CAC. Then $\mathcal{A}$ is isomorphic to an automaton of the form $\mathcal{R}(na(\mathcal{N}))$ for some CPN $\mathcal{N}$ if and only if $\mathcal{A}$ is rich and each state of $\mathcal{A}$ is reachable.*
In this case, $(\sigma_0, id_E) : \mathcal{A} \to \mathcal{R}(na(an(\mathcal{A})))$ is an isomorphism.

Proof. By Corollary 6.6, each automaton of the form $\mathcal{R}(na(\mathcal{N}))$ is rich, and trivially each state is reachable. For the converse, by Theorem 6.4, (σ_0, id_E) embeds $\mathcal{A}$ into $na(an(\mathcal{A}))$. Now apply Lemma 6.7   □

Let $\underline{CAC}^{\mathcal{R}}$ be the full subcategory of $\underline{CAC}$ consisting only of rich CAC's in which each state is reachable. Clearly, $\mathcal{R} : \underline{CAC} \to \underline{CAC}^{\mathcal{R}}$ as defined above is a functor. Then the composition $na^{\mathcal{R}} := \mathcal{R} \circ na : \underline{CPN} \to \underline{CAC}^{\mathcal{R}}$ is also a functor. Let $an^{\mathcal{R}} : \underline{CAC}^{\mathcal{R}} \to \underline{CPN}$ just be the restriction of the functor an to $\underline{CAC}^{\mathcal{R}}$. Now we obtain our desired coreflection result:

Theorem 6.9. *The functors $an^{\mathcal{R}} : \underline{CAC}^{\mathcal{R}} \to \underline{CPN}$ and $na^{\mathcal{R}} : \underline{CPN} \to \underline{CAC}^{\mathcal{R}}$ form a coreflection with $an^{\mathcal{R}}$ the left adjoint of $na^{\mathcal{R}}$; the components of the unit of this adjunction are the (iso-)morphisms (σ_0, id_E) given in Lemma 5.1.*

Proof. By Theorem 5.3 it only remains to show that $(\sigma_0, id_E) : \mathcal{A} \to na^{\mathcal{R}} \circ an^{\mathcal{R}}(\mathcal{A})$ is an isomorphism for each $\mathcal{A} \in \underline{CAC}^{\mathcal{R}}$. But this is immediate by Corollary 6.8. $\square$

For later use in section 8, we also include the following remark.

Lemma 6.10. *Let $\mathcal{A}, \mathcal{A}' \in \underline{CAC}$ such that each state in $\mathcal{A}$ is reachable. Let $(\sigma, \eta), (\sigma', \eta') : \mathcal{A} \to \mathcal{A}'$ be two morphisms. If $\eta = \eta'$, then $\sigma = \sigma'$.*

Proof. Let $\mathcal{A} = (S, E, T, s_0, \|)$ and $\mathcal{A}' = (S', E', T', s_0', \|')$. Choose any $s \in S$. Then $s = T(s_0, t, c)$ for some $t \geq 0$ and some control path $c \in \mathcal{C}(E)$. Hence $\sigma(s) = T'(s_0', t, \overline{\eta} \circ c) = T'(s_0', t, \overline{\eta'} \circ c) = \sigma'(s)$. $\square$

7   Products and Coproducts

In this section, we wish to consider the product and coproduct construction for the categories of continuous Petri nets and continuous automata with concurrency.

First, since these constructions naturally involve the direct product of event sets, the canonical projections forming the event mappings of the morphisms have to be countable-to-one. Therefore we require subsequently all event sets to be countable. From the point of view of applications (with events representing executions of controls), this is no essential restriction. Therefore, let $\underline{CPN}_0, \underline{CAC}_0, \underline{CAC}_0^{\mathcal{R}}$ denote the full subcategories of $\underline{CPN}, \underline{CAC}$ and $\underline{CAC}^{\mathcal{R}}$, respectively, whose objects have only countable event sets.

Since for our Petri nets continuity is important for the definition of the dynamic behaviour, but not for the net structure and the morphisms, here we can use the same product and coproduct construction as in the discrete case, as developed in [10]. For the convenience of the reader, we sketch these constructions here, since they will be used for the category $\underline{CAC}_0^{\mathcal{R}}$.

Definition 7.1. Let $\mathcal{N}_i = (P_i, E_i, F^i, M_{in}^i)$ $(i = 1, 2)$ be two disjoint Petri nets in $\underline{CPN}_0$. We define the Petri net $\mathcal{N} = (P, E, F, M_{in})$ as follows.

1. $P = P_1 \cup P_2$ and $E = E_1 \times E_2$
2. For $p \in P_i$ $(i = 1, 2)$, $(e_1, e_2) \in E$ and $j = 1, 2$, let $F_j(p, (e_1, e_2)) = F_j^i(p, e_i)$
3. $M_{in}(p) = M_{in}^i(p)$ if $p \in P_i$ $(i = 1, 2)$.

Define morphisms $\pi_i = (id_{P_i}, \pi_{E_i}) : \mathcal{N} \to \mathcal{N}_i$, where $id_{P_i} : P_i \to P$ is the identity mapping and $\pi_{E_i} : E \to E_i, \pi_{E_i}(e_1, e_2) = e_i$, is the projection mapping onto E_i $(i = 1, 2)$.

Lemma 7.2. *With the above notation, the product of $\mathcal{N}_1$ and $\mathcal{N}_2$ in the category $\underline{CPN}_0$ is the net $\mathcal{N}$ equipped with the projections π_1, π_2.*

Proof. We first have to show that $\mathcal{N}$ actually belongs to $\underline{CPN}_0$. Indeed, E is countable and for any $p \in P_i$ and $i, j \in \{1, 2\}$ we have $\|F_j(p)\|_\infty = \|F_j^i(p)\|_\infty$. Hence $\mathcal{N} \in \underline{CPN}_0$. Since all event sets are countable, $\pi_i : \mathcal{N} \to \mathcal{N}_i$ is a morphism $(i = 1, 2)$. The fact that $\mathcal{N}$ is the categorial product of $\mathcal{N}_1$ and $\mathcal{N}_2$ can be shown easily as in [10], Lemma 4.2. $\square$

Next we turn to the coproduct. As in [10], in general the coproduct of two nets $\mathcal{N}_1, \mathcal{N}_2$ does not exist. For example, assume that the initial markings of $\mathcal{N}_1$ and $\mathcal{N}_2$ have different values, i.e. $M_{in}^1(p_1) \neq M_{in}^2(p_2)$ for all places $p_1 \in P_1$, $p_2 \in P_2$. Then there is *no* Petri net $\mathcal{N}$ with morphisms $(\pi_i, \eta_i) : \mathcal{N}_i \to \mathcal{N}$ $(i = 1, 2)$: choose any $p \in P$, then $M_{in}^1(\pi_1(p)) = M_{in}(p) = M_{in}^2(\pi_2(p))$, a contradiction. This leads to the following concept of *conditional coproducts*.

Definition 7.3. Let $\mathcal{C}$ be a category. We say that $\mathcal{C}$ has *conditional coproducts*, if for any two objects A, B of $\mathcal{C}$ the following holds: If there is an object $C \in \mathcal{C}$ with morphisms $\varphi : A \to C$ and $\psi : B \to C$, then there is a coproduct of A and B in $\mathcal{C}$.

We just remark that this notion is similar to the concept of bounded completeness (consistent completeness) in the theory of Scott-domains: a partially ordered set is bounded complete, if any subset which has an upper bound has a supremum (= least upper bound). Now we will show that $\underline{CPN_0}$ has conditional coproducts. The construction of the coproduct itself is almost dual to the one of products: for the event sets we take the disjoint union, but for the state sets only a subset of the direct product.

Definition 7.4. Let $\mathcal{N}_i = (P_i, E_i, F^i, M_{in}^i)$ be two Petri nets in $\underline{CPN_0}$, and assume there is a Petri net $\mathcal{N}^* = (P^*, E^*, F^*, M_{in}^*)$ with morphisms $\varrho_i = (\pi_i, \eta_i) : \mathcal{N}_i \to \mathcal{N}^*$ $(i = 1, 2)$. We define $\mathcal{N} = (P, E, F, M_{in})$ as follows:

1. $P = \{(p_1, p_2) \in P_1 \times P_2 \ : \ M_{in}^1(p_1) = M_{in}^2(p_2)\}$,
2. $E = E_1 \dot\cup E_2$ (disjoint union),
3. $F_j((p_1, p_2), e_i) = F_j^i(p_i, e_i)$ for $i, j \in \{1, 2\}$,
4. $M_{in}(p_1, p_2) = M_{in}^1(p_1) = M_{in}^2(p_2)$.

Furthermore, let $in_i = (\pi_{P_i}, id_{E_i}) : \mathcal{N}_i \to \mathcal{N}$ where $\pi_{P_i} : P \to P_i$ is the projection and $id_{E_i} : E_i \to E$ is the identity mapping.

Proposition 7.5. *Under the above assumptions, $\mathcal{N}$ together with the inclusion morphisms in_1, in_2 is the coproduct of $\mathcal{N}_1$ and $\mathcal{N}_2$.*

Proof. We first show that P is non-empty. By assumption, choose any $p^* \in P^*$. Then $M_{in}^1(\pi_1(p^*)) = M_{in}^*(p^*) = M_{in}^2(\pi_2(p^*))$, so $(\pi_1(p^*), \pi_2(p^*)) \in P$. Clearly, $\|F_j(p_1, p_2)\|_\infty = \max_{i=1,2} \|F_j^i(p_i)\|_\infty$ for any $(p_1, p_2) \in P$ and $j = 1, 2$. Hence $\mathcal{N} \in \underline{CPN_0}$. Also, in_i is a morphism, and the categorial coproduct properties of $\mathcal{N}$ follow easily as in [10] □

Now we turn to the product and coproduct construction for the categories $\underline{CAC_0}$ and $\underline{CAC_0^\mathcal{R}}$, where the situation is more complicated. First we consider the product. If E_1, E_2 are sets, we let $\pi_{E_i} : E_1 \times E_2 \to E_i$ denote the canonical projection (i = 1,2). Recall then the linear transformation $\overline{\pi_{E_i}} : \ell(E_1 \times E_2) \to \ell(E_i)$.

Definition 7.6. Let $\mathcal{A}_i = (S_i, E_i, T_i, s_0^i, \|\cdot\|^i) \in \underline{CAC_0}$ $(i = 1, 2)$. We define $\mathcal{A} = (S, E, T, s_0, \|\cdot\|)$ as follows:

1. $S = S_1 \times S_2, E = E_1 \times E_2$ and $s_0 = (s_0^1, s_0^2)$.
2. $T((s_1, s_2), t, c) := (T_1(s_1, t, \overline{\pi_{E_1}} \circ c), T_2(s_2, t, \overline{\pi_{E_2}} \circ c))$, provided this pair exists for any $(s_1, s_2) \in S, t \in \mathbb{R}_+$ and $c \in \mathcal{C}(E)$,
3. $c \|_{(s_1, s_2)} c'$ iff $\overline{\pi_{E_i}} \circ c \|_{s_i}^i \overline{\pi_{E_i}} \circ c'$ in $\mathcal{A}_i$ for $i = 1, 2$, for any $(s_1, s_2) \in S$, and $c, c' \in \mathcal{C}(E)$.

Furthermore, define $\pi_i = (\pi_{S_i}, \pi_{E_i}) : \mathcal{A} \to \mathcal{A}_i$ as componentwise projection, i.e.

$$\pi_{S_i}(s_1, s_2) = s_i \text{ and } \pi_{E_i}(e_1, e_2) = e_i \quad (i = 1, 2).$$

Lemma 7.7. *With the above notation, the following hold:*

(a) $\mathcal{A}$ *together with the projections* π_1, π_2 *is the product of the automata* $\mathcal{A}_1$ *and* $\mathcal{A}_2$ *in the category* $\underline{CAC}_0$.
(b) *If* $\mathcal{A}_1$ *and* $\mathcal{A}_2$ *are rich, then* $\mathcal{A}$ *is rich.*
(c) *Let* $\mathcal{A}_1, \mathcal{A}_2 \in \underline{CAC}_0^R$. *Then the product of* $\mathcal{A}_1$ *and* $\mathcal{A}_2$ *in the category* $\underline{CAC}_0^R$ *is* $R(\mathcal{A})$, *together with the restrictions of the projections* π_1, π_2.

Proof. (a) We first show that $\mathcal{A} \in \underline{CAC}_0$. To check condition 4.1.2(i), let $c, c' \in \mathcal{C}(E), s = (s_1, s_2) \in S$ and $t, t' \in \mathbb{R}_+$ such that $s' = (s_1', s_2') = T(s, t, c)$ and $s'' = (s_1'', s_2'') = T(s', t', c')$ exist. Then $s_i' = T_i(s_i, t, \overline{\pi_{E_i}} \circ c)$ and $s_i'' = T_i(s_i', t', \overline{\pi_{E_i}} \circ c')$ in $\mathcal{A}_i$, so $s_i'' = T(s, t + t', \overline{\pi_{E_i}} \circ c^t + \overline{\pi_{E_i}} \circ c_t')$ $(i = 1, 2)$. Since $\overline{\pi_{E_i}} \circ c^t + \overline{\pi_{E_i}} \circ c_t' = \overline{\pi_{E_i}} \circ (c^t + c_t')$ by linearity of $\overline{\pi_{E_i}}$, we get $s'' = T(s, t+t', c^t + c_t')$ as required. Similarly, we obtain the rest of condition 4.1.2 and 4.1.3. It follows immediately from the definitions that $\pi_i : \mathcal{A} \to \mathcal{A}_i$ are morphisms (i = 1,2).

To check the universal product properties, let $\mathcal{A}' = (S', E', T', s_0', \|') \in \underline{CAC}_0$ and $(\sigma_i, \eta_i) : \mathcal{A}' \to \mathcal{A}_i$ $(i = 1, 2)$ be two morphisms. We claim that the usual pairing $(\sigma_1 \times \sigma_2, \eta_1 \times \eta_2) : \mathcal{A}' \to \mathcal{A}$ is a morphism. Indeed, since $\pi_{E_i} \circ (\eta_1 \times \eta_2) = \eta_i$, by the remarks in section 2 we have

$$\overline{\pi_{E_i}} \circ \overline{\eta_1 \times \eta_2} = \overline{\pi_{E_i} \circ (\eta_1 \times \eta_2)} = \overline{\eta_i} \quad , \text{ and so}$$

$$\overline{\pi_{E_i}} \circ \overline{\eta_1 \times \eta_2} \circ c = \overline{\eta_i} \circ c$$

for each path $c \in \mathcal{C}(E')$ and $i = 1, 2$. Now if $s' \in S', t \in \mathbb{R}_+$ and $c \in \mathcal{C}(E')$ such that $s'' = T'(s', t, c)$ is defined, then $\sigma_i(s'') = T_i(\sigma_i(s'), t, \overline{\eta_i} \circ c)$ in $\mathcal{A}_i$ (i = 1,2), and so $\sigma_1 \times \sigma_2(s'') = (\sigma_1(s''), \sigma_2(s'')) = T(\sigma_1 \times \sigma_2(s''), t, \overline{\eta_1 \times \eta_2} \circ c)$ in $\mathcal{A}$ as required. Similarly, $c \|_{s'}' c'$ in $\mathcal{A}'$ implies $\overline{\eta_i} \circ c \|_{\sigma_i(s')}^i \overline{\eta_i} \circ c'$ in $\mathcal{A}_i$ $(i = 1, 2)$, showing $\overline{\eta_1 \times \eta_2} \circ c \|_{\sigma_1 \times \sigma_2(s')} \overline{\eta_1 \times \eta_2} \circ c'$ in $\mathcal{A}$. This proves our claim. The rest is straightforward.

(b) First we show for any fixed $i \in \{1, 2\}$ that if $\varphi_i = (\varphi_{S_i}, \varphi_{E_i})$ is a place of $\mathcal{A}_i$, then $\varphi = (\varphi_S, \varphi_E)$ defined by $\varphi_S = \varphi_{S_i} \circ \pi_{S_i}, \varphi_E = ((\varphi_{E_i})_1 \circ \pi_{E_i}, (\varphi_{E_i})_2 \circ \pi_{E_i})$ is a place of $\mathcal{A}$. Indeed, suppose that $(s, t, p) \in S \times \mathbb{R}_+ \times \mathcal{C}(E)$ and $(s_1', s_2') = T(s, t, c)$ exists in $\mathcal{A}$, where $s = (s_1, s_2)$. Then $s_i' = T_i(s_i, t, \overline{\pi_{E_i}} \circ c)$ in $\mathcal{A}_i$ and so

$$\varphi_S(s_1', s_2') = \varphi_{S_i}(s_i')$$
$$= \varphi_{S_i}(s_i) - \langle (\varphi_{E_i})_1, w(\overline{\pi_{E_i}} \circ c)(t) \rangle + \langle (\varphi_{E_i})_2, w(\overline{\pi_{E_i}} \circ c)(t) \rangle$$
$$= \varphi_S(s) - \langle (\varphi_E)_1, w(c)(t) \rangle + \langle (\varphi_E)_2, w(c)(t) \rangle$$

using Proposition 2.1 and formula (2). This proves condition 4.5(I), and 4.5(II) follows similarly.

Now we show that $\mathcal{A}$ is rich. To check condition 6.1.1, let $s = (s_1, s_2), s' = (s'_1, s'_2) \in S$ such that $\varphi(s) = \varphi(s')$ for each place φ of $\mathcal{A}$. By the above, we obtain $\varphi_i(s_i) = \varphi_i(s'_i)$ for each place φ_i of $\mathcal{A}_i$ and, since $\mathcal{A}_i$ is rich, thus $s_i = s'_i$ $(i = 1, 2)$. Hence $s = s'$. Similarly, we can check conditions 6.1.2 and 6.1.3.

(c) Let $\mathcal{A}' = (S', E', T', s'_0, \|')\in \underline{CAC}_0^{\mathcal{R}}$ and let $(\sigma_i, \eta_i) : \mathcal{A}' \to \mathcal{A}_i$ be morphisms $(i = 1, 2)$. By (b), $\mathcal{A}$ is rich. Since $R(\mathcal{A})$ is a subautomaton of $\mathcal{A}$, Lemma 6.5 shows that $R(\mathcal{A}) \in \underline{CAC}_0^{\mathcal{R}}$. We define $(\sigma, \eta) : \mathcal{A}' \to \mathcal{A}$ by $\sigma = (\sigma_1, \sigma_2)$, i.e. $\sigma(s') = (\sigma_1(s'), \sigma_2(s'))$, and $\eta = (\eta_1, \eta_2)$ (correspondingly). As in (a), (σ, η) is a morphism and $(\sigma_i, \eta_i) = \pi_i \circ (\sigma, \eta)$ $(i = 1, 2)$. It only remains to show that (σ, η) is a morphism from $\mathcal{A}'$ actually to $R(\mathcal{A})$, i.e. that each state of $\sigma(S')$ in $\mathcal{A}$ is reachable. Indeed, if $s' \in S'$, then $s' = T'(s'_0, t, c)$ in $\mathcal{A}'$ for some $t \geq 0$ and some control path $c \in \mathcal{C}(E')$. Then $\sigma(s') = T(s_0, t, \bar{\eta} \circ c)$ is reachable in $\mathcal{A}$.   □

Next we turn to coproducts in $\underline{CAC}_0$. Here, the situation is easy:

Proposition 7.8. *The category $\underline{CAC}_0$ has coproducts.*

Proof. Given $\mathcal{A}_1, \mathcal{A}_2$, define $\mathcal{A}$ as usual as the disjoint union of $\mathcal{A}_1$ and $\mathcal{A}_2$, identifying the initial states. Then together with natural inclusions, $\mathcal{A}$ is the coproduct of $\mathcal{A}_1$ and $\mathcal{A}_2$.   □

By the coreflection of Theorem 6.9, the category $\underline{CAC}_0^{\mathcal{R}}$ is 'embedded' into $\underline{CPN}_0$. This will enable us to form a conditional coproduct for CAC's inside $\underline{CPN}_0$. In fact, the following argument is analogous to the proof of the corresponding result for discrete Petri nets and automata with concurrency, cf. [10], Proposition 4.9.

Proposition 7.9. *The category $\underline{CAC}_0^R$ has conditional coproducts.*

Proof. Let $\mathcal{A}_1, \mathcal{A}_2 \in \underline{CAC}_0^R$, and assume there are morphisms from $\mathcal{A}_1, \mathcal{A}_2$ to some $\mathcal{A}^* \in \underline{CAC}_0^R$. Then in $\underline{CPN}_0$ there are morphisms from $\mathcal{N}_1 = an(\mathcal{A}_1)$ and $\mathcal{N}_2 = an(\mathcal{A}_2)$ to $an(\mathcal{A}^*)$. By Proposition 7.5, we form the coproduct $\mathcal{N}$ with inclusions $in_i : \mathcal{N}_i \to \mathcal{N}$ $(i = 1, 2)$ of $\mathcal{N}_1$ and $\mathcal{N}_2$. Choose the isomorphisms $\iota_i : \mathcal{A}_i \to na^R \circ an(\mathcal{A}_i)$ given by Corollary 6.8. We claim that $na^R(\mathcal{N})$, together with the morphisms $na(in_i) \circ \iota_i : \mathcal{A}_i \to na^R(\mathcal{N})$ $(i = 1, 2)$, is the coproduct of $\mathcal{A}_1$ and $\mathcal{A}_2$. Choose any $\mathcal{A}' \in \underline{CAC}_0^R$ and morphisms $\rho_i : \mathcal{A}_i \to \mathcal{A}'$. We have to show that there is a unique morphism $\sigma : na^R(\mathcal{N}) \to \mathcal{A}'$ such that $\sigma \circ na(in_i) \circ \iota_i = \rho_i$ for $i = 1, 2$.

Since $an(\rho_i) : \mathcal{N}_i \to an(\mathcal{A}')$ are morphisms and $\mathcal{N}$ is the coproduct of $\mathcal{N}_1$ and $\mathcal{N}_2$, there is a (unique) morphism $\rho : \mathcal{N} \to an(\mathcal{A}')$ such that $\rho \circ in_i = an(\rho_i)$ $(i = 1, 2)$. Then $na(\rho) : na^R(\mathcal{N}) \to na^R \circ an(\mathcal{A}')$ and $na(\rho) \circ na(in_i) = na \circ an(\rho_i)$ $(i = 1, 2)$. Furthermore, choose the isomorphism $\iota : \mathcal{A}' \to na^R \circ an(\mathcal{A}')$ given by Corollary 6.8, see Fig. 4.

Now suppose we can show that

$$(na \circ an(\rho_i)) \circ \iota_i = \iota \circ \rho_i \text{ for } i = 1, 2. \tag{5}$$

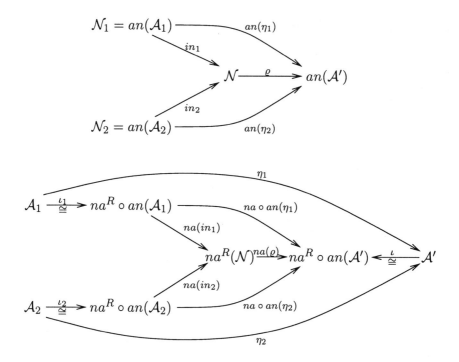

Fig. 4. coproducts

Then the whole diagram of Fig. 4 commutes, so $\iota^{-1} \circ na(\rho) \circ na(in_i) \circ \iota_i = \iota^{-1} \circ (na \circ an(\rho_i)) \circ \iota_i = \rho_i$ for $i = 1, 2$, and we may put $\sigma = \iota^{-1} \circ na(\rho)$.

To show (5), observe that both ι_i and ι act like the identity on the event sets, hence the event mappings of the left hand side resp. the right hand side of (5) are equal. Since $\mathcal{A}_i$ has only reachable states, by Lemma 6.10 the two corresponding state mappings have to be equal, too. This proves (5).

To show that σ is unique, let $\sigma' : na^R(\mathcal{N}) \to \mathcal{A}'$ satisfy $\sigma' \circ na(in_i) \circ \iota_i = \rho_i$ for $i = 1, 2$. The event mapping of $na(in_i) \circ \iota_i$ is the identity, so σ' and ρ_i have the same event mapping on the event set of $\mathcal{A}_i$, and the event mapping of ρ_i equals the corresponding restriction of $\sigma = \iota^{-1} \circ na(\rho)$. Hence σ' and σ have the same event mapping and hence, again by Lemma 6.10, we obtain $\sigma' = \sigma$ as required.   □

Now we can summarize our results:

Theorem 7.10. *The category* $\underline{CAC_0}$ *has products and coproducts and the categories* $\underline{CPN}$ *and* $\underline{CAC_0^R}$ *have products and conditional coproducts.*

8   Conclusion

We have presented a model of Petri nets in which places may carry continuous amounts of tokens and events can act by control paths continuously on them.

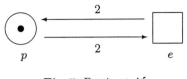

Fig. 5. Petri net $\mathcal{N}$

The dynamic behaviour of these continuous Petri nets could be described by continuous transition systems with concurrency, and we obtained an adjunction between the associated categories using the concept of regions from 'discrete' Petri net theory. By a coreflection we characterized which continuous transition systems correspond to the dynamic behaviour of continuous Petri nets. We showed that suitable categories of continuous nets and continuous automata, respectively, have products and conditional coproducts.

These results were obtained, taking into account some observations on control paths, quite analogously to the corresponding results for discrete Petri nets and automata with concurrency relations, cf. [10]. The question arises about the relationship between these two types of similar results. In section 3, we related the discrete model to the present one by associating with each event e a pure control path c^e such that if $M \xrightarrow{e} M'$ in the discrete case, then c^e transforms M into M' also in the continuous case in one unit of time. However, we note that the converse *fails*. Consider the Petri net $\mathcal{N}$ of Fig. 5.

We have $M_{in}(p) = 1$ and $F_1(p,e) = F_2(p,e) = 2$. Clearly, $M_{in} \xrightarrow{c^e} M_{in}$ as is easy to check (for inequality (4) take any $\Delta t \leq 0.5$). But, in the discrete case, e cannot fire at marking M_{in}. Hence the results on the discrete case of [10] mentioned above do not follow from the present ones. Further similar discrepancies between the discrete and the continuous model have been noted in Recalde, Teruel and Silva [14].

This is due to the fact that when determining the weight function, we chose the usual Lebesgue (or Riemann) integration. We could incorporate the discrete case into the present setting by integrating with respect to Dirac (point) measures. Then, indeed, we obtain $M \xrightarrow{e} M'$ in the discrete case if and only if $M \xrightarrow{c^e} M'$ in this continuous fashion with respect to the Dirac measure after, say, one unit of time. A generalization containing both results would require a bit more measure theory and for the sake of clarity we preferred just to deal with the present case.

Also, the question arises whether we could not model a system with both continuous *and* discrete components. This can be done, of course, by approximating e.g. the discrete components continuously (or by measurable functions); but if we want to do this exactly, one could use Riemann integration as here and model the discrete part as in physics by Dirac functions, i.e. resort to the theory of distributions. One could probably develop similar results as here (we have not checked the details), but the present restricted approach is mathematically much simpler.

In any case, since the present approach using just measurable functions is very general, it could be interesting to try to specialize the duality $CPN \leftrightarrow CAC$ to more concrete versions of continuous Petri nets investigatedd in the literature. Possibly, then some of the consequences of this duality developed for the discrete case could also be utilized in the continuous setting.

References

1. R. Alur, C. Courcoubetis, N. Halbwachs, T. A. Henzinger, P. -H. Ho, X. Nicollin, A. Olivero, J. Sifakis, S. Yovine: The algorithmic analysis of hybrid systems *Theoretical Computer Science* <u>138</u> (1995) 3–34.
2. H. Alla and R. David: Continuous and hybrid Petri nets. *Journal of Circuits, Systems and Computers,* <u>8</u>, (1998) 159–188.
3. E. Badouel and P. Darondeau: Trace nets and process automata. *Acta Informatica* <u>32</u> (1995), 647–679.
4. E. Badouel and P. Darondeau: Theory of regions *in: Third Advanced Course on Petri Nets,* Dagstuhl Castle, Lecture Notes in Computer Science vol. 1279, Springer, 1998, 529–586.
5. R. David and H. Alla: Petri nets for modeling of dynamic systems – a survey. *Automatica* <u>30</u> (1994), 175–202.
6. E. Dubois, H. Alla and R. David: Continuous Petri net with maximal speeds depending on time. In: *4th Int. Conf. RPI, Computer Integrated Manufacturing and Automation Technology,* Troy, USA, Oct. 1994.
7. I. Demongodin and N. T. Koussoulas: Differential Petri nets: Representing continuous in a discrete-event world. *IEEE Trans. on Automatic Control* <u>43</u> (1998), 573–579.
8. M. Droste: Concurrent automata and domains. *Intern. J. of Found. of Comp. Science* <u>3</u> (1992), 389–418.
9. M. Droste and R. M. Shortt: Petri nets and automata with concurrency relations – an adjunction. In: *Semantics of Programming Languages and Model Theory* (M. Droste, Y. Gurevich, eds.), Gordon and Breach Science Publ., 1993, 69–87.
10. M. Droste and R.M. Shortt: From Petri nets to automata with concurrency. *Applied Categorical Structures,* to appear
11. A. Ehrenfeucht and G. Rozenberg: Partial 2-structures, *Acta Informatica* <u>27</u> (1990), 315–342 and 343–368.
12. M. Mukund: Petri nets and step transition systems. *Intern. J. of Found. of Comp. Science* <u>3</u> (1992), 443–478.
13. M. Nielsen, G. Rozenberg and P.S. Thiagarajan: Elementary transition systems. *Theor. Comp. Science* <u>96</u> (1992), 3–33.
14. L. Recalde, E. Teruel and M. Silva: Autonomous continuous P/T systems. In: *Proc. 20th Int. Conf. on Applications and Theory of Petri Nets,* Lecture Notes in Computer Science, vol. 1639, Springer, 1999, 107–126.
15. G. Winskel and M. Nielsen: Models for concurrency. In: *Handbook of Logic in Computer Science vol. 4* (S. Abramsky, D.M. Gabbay, T.S.E. Maibaum, eds.), Oxford University Press, 1995.

Author Index

Lecture Notes in Computer Science

For information about Vols. 1–2175
please contact your bookseller or Springer-Verlag